SURFACE WATER HYDROLOGY

PROCEEDINGS OF THE INTERNATIONAL CONFERENCE ON WATER RESOURCES MANAGEMENT IN ARID REGIONS (WaRMAR), MARCH 23–27, 2002, KUWAIT

Surface Water Hydrology

Edited by

V.P. Singh
Department of Civil and Environmental Engineering
Louisiana State University, Baton Rouge, LA 70803-6405, USA

M. Al-Rashed
Water Resources Division
Kuwait Institute for Scientific Research, Kuwait

M.M. Sherif
Department of Civil Engineering
United Arab Emirates University, Al Ain, UAE

VOLUME 1

A.A. BALKEMA PUBLISHERS   LISSE / ABINGDON / EXTON (PA) / TOKYO

Published by: A.A. Balkema, a member of Swets & Zeitlinger Publishers
          www.balkema.nl and www.szp.swets.nl

For the complete set of four volumes: ISBN 90 5809 362 X
Volume 1: 90 5809 363 8
Volume 2: 90 5809 364 6
Volume 3: 90 5809 365 4
Volume 4: 90 5809 366 2

Printed in The Netherlands

Surface Water Hydrology, Singh, Al-Rashed & Sherif (eds)
© 2002 Swets & Zeitlinger, Lisse, ISBN 90 5809 363 8

Table of contents

Section 7: Flood and drought frequency analysis

Section 8: Hydrologic applications

Surface Water Hydrology, Singh, Al-Rashed & Sherif (eds)
© 2002 Swets & Zeitlinger, Lisse, ISBN 90 5809 363 8

Preface

About a third of the land surface in the world is either arid with less than 250 mm of annual precipitation or semi-arid with annual precipitation between 250 mm and 500 mm. The lack of freshwater resources in these regions constitutes a major deterrent to their sustainable development. On the other hand, growing population, rising standard of living, and expanding opportunities exert increasing demands for varied needs for water. These needs may be domestic, agricultural, industrial, waste disposal, power generation, navigational, transportation, recreational, and so on. To meet demands for water for a multitude of such needs is a continuing struggle.

The water shortage has already become commonplace in many parts of the world, including those that are not arid or semi-arid. In many of those parts where water is plentifully available now, it is expected that by the middle of this century they will start experiencing severe water shortages. The per capita share of freshwater has dropped significantly due to burgeoning population and the attendant increasing need for water. Notwithstanding these shortages, water continues to be used unwisely, wasted and polluted. Insufficient water at the right place at the right time with the right quality requires, more than ever before, improved management, efficient utilization, and increased conservation of limited freshwater resources. These demands can only be met if water resources are properly managed. This constituted the primary motivation for organizing this international conference on water resources management in arid regions.

Kuwait is a cultural oasis in the Arab world. Its main source of revenue is oil. Its people enjoy a high standard of living. Kuwait typifies an arid area and has very limited freshwater resources. Its main source of water is through desalination which is quite expensive. In order to sustain its development it needs a lot of freshwater. The management of its water resources has long been recognized as fundamental to its sustained development. In fact, the same can be said about many other countries of the world. Thus, particular attention needs to be paid to the development and management of its precious freshwater resources. In this vein, the venue of the conference could not have been more appropriate. This is probably the first international conference on its kind in Kuwait.

The main objective of this conference was, therefore, to bring together educators, researchers, practitioners, managers, and policy makers from all over the world involved in various aspects of water resources in arid and semi-arid regions. More specific objectives were to (1) assess the current state of art of water resources management in arid regions, with particular emphasis on developing countries, (2) promote interdisciplinary dialog and interaction, (3) discuss the applicability of interdisciplinary approaches and models, (4) discuss the future needs, and (5) determine the direction of future research.

We received an overwhelming response to our call for papers. The number of abstracts received exceeded 270. Each abstract was reviewed and about two thirds of them, deemed appropriate to the conference theme, were accepted. That led to submission of about 180 full-length papers. These papers were reviewed by the conference organization and about 15% of them were found unsuitable and were, therefore, excluded from the conference proceedings books. The subject matter of the conference was divided into 25 topics encompassing virtually all major facets of water resources management in arid and semi-arid regions. Each topic comprised a number of contributed papers and in some cases a state-of-the-art paper. These papers provided a natural blend to reflect a body of knowledge on that topic.

The papers contained in this book, *Surface Water Hydrology,* represent one part of the conference proceedings. The other parts are embodied in three separate companion books entitled, *Groundwater Hydrology, Environmental and Groundwater Pollution,* and *Water Resources Development and Management*, which are being published simultaneously. Arrangement of the contributions in these four books under different titles was a natural consequence of the diversity of the papers presented at the conference and the topics therein. These books can be treated almost independently, although considerable interconnectedness exists amongst them.

This book contains eight sections encompassing major aspects of surface water hydrology. Section 1 deals with data extraction, processing and development as well as hydrometric networks. There has been and continues to be a considerable attention on climate change and its effect on water resources. This topic has assumed heightened significance in Kuwait due to burning of oil and gas wells which lasted nearly a year during and subsequent to the Iraqi invasion. The effect of climate change constitutes the subject of Section 2. The effects on river runoff, flood flows, crop evapotranspiration, seasonal drought, and water resources are discussed in this section.

Rainfall constitutes the subject of Section 3. Rainfall simulators, extreme precipitation events, and probabilistic modeling of rainfall, regional rainfall, and seasonal rainfall are discussed. Section 4 goes on to discuss evaporation and evapotranspiration. Infiltration is discussed in Section 5. Watershed modeling is the subject of Section 6. Beginning with the state of art, the papers go on to discuss runoff modeling, transport of sediment and agricultural chemicals, determination of mean velocity in channels, and how deserts are kept alive.

Section 7 deals with flood and drought frequency analysis. Beginning with a discussion of the effect of climate and soil moisture characteristics on the probability distribution of floods, the papers go on to discuss nonstationarity in hydrologic time series, multivariate estimation, bias estimation, and discontinuous distributions. The concluding section 8 presents hydrologic applications, including flood mitigation, drought alleviation, sand dams, and canal controls.

The book will be of interest to researchers and practitioners in the field of hydrology, environmental engineering, agricultural engineering, and watershed and range sciences, as well as to those engaged in water resources planning, development and management in arid and semi-arid areas. Graduate students and those wishing to conduct research in hydrology, environmental science and engineering, and water resources will find the book to be of value.

We wish to take this opportunity to express our sincere appreciation to all the members of the Organizing Committee and the International Technical and Scientific Committee, Kuwait Institute for Scientific Research (KISR), and Louisiana State University. Dr. Abdul Hadi Al-Otaibi, the Director General, and Dr. Nader Al-Awadi, the Deputy Director General for Research Affairs of KISR, provided support, encouragement and facilities for hosting the conference. Mr. Tariq Rashid, Mrs. Ruby Crasta, Ms. Nancy Carmen De Souza, and Mr. Mohamed Tufail helped with preparation of papers in camera-ready form. Without their assistance, the proceedings books would not have been of the quality that they are. Numerous other people contributed to the conference in one way or another, and a lack of space does not allow us to list all of them by name here. The authors, including the invited keynote speakers, contributed to the conference technically and made the conference what it was; this book is a result of their collective efforts and contributions. The session chairmen and co-chairmen administered the sessions in a positive and professional manner. We owe our gratitude to all of these individuals.

WaRMAR attracted a large number of nationally and internationally well-known people who have long been at the forefront of hydrologic, groundwater, environmental, and water resources research. More than 50 countries, covering five continents and many of the major countries of the world active in hydrologic, groundwater, environmental, and water resources were represented. It is hoped that long and productive personal associations and friendships will develop as a result of this conference.

V.P. Singh, Technical and Scientific Committee Chair
M. Al-Rashed, Conference Chair
M.M. Sherif, Conference General Secretary

Acknowledgments

The International Conference on Water Resources Management was sponsored by a number of organizations. The sponsors offered financial support to the conference, without which the conference would not have come to fruition. Their financial support is gratefully acknowledged. The co-sponsors offered their support by promoting the conference through their periodicals, newsletters, and magazines. The following is a list of conference sponsors and co-sponsors:

Sponsors

Kuwait Institute for Scientific Research (KISR)
Mohammed Abdulmohsin Al Kharafi & Sons
Ministry of Electricity and Water (MEW)
Kuwait Foundation for the Advancement of Sciences (KFAS)
Chevron Overseas Petroleum
International Hydrological Programme (IHP), UNESCO
Islamic Development Bank (IDB), Jeddah, Saudi Arabia
Kuwait Airways (KAC)
UNESCO Cairo Office
Arab League Educational Cultural and Scientific Organization (ALECSO)
Islamic Educational Scientific and Cultural Organizations (ISESCO)

Co-Sponsors

Food and Agriculture Organization of the United Nations (FAO)
European Water Resources Association (EWRA)
International Association of Hydrogeologists (IAH)

Section 1: Hydrologic data

Surface Water Hydrology, Singh, Al-Rashed & Sherif (eds)
© 2002 Swets & Zeitlinger, Lisse, ISBN 90 5809 363 8

Status of hydrometric network in Oman

Abdul Baqi Al Khabouri and Tariq Helmi Abo Al A'ata

Ministry of Regional Municipalities, Environment and Water Resources
P.O. Box 2575 Ruwi, PC 112 Ruwi
Sultanate of Oman
e-mail: hayatabd@omantel.net.om, e-mail: tariq_h40@hotmail.com

Abstract

Oman is situated in the southeastern part of the Arabian Peninsula. Due to its location in the Arid and Semi-Arid region, the water resources are limited. Special attention was given to establish an up to date hydrometric monitoring network across the entire country. This paper has reviewed the current status and compared the distribution and density of the network with WMO standards. The overall density of rainfall, wadi and aflaj stations meet WMO standards in coastal and plain areas. Attention should be given to the less developed desert regions. The groundwater monitoring shows a good distribution across the entire country, averaging 38 Km^2/ well in the non-desert areas. Although the regional distribution of the hydrometric network shows a good coverage, the larger scale density i.e. catchment scale may not be.

Introduction

Sultanate of Oman is situated in the south eastern part of Arabian peninsula, surrounded by United Arab Emirates (U.A.E) from north west, Saudi Arabia from the west, Gulf of Oman and Arabian sea from the east and south east, Fig. 1.

Due to its location on the Arid and Semi-Arid region water resources are very scarce. Since the establishment of Ministry of Regional Municipalities, Environment and Water Resources (MRMEWR), systematic hydrometric network started. Currently there are over 3,000 Monitoring stations for climate, rainfall, wadi flow, sediments, aflaj, groundwater levels and groundwater quality. Data are collected using electric water level indicator (Groundwater) once a month, where rainfall and wadi flow stations are equipped with dataloggers or autagraphic water level recorders. Based on topographical, geological and hydrological features, Oman has been divided into seven (7) main assessment regions, Fig. 1. These are Muscat, Batinah, Musandam, Dakhliyah, Dhahirah, Sharqiyah and Dhofar & Al wusta. Due to Oman's rapid development since 1970, demands for water resources data are increased. The distribution of current hydrometric network across the entire country is discussed and the density of MRMEWR stations is compared with WMO standards for Arid countries and for the main physiographic zones (desert, coast, mountains, interior plain).

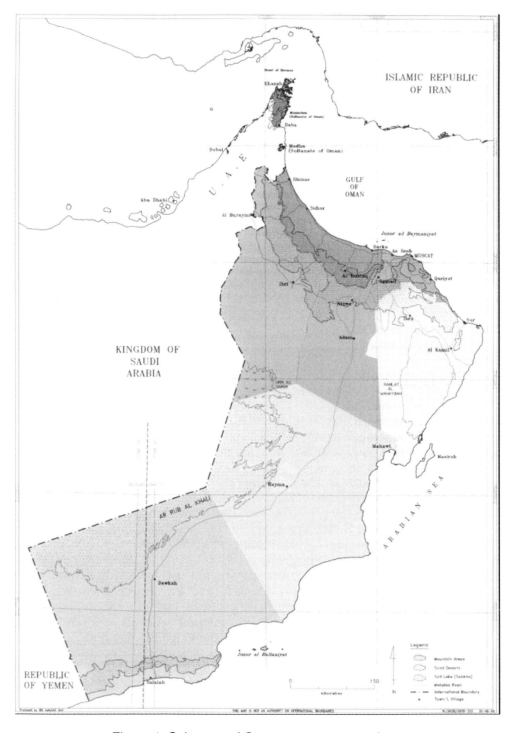

Figure 1. Sultanate of Oman assessment regions.

Main Objectives of Monitoring Stations

All stations are designed to provide basic water resources data which can be used to better assess and manage Oman's water resources. Long term records are necessary to produce frequency and trend data for planning and engineering purposes. Basic data requirements are summarized below:

Wadi flow stations: Data are used for water balance assessments, water management planning and projects, peak flows and volume data for design of dams, bridges, culverts and flood protection.

Rainfall: Primary network - data are necessary for water balance assessments, water management planning and intensity calculations for engineering designs.

Aflaj flow: Data are used to estimate the water supply, water balance assessments, development and agricultural planning.

Groundwater levels: Data are used to estimate the water situation, water balance assessments, development and agricultural planning.

Groundwater salinity: Data are used for water supply projects, water balance assessments, assessment of saline intrusion, development and agricultural planning.

Surface Water Network

There are 1071 Surface Water Monitoring stations (rainfall, wadi flow, aflaj, springs/khawrs) across the entire country, Table 1. Rainfall data measured using automatic recorders, data are recorded on a daily and hourly bases. A small portion of gauges are standard devices and chart autographic recorders where data are read on a daily bases. Wadi flow gauges are mostly automatic with few chart recorders. Aflaj flow and water quality data with springs and khowrs are collected once a month.

Density of Surface Water Network

Table 1 shows the density and distribution of Surface Water Network for all the assessment regions. As it could be seen, that the current network generally meets the WMO guidelines in term of number of gauges, the highest density of monitoring is with aflaj, followed by rainfall and wadis. This fits with Oman's unique water situation where aflaj, springs and groundwater form the most common and reliable source of water. The desert areas are not gauged enough to meet WMO standards where the non-dessert areas are well covered meeting WMO standard in most regions.

Figs. 2 to 4 show the distribution of the Monitoring stations in different regions of the country.

5

Assessment Region	Area (b) (Km²)	Rainfall Density	Rainfall Number	Wadi* & Dams Density	Wadi* & Dams Number	Aflaj/Springs Density	Aflaj/Springs Number	(km² per station) WMO Guidelines Rainfall	(km² per station) WMO Guidelines Wadi flow	(km² per station) WMO Guidelines Aflaj
Muscat	7,641	174	44	347	22	273	28	250-900	300-1,000	
Dakhliyah	7,488	250	30	357	21	70	107	250-575	300-1,000	
Dakhliyah - desert	25,741	12,871	2	8,580	3	0	4	10,000	5,000-20,000	N
Dhahirah	14,509	468	31	764	19	165	88	250-575	300-1,000	o
Dhahirah - desert	19,491	9,746	2	0	0	0	0	10,000	5,000-20,000	
South Batinah	6,925	204	34	266	26	95	73	250-900	300-1,000	W
North Batinah	7,696	151	51	175	44	428	18	250-900	300-1,000	M
Salalah Plain Assessment	1,420	36	40	284	5	118	12	575-900	300-1,000	o
Nejd region	91,415	6,094	15	15,236	6	3,657	25	10,000	5,000-20,000	G
Sharqiyah	12,774	278	46	511	25	58	219	250-575	300-1,000	u
Sharqiyah-desert	15,296	2,185	7	7,648	2	0	0	10,000	5,000-20,000	i d
Musandam and Madha	1,744	125	14	349	5	1,744	1	250	300-1,000	e
Al Wusta coast	18,418	0	0	0	0	0	0	250-900	300-1,000	l i
Al Wusta desert	78,582	39,291	2	0	0	0	0	10,000	5,000-20,000	n e
Total	**309,140**	972	318	1,737	178	538	575			
Desert area	230,525	8,233	28	20,957	11	7,949	29			
Non-desert	78,615	271	290	471	167	144	546			

Notes:
(a) WMO guidelines based on minimum density - a rough guide.
Rainfall gauge density shown is based on standard gauges in the jabals and plains areas because of the mixed standard and automatic; and for the automatic gauges in the desert. WMO allows the density to be 10 times less for automatic gauges than standard.

(b) Regional name indicates the area where the main habitations and villages are near the jabals and plains.
Desert indicates marginal rainfall zone away from main habitations and villages, with less than 75 mm annual rainfall.
Areas based on 1996 MWR report, "Mean Annual Isohyets and Mean Annual Rainfall Yield in the Sultanate of Oman", and adjusted to fit the assessment boundaries where necessary.

* recording stations only

References: UNESCO/WMO (1982) Guidelines for Evaluation of Water Resource Assessment Programmes based on the Handbook for National Evaluation. WMO (1981) Guide to Hydrological Practices, Publication No. 168, Geneva.

Rainfall Gauges (Fig. 2)

For desert areas WMO standards specify 10,000 km² should be covered by a station. As it could be seen from Fig. 2, most of the desert areas in Oman are covered. Special attention should be given to Dakhliyah desert (including Andam/Halfayn), Al Wusta desert and Buraimi desert to meet WMO standard. The coastal plain and Jabal regions are very well covered meeting WMO standard.

Coastal plains should be covered every 900 km², all of Oman's coastal plains has been covered with less than 200 km², except Al Wusta coast. The Jabal regions are also covered with less than 400 km² meeting WMO standard. However, Dhahirah region still need to be covered particularly Buraimi desert.

Wadi Flow Gauges (Fig. 3)

Most of the main wadi catchments in the coastal plains and Jabal regions are very well covered with wadi flow stations. Coastal plains are covered with flow gauges density of less than 400 km² meeting WMO standard of 2,500 km². The Jabal and

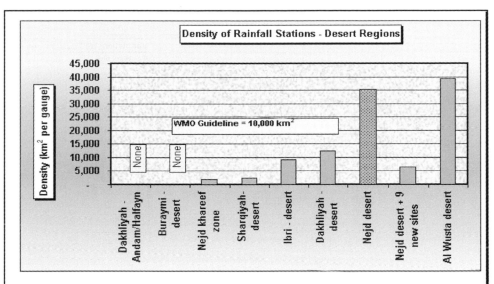

Once the 9 new raingauges are installed in Nejd, the areas that need the
most attention to meet the WMO Guidelines are Andam/Halfayn, Buraymi desert,
and Al Wusta.

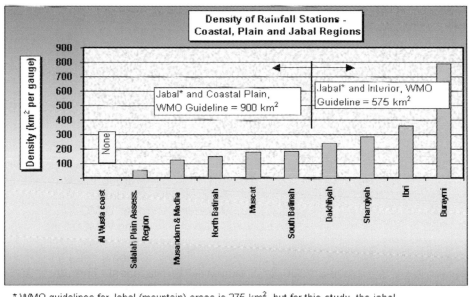

* WMO guidelines for Jabal (mountain) areas is 275 km², but for this study, the jabal
and adjacent area are combined.
The areas needing most attention are Al Wusta coast and Buraymi.

Figure 2. Density of rainfall stations.

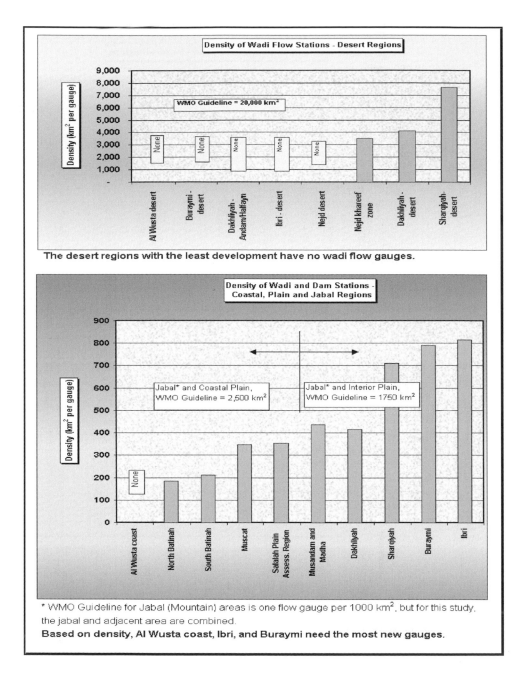

Figure 3. Density of Wadi flow and dams stations.

interior regions has a density of less than 800 km^2 compared to WMO recommendation of 1750 km^2. However Al Wusta coast is still short of wadi flow gauges. As most of the desert regions in Oman form a significant recharge area, special at-

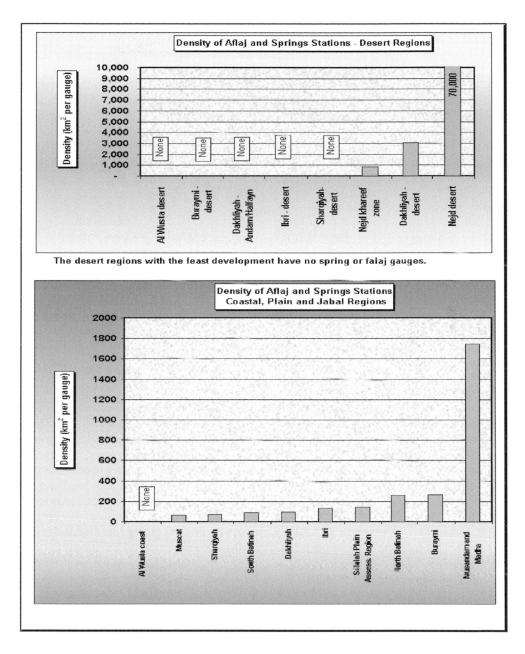

The desert regions with the least development have no spring or falaj gauges.

Figure 4. Density of Aflaj and Springs stations.

tention should be given to gauge the wadi flow to better assess the amount of re-charge. It could be seen from the Figure that dessert regions with least develop-ment have no wadi flow gauging stations. An increase in gauging stations should be considered as development increases.

Aflaj/Springs (Fig. 4)

Falaj (plural is Aflaj) is a man made stream which intercepts groundwater at the footslopes of mountains (Jabals) and brings it to surface at a lower level for irrigation and drinking purposes. Aflaj forms a unique feature of Oman's water resources, but unfortunately WMO has no standards for monitoring them. Most of the regions near Jabals have a large number of aflaj and springs, therefore the density of monitoring is high (less than 300 km^2 except for Musandam). Most of the desert areas do not have springs or aflaj and therefore show poor density. Special attention should be given in monitoring Musandam springs and aflaj. It should be noted howevre that the currunt monitoring of Aflaj is considered adequate.

Khawr Monitoring Stations

Khawrs are the brackish water estuary of some sea bound wadis, usually separated from the sea by a sandbar except during heavy flooding. They represent a potential water source, some are drinkable without treatment. Their water quality is representative of the mixing zone between the sea and the groundwater and they form an important link in salt-water intrusion studies. Water levels of the khawrs are measures of the water table where the salt water and fresh water mix. MRMEWR monitors water quality and water levels at 11 khawrs in Salalah region. Monthly records are collected for water levels, EC and general minerals water quality.

Groundwater Monitoring Network

In Oman, reliable fresh water resources occur only as groundwater, typically in shallow alluvial unconfined aquifers of the wadi systems that drain the mountains (Duncan et al. 1995). That is true in most of the 7 assessment regions. Notable exceptions are Batinah region (coastal plain) where the sediments are in excess of over 350 meters thick in places, and in Dhofar region (Nejd artesian multi-aquifer system in the south of the country). Since 1989, a great efforts were made to expand and improve the efficiency of the monitoring network particularly, the groundwater network. The network includes water level measurements and water quality samples. The existing overall national groundwater network was designed to cover all the large main exploited aquifers in the Sultanate, to monitor water levels changes and groundwater contamination including both pollution and sea water intrusion in coastal areas, and to evaluate the natural and artificial replenishment.

Water Levels

A total of 2531 wells are measured, 1797 measured every month and 734 measured every 3 months. The seven assessment regions are covered by 10 Water Resources Regional Offices where technicians are routinely collect water level measurements from a nation wide network of boreholes and dug wells. Most are measured once every month using an electric water level indicators, a small proportion are equipped with dataloggers. The oldest groundwater level data began in the early seventies, mainly in the coastal areas, routine long-term data collection expanded in 1982-83. The water level data are compiled in a major database after general checking. The MRMEWR database contains 261850 water level records (March 1999) increasing at a rate of about 1800 records per month.

Groundwater Quality Network

Monitoring of groundwater quality is a vital issue in Oman. Specially as groundwater represent the dominant water sources. In many cases, however, the chemical nature of the groundwater, not the volume of water obtained, may be the major factor determining the water value. In 1998, the Ministry expand its groundwater quality program to meet the following requirements, 1) detecting any local salinity changes in the main aquifer on each assessment region, 2) monitor salinity of coastal aquifers, and 3) locate groundwater contamination from any pollutant source. Consequently, the network includes all the available water resources such as observation wells; private dug wells, aflaj and springs. In order to meet these requirements, The Ministry implemented a special monitoring program at each assessment region according to geographic setting and hydrogeological conditions prevailing on each region. The groundwater quality network has a total of 1280 water samples collected every year. In addition, since 1995, there had been a special program at Al Batinah Assessment Region in order to monitor the sea water intrusion where over abstraction is recorded at Batinah coastal plain. Over 3000 wells are measured every two years to evaluate the salinity changes at coastal wells.

Table 2. MRMEWR Groundwater Monitoring Network.

Assessment Region	Area (Km²)	Number of Wells Measured at			Density (Km² per well)
		1 month	3 month	Total	
Muscat	7,641	132	73	205	37
Dakhliyah	7,488	302	52	354	21
Dakhliyah desert	25,741	0	74	74	348
Dhahirah	14,509	469	80	549	26
Dhahirah desert	19,481	0	108	108	180
South Batinah	6,925	111	0	111	62
North Batinah	7,696	190	34	224	34
Salalah Plain	1,420	133	62	195	7
Nejd Region	61,115	39	10	49	1866
Sharqiyah	12,774	306	19	325	344
Sharqiyah desert	15,296	33	19	52	294
Masandam	1,744	82	10	92	19
Al Wusta Coast	18,418	0	0	0	na
Al Wusta desert	78,582	0	193	193	407
Total	309,130	1797	734	2531	3647
Desert area	230,515	72	404	476	484
Non-desert	78,615	1725	330	2055	38

Table 3. Density of MRMEWR Monitoring Network at Batinah Assessment Region.

Catchment / Wadi	Area	Density of MRMEWR Monitoring Stations (km^2 per station)							
		Rainfall		Wadi & Dams		Aflaj		Wells	
	(km^2)	Density	Number	Density	Number	Density	Number	Density	Number
South Batinah									
Wadi Taw	366	183	2	183	2	52	7	18	20
Wadi ma'awil	1,098	157	7	183	6	137	8	27	41
Wadi Bani Kharus	1,180	148	8	169	7	98	12	59	20
Wadi Al Fara'	1,173	130	9	196	6	59	20	73	16
Wadi Bani Ghafir	1,384	173	8	346	4	115	12	115	12
Al Mayhah-Al Hajir System	1,450	na	0	1,450	1	161	9	na	0
Wadi mashin	274	na	0	na	na	55	5	137	2
Total	6,925	204	34	266	26	95	73	62	111
North Batinah									
Wadi Al Hawasinah	980	89	11	140	7	327	3	65	15
Wadi Shafan	755	252	3	755	1	na	0	63	12
Wadi As Sarami	406	203	2	203	2	135	3	81	5
Wadi Sakhin	366	366	1	na	0	na	0	73	5
Wadi Ahin	1,004	100	10	126	8	335	3	18	56
Wadi Al Jizi	1,151	128	9	230	5	384	3	30	38
Wadi Hilti	648	81	8	72	9	na	0	26	25
Wadi Suq	211	211	1	na	0	211	1	8	25
Wadi Bani Umar Al Gharbi	480	480	1	160	3	240	2	96	5
Wadi Fizh	294	294	1	98	3	147	2	147	2
Wadi Rajma	473	473	1	na	0	na	0	118	4
Wadi Al Bid'ah	154	na	0	na	0	na	0	77	2
Wadi Faydh	166	166	1	na	0	na	0	21	8
Wadi Hatta (Oman)	373	373	1	124	3	na	0	23	16
Wadi Al Hawarim (Oman)	142	na	0	na	0	na	0	na	0
Wadi Al Qwr (Oman)	63	63	1	21	3	63	1	11	6
Wadi Malahah (Oman)	30	na	0	na	0	na	0	na	0
Total	7,696	151	51	175	44	428	18	34	224

na : not applicable

Distribution and Frequency

The main 7 assessment regions are subdivided into sub-regions controls by the characteristics of the hydrological and geological units. The 2531 monitoring wells for water levels are shown in Table 2 with their monitoring frequency and density over each Assessment Region, of which 1280 wells for water quality. It should be noted that 734 wells are monitored once every three months.

The groundwater monitoring stations covers most of the Sultanate with reasonable density. The main assessment regions and the heavily populated areas were moni-tored with an average density of 38 km^2. Remote and less populated areas mainly deserts are less monitored with an average of 484 km^2. Salalah Plain at Southern Oman has the highest well density 1 well per 7 km^2 where Al Wusta coastal region

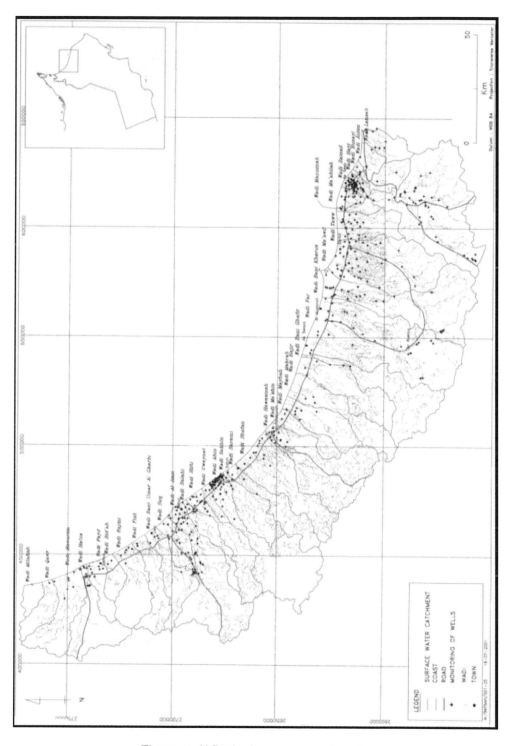

Figure 5. Al Batinah assessment region.

13

has no wells. To best illustrate the well distribution and density, the Hydrometric Network at Al Batinah Assessment Region is shown in details, Table 3 and Fig. 5. The hydrometeric network at north Batinah assessment region is well distributed where rainfall stations has an average of 151 km^2, wadi gauge stations averaging 175 km^2 and groundwater network (boreholes/wells) has an average of 34 km^2.

A concept of integrated surface and groundwater monitoring at "primary monitoring site" has been implemented within the hydrometric network, particularly for the Batinah assessment region. Consideration has been given to the planning of integrated monitoring sites where wadi flow and groundwater throughflow co-exist to assess recharge and throughflow, i.e. Al Batinah Plain. Braiding of the main channels, variable abstraction, a broad groundwater discharge boundary and a saline interface complicate the monitoring of outputs from the aquifer, and should only be attempted at selected study catchments.

Conclusions

- In general, the current hydrometric Network density in Oman is very well distributed.

- The surface water network covers most of Oman's main assessment regions, special attention should be given to remote areas as the infra-structural development reach them.

- The groundwater network is considered more than adequate and very well distributed in regional scale. On the other hand, few sub-regions need to be upgraded.

- Although the regional network coverage is considered adequate, the distribution of the stations in a particular catchment area is currently considered by MRMEWR.

References

Duncan Storey, et al., 1995. "An Approach to Water Resources Assessment and Management using an Integrated Catchment Model: A case Study from Northern Oman", International Conference on Water Resources Management in Arid Countries, Vol. 2, p. 401-410, 12-16 March 1995, Muscat, Oman.

MWR report, 1993. "Mean Annual Isohyets and Mean Annual Rainfall Yield in the Sultanate of Oman", Unpublished Report, March 1996, Sultanate of Oman.

UNESCO/WMO (1982) Guidelines for Evaluation of Water Resources Assessment Programmes based on the Handbook for National Evaluation. WMO (1981) Guide to Hydrological Practices, Publication No. 168, Geneva.

Surface Water Hydrology, Singh, Al-Rashed & Sherif (eds)
© 2002 Swets & Zeitlinger, Lisse, ISBN 90 5809 363 8

Automatic extraction of boundary and channel network of mountainous watershed

Abdullah S. Al-Wagdany
Department of Hydrology, Faculty of Meteorology, KAAU
P.O. Box 80208, Jeddah 21589, Saudi Arabia
e-mail: awagdani@yahoo.com

Abstract

Digital elevation model was created for the mountainous region located between the cities of Makkah AlMukaramah and Taif. The GIS software Idrisi was utilized to create the DEM of the region through digitizing contour lines of topographic maps of the region and interpolating the elevation over the surface. The Watershed Management System software was then used to automatically extracting the boundary and the channel network of Wadi Namman watershed from the created DEM. Both boundary and channel network of the wadi were extracted manually from topographic maps of the region. Then, automatically and manually extracted boundary and network were compared. The automatic procedure produced reasonable estimate of drainage basin characteristics. The automatic procedure was recommended since it is faster, has less error due to human mistakes and cost less whenever the digital map of the studied region is available.

Introduction

Digital elevation models (DEMs) are raster digital maps that are used to delineate stream channel and watershed divide. Considerable research has been carried out in this field by hydrologists and geologists such as O'Callaghan and Mark, 1984; Yuan and Vanderpool, 1986; Jenson and Domingue, 1988 and Band, 1988 and 1989. Square-grid digital elevation models have emerged as the most widely used elevation data structure during the past decade because of their simplicity and ease of computer implementation (Wilson and Gallant, 2000).

The structure of digital map of the DEM usually consists of grid network. Each grid in these networks holds a numerical value of elevation and known as a raster cell. The resolution of the elevation data is determined by the size of the cell. For instance, data that are stored in cells of 50×50-meter have finer resolution than that stored in 100×100-meter cells.

The majority of published algorithms that are used to extract drainage boundary and channel network of watersheds utilized the eight neighbors' method (Fairfield and Leymarie, 1991). In this procedure, the eight neighbors of the cell are examined in order to determine the direction of the flow. Hence, the direction of the flow is assumed to be in the direction of the cell toward which the slope is steepest.

The accuracy and details of automatically extracted channel networks depend on the quality and resolution of the digital elevation maps. Watersheds extracted from

DEMs with fine resolutions are expected to be closer to those of actual basins than those extracted from DEMs with course resolution. However, to check the validity of an automatically extracted drainage network, it should be compared with a drainage network manually extracted from topographic maps. Chorowicz et al. (1992) stated that "any method yields a drainage network comparable to that extracted from topographic maps should be regarded as reliable.

Study Area

The study area is located in the western region of Saudi Arabia east of Makkah and west of Taif. It extends between longitudes $40°00'$ and $40°20'$ E and latitudes $21°07'$ and $21°30'$ N. The area of study contains Wadi Namman which has a drainage area of about 670 km^2. Namman is an important wadi since it is close to the holly places of Arafat and a major source of water during the Hajj season. Its groundwater is the source of water for the historic underground gallery known as "Ain Zubidah" which used to supply water to Arafat, Mena and Makkah for about twelve centuries. Most of the area of Namman is covered with mountains. The remaining areas of the wadi are covered with alluvial deposits.

The boundary and channel network of Namman Basin was manually delineated from four topographic maps at the scale of 1:50,000. Results of the manual delineation are shown in Fig. 1. The delineated channels were then classified according to Strahler (1957) ordering systems. The basin was found to be a fifth order basin. Lengths, slope and area draining to all channels in the basin were measured. Values of channel orders, mean lengths, mean slopes and mean areas draining to channels are summarized in Table 1.

Automatic Extraction of Boundary and Channel Network

Digital elevation model for the study area was unavailable. Therefore, it should be created from the available topographic data for the region. The contour lines of the region of Wadi Namman were identified on the four 1:50,000 topographic maps used for manual extraction of the channel network. The contours were digitized using an A1 size digitizing board and Cartalinx software. Cartalinx is spatial data builder software developed by Clark Labs at Clark University. The process of maps digitizing required map registration and then the use of a cursor that is moved over the contour lines to capture the coordinates of the points representing the contour lines. The coordinates are then saved on an output files. The information from the four maps was combined in a single file using the Append module which is a specialized module within Cartalinx software.

The DEM file for the area of the study was produced using Idrisi GIS software. Idrisi is a raster-based GIS and image processing software system developed also by Clark Labs at Clark University. It consists of a main interface program and a collection of more than 150 program modules (Idrisi, 1997). These modules provide facilities for the input, display, and analysis of spatial data. INTERCON is one of the modules of Idrisi and is used to produce digital elevation models. The input for the INTERCON should be a raster file which contains coordinate of the contour lines.

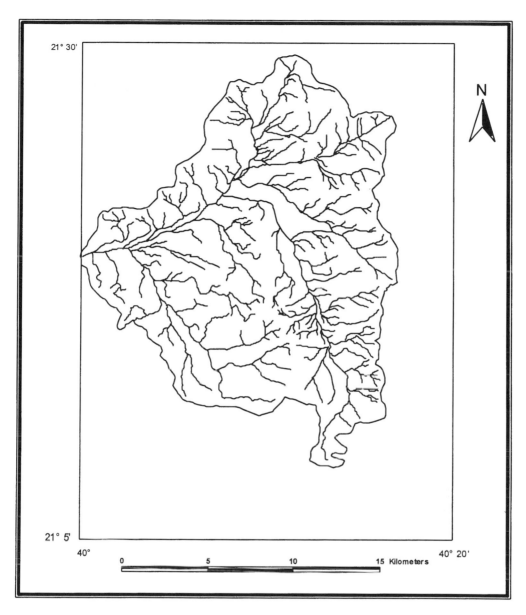

Figure 1. Drainage network of Wadi Namman delineated manually.

The vector files for the contour line produced by Cartalinx was converted to raster file using the LINERAS module of Idrisi software. The produced raster file was used as an input file to the INTERCON module to produce a DEM for the region. The produced DEM consists of 650 rows and 500 columns, which means a total of 325000 cells. The range of values of elevation for these cells is between 500 and 3000 meters above sea level.

Table 1. Comparison of the drainage properties of Wadi Namman.

Property	Method of delineation	
	Manual	Automatic
Drainage Area (km^2)	671.1	671.23
Length of Main stream (km)	49.95	48.89
Channel order	Number of Channels	
1	215	387
2	62	85
3	17	19
4	5	4
5	1	1
Channel order	Mean Length of Channels (km)	
1	1.81	1.07
2	2.32	2.15
3	4.51	4.13
4	8.23	12.98
5	19.25	18.5
Channel order	Mean Area Draining to Channel (km^2)	
1	1.82	1.07
2	2.17	1.70
3	4.74	3.34
4	9.20	10.64
5	17.80	11.05
Channel order	Mean Slope of Channels (m/m)	
1	0.198	0.1490
2	0.0889	0.0742
3	0.0332	0.0388
4	0.0259	0.0214
5	0.0109	0.0093

Watershed modeling system (WMS) software were used to automatically extract the boundary and the stream network for Wadi Namman from the produced DEM. WMS software is a comprehensive environment for hydrologic analysis. It was developed by the Engineering Graphic Laboratory of Brigham Young University in corporation with the U.S. Army Corp of Engineers Waterways Experiment Station (WMS, 1998). It is distinguished by its ability to take the advantage of digital terrain data for hydrologic model development. WMS uses three data sources for model development, which are GIS data, DEMs and triangular irregular networks (TINs).

The format of the developed DEM of the Wadi Namman region was Idrisi raster image format, which is not among the format that can be directly imported by the WMS software. However, WMS accept USGS DEMs format in which elevation data are arranged in an array of rows and columns. The USGS file contains a header, which give information about the reference geographic system and number of rows and columns of the elevation data array. On the other hand, the data in an Idrisi raster file are arranged in a single column and its introductory information are stored in another file called document file. A short FORTRAN program was used to read elevation data from the Idrisi raster file and rewrite them as an array of rows and columns. The introductory information was copied from the Idrisi document file and added as a header of the array.

The resulted file is a DEMs for the study area, which is arranged according to USGS format and can be imported by WMS software. The DEMs file is shown in Fig. 2. It consists of 650 rows and 500 columns covering the study area and has a 70 m square cell size. The geographic reference system of the file is Universal Transverse Mercator (UTM 37-N).

In order to extract basin boundary and channel network the WMS software required the creation of a flow direction grid and flow accumulation files. The two files can be produced using the TOPAZ (Topographic parameterization) model distributed with WMS. The flow direction file can also be produced in GRASS or ArcView GIS software. These programs all use variation of eight-point pour model (WMS, 1998; Garbrecht and Martz, 1995).

In the flow direction file, a flow direction value is assigned to each cell in the DEM file. The flow direction value of a cell is computed by determining which of the eight neighboring DEM cells has the lowest elevation. The values are integer numbers between 1 and 9 depending on the direction of the flow. The file can be displayed as an array of arrows representing the flow direction of each cell in the DEM.

The flow accumulation file is created by computing a value of flow accumulation for each cell in the DEM file using the flow direction file. This value equals to the number of cells whose flow paths eventually pass through that cell. Therefore, DEM cells that are part of a channel have high flow accumulation values since the flow paths of all upstream cells will pass through them. The outlet of the basin should have the highest flow accumulation value since the flow paths all cells in the basin will eventually pass through the outlet cell.

The flow accumulation file is used to extract channel network of a basin. Channels are created for all DEM cells that have a flow accumulation value larger than a user specified threshold value. The channel is traced upstream by noting the neighboring DEM cells with the next highest accumulation value. This process is repeated until no neighboring cell has a accumulation value larger than the threshold value. The boundary of the basin is determined through selecting a cell as the outlet of the basin and then assigning the same value to all cells that eventually drain to the outlet cell using the flow accumulation file.

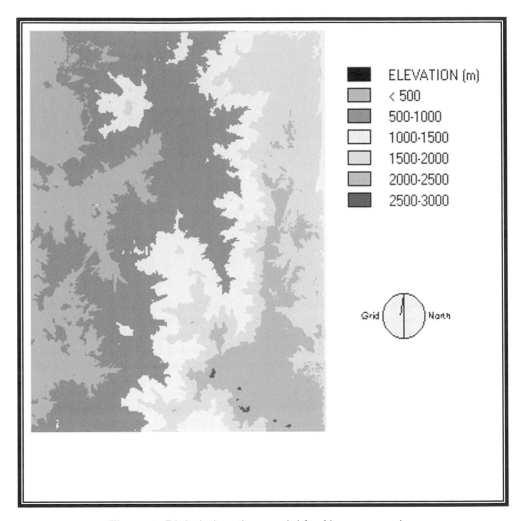

Figure 2. Digital elevation model for Namman region.

The DEM of the study region was exported to the WMS software. The TOPAZ module was used to create both of the flow direction and flow accumulation files. Basin boundary was extracted by WMS as shown in Fig. 3. Several trails were conducted in order to select a threshold value that gives a channel network closer to the manually delineated one. A threshold value equals to 150 produced a channel networks which was very similar to those delineated from the topographic maps. These channels are consisted from all the cells in the DEM which have an accumulation value larger than 150. The extracted channel network according to the specified threshold is also shown in Fig. 3.

The automatically extracted basin boundary and channel network were compared to those delineated manually. The results of the comparison are presented in Table 1. The drainage boundaries of the two maps are very similar. Numbers of both first

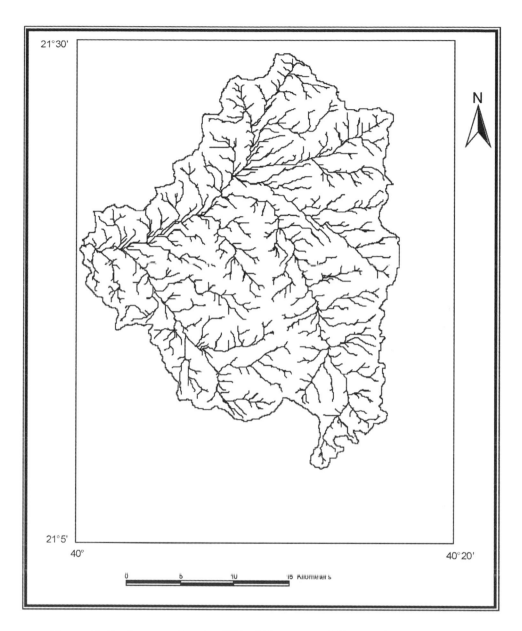

Figure 3. Drainage network of Wadi Namman delineated automaticaly.

and second order channels were significantly larger in the automatically extracted drainage map compare to the manually delineated channels. However, the differences between the number of higher order channels in both maps were not significant. In fact, the order of the basin was the same in both cases. The high difference between the values of mean length of first order channel was due to the large number of very short first order channels produced by the automatic procedure.

The results also show that values of mean area draining to channel and mean slope of channels were comparable between the two procedures.

Conclusions and Recommendations

In comparison to manual extraction procedure, the automatic derivation of watershed boundary and channel network significantly reduces the time and cost required for extracting them provided that DEM is available to the study region. In addition, the automatic technique has less error due to human mistakes. Unfortunately, DEMs are not available for most of the regions in Middle East with the proper resolutions that is suitable for hydrologic studies on basin scale. Therefore, It is recommended to develop DEMs for the region. This is step is expected to particularly promote the applications of hydrologic distributed models in the region. Moreover, DEMs are essential for various applications of GIS in large number of scientific fields that use DEMs as input maps.

References

Band, L.E. (1986). "Topographic partition of watersheds with digital elevation models". Water Resources Research, 22, 25-24.

Band, L.E. (1989). "A terrain-based watershed information system". Hydrological Processes, 3, 151-162.

Chorowicz, J., C. Ichoku, S. Riazanoff and Y. Kim (1992). "A combined algorithm for automated drainage network extraction". Water Resources Research, 28(5), 1293-1302.

Fairfield, J. and P. Leymarie (1991). "Drainage networks from digital elevation models". Water Resources Research, 27(5), 709-717.

Garbrecht, J. and L.W. Martz (1995). "Topaz: An Automated Digital Landscape Analysis Tool for Topographic Evaluation, Drainage Identification, Watershed Segmentation, and Subcatchment Parameterization: Overview," U.S. Department of Agriculture, Agriculture Research Service, ARS Publication No. NAWQL 95-1.

Idrisi Reference Manual, version 2 (1997). Clark Labs, Clark University, Worcester, Masochists, USA.

Jenson, S.K. and J.O. Dominque (1988). "Extraction topographic structure from digital elevation data for geographic system analysis". Photogramm, Engineering, Remote Sensing, 54(22), 1593-1600.

O'Callaghan, J. and D.M. Mark (1984). "The extraction of drainage networks from digital elevation data". Computer Vision Graphics Image Process, 28, 323-344.

Strahler, A.N. (1057). "Quantitative analysis of watershed geomorphology". Transactions of American Geophysical Union, 38, 913-920.

Wilson, J. and J.C. Gallant (2000). Terrain Analysis, New York, USA.

WMS Reference Manual, version 5.1 (1998). Brigham Young University, Provo, Utah, USA.

Yuan, L.P. and N.L. Vanderpool (1986). "Drainage network simulation". Computational Geoscience, 12(5), 653-665.

Surface Water Hydrology, Singh, Al-Rashed & Sherif (eds)
© 2002 Swets & Zeitlinger, Lisse, ISBN 90 5809 363 8

Database development for an experimental watershed

Allen T. Hjelmfelt, Jr.
USDA-Agricultural Research Service
268 Agricultural Engineering, UMC, Columbia, MO 65211, USA
e-mail: HjelmfeltA@missouri.edu

Teri L. Oster and Michelle Pruitt
Biological Engineering Department
University of Missouri, Columbia, MO 65211, USA
e-mail: OsterT@missouri.edu & PruittM@missouri.edu

Abstract

A relational database is being developed to facilitate data storage and information retrieval for scientists and technicians working on an experimental watershed. The project includes (1) migrating legacy hydrologic and climatic records to a modern data storage system; (2) developing new systems for extracting information from data logger records and moving that information to our data storage system; (3) developing an environmental sampling data-entry system to track samples along with QA/QC information from collection through laboratory processing, and into our data storage system; (4) updating field operation data entry to incorporate hand-held devices to record and store all agronomic observations associated with research fields and plots; and (5) providing a graphical interface that allows easy access to the many data layers.

Introduction

Contamination of surface water with sediment and agricultural chemicals is a major public concern for water quality in the central United States. The Goodwater Creek watershed, a 7250 ha agricultural watershed, was instrumented as an experimental erosion and nutrient study site by the U.S. Department of Agriculture, Agricultural Research Service in 1971. In 1990 the Missouri Management System Evaluation Area (MSEA) was established at that location to determine the impact of prevailing farming systems' pesticide use on surface and ground water quality. The watershed topography consists of broad, nearly flat divides, gentle side slopes, and broad alluvial valleys that are dissected by small streams. The soils were formed in shallow deposits of Wisconsin loess over glacial till. Soils are the Udollic Ochraqualfs, Albaquic Hapludalfs, and Vertic Ochraqualfs of the Mexico, Adco, and Leonard series. The most significant characteristic of these soils is a naturally occurring argillic claypan horizon located 0.15 to 0.30 m below the soil surface. The clay content of the argillic horizon is generally greater than 50% and consists

---

Mention of trade names or commercial products in this article is solely for the purpose of providing specific information and does not imply recommendation or endorsement by the U.S. Department of Agriculture.

mostly of montmorillonite. This slowly permeable claypan together with the nearly level topography causes wetness problems during the rainy season. Annual precipitation averages 950 mm. May to June rains, during seedbed preparation and planting, are short but intense. Mean maximum and minimum daily temperatures are 17.1 °C and 6.3 °C, respectively. The soil at the surface and in the rooting zone dries and cracks during the summer growing season.

Site Characteristics and Instrumentation

Investigations are being conducted on three scales: small watershed, field, and plot. The small 7250 ha watershed is instrumented as three nested watersheds with areas of 7250, 3150, and 1215 ha. Weirs provide control for flow measurement during low flow, and bridge openings and the channel provide control for high flows. Ground water quality has been sampled quarterly from 105 ground water wells. Piezometers have also been installed to measure fluctuations in ground water level. As part of the MSEA project, three field-sized watersheds were established within the Goodwater Creek watershed. These sites were instrumented to evaluate the effect of specific farming systems on surface and ground water quality. Surface runoff is measured using V-notch weirs. Ground water is monitored with five well nests on each field. Each nest consists of two to five wells. One well of each nest was drilled to bedrock, which ranged from 4.7 to 16.4 m. The other wells were drilled to successively shallower depths. Ground water elevation has been measured biweekly, and the wells have been sampled quarterly for water quality. Soil water access tubes were installed within each field to monitor changes in soil water content within the unsaturated zone. Thirty research plots were established adjacent to one of the fields for more carefully controlled field studies. Each plot is 18 m wide by 189 m long and has an area of 0.35 ha. Some plots are instrumented with Parshall flumes. Rainfall over the watershed is measured with an array of 17 recording raingages. In addition, a weather station, which includes an evaporation pan, is maintained on the watershed.

Surface water runoff events are sampled for water quality evaluation with flow-proportional pumping samplers installed at each of the watershed weir sites and at the field weirs. In addition, weekly grab samples are collected at the weir sites when adequate flow is available. During 1993 and 1994, grab samples were collected at selected times during the year at stream-road crossings throughout the watershed.

Legacy System

Our legacy data collection system for field information consisted of float-actuated chart recorders marking water level and weighing raingages with chart recorders marking rainfall. These recorders gave a continuous pen trace on a preprinted chart, which could record up to seven days worth of data, each revolution of the drum representing one day. The policy of USDA-Agricultural Research Service has been to express the chart trace using a series of straight-line segments that closely approximate the curve on the chart. The time duration of each line segment varies as needed. Thus, only the points at which a break in the slope occurs are recorded, which produce what are termed "breakpoint" records of rainfall and runoff. These breakpoints were marked manually on each chart, the values recorded on prepared coding forms then transferred to punched cards for processing. Later the punched cards were replaced by simple dBase® data entry tables that were then copied to text files which were uploaded before processing. Daily values for other climatic data, such as temperature and pan evaporation, were handled in a similar manner.

Typically this field information was uploaded and processed on an annual basis. The processing programs resided on mainframe disk and were written in FOR-TRAN. Once all the data for a calendar year had been uploaded, the cyclic execution-correction-reexecution process began. In later years, we were able to stream-line this process with a collection of pre-processing routines that checked the input files for various errors. After errors had been corrected, a final run was executed with output going to both paper and disk. The disk datasets were then written to magnetic tape and erased from the disk. Requests for data then required accessing the magnetic tape files to retrieve desired datasets, a time-consuming task.

Impact of New Instrumentation

The transition to electronic sensors was slow as we tested many devices for stability under extremes of temperature and during thunderstorms. Eventually we modified our raingages with a load cell sensor and data logger. The weighing mechanism and chart recorder remains intact so comparison testing can be carried out as desired. Water levels are sensed using bubbler or linear-differential-transformer pressure transducers connected to data loggers. The water level sensors are integrated with our automatic water quality sampling devices. The recording frequency results in a large number of data points that require a technician to download data weekly to avoid overwriting the data buffer of the data logger.

Water quality samples are obtained as grab samples and from automatic samplers. Grab samples are taken weekly by hand, if flow is available, when the samplers are serviced. The water quality samplers are programmed to take flow-weighted samples. The water quality samples are brought to the laboratory for processing. The data downloaded from the data loggers also requires processing. The binary data are converted to ASCII text format and uploaded to a folder on the Local Area Network for processing. Rainfall and runoff files are used in hydrologic programs whereas the sample start and stop times are incorporated into the chemical database.

The data loggers collect rainfall and runoff at two- and five-minute intervals, respectively, producing annual datasets of 262,800 and 105,120 records each. Most of these records are zero readings or other unnecessary repetitive values. Removing these superfluous readings reduces the mass of data and makes the information more accessible, while not changing the resulting information. This process was carried out by hand on the paper charts. For data logger runoff files, a program called PointBreaker is used to graph the data points on the computer monitor, simulating the recording chart. The technician then uses the mouse to select the breakpoints. For rainfall data, the selection process has been automated as a function of another program called the Input Manager.

Figure 1 shows a screen shot of the Input Manager's graphical interface that allows the user to select and unselect data points. On the left is a grid displaying the original data and on the right is a chart of the original data plus the selected data points. The selected data points are marked with asterisks on the chart. No points are selected on the declining portions of the curve, as those represent periods of no precipitation, when evaporation is occurring. The steep decline in the center of the graph indicates when the data were gathered and the collection pan was emptied.

Since data are collected on a weekly basis, the files for each collection site must be merged to form the annual dataset. As we currently have eleven rainfall gages from which data are collected, manual merging was fraught with error opportunities. The Input Manager merges the files by loading the text data into Paradox tables. Given user-supplied file specifications and date ranges, the program searches for breakpoint data files based on header information contained in the first three records of each dataset.

Data Management for the Chemical Laboratory

The Chemical Laboratory measures concentrations of a suite of agricultural herbicides and inorganic nutrients using a Beckman® High Performance Liquid Chromatography (HPLC) instrument, a Varian® Saturn® Gas Chromatograph/ Mass Spectrometer (GCMS), a LACHAT® QuikChem® Automated Ion Analyzer and ELISA (enzyme-linked immunoassay).

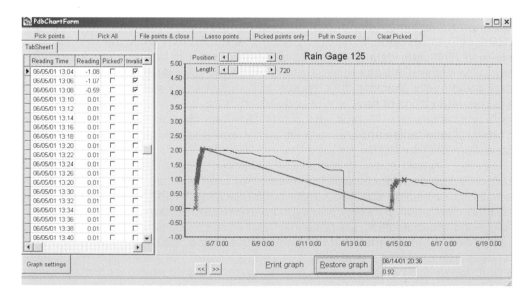

Figure 1. Screen image of Input Manager rainfall selection.

Prior to the development of a Laboratory Information Management System (LIMS), sample custody and analysis data were manually entered in Corel® Quattro Pro® spreadsheets. Quality assurance (QA) calculations were also done in Quattro spreadsheets and required nearly a week's effort by the overseeing scientist to complete each calendar quarter. The LIMS was developed in Borland® Delphi® with Corel® Paradox® database tables. Although between 1991 and 1997 the laboratory averaged only 800 samples per year and since then has averaged around 2200 samples per year, data reports can now usually be prepared the same day they are requested.

Lab personnel, both technicians and students, and overseeing scientists use the LIMS for data entry, performing and reviewing QA/QC calculations, and preparing and viewing data reports. The LIMS was developed with three major goals in mind:

1. Assurance of data accuracy, by facilitating data entry and minimizing manual entry of instrument data to avoid transcription errors, as well as implementing systematic "double checking";
2. Improved data access for lab personnel and scientists;
3. Documentation and reproducibility of results, including automation of QA calculations.

All user interface modules are built over Paradox tables in a relational database.

The LIMS user interface simplifies data entry of custody information in several ways. It eliminates repetitive entry of information that applies to an entire track, such as the received date, by copying fields from the current record into new records when they are inserted. The LIMS automatically increments sample number on a new record and automatically increments track number on a new track. In addition, samples are categorized by 'Project'; relevant fields such as scheduled analyses are automatically filled, and the site list is filtered to show only sites associated with the project. Whenever possible within the limitations of laboratory instrument software, data files (including analysis results and data logger records) are saved as ASCII files to the network file server and imported electronically with appropriate error checking such as automatic range checking of concentration values and verification of sample and track numbers.

Both database tables and raw data (analysis results, data logger records, etc.) are stored on a networked file server. Because lab data are available on the file server, lab personnel and scientists no longer must refer to lab notebooks to determine sample status or obtain results. Added benefits of saving data files to the file server are the daily backup done as standard maintenance by the server administrator, and the elimination of the need to synchronize multiple copies of data files and database tables. In addition, standard Windows tools can be used from any computer on the network to ascertain the location of original data files.

Because the number and output data format of the laboratory instruments determine the data stream, ease of incorporating data from new instruments and analyses into the LIMS was a primary design consideration. To facilitate the transition period during which the LIMS was under development, to accommodate existing laboratory routines, and/or to retain the independence of the database from the LIMS user interface, some tables are not strictly relational.

The QA notebook of the LIMS is primarily concerned with giving lab personnel and scientists access to spike and standard recoveries, recovery fractions, and corrected data. Forming the basis for the LIMS QA notebook, which was developed in Delphi, are Paradox QBE's (query by example) that calculate average recovery fractions and then correct field samples in a recursive calculation. Besides database tables holding raw concentration and volume data and custody information for samples, QA calculations require tables for field spike recoveries that can be reviewed by scientists before corrections are made and documentation of how the calculations were done to ensure reproducibility of results.

Farming System Information

In addition to climatic and chemical information, the MSEA project saw the advent of a farming systems database that manages data concerning various farming operations and events. Included in this collection are planting dates and populations, cultivation operations, fertilizer and pesticide applications, harvest dates and yields. Data are currently stored in relational Paradox tables that reside on our LAN.

Initially this information was gleaned from field notes taken by field technicians and other personnel. To simplify and clarify this information for data entry, a set of forms was designed for organized data recording. After one season using these forms, the technician obtained a handheld device for capturing this data and minimizing the data entry tasks.

The handheld device is a Compaq iPaq that runs Windows CE. We purchased a software package that runs on Windows CE, called FieldWorker, for collecting field data and recording GPS information. The FieldWorker application uses embedded Oracle tables from which we will eventually directly transfer information into our main Oracle data storage tables. For now, the information is exported from FieldWorker as tab-delimited ASCII text then imported into the Paradox tables for report generation.

Data Storage

The database system was constructed using an n-tier architecture. The main data storage is in Oracle® tables on the University of Missouri central data hub. Intermediate storage and middleware reside on a LAN file server. The decision to use Oracle tables for storing finalized data was straightforward, primarily based on one consideration: this database management system was available through the University of Missouri. Data entry and information retrieval are achieved through networked computer workstations, laptops, and handheld devices.

Several issues were considered when designing the structure of our database tables. For data integrity, the database needed to be relationally sound without any

30

redundant elements. Speed of retrieval was also a significant factor. Relative simplicity was also important so that, given the basic table structure, a layperson could construct a query. In addition, when naming elements of the database, highly descriptive names were used, including units in field names.

The data were sorted by type (runoff or rainfall) and frequency (breakpoint or daily), so that each table contains data for every location that collects that particular type and frequency of data. Identical physical tables for each location were considered, but quickly discarded as that design would make the database too unwieldy and yield negligible speed advantages. For ease of use, views were constructed that mimicked individual tables by retrieving data from individual locations.

We decided to calculate and store daily accumulation values for any breakpoint data since most data requests and analysis are based on that frequency. This will also facilitate the computation of long-term averages. The monthly and yearly accumulation values are then accessible through various views constructed using the daily data.

One of our concerns in choosing to store our data in relational database system was the highly sequential nature of our data. As a result, all time-sequenced data are keyed on site location and date and time or date only. When creating views and queries, we are careful to order our resulting datasets by date and time.

Using the Oracle Database

Currently the Oracle datasets are updated using an uploading program that was custom written in Borland Delphi. A spreadsheet application with ODBC connectivity is used to query the database. A client-server application is being constructed to facilitate essential query access to the data by our Unit scientists and personnel. The client application runs on Windows and allows the user to select locations and date ranges using a map of the Goodwater Creek watershed on the main window. The client then sends the data requests to the middleware data server running on a Windows NT/2000 server. The middleware accesses the Oracle tables, processing the request and returning the resulting dataset to the client. Our plan is to provide access to select parts of our database via web browser in the near future.

We are also building analysis applications that use the Oracle data as input. One such application calculates sediment and nutrient transport using merged breakpoint runoff and water sample data. Because each sample is analyzed for multiple pollutants, runoff breakpoint times are merged once with sampling times then individual analyte concentrations are applied to calculate the load transport of each pollutant.

Conclusion

The legacy hydrologic and climatic data have been moved into the Oracle system, decreasing retrieval time immensely. The process for extracting information from rainfall and runoff data logger records is operational, and we continue to streamline that process. The water quality data management system provides lab personnel and scientists efficient access to its data and processes, saving time and money. Soon the water quality results will be available from the Oracle tables as well. The agronomic data system is continually evolving, as new technologies are adopted to simplify the data gathering process. Data access is available to unit scientists through spreadsheet import queries while the graphical user interface is under development. Completion of this work seems continually delayed by related projects. One of the most recent related projects involved merging runoff breakpoint records with water quality sampling records to determine pollutant load values.

Surface Water Hydrology, Singh, Al-Rashed & Sherif (eds)
© 2002 Swets & Zeitlinger, Lisse, ISBN 90 5809 363 8

Hydrological data pre-processing for reservoir planning analysis: Case study using data from England and Iran

A.J. Adeloye
Department of Civil and Offshore Engineering, Heriot-Watt University
Edinburgh EH14 4AS, UK
e-mail: a.j.adeloye@hw.ac.uk

M. Montaseri
Department of Irrigation Engineering, Urmia University, Iran

Abstract

The analysis and use of hydrological data for decision making in water resources can only be meaningful if the data possess the right characteristics. In general, it is customary that data being analysed are consistent, free of trend, constituting a random series and are following the appropriate probability distribution hypothesis. These characteristics were investigated during the course of an extensive water resources planning study recently carried out. The results and the experience gained during the exercise form the subject of this paper.

Introduction

Most statistical analyses of hydrological time series data are based on a set of fundamental assumptions, i.e. that the series is consistent, is trend-free, constitutes a random series, and follows the appropriate probability distribution function. Data exhibiting these properties are much easier to analyse than data which do not. On the other hand, it is wrong to press ahead with analysing data based on the premise of these assumptions without first establishing, through proven statistical or other tests, that the assumptions are valid or, more appropriately, cannot be rejected on the basis of statistical evidence. In general therefore, it is important that appropriate tests are carried out to establish these characteristics prior to any further analyses of the time series data.

Consistency implies that all the collected data belong to the same statistical population. This would mean that the data set has been collected without changes in equipment, observation procedures or gauge location. If any of these had happened, then the data are no longer consistent and the cause of the inconsistency must be identified and the inconsistency corrected before any further analyses of the data take place.

Trend exists in a data set if there is a significant correlation (positive or negative) between the observations and time. Trend or non-stationarity is normally introduced through human activities such as land use changes or the human-induced

33

climate change. The presence of trend complicates data analysis, if such a trend must be explicitly accommodated in the analysis. Therefore it is often much simpler if the trend is quantified and then removed from the data. Following analysis of the de-trended data, any quantified trend can then be added back.

Randomness in a hydrologic time series in general means that the data arise from natural causes. If the data series is not random, then it is persistent; this persistence is normally quantified in terms of the serial correlation coefficient (see McMahon and Mein, 1986). Most statistical analyses in water resources and hydrology assume a random sample and it is therefore important to establish this characteristic before subjecting the hydrological time series to statistical analysis. However, if an otherwise significant serial dependence is ignored, this could have consequences for decisions based on the outcome of the data analysis. For example, the presence of serial dependence in streamflow data causes reservoir storage requirements to increase for a given demand (see Adeloye et al., 2001). If the serial dependence is ignored and the streamflow data are assumed to be random, the storage requirement will be underestimated, thus resulting in possible serious under-performance of the reservoir in meeting the demand placed on it. Finally, the choice of the appropriate probability distribution for describing the random sample should also be based on a robust statistical approach.

Commonly used tests for establishing each of the above characteristics are described in the following Sections.

Consistency Test

The most widely used technique for evaluating a time series data, such as rainfall or streamflow, for consistency is the double mass curve (Bras, 1990). The double mass curve procedure is based on the comparison of cumulative values of two data sets in a diagram form; hence the name double mass. One of the data records is consistent or reliable while the other is suspect. When plotted, the double mass diagram should show a linear relationship when the suspected data set is consistent; otherwise, there will be a departure from linearity, which is an indication of change or inconsistency. Usually, this change is revealed by a change in slope and the point in time where this occurs is the start of the change.

Since the mean data for a number of sites is unlikely to be affected by changes at one of the component sites, it is customary to use the mean data as the basis of the consistent or reliable data. Alternatively, data at a long-term station in the region with no known human influences could be used as the consistent set.

A schematic illustration of a double mass curve plot is shown in Fig. 1. The accumulated data at the suspect site are represented by the ordinate and the accumulated data based on the mean data at N_s sites in the region constitute the abscissa. Two possible changes occurring in time T are circled, one resulting in a steeper slope (S_1) and the other resulting in a flatter slope (S_2) in comparison with the slope prior to time T.

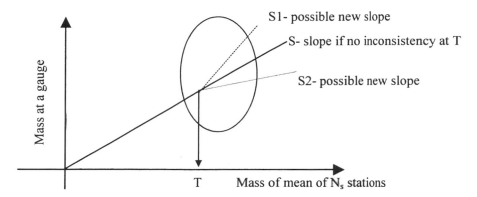

Figure 1. Schematic illustration of double mass curve and inconsistency in hydrological data.

The post-T record at the suspect gauge cannot be used with its pre-T record without first carrying out some adjustments. The precise adjustment will depend on which segment of the suspect record, pre-T or post-T, is thought to be more reliable. For example, if more confidence is placed on the post-T measurements, then the pre-T records can be corrected for current slope S_1 by multiplying by (S_1/S), or for current slope S_2 by multiplying by (S_2/S). S is the pre-T slope and S_1 and S_2 are as defined previously. Similar corrections can be made if the pre-T records were thought to be more reliable.

Trend Test

As remarked earlier, trend occurs when there is correlation between the data values and time, which distinguishes it from persistence which is measure of the correlation between adjacent data values (Hameed et al., 1998). Thus, the most commonly used method of trend detection is to formulate a linear model between the data values and time of the form (Vogel and Kroll, 1991; Hameed et al., 1998):

$$Y_i = a + b_i + v_i \qquad (1)$$

where Y_i, i= 1,2....n, is the data value at time i, a and b are regression coefficients and v_i is a random error (white noise) with a mean of zero and variance of S_v^2. If there is a linear trend in the data set, then the regression estimate of the slope parameter b will be statistically different from zero; otherwise it becomes difficult to justify the existence of trend.

Hameed et al. (1998), however, caution that the methodology in Eq. 1 does not discriminate between the existence of a long-term trend, which is of interest, and persistence (or stochastic trend – see Box and Jenkins, 1976; Cryer, 1986). Persistence merely represents the correlation between adjacent observed values in a data record. When such a neighbourly correlation or persistence is present in the data, it becomes impossible for the residual component v_i to be a white noise, thus violating one of the cornerstone assumptions of the regression model in Eq. 1

(Draper and Smith, 1966). (Another cornerstone assumption of Eq. 1 is that v_i must have a normal distribution.) To avoid the confounding effect of the neighbourly correlation on the long-term trend, Hameed et al. (1998) then suggest modelling the time series of the v_i using the auto-regressive (AR) model.

To avoid all of the above complications, the Spearman Rank Order Correlation (SROC) non-parametric test was used to investigate the correlation between data values and time and hence to establish whether or not a long-term trend exists in the data. This test has been recommended by the World Meteorological Organization for trend detection (WMO, 1988). Being non-parametric, the SROC test is sufficiently robust because it makes no *a priori* assumption about the distribution hypothesis of the time series. The SROC test is fully described by Kendall and Stuart (1976) (see also Kendall and Gibbons, 1990). We adopt the presentation by McGhee (1985) to illustrate the methodology as follows.

Let data series Y_i, i =1,....,n be observed in time i.

1. Assign ranks R_{yi} to Y_i, such that the least Y_i has R_{yi} =1 and the largest Y_i has a rank = n. Where there are ties in the Y_i, then assign to each of the ties a rank equal to the mean of the ranks that would have been used had there been no ties.

2. Compute the difference $d_i = R_{yi} - i$.

3. Compute the coefficient of trend r_s as:

$$r_s = 1 - \frac{6 \sum_{i=1}^{n} d_i^2}{n(n^2 - 1)} \qquad (2)$$

Under the null hypothesis (H_o) that the time series has no trend, i.e. that there is no correlation between the Y_i and i, it can be shown that the variable:

$$t = r_s \sqrt{\frac{n-2}{1-r_s^2}} \qquad (3)$$

has a Student's t-distribution with n–2 degrees of freedom.

4. Obtain critical values of the t-distribution for the chosen significance level α and n–2 degrees of freedom. For a two-tailed test, denote these critical values by $\pm t_{\alpha, n-2}$.

5. Compare the t obtained in Eq. 3 with the critical values. Reject H_o if $t > t_{\alpha, n-2}$ or $t < -t_{\alpha, n-2}$.

We illustrate the SROC methodology with a simple hypothetical example adapted from McGhee (1985).

Example

The annual runoff data (mm) recorded at a gauging site on a medium-sized river basin over a 12-year period are as follows:

i (yr)	1	2	3	4	5	6	7	8	9	10	11	12
y_i (mm)	68	21	88	92	32	5	50	81	64	70	26	84

Using the SROC test, establish whether or not a trend exists in the runoff data.

Solution

The solution is set out in the table below.

i	1	2	3	4	5	6	7	8	9	10	11	12
y_i	68	21	88	92	32	5	50	81	64	70	26	84
R_{yi}	6	11	2	1	9	12	8	4	7	5	10	3
d_i	5	9	-1	-3	4	6	1	-4	-2	-5	1	9
d_i^2	25	81	1	9	16	36	1	16	4	25	1	81

$$r_s = 1 - \frac{6 \times 296}{12(144 - 1)} = -0.03$$

The negative sign of r_s is an indication that when one of either the runoff or time increases, the other should reduce. However, given that its numerical value (absolute) is rather low, being only 0.03, this association is likely to be very weak. Using Eq. 3, t = −0.095. For a 5% significance level and degrees of freedom = 12 − 2 = 10, critical values from the t-distribution are ±$t_{0.025,10}$ = ±2.228. Since −0.095 is greater than −2.228, the null hypothesis is not rejected. This is to say that there is no statistical evidence of a long-term trend in the data record.

Randomness Test

As noted by Durrans and Tomic (1996), it is often difficult to find a satisfactory test for randomness when the data length is less than 100. They therefore recommend using simple visual inspection of the time series plot, in combination with any background knowledge about possible human influences on the basin, to decide on whether or not the data series is random. However, such an approach is highly subjective, meaning that its outcome will vary from one analyst to another.

Because of this, we have employed the non-parametric Run Test to evaluate our data series for randomness. This test is described by McGhee (1985) which has been adapted here. Let the objective be to test whether the data sample Y_i, i = 1,....,n is random based on the runs of the data with respect to the median of the observation. The procedure is therefore as follows:

1. Determine the median of the observation. To do this, sort the sample in increasing order of magnitude such that $y_1 \leq y_2 \leq .... \leq y_n$. Then for an integer k, such that n = 2k (even) or n = 2k+1 (odd), the sample median denoted by $\bar{y}_{0.5}$ is estimated as (Loucks et al., 1981, p. 102):

$$\bar{y}_{0.5} = \begin{cases} y_{k+1} & \text{for } n = 2k \\ 0.5(y_k + y_{k+1}) & \text{for } n = 2k + 1 \end{cases} \tag{4}$$

2. Examine each data item in turn to see whether or not it exceeds the median. If a data item exceeds the median, then this is a success case (replaced by letter S) but if it does not exceed the median, it is a failure case (denoted by letter F). Cases that are exactly equal to the median are excluded.
3. Count the number successes and denote this by n_1; similarly denote the number failures by n_2. In general, $n = n_1 + n_2$ except where some of the values are omitted as explained in step 2 above.
4. Determine the total number of runs in the data. A run is a continuous sequence of S's until it is interrupted by an F and vice versa. Let the total number of runs be denoted by R.
5. Compute the test statistic

$$z = \frac{R - \left(\frac{2n_1 n_2}{n_1 + n_2} + 1\right)}{\sqrt{\frac{2n_1 n_2 (2n_1 n_2 - n_1 - n_2)}{(n_1 + n_2)^2 (n_1 + n_2 - 1)}}} \tag{5}$$

6. Under the null hypothesis H_o that the sequence of S's and F's is random, z has a standard normal distribution. Hence obtain critical values of the standard normal distribution for the chosen significance level α and denote these by $\pm z_{\alpha/2}$.
7. Compare the z obtained in step 5 (see Eq. 5) with the critical values $\pm z_{\alpha/2}$. Reject H_o if $z < -z_{\alpha/2}$ or $z > z_{\alpha/2}$.

We also illustrate the run test with a hypothetical example adapted from McGhee (1985).

Example

The following are the annual rainfall measurements (mm) for a river basin over a 27-year period.

8,14,13,12,14,8,6,7,15,16,14,7,11,10,
9,9,12,13,12,14,15,9,8,9,10,11,11

Using the run test, decide whether or not the data can be considered as a random sample with respect to the median.

Solution

The median of the 27 observations is 11. Replacing measurements below 11 by F and above 11 by S, we have the series in terms of S and F as:

F SSSS FFF SSS FFFF SSSSS FFFF

The runs have been underlined. Note that the three 11's have been ignored since these are exactly equal to the median. Thus in this example, $n_1 = 12$; $n_2 = 12$; and R = 7. Using these, z = −2.50. At the 5% significance level for a two-tailed test, $\pm z_{\alpha/2} = \pm1.96$. Since −2.50 < −1.96, we reject the null hypothesis. There is therefore no statistical evidence of randomness in the data.

Probability Distribution Goodness-of-Fit Tests

There are several commonly used formal methods of testing the goodness-of-fit of time series data to postulated theoretical probability distribution functions (see Stedinger et al., 1993). These include the chi-squared test, the kolmogorov-Smirnov test, the probability plot correlation coefficient (PPCC) test, and the moment (l-moments and p-moments) ratio diagrams test.

Apart from the formal tests, it is also possible to infer the form of the probability distribution by plotting the histogram of the data. This histogram may either be symmetrical, indicating the use of the normal distribution, or may be unsymmetrical indicating a non-zero skew which cannot be modelled with the normal distribution. To reinforce the indication of a normal distribution when using the histogram plot methodology, a simple statistical test can be carried out to determine whether or not the estimated sample skew coefficient is statistically zero.

To test the significance of the skew coefficient, consider again the random variable y_i, i =1,2,3....n, where n is the sample size. The skew coefficient of y, g_y, which is the third moment, can be estimated using:

$$g_y = \frac{n\sum_{i=1}^{n}(y_i - \bar{y})^3}{(n-1)(n-2)S_y^3} \tag{6a}$$

$$\bar{y} = \frac{\sum_{i=1}^{n}y_i}{n} \tag{6b}$$

$$S_y = \sqrt{\frac{\sum_{i=1}^{n}(y_i - \bar{y})^2}{n-1}} \tag{6c}$$

where S_y is the sample standard deviation and $\bar{y}$ is the sample mean. The standard error of estimate of g_y is only a function of n and is given by (McMahon and Mein, 1986):

$$S_{g_y} = \sqrt{\frac{6n(n-1)}{(n-2)(n+1)(n+3)}} \tag{7}$$

Assuming that the distribution of the skew coefficient is normal, then under the hypothesis that the skew coefficient is zero, the 95% confidence limits for zero skew become:

$$CL = 0 \pm 1.96 S_{g_y} \tag{8}$$

If the skew coefficient estimated using Eq. 6a is within the range of Eq. 8, then the null hypothesis is not rejected; otherwise the data is assumed to have a non-zero skew coefficient. When the skewness is statistically non-zero, it becomes imperative to use one of the formal goodness-of-fit tests to discriminate between competing distribution hypotheses.

As noted by Stedinger et al. (1993), the available formal tests mentioned above vary in their complexity and in their power to discriminate between competing distributions. Following a review of all the relevant issues, we have adopted the PPCC test, originally due to Filliben (1975), which is both simple to use and also powerful in discriminating between competing probability distribution functions. The test is based on the correlation coefficient between the ordered sample y_i, i.e. such that $y_1 \le y_2 \le y_3 \le \ldots \le y_n$, and their corresponding fitted quantiles.

The fitted quantiles are obtained first by assigning probability to the ranked observations using an appropriate plotting position formula. Let the ordered data y_i have a rank R_{yi}, such that y_1 has $R_{y1} = 1$ and y_n has a rank $R_{yn} = n$. The most commonly used plotting position formula, because of its unbiasedness, is (Cunnane, 1978):

$$p_i = \frac{R_{yi} - 0.4}{n + 0.2} \tag{9}$$

where p_i is the probability of non-exceedance for observation y_i. The fitted quantile corresponding to p_i then becomes $w_i = G^{-1}(p_i)$, where $G^{-1}(.)$ is the inverse of the cumulative distribution function (cdf), corresponding to the argument in the bracket, of the probability distribution function being tested. The correlation coefficient between the y_i's and w_i's is then obtained using:

$$r = \frac{\sum_{i=1}^{n}(y_i - \bar{y})(w_i - \bar{w})}{\left[\sum_{i=1}^{n}(y_i - \bar{y})^2 \sum_{i=1}^{n}(w_i - \bar{w})^2\right]^{0.5}} \tag{10}$$

In Eq. 10, both the sample means $\bar{y}$ and $\bar{w}$ are obtained using the expression in Eq. 6b.

To decide on whether or not a correlation is significant, the calculated r is compared with critical points of the PPCC. These critical points are provided by Vogel (1986) for the normal, lognormal and gumbel probability distribution functions. Vogel and McMartin (1991) also provide the critical values for the Pearson type 3 distribution; these can be used for logarithm-transformed variates if the data are being tested for the log Pearson type 3 distribution.

It is obvious from the foregoing discussions that the PPCC test is most suited for distributions whose inverse cdf, i.e. $G^{-1}(.)$, can be readily obtained. This is more or less the case for the commonly used probability distribution functions for modelling skewed data series in hydrological analysis, such as the log-normal distribution (2-parameter, LN2, and 3-parameter, LN3), the gamma distribution and the log-Pearson III (LP3) distribution.

In general, a variable Y has a 3-p log-normal distribution if the variable $X = \log_e(Y - \vartheta)$ has a normal distribution, where ϑ is the lower limit. The pdf of the 3-p log-normal distribution is:

$$f_Y(y) = \frac{1}{\sqrt{2\pi}\sigma_x y} \exp\left[-\frac{1}{2}\left(\frac{\ln(y - \vartheta) - \mu_x}{\sigma_x}\right)^2\right]; \qquad 0 \le y < \infty \qquad (11)$$

where μ_x and σ_x are respectively the mean and standard deviation of $x = \log_e(y - \vartheta)$. Together with ϑ, μ_x and σ_x constitute the three parameters that completely describe the LN3 distribution function. Nearly maximum likelihood estimates of the parameters can be found using (Stedinger et al., 1993):

$$\vartheta = \frac{y_{max}y_{min} - y_{0.5}^2}{y_{max} + y_{min} - 2y_{0.5}}; \quad y_{max} + y_{min} > 2y_{0.5} \qquad (12)$$

$$\mu_x = \left(\frac{1}{n}\right)\sum_{i=1}^{n} x_i \qquad (13)$$

$$\sigma_x^2 = \frac{1}{(n-1)}\sum_{i=1}^{N}[x_i - \mu_x]^2 \qquad (14)$$

where y_{max}, y_{min} and $y_{0.5}$ are the largest observed, smallest observed and sample median of variable y. With the parameters estimated, quantiles w_i can then be estimated using:

$$w_{i,LN3} = G^{-1}(p_i) = \vartheta + \exp\left(\mu_x + z_{p_i}\sigma_x\right) \qquad (15)$$

41

where $w_{i,LN3}$ is the value of y at the $p_i\%$ probability of non-exceedance as predicted by the LN3 distribution, and Z_{pi} is the standardised normal variate at $p_i\%$. A good approximation for Z_{pi}, given p_i, is (Stedinger et al., 1993):

$$Z_{pi} = \frac{(p_i/100)^{0.135} - (1-p_i/100)^{0.135}}{0.1975} \quad (16)$$

The LN2 distribution is a special case of the log-normal distribution in which the lower limit $\vartheta = 0$. However, while it is also possible to model skewed data with the LN2 distribution, this is often unjustifiable because of the unique relationship between the coefficient of variation (CV) and the skew coefficient (g_y) of the LN2:

$$g_y = CV^3 + 3CV \quad (17)$$

The CV is the ratio of the standard deviation (Eq. 6c) to the mean (Eq. 6b) and g_y can be obtained using Eq. 6a.

To fit the gamma distribution and obtain its quantiles, Loucks et al. (1981) recommend using the Wilson-Hilferty transformation (Wilson and Hilferty, 1931). Using this transformation, the quantiles for the gamma distribution, $w_{i,g}$, are given by:

$$w_{i,g} = G^{-1}(p_i) = \mu_y + S_y \left[\frac{2}{\gamma_y} \left(1 + \frac{\gamma_y Z_{pi}}{6} - \frac{\gamma_y^2}{36} \right)^3 - \frac{2}{\gamma_y} \right] \quad (18a)$$

$$\gamma_y = \frac{g_y(1-\rho_1^3)}{(1-\rho_1^2)^{1.5}} \quad (18b)$$

In Eq. 18, ρ_1 is the lag-1 auto-correlation coefficient and all the other symbols are as defined previously. An efficient algorithm for obtaining ρ_1 is provided by McMahon and Mein (1986). The Wilson-Hilferty transformation is usually satisfactory when the modified skew coefficient obeys the inequality $-3 \le \gamma_y \le 3$ (Kirby, 1972).

The gamma distribution, like the LN2, is a 2-parameter function with a unique theoretical relationship between its CV and skew coefficient given by:

$$g_y = 2CV \quad (19)$$

Therefore, where there must be flexibility in the skew to be modelled, a more general 3-parameter version of the gamma distribution, known as the Pearson type 3 (P3) distribution, must be used. The fitting of the P3 distribution follows that of the gamma distribution by working with the variable $x_i = y_i - \xi$, $i = 1,2,...,n$, where ξ is the third parameter of the P3 distribution known as the location parameter. For the gamma distribution, $\xi = 0$. Stedinger et al. (1993, Table 18.2.1) provide efficient algorithms for estimating the three parameters, i.e. location, scale and shape, of the

42

P3 distribution. To model the LP3 distribution, Stedinger et al. (1993) suggest taking the natural logarithms of P3 distributed y and treating such logarithms as if they were distributed as P3.

Application

The above data analysis procedures were implemented as part of an extensive water resources planning study whose aim was to further understand the planning characteristics of within-year and over-year reservoir systems. The planning analysis was implemented within a Monte Carlo framework and the associated statistical analyses meant that the reservoir inflow data had to possess the right characteristics of consistency, randomness, no trend and should be modelled with the right probability distribution function.

The analysis made use of monthly river flow data from a total of seven sites- three sites in Iran and four sites in Yorkshire region, England. The relevant physical characteristics as well as the summary statistics of the annual flow data at the seven sites are shown in Table 1. On the basis of the coefficient of variation (CV) of the annual flows shown in Table 1, all seven sites would be characterised as low to medium variability streams (see McMahon et al., 1992). The corresponding summary statistics for the monthly streamflow data records at the sites are shown in Table 2. All the summary statistics shown in Tables 1 and 2, i.e. the mean, CV, skewness and serial correlation coefficients were obtained using sample estimator expressions presented by Stedinger et al. (1993, page 18.4). Some of the results of the water resources planning studies have been published elsewhere (see Montaseri and Adeloye, 1999; Adeloye et al., 2001); consequently, only the results of the preliminary data analyses are reported in this paper.

Table 1. Summary characteristics of the river basins analysed.

Site	Reservoir	Area	Annual flow			
			Mean	CV	Skewness	Serial correlation (lag-1)
		km^2	(10^6 m^3)			
England						
1	Gorpley	2.8	2.6	0.19	0.20	0.04
2	Hebden	26.4	25.1	0.20	0.12	0.17
3	Luddenden	5.4	5.1	0.18	0.26	0.19
4	Ogden	5.4	4.3	0.22	0.01	-0.02
	Group total	40.0	37.1	0.19	0.16	0.14
Iran						
1	Baranduz	618	272.7	0.34	1.21	0.34
2	Shahr	418	170.3	0.38	1.04	0.38
3	Nazlu	1715	390.3	0.41	1.03	0.41
	Group total	2751	833.3	0.35	0.96	0.22

Table 2. Summary statistics of monthly flow records at the sites.

Site	Parameter	Jan	Feb	Mar	Apr	May	Jun	Jul	Aug	Sep	Oct	Nov	Dec
Gorpely	Mean (10^3 m^3)	371.3	277.5	235.2	159.9	109.9	79.8	88.6	145.4	170.8	248.8	342.2	382.6
	CV	0.45	0.54	0.57	0.50	0.66	0.74	0.72	0.85	0.72	0.64	0.47	0.45
	Skew.	0.56	0.49	1.48	0.36	1.20	1.31	0.76	1.34	0.90	1.18	0.19	0.73
Hebden	Mean (10^3 m^3)	3554.7	2628.1	2225.9	1624.3	1006.7	797.5	922.0	1492.3	1692.4	2385.9	3153.4	3596.1
	CV	0.40	0.46	0.58	0.54	0.65	0.67	0.67	0.77	0.69	0.60	0.48	0.40
	Skew.	0.72	0.55	1.28	0.67	1.39	1.07	0.97	1.48	0.90	1.07	0.17	0.47
Luddenden	Mean (10^3 m^3)	625.8	521.6	473.6	399.8	316.4	273.6	251.2	322.2	345.0	442.6	544.0	601.0
	CV	0.29	0.37	0.37	0.35	0.41	0.42	0.39	0.53	0.57	0.51	0.42	0.37
	Skew.	0.36	0.32	0.88	0.77	1.27	1.16	1.06	1.22	0.93	1.02	0.20	0.58
Ogden	Mean (10^3 m^3)	609.5	481.9	376.5	290.7	177.8	141.0	125.0	224.3	276.8	403.9	554.6	619.5
	CV	0.40	0.49	0.57	0.61	0.80	0.83	0.90	0.97	0.87	0.72	0.53	0.49
	Skew	0.58	0.60	1.00	0.96	1.86	1.37	1.58	1.91	1.37	1.15	0.30	0.71
Baranduz	Mean (10^6 m^3)	12.73	12.45	17.42	37.77	61.19	51.87	26.86	11.14	5.70	7.48	13.53	14.57
	CV	0.26	0.29	0.45	0.41	0.39	0.45	0.59	0.81	0.55	0.55	0.44	0.32
	Skew	1.20	1.54	1.87	1.01	0.88	0.73	1.33	2.41	1.07	1.21	0.93	0.92
Shahr	Mean (10^6 m^3)	3.22	3.46	7.10	25.25	47.43	46.94	18.00	5.80	2.65	2.51	4.02	3.88
	CV	0.49	0.44	0.54	0.46	0.38	0.44	0.62	0.71	0.61	0.63	0.71	0.56
	Skew	1.32	1.20	1.37	0.89	1.02	0.72	1.18	1.34	0.81	1.48	1.36	2.19
Nazlu	Mean (10^6 m^3)	13.07	12.87	19.99	60.20	115.76	83.46	30.33	11.18	7.73	8.99	13.55	13.22
	CV	0.31	0.31	0.46	0.44	0.46	0.60	0.76	0.80	0.72	0.54	0.45	0.32
	Skew	1.34	0.95	1.68	0.95	0.66	0.94	1.60	1.13	1.14	1.15	0.98	1.02

Results and Discussion

Consistency Test

The double mass curves at the English sites are shown in Fig. 2 while those for the Iranian sites are shown in Fig. 3. In each of the graphs, the ordinate is the accumulated annual flow at the particular site under consideration while the abscissa is the sum of the annual flow at all the sites in the region, i.e. four sites in England and three sites in Iran. As can be seen in the plots, the double mass curves indicate a reasonable degree of consistency, with no discernible change in slope for any of the sites. This would imply that all the collected data belong to the same population and could be used together as such.

Trend Test

The values of the SROC test statistic for the annual data at each of the sites are compared with the critical points of the t-distribution at the 5% significance level in Table 3. On the basis of this information, it can be concluded that there is no statistically significant long-term trend in the annual runoff data records at any of the seven sites. It can therefore be safely assumed that all the annual runoff data series are stationary. A different outcome would have been surprising for these data sets because all the flow measurement sites are upstream of any major flow regulation which could impart non-stationarity (or trend) into the data. Indeed, the three sites in Iran are sites of planned reservoir development and so even here, the influence of human intervention would be much less than at the English sites. Another possible source of non-stationary behaviour is the possible impact of climate change; however, given the moderate lengths of the data records considered in the analysis, this is not likely to be significant.

Randomness Test

The results of the run tests are also presented in Table 3. There is obviously no evidence to reject the null hypothesis that the annual time series record of runoff at each of the seven sites is random at the 5% significance level. This should come as no surprise given the very low values of the annual lag-1 auto-correlation coefficient for the seven data sets shown in Table 1. It should be noted that although the annual data are random on the evidence of the run test, the stochastic modelling of the annual flows for the subsequent Monte Carlo water resources planning analysis still utilised the lag-1 auto-regressive (AR) model (see Montaseri and Adeloye, 1999). Thus the low to moderate lag-1 annual auto-correlation coefficients shown in Table 1 were actually explicitly incorporated in the stochastic models for the annual flows. This was to forestall any of the previously outlined undesirable impacts of ignoring the auto-correlation on estimated reservoir planning characteristics.

Goodness-of-Fit Tests for Probability Distribution

As remarked previously, the water resources planning studies, which employed the data series, were implemented within a Monte Carlo Simulation (MCS) framework.

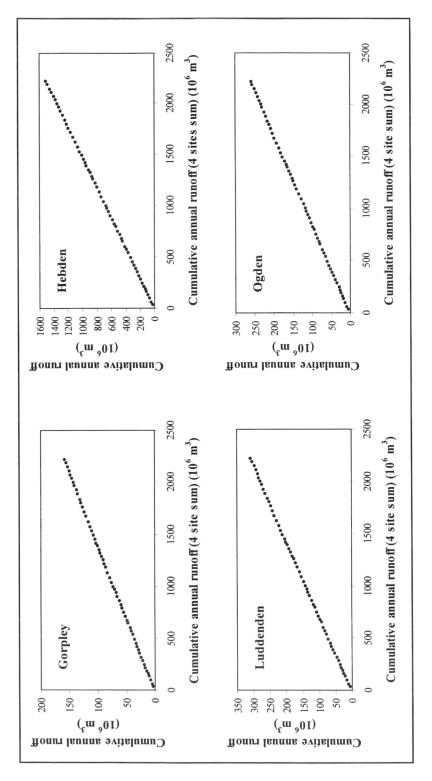

Figure 2. Double mass curves for annual flows at the English sites.

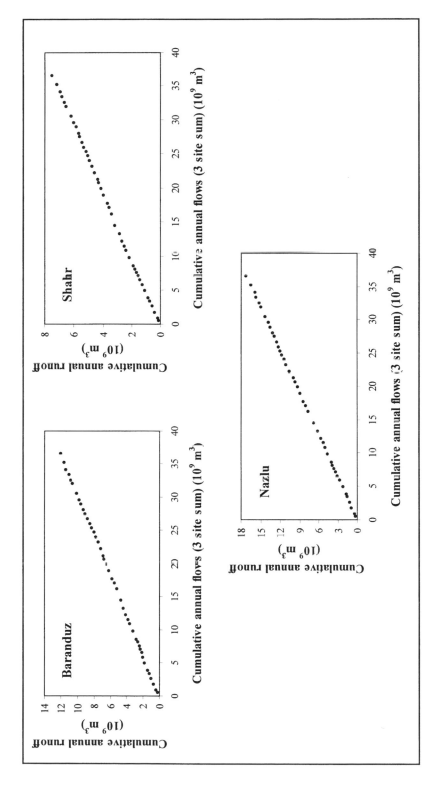

Figure 3. Double mass curves for annual flows at the Iranian sites.

Table 3. Results of the SROC and Run tests for trend and randomness respectively at the seven sites.

Site	Reservoir	Trend		Randomness	
		Test statistic	Critical value at 5% level (t-distribution)	Test statistic	Critical value at 5% level (normal distribution)
England					
1	Gorpley	-1.15	±2.00	+0.01	±1.96
2	Hebden	-1.99	±2.00	-0.26	±1.96
3	Luddenden	-1.26	±2.00	-1.32	±1.96
4	Ogden	-1.17	±2.00	+1.32	±1.96
Iran					
1	Baranduz	-1.55	±2.02	-0.31	±1.96
2	Shahr	-0.74	±2.02	-0.94	±1.96
3	Nazlu	-1.47	±2.02	-0.94	±1.96

In MCS, several realisations of the single historic data records must be generated through stochastic models whose parameters are estimated from the historic records. For such stochastic models to be effective, it is important that the models are formulated using the right probability distribution function. In the water resources studies, the interest covered both within-year and over-year reservoir system behaviours (see Montaseri and Adeloye, 1999); consequently, both the probability distribution of the annual streamflow series as well as those for each of the twelve months in the year must be specified. Consequently, some of the procedures described in the previous section were applied to test competing probability distribution functions at the sites.

Histogram and Skew Tests

The histograms of the annual flows at the English and Iranian sites are shown in Figs 4 and 5 respectively. It can be seen from these Figs that while the histograms for the English sites are almost symmetrical in shape, those for the Iranian sites are skewed to the right. This evidence represents a preliminary indication that the annual flows at the Iranian sites, unlike their English counterparts, possess a positive skew and may not be modelled with the normal distribution function. The sample skew coefficients for the annual flows are presented in Table 1.

To further prove the significance of the skew coefficients at the Iranian sites when compared with the English sites, we performed the simple skew test described previously (see Eqs 6–8) by constructing approximate confidence intervals for zero skew at the sites. For n = 60 at the English sites, the approximate 95% confidence limits about zero for the sample skew are ± 0.605 (= ± 1.96 × 0.31, where 0.31 is the standard error of the skew coefficient as estimated from Eq. 7, based on n = 60). As seen in Table 1, all the annual skew estimates for the English sites are within this range, implying that the estimated skew coefficients at the sites are not statistically different from zero at the 5% level. The corresponding limits at the Iranian sites, with n = 44 years, are ± 0.701 (= ± 1.96 × 0.36); the skew estimates for the Iranian sites in Table 1 are well outside this range, implying that they are significant.

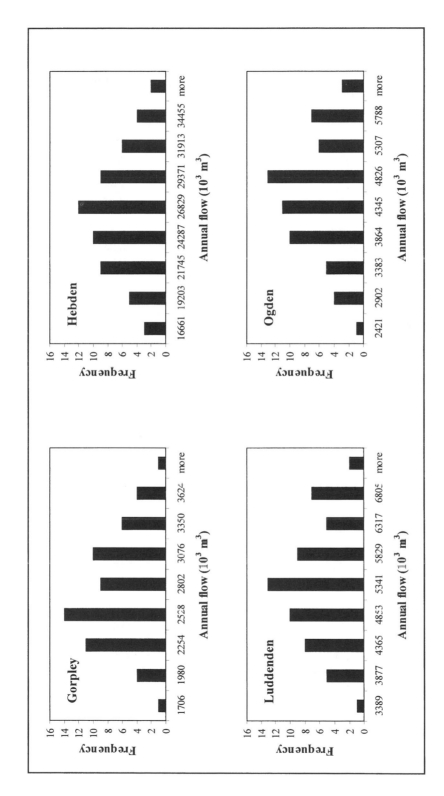

Figure 4. Histograms showing frequency distribution of historical annual flows for the English sites.

49

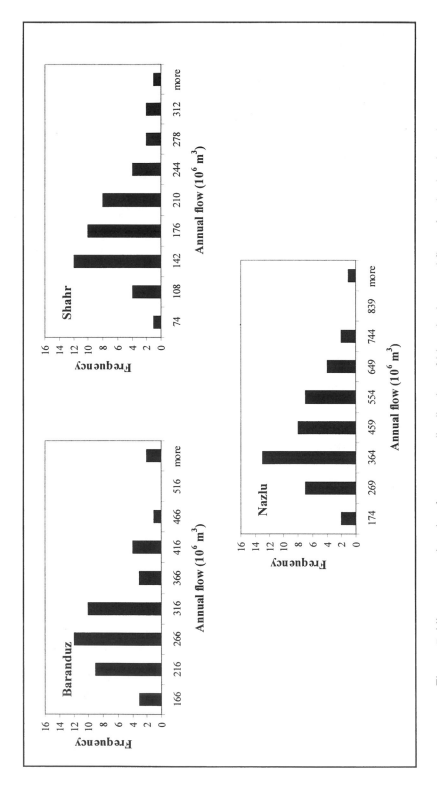

Figure 5. Histograms showing frequency distribution of historical annual flows for the Iranian sites.

The histograms of the flows for each of the twelve months have not been shown here for lack of space. However, for both the English and Iranian sites, most of the sample skew coefficients for the monthly flows shown in Table 2 are out of the range of their appropriate confidence limits; consequently the assumption of normality for these monthly data records does not hold.

PPCC Goodness-of-Fit Probability Distribution Test

Table 4 presents the correlation coefficients obtained from the PPCC method for the historical annual and monthly streamflow data at one of the English sites (Gorpley). Table 5 contains the corresponding results at one of the Iranian sites (Baranduz). Individual results for the other five sites are not presented here for lack of space; they are, however, similar to those in Table 4 (English sites) and Table 5 (Iranian sites) (see Montaseri, 1999 for details). For completeness, the results in Tables 4 and 5 also include the normal distribution.

In an earlier section where the PPCC methodology was presented, it was stated that the way to test the appropriateness of a distribution hypothesis is to compare the estimate of its PPCC with the appropriate critical points for that distribution. However, rather than adopt this approach in the study, we based our selection of distribution on the magnitude of the PPCC and we chose the distribution whose PPCC was the highest. The advantage of this approach over the more formal one is that it is not necessary to have the table of critical points for the various distributions to make a decision on the most appropriate probability distribution function. In addition, since the PPCC represents the degree of linearity between the observed

Table 4. Correlation coefficients obtained from the PPCC test for monthly and annual historical streamflow data record in the Gorpley basin (maximum PPCC is underlined).

Month	Normal	Gamma	LN2	LN3	LP3
Jan.	.9863	.9949	.9753	<u>.9952</u>	.9949
Feb.	.9787	.9855	.9743	.9892	<u>.9940</u>
Mar.	.9414	.9931	.9936	<u>.9942</u>	.9939
Apr.	.9853	.9904	.9803	.9903	<u>.9932</u>
May.	.9493	.9906	.9356	.9925	<u>.9959</u>
Jun.	.9321	<u>.9887</u>	.9860	.9871	.9886
Jul.	.9416	.9671	.9788	.9766	<u>.9832</u>
Aug.	.9398	<u>.9914</u>	.9702	.9786	.9906
Sep.	.9677	.9920	.9635	.9872	<u>.9927</u>
Oct.	.9561	.9917	.9677	<u>.9955</u>	.9937
Nov.	.9777	.9797	.9734	.9789	<u>.9818</u>
Dec.	.9817	<u>.9958</u>	.9828	.9946	.9948
Total score[*]	0	3	0	3	6
Annual	<u>.9947</u>	.9940	.9944	.9938	.9942

* Number of underlined PPCC.

Table 5. Correlation coefficients obtained from the PPCC test for historical stream-flow data record in the Baranduz basin (maximum PPCC is underlined).

Month	Normal	Gamma	LN2	LN3	LP3
Jan.	.9541	.9949	.9951	.9955	.9954
Feb.	.9625	.9881	.9913	.9912	.9913
Mar.	.9737	.9943	.9942	.9946	.9943
Apr.	.9519	.9777	.9808	.9841	.9816
May.	.9373	.9937	.9818	.9943	.9933
Jun.	.9045	.9878	.9763	.9936	.9935
Jul.	.9637	.9879	.9868	.9913	.9867
Aug.	.9692	.9909	.9938	.9938	.9928
Sep.	.9722	.9889	.9885	.9909	.9908
Oct.	.9497	.9923	.9941	.9965	.9966
Nov.	.8811	.9885	.9954	.9969	.9967
Dec.	.9633	.9953	.9508	.9928	.9648
Total score*	0	1	2	8	1
Annual	.9561	.9914	.9902	.9939	.9923

* Number of underlined PPCC.

and probability-distribution-predicted quantiles, it seems reasonable to accept the distribution with the highest numerical value of the PPCC as being the most appropriate for modelling a particular data set.

As see in Tables 4 and 5, all the correlation coefficients are positive which is plausible since one expects the observed quantiles to increase as the probability-distribution-predicted quantiles increase. The maximum values of correlation coefficients for the various probability distributions have been underlined in Tables 4 and 5. Based on the approach we adopted for selecting the best distribution, it can be seen that the normal distribution is the most appropriate distribution for annual flows at Gorpley (and at the other English sites). This further reinforces the earlier evidence obtained both through the histograms and skewness test. For Baranduz (and all the other Iranian sites), the LN3 distribution produced the maximum PPCC for the annual flows. The histogram and skew test carried out for the Iranian annual data sets revealed that the data have non-zero skews, which has been further confirmed by the formal PPCC test.

The probability plots of the observed annual runoff data are shown in Figs 6 and 7 for the English and Iranian sites respectively. The probability axis in both plots is based on the normal distribution; the ordinate axis in Fig. 7 for the Iranian data is logarithmic to account for the use of the LN3 distribution. The upper and lower boundary limits of the confidence interval at 5% level of significance are also included in the plots. These probability plots indicate that the historical annual flows all lie inside their respective confidence limits at the 5% level of significance and show that there is reasonable agreement between the observed and theoretical

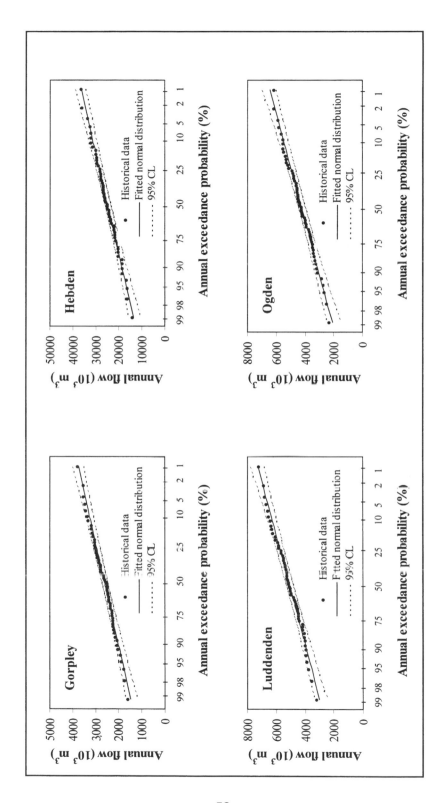

Figure 6. Normal probability plots of historical annual flows for the English sites (CL = Confident Limits).

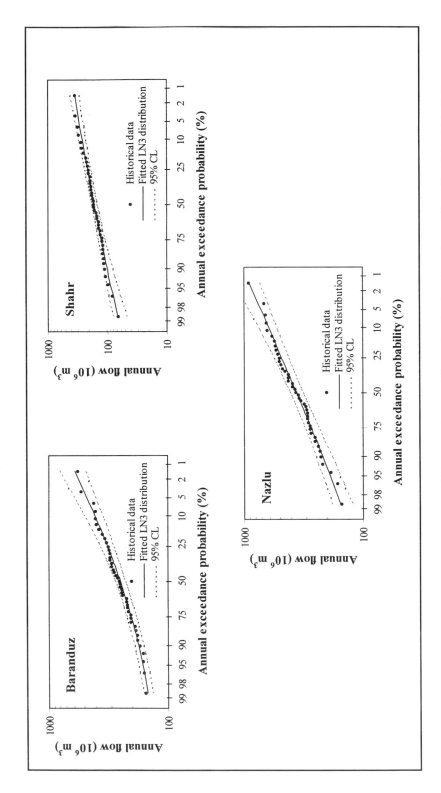

Figure 7. Log-normal probability plots of historical annual flows for the Iranian sites (CL = Confident Limits).

values. This further reinforces the decision to use the normal and LN3 distributions for modelling the annual flows at the English and Iranian sites, respectively.

As for the monthly flows, it is apparent in Tables 4 and 5 that the lowest correlation coefficient is associated with the normal distribution, which is not surprising given their high skew estimates reported in Table 2. This rules out any further consideration of the normal distribution. However, because different months tended to have different 'best' distribution on the basis of the maximum correlation coefficient of the PPCC test, a slightly different approach was used for selecting the best of the remaining four distributions for the monthly flows. Rather than select the best distribution for each month, the distribution having the highest score of total number of occasions in which it represented the best distribution was selected for use in all the months. Tables 4 and 5 contain the total score for each of the probability distribution functions. Clearly the LP3 produced the most number of best distributions for Gorpley (and all the other English sites) whereas the LN3 produced the highest score for Baranduz (and the other sites in Iran). Consequently, all the monthly flows at the English sites were modelled using the LP3 distribution while all the monthly flows at the Iranian sites were modelled using the LN3 distribution function. The use of a universal distribution function across all the twelve months simplified considerably the subsequent stochastic modelling of the flows to replicate the data for the Monte Carlo water resources planning analysis.

Conclusions

This study has used data from Iran and England to illustrate some of the relevant preliminary analyses which must be carried out on data prior to using them for decision-support purposes in water resources planning and management. The preliminary tests cover a check on the consistency of the data, its randomness, whether or not the data set possesses a long-term trend and the selection of the most appropriate probability distribution function for modelling the data. As it would turn out, all the seven data sets analysed exhibited the right attributes but this may not always be true in some situations, in which case some of the ideas discussed in the paper about how the undesirable characteristics can be accommodated will become relevant. By the same token, the PPCC goodness-of-fit test had sufficient power to discriminate between candidate probability distribution functions for modelling the data sets. The PPCC test is a very simple one to use and as seen in this paper can be employed even where tables of critical points are not available. It has been recommended by Stedinger et al. (1993) and we would like to endorse the approach given the ease and huge success with which it was implemented in our study.

Finally, it should be realised that not all pre-processing requirements likely to confront the practising hydrologist have been covered in this study. Indeed, there may be other pre-processing needs dictated by the nature and state of the historic data record. In particular, the present study has been fortunate that the available historical data were of reasonable lengths to warrant their being used without the need for extension. The data records also had no gaps in them that must be filled. However, both of these are often not true in practice; so another pre-processing analysis commonly carried out is data record augmentation and extension. Hirsch et al. (1993) discuss record extension methodologies, while Salas (1993) describes procedures for data filling and extension. Relatively more recently, Wang (2001) discusses a Bayesian approach for augmenting flood information at sites with insufficient data records. The data record may also contain abnormally extreme entries, termed outliers, which must be objectively identified, removed and replaced by more plausible values (see Kottegoda, 1984). All of this kind of analysis is often necessary to improve the information content of the data and so ultimately to enhance decision-support based on the outcome of the data analysis.

References

Adeloye, A.J., Montaseri, M. and Garmann, C. (2001). "Curing the misbehaviour of reservoir capacity statistics by controlling shortfall during failures using the modified sequent peak algorithm". Wat. Resour. Res., 37(1), 73-82.

Box, G.E.P. and Jenkins, G.M. (1976). "Time series analysis, forecasting and control". Holden-Day Inc. CA, USA.

Bras, R.L. (1990). "Hydrology: An introduction to hydrologic science". Addison-Wesley Publishing Co.

Cryer, J.D. (1986). "Time series analysis". Duxbury Press, Boston, MA, USA.

Cunnane, C. (1978). "Unbiased plotting positions – a review". J. Hydrol., 37, 205-222.

Draper, N.R. and Smith, H. (1966). "Applied regression analysis". John Wiley & Sons, New York.

Durrans, S.R. and Tomic, S. (1996). "Regionalization of low-flow frequency estimates: An Alabama case study". Wat. Resour. Bull., 32(1), 23-37.

Filliben, J.J. (1975). "The probability plot correlation test for normality". Technometrics, 17(1), 111-117.

Hameed, T., Marino, M.A., DeVries, J.J. and Tracy, J.C. (1998). "Method for trend detection in climatological variables". J. Hydrologic Engrg., ASCE, 2(4), 154-160.

Hirsch, R.M., Helsel, D.R., Cohn, T.A. and Gilroy, E.J. (1993). "Statistical analysis of hydrologic data". In: Handbook of applied hydrology (ed. D.R. Maidment), Chapter 17, McGraw-Hill, New York.

Kendall, M. and Gibbons, J.D. (1990). "Rank correlation methods". Edward Arnold, London.

Kendall, M. and Stewart, A. (1976). "The advanced theory of statistics, Volume 3. Design and analysis, and time series". Charles Griffin & Co Ltd., London.

Kirby, W. (1972). "Computer-orientated Wilson-Hilferty transformation that preserves the first three moments and the lower-bound of the Pearson type III distribution". Wat. Resour. Res., 8(5), 1251-1254.

Kottegoda, N.T. (1984). "Investigation of outliers in annual maximum flow series". J. Hydrol., 72, 105-137.

Loucks, D.P., Stedinger, J.R. and Haith, D.A. (1981). "Water resources systems planning and analysis". Prentice-Hall, Englewod Cliffs, New Jersey.

McGhee, J.W. (1985). "Introductory statistics". West Publishing Co., New York.

McMahon, T.A., Finlayson, B.L., Srikanthan, R. and Haines, A.T. (1992). "Global runoff: Continental comparisons of annual flows and peak discharges". Catena Verlag, Gremlingen-Desteedt, Germany.

McMahon, T.A. and Mein, R.G. (1986). "River and reservoir yield". Water Resources Publication, Littleton, CO, USA.

Montaseri, M. (1999). "Stochastic investigation of the planning characteristics of within-year and over-year reservoir systems". PhD thesis, Heriot-Watt University, Edinburgh, UK.

Montaseri, M. and Adeloye, A.J. (1999). "Critical period of reservoir systems for planning purposes". J. Hydrol., 224, 115-136.

Salas, J.D. (1993). "Analysis and modeling of hydrologic time series". In: Handbook of applied hydrology (ed. D.R. Maidment), Chapter 19. McGraw-Hill, New York.

Stedinger, J.R., Vogel, R.M. and Foufoula-Georgiou, E. (1993). "Frequency analysis of extreme events". In: Handbook of applied hydrology (ed. D.R. Maidment), Chapter 18. McGraw-Hill, New York.

Vogel, R.M. (1986). "The probability plot correlation coefficient test for the normal, log-normal, and gumbel distribution hypotheses". Wat. Resour. Res., 22(4), 587-590.

Vogel, R.M. and Kroll, C.N. (1991). "The value of streamflow record augmentation procedures in low-flow and flood-flow frequency analysis." J. Hydrol., 125, 259-276.

Vogel, R.M. and McMartin, D.E. (1991). "Probability plot goodness-of-fit and skewness estimation procedures for the Pearson Type III distribution". Wat. Resour. Res., 27(12), 3149-3158.

Wang, Q.J. (2001). "A Bayesian joint probability approach for flood record augmentation". Wat. Resour. Res., 37(6), 1707-1712.

Wilson, E.B. and Hilferty, M.M. (1931). "Distribution of chi-square". Proceedings, Nat. Aca. Sci, 17, 684-688.

WMO (1988). "Analysing long time series of hydrological data with respect to climate variability". WCAP-3, WMO/TD- no. 224, Geneva.

Section 2: Effect of climate change

Surface Water Hydrology, Singh, Al-Rashed & Sherif (eds)
© *2002 Swets & Zeitlinger, Lisse, ISBN 90 5809 363 8*

Investigation of trends and variability in rainfall and temperature data in the Arid Arab region

Khaled H. Hamed
Department of Irrigation and Hydraulics, Faculty of Engineering,
Cairo University, Orman, Giza, Egypt
e-mail: hamedkhaled@hotmail.com

Abstract

Recently, the study of temperature and rainfall data on a global scale show the existence of significant trends, which are linked to the global warming phenomena. Studies also show that the variability of meteorological and hydrologic data can be attributed to some global phenomena such as the sunspot cycle and El-Nino/Southern oscillations, among others. In this paper, a number of temperature and rainfall data series in Egypt and the surrounding Arab countries are investigated for the existence of trends, their significance, and directions of change. Also, the variability in the characteristics of these time series and their relationship to global variations are studied. A number of new trend analysis techniques and modified statistical tests as well as improved spectral analysis methods are used.

Introduction

In the past few decades, attention has been rising regarding changes in the climate of the earth. These changes are linked to natural as well as manmade changes in the environment. Natural changes are manifested in the form of cycles or recurring weather patterns, which are mainly due to natural origins. These changes are usually linked to natural phenomena such as the sunspot cycle, the sun magnetic cycle, El-Nino/Southern oscillations, and lunar cycles, among others. Other changes show in the form of trends, either increasing or decreasing. Those changes are linked to such phenomena as the global warming due to the greenhouse effect of Carbon Dioxide produced by the industrialized world. These effects are usually studied on a global scale. This paper investigates trends and variability in tempera ture and rainfall data in the Arid Arab region. The goal is to compare recent changes in this region with the global trends of climatic changes and global climate variability.

Data Used in the Study

In this study, temperature and rainfall data from GHCN, the Global Historical Climatology Network (WMO, 2001) are used. Two sets of time series are used. The first set contains data from individual stations. This data set is not up-to-date (extent up to 1987 only) and so will not be used in trend analysis, which needs more recent data. However this data set will be used in spectral analysis. The time series from individual stations are chosen from the GHCN database such that the length of the time series is not less than 30 years in order to get meaningful results. The missing

Table 1. Temperature and rainfall individual stations from GHCN.

Country Code	Country	No. of Temp. Stations	No. of Rain Stations
101	Algeria	10	23
115	Egypt	8	24
124	Libya	4	15
128	Mauritania	6	14
130	Morocco	6	23
152	Tunisia	2	9
169	West Sahara	—	1
202	Bahrain	1	1
209	Iraq	6	2
212	Kuwait	—	1
218	Oman	—	3
220	Qatar	1	1
223	Saudi Arabia	5	1
233	Yemen	—	1
624	Jordan	3	5
627	Lebanon	2	2
647	Syria	6	6

monthly data (a few are missing in some stations) are replaced with the monthly mean. The details of the chosen time series are given in Table (1).

The second data set is a gridded data set at 5 degrees resolution (GHCN, 2001; CRU, 2001). The value for a given month is the average of a number of stations that lie in given grid cell. For rainfall data, Theissen polygon weights were used to average gauge data within each grid cell. Missing data were estimated from the surrounding stations. The data is also corrected for gross outliers. The use of this gridded data set provides two advantages over the use of individual station records. First, in addition to the difficulty of obtaining the records, most individual datasets are incomplete, have missing data, discontinued, or simply too short, and tedious work is needed to render them suitable for analysis, a work that has already been carried out to produce the gridded set. Second, the study of individual datasets gives only a local look at changes, which will be affected by various factors, such as relocation of the station. Also, point measurements may be misleading in indicating variations and trends especially in the case of rainfall, which is characterized by its large spatial variability. Spatial averaging in this case is expected to reduce much of the spatial randomness of rainfall data to give a clear picture of trends and variations on a regional scale. At a grid size of 5 by 5 degrees, each country will be represented by as few as one grid cell in the case of small countries (Kuwait, Bahrain, ... etc.) up to 6 or more grid cells for large countries (Sudan, Libya, Algeria, ... etc.). Due to the lack of data in some areas, data are not available for some Arab countries, but all available data are studied here. Available monthly temperature data are nine time series covering more or less Mauritania, Morocco, Libya, Egypt, Jordan, Sudan, Oman, Qatar and Bahrain. The annual av-

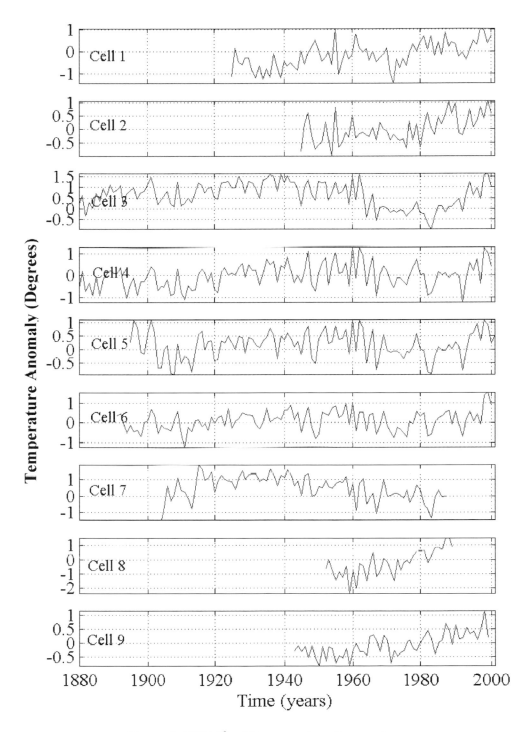

Figure 1. HCN 5° grid temperature time series.

Table 2. Grid cells for temperature data.

Cell No.	Country
1	Morocco
2	Libya
3	Egypt
4	Jordan
5	Egypt
6	Bahrain and Qatar
7	Sudan
8	Sudan
9	Oman

erage temperature time series from these grid cells are shown in Figure (1). Table (2) gives the cell numbers and the corresponding countries covered by these grid cells. The numerical values in the temperature time series are anomalies from the long-term average at these grids.

Available monthly rainfall time series are from 26 gridded cells covering most of the Arab countries. Samples of these data series are shown in Figure (2). Table (3) gives the cell numbers and the corresponding countries covered by these grid cells. The numerical values in these time series are rainfall depth in millimeters.

Another more complete dataset on a 2.5° grid, but with extent from 1986 to 2000 only (GPCC, 2001; Rudolf et al., 1994) is used in the analysis. This dataset provides 79 grid cells covering most of the Arab countries. The extent of this dataset is short but gives a closer look at the recent trends and variations.

Table 3. Grid cells for rainfall data.

Cell No.	Country	Cell No.	Country
1	Djibouti / Yemen	14	Morocco
2	Sudan	15	Morocco
3	Sudan	16	Algeria
4	Oman / Yemen	17	Algeria
5	Mauritania	18	Libya
6	Mauritania	19	Libya
7	Sudan	20	Libya
8	Sudan	21	Egypt
9	Oman	22	Egypt
10	Mauritania	23	Jordan
11	Algeria	24	Iraq
12	Egypt	25	Syria
13	Jordan / Saudi Arabia	26	Syria

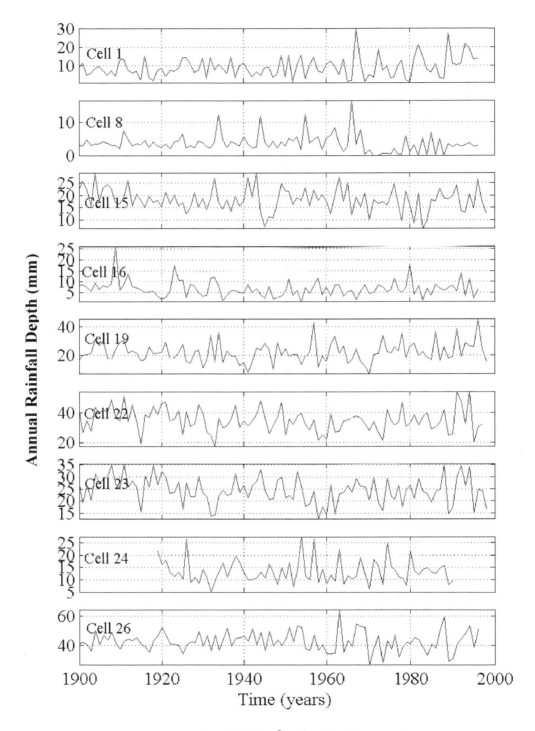

Figure 2. Samples of HCN 5° grid rainfall time series.

Trend Analysis

Trend analysis is performed on both temperature and rainfall time series. The statistical test used in this paper is the modified Mann-Kendall trend test (Hamed and Rao, 1998). This test is suitable for the data since autocorrelation is expected to exist in the data. Also, the original Mann-Kendall trend test (Kendall, 1970) and the modified test are nonparametric tests, which do not require the condition of normality of the data, which is violated in most of the natural meteorologic and hydrologic time series. The original test is based on the ranks of the data rather than the data values themselves. For a time series x_1, x_2, ..., x_n with ranks R_1, R_2, ..., R_n the test statistic is

$$S = \sum_{i<j} sign(R_j - R_i) \tag{1}$$

This statistic is asymptotically normally distributed under the null hypothesis of zero trend. The variance of the statistic is given by

$$var(S) = \frac{1}{18}\left[n(n-1)(2n+5) - \sum_t t(t-1)(2t+5)\right] \tag{2}$$

where n is the total number of observations and t is number of x values involved in a given tie of ranks, and the summation is over all ties. The standardized normal test statistic is thus given by

$$Z = S/[var(S)]^{1/2} \tag{3}$$

The sign of Z indicates the direction of trend, while its value indicates the significance of the trend based on the Normal distribution. Hamed and Rao (1998) have made a correction to the test, which accounts for the autocorrelation in the data, which gives false indication of significant trends. The correction involves the modification of the variance of the statistic to become

$$var(S) = \frac{1}{18}\left[n(n-1)(2n+5) - \sum_{tt} t(t-1)(2t+5)\right] \times \frac{n}{n_S^*} \tag{4}$$

where n_S/n_S^* is given by

$$\frac{n}{n_S^*} = 1 + \frac{2}{n(n-1)(n-2)}\sum_{i=1}^{p}(n-i)(n-i-1)(n-i-2)\rho_S(i) \tag{5}$$

where $\rho_S(i)$ is the autocorrelation between the ranks of the observations at lag i, and p is the number of significant autocorrelations which is usually small (1 to 3).

The test is performed based on a suitable significance level for the statistic Z, in this study the significance level is taken as 0.05.

Trends in Temperature Data

The results of applying the modified Mann-Kendall test to the GHCN full-length annual average 5° gridded temperature time series are given in Table (4).

Table 4. Results of trend test on full-length average temperature time series.

Grid Cell	1	2	3	4	5	6	7	8	9
Trend Direc.	+	+	-	+	+	+	-	+	+
Significant	Y	Y	N	Y	N	N	N	Y	Y

From Table (4) it can be concluded that seven cells out of 9 show positive (increasing) trend, five of which are statistically significant. Two cells (3 and 7), however, show an insignificant negative trend.

The analysis based on the full-length time series indicates positive trends in most time series. The number of significant positive trends (5) suggests an increasing trend in the region both in mean and maximum temperature. This is based on the full record.

A closer look at the time series in Figure (1) reveals that the time series are not all of equal length, start, or end date, especially cells 7 and 8, which stops at the year 1985. The difference in length, start and end dates of the studied time series will certainly give misleading conclusions. In this case, a better way to identify trends that are comparable in different time series is to look at significant trends at different periods in the record. This is done by progressively testing segments of different length starting at different points in each time series. Plotting the significant trends side by side gives a clearer picture about the variation in trends relative to time in different time series.

Figure (3) shows the significant trends in mean temperature time series at different segments in the time series at the 0.05 level. In this figure, significant positive trends are shown as black lines and significant negative trends are shown as white lines. Each time series is represented as a cluster of black and (possibly overlapping) white lines indicating significant trends in the time series. Although the number of time series analyzed is not large, one can distinguish three different trend periods that are more or less consistent in almost all the temperature time series. The first significant trend is an increasing trend from as early as 1880 or 1900 in the longest records up to around 1960. Superimposed on that is a decreasing trend from around 1920 to around 1985. Again, a significant increasing trend exists starting around 1970 and extending to the end of the record at 2000. This analysis shows that if the time series are investigated for example in the period 1920 to 1985, the analysis would indicate a significant negative trend for that period in all time series, as opposed to the period from 1960 to 2000 which would indicate a significant increasing trend for all the time series.

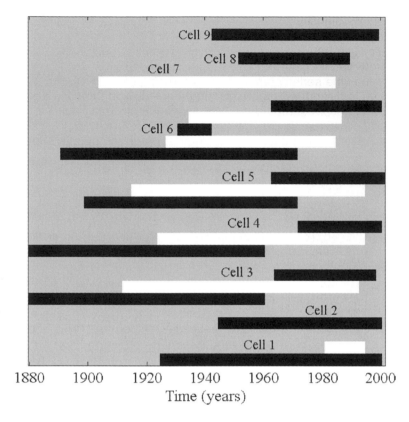

Figure 3. Trends in segments of 5° grid temperature time series.

Trends in Rainfall Data

The results of applying the modified Mann-Kendall trend test to the full-length rainfall time series for the 26 grid cells are given in Table (5).

The results show that 20 out of the 26 cells indicate a negative trend, eight of which are significant. On the other hand, six cells indicate a positive insignificant trend. These six cells are in countries Yemen, Djibouti, Oman, and Libya.

Table 5. Results of trend test on full-length rainfall time series.

Grid Cell	1	2	3	4	5	6	7	8	9	10	11	12	13
Trend Dir.	+	-	-	+	-	-	-	-	+	-	-	-	-
Significant	N	N	Y	N	Y	Y	N	Y	N	Y	N	N	Y
Cell	14	15	16	17	18	19	20	21	22	23	24	25	26
Trend Dir.	-	-	-	0	+	+	-	-	-	-	-	-	-
Significant	N	N	N	N	N	N	N	N	Y	Y	N	N	N

Figure (4) shows significant trends in segments of rainfall time series. Significant positive trends are shown as black lines and significant negative trends are shown as white lines. The trends in Figure (4) are not as clear as in the case of temperature data. However, in general, one can distinguish a significant negative trend in almost all of the time series starting at the beginning of the record and extending to around 1980. On the other hand, about half of the stations show a significant positive trend starting from somewhere in the 1980's and extending to the end of the record in 1999, while the other half shows a negative trend in the same period.

Figure 4. Trends in segments of 5° grid rainfall time series.

On the other hand, the results of testing the 79 GPCC 2.5° grid cells show that out of those 79 grid cells, 31 series show a positive trend, 36 show a negative trend, and 12 show no trend at all. It is worth noting that decreasing trends are mainly in the west along the Mediterranean (Morocco, Tunisia, Algeria, Egypt) except Libya, while increasing trends are to the east (gulf countries, Syria, Iraq, and Yemen). These countries coincide with some of those identified in the 5° grid cell data. However, only one time series of the 79 gives a significant increasing trend and two time series give significant decreasing trend.

Discussion of Trend Results

Based on the statistical testing of the full-length temperature data it was found that five out of nine mean temperature series studied indicate a significant increasing trend. Further analysis of segments of the time series indicates that eight out the nine time series show a significant positive trend in the period starting from 1970 up to the present in 2000. These results agree with the global increase in temperature known as global warming. Figure (5) shows the annual global temperature in the period from 1958 to 1999 (NCDC, 2001) expressed as deviations from the 20-year mean of 1958-1977. It is clear from Figure (5) that an increasing trend starts around 1965 up to the present. These results also are similar to those obtained by Hamed (1997) for stations and regional data in the Midwest USA, which indicated a significant increasing trend in the period from 1970 towards the end of the record at 1997. The effect of global warming thus is in agreement with the global trends in the northern hemisphere.

For rainfall data, the full-length time series indicate a decreasing trend in 20 out of 26 time series, with eight being significant. However, segment analysis indicates that in about half of the time series a significant decreasing trend extends from the beginning of the record up to 1980, but followed by a significant positive trend from 1980 up to 1999, an extent much smaller than that of the negative trend. On the other hand, the data from the GPCC database from 1986 to 2000 give only one significant positive trend and two significant negative ones. Non-significant trends, on the other hand are divided between the countries. Therefore trends in rainfall time series are not as clear as those in temperature data.

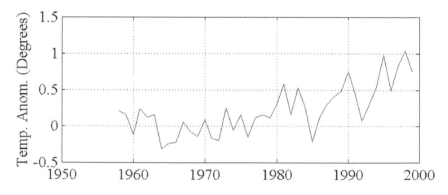

Figure 5. Temperature anomalies in the northern hemisphere (1958-1999).

Spectral Analysis

In this analysis the multi-taper spectral analysis method (Thomson, 1982) is used. As opposed to conventional spectral estimation (periodogram or correlogram and conventional windowing), this method considers a process with a background continuum (the underlying stationary process) plus a number of periodic components. The method provides also a likelihood ratio test (F-test), which can be used to assess the significance of periodic components. Usually, a significance level of $1/n$ is recommended (Thomson, 1990a), where n is the number of observations. However, since the goal of this analysis is to find similar components in different time series, the significance is relaxed to 90% or above probability. The multi-taper method also provides a procedure for removing the mean of the process. Simply subtracting the mean creates a hole in the spectrum around zero frequency and often leads to the appearance of a spurious peak commensurable with the length of the time series. The details of the application of the method can be found in Thomson (1990a and b) and Percival and Walden (1993).

Cycles in Temperature Data

Figure (6) is a representation of the significant cycles in mean temperature series from the 5° grid cells. Circles are sized according to their significance (probabilities between 90% and 100%). The plus sign means that the component is also significant at a level $1/n$. In Figure (6), the most similar cycles between series are between 53.9 and 120 years. However, the values of these cycles are close to the length of the time series, and therefore are questionable, since only one cycle would appear in the record, although a period of 102 ± 15 years is reported in longer time series such as the central England temperature data from 1659 to 1990 (Baliunas et al., 1997). The other feature in Figure (6) is the existence of periods of 2 to 2.7 years in most of the time series. These are usually a result of the Quasi-Biannual Oscillations (QBO), which exist in most natural time series with periods from 2 to 3 years. Some scattered periodicities of 3 to 5.6 years appear in most time series. These are usually of an El-Nino/Southern Oscillations (ENSO) origin. One observation about the results in Figure (6) is that the 11-year cycle, which appears in most temperature data such as North America and Europe (Hamed, 1997), does not show in this dataset. This periodicity is often attributed to the Sunspot cycle of 11 years. However, Mann and Park (1994) indicate that these cycles although equal to the sunspot cycle, they correlate poorly with the sunspot cycle. They seem to be related to a circulation anomaly in the north Atlantic (Mann and Park, 1996).

Figure (7) shows the significant cycles in the GHCN individual stations time series. Again, Periodicities of 40 to 80 years show at the bottom of the figure. The next group of cycles is between 8 and 9 years. This periodicity has not been identified in other temperature series. However, the same cycles are show in the 5° grid data, Figure (6), at three cells, which are also significant at the 1/n level. The origin of this cycle is not known. The next group is the 3 to 4 years, which also shows in most stations, suggesting the effect of ENSO. Periodicities of 2.5 to 2.8 years also show in many stations. These periodicities do not seem to be related to natural

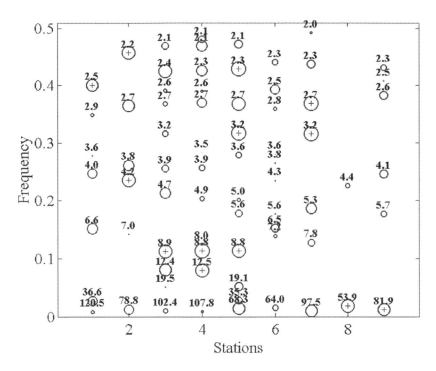

Figure 6. Significant cycles in 5° grid temperature time series.

phenomena, except if it is an interaction between ENSO and QBO. The QBO signal (2 to 3 years) shows clearly in most of the stations.

A general observation, however, from Figures (6) and (7) is the inconsistency of frequencies between different stations. The bands over which the periodicities are spread are large and it is inevitable that the grouping of frequencies is rather subjective. In another study by the author in the Midwest USA (Hamed, 1997) the variations in cycles between stations was more or less confined, with noticeable spatial drift of frequencies. The quality of data, missing observations, as well as the scale of the study may have contributed to such a spread of the results.

Cycles in Rainfall Data

Figure (8) shows the significant periodicities in the GHCN 5° grid rainfall time series. It is clear from Figure (8) that the spread of periodicities is much more than the in temperature data, which is typical of rainfall data due to increased temporal and spatial randomness. Periods of 2 to 2.3 years at the top of Figure (8) suggest a QBO origin. The largest periods detected as significant are in the range of 30 to 136 years, with no consistency between different time series. Other frequencies are hard to group.

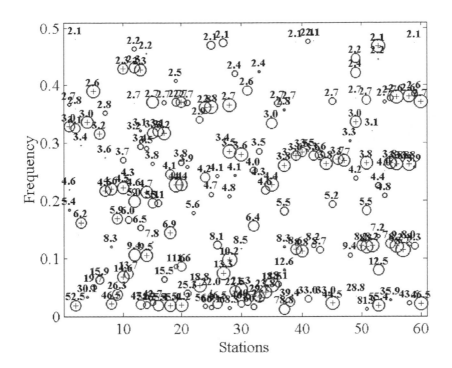

Figure 7. Significant Cycles in GHCN individual stations temperature data.

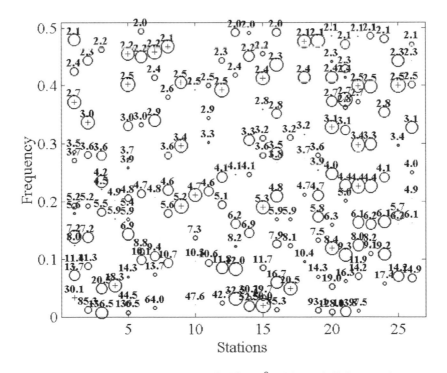

Figure 8. Significant cycles in GHCN 5° grid rainfall time series.

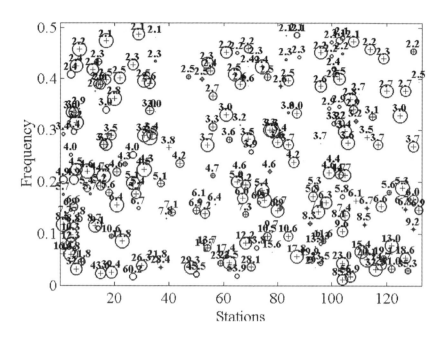

Figure 9. Significant cycles in GHCN individual rainfall time series.

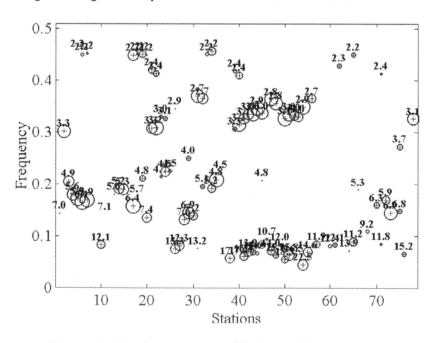

Figure 10. Significant cycles in GPCC rainfall time series.

Figure (9) shows the significant cycles in the GHCN individual stations rainfall time series. The cycles are further more spread and no grouping for the frequencies is possible, except for the QBO signal (2 to 2.3 years), which exits in most stations. The largest detected cycles are in the range between 40 and 85 years, but are inconsistent between different time series.

Figure (10) shows significant cycles in the GPCC 2.5° grid rainfall time series. A QBO signal (2 to 2.4 years) shows in a few stations. The largest cycles detected are 12 to 22 years, but again, these are inconsistent between time series and are also close to the length of the time series (15 years).

An overall observation, similar to that made concerning temperature data analysis, is the inconsistencies between different time series, which makes an overall statement about the variations in the analyzed time series rather difficult.

Conclusions

1. There is evidence of a statistically significant increasing trend in temperature time series in the period from 1970's and 1980's towards 2000 in the Arab region. This trend follows a decreasing trend from 1920 through the 1980's and an increasing trend from 1880 through 1960.

2. A corresponding change in rainfall time series is however questionable. Some time series show increasing trends, while the others show a decreasing trend. The time series show an overall decreasing trend over the period 1900 to 1999. Note that the same period in temperature time series have a decreasing then an increasing trend. Some stations show a positive trend in the period 1985 to 1999 but the period is not long enough to judge such a trend. More rainfall data is needed before a definite statement about trends in rainfall can be made.

3. Although not significant, there is an indication that increasing rainfall trends are for the eastern side Arab countries, gulf countries, Syria, Iraq, and Yemen in addition to Libya. Decreasing trends, on the other hand are observed in the western Mediterranean countries Morocco, Tunisia, Algeria, and Egypt, excluding Libya.

4. Concerning variability and cycles in temperature data, short-term variability shows in the order of the ENSO and QBO variations. Long-term cycles are close to the lengths of the records and therefore cannot be confirmed. The consistency of cycles between stations is however poor. This may be attributed to the quality of the data and the large extent of the region.

5. Cycles in rainfall data do not show clearly in the region. Some QBO signals are detected, but longer-term cycles are hard to distinguish or group. This also may be attributed to the quality of the data and the extent of the region.

References

Baliunas, S., Frick, P., Sokoloff, D., and Soon, W. (1997). "Time scales and trends in the central England temperature data (1659-1990): A wavelet analysis." *Geophysical Research Letters*, v 24, no. 11, pp. 1351-1354.

CRU (2001). *Climate Research Unit database*. University of West Anglia Norwich, UK, Internet site: http://www.cru.uea.ac.uk.

GHCN (2001). *Global Historical Climatology Network (GHCN) gridded database*. Internet site: http://lwf.ncdc.noaa.gov/oa/climate/research/ghcn/ghcngrid.html.

GPCC (2001). *The Global Precipitation Climatology Centre*. Internet site: http://www.dwd.de/research/gpcc.

Hamed, K.H. (1997). *Investigation of stationarity of hydrologic and climatic time series*. Ph. D. Thesis, Purdue University, West Lafayette, IN, USA.

Hamed, K.H. and Rao, A.R. (1998). "A modified Mann-Kendall trend test for autocorrelated data." *Journal of Hydrology*, 204, pp. 182-196.

Kendall, M.G. (1970). *Rank correlation methods*. Griffin, London.

Mann, M.E. and Park, J. (1994). "Global-scale modes of surface temperature variability on interannual to century time scales." *Journal of Geophysical Research*, 99, pp. 25819-25834.

Mann, M.E. and Park, J. (1996). "Joint spatiotemporal modes of surface temperature and sea-level pressure variability in the northern hemisphere during the last century." *Journal of Climate*, 9, pp. 2137-2162.

NCDC (2001). *National Climatic Data Center database*. Internet site: http://lwf.ncdc.noaa.gov/oa/climate/climatedata.html.

Percival, D.B. and Walden, A.T. (1993). Spectral analysis for physical applications: multi-taper and conventional univariate techniques. Cambridge University Press, Great Britain.

Rudolf, B., H. Hauschild, W. Rueth and U. Schneider (1994). Terrestrial Precipitation Analysis: Operational Method And Required Density Of Point Measuremenets. In: *Global Precipitations and Climate Change* (Ed. M. Desbois, F. Desalmond).

Thomson, D.J. (1982). "Spectrum estimation and harmonic analysis." *Proc. IEEE*, 70, pp. 1055-1096.

Thomson, D.J. (1990a). "Time series analysis of Holocene climate data." *Phil. Trans. Roy. Soc. London,* Series A, 330, pp. 601-616.

Thomson, D.J. (1990b). "Quadratic-inverse spectral estimates: Applications to paleoclimatology." *Phil. Trans. Roy. Soc. London*, Series A, 332, pp. 593-597.

WMO (2001). *Global Historical Climatology Network (GHCN) V2 database*. Internet site: http://www.wmo.ch/index-en.html.

Surface Water Hydrology, Singh, Al-Rashed & Sherif (eds)
© *2002 Swets & Zeitlinger, Lisse, ISBN 90 5809 363 8*

Climate and land use change impacts on flood flows in Greece

M.A. Mimikou, E. Varanou and M. Pikounis

Department of Water Resources, Hydraulic and Maritime Engineering
Laboratory of Hydrology and Water Resources Management
National Technical University of Athens
5, Iroon Polytechniou, 157 80 Athens, Greece
e-mail: Mimikou@chi.civil.ntua.gr

Abstract

Climate and land use changes are considered to be the most important anthropo-genic factors that influence the flow and consequently the flood regime of river basins. The paper herein is focusing on the Pinios catchment located in Central Greece. Being an agricultural area with important annual crop production, the need to assess the magnitude of flooding in environmentally changed conditions is straightforward. The study uses a scenario based approach for both the climate and the land use change conditions. The flows with a return period of 10, 20, 100 and 1000 years are examined. The altered flows due to climate and land use changes, indicate an increase of the flood magnitude for all the return periods examined. In any case, the study outlines the need for revised flood mitigation procedures that will take into account the environmentally changed conditions.

Introduction

The growing impact of the flooding hazard throughout the world and the potential of even greater flooding disasters under a variable and uncertain environment under-scores the need for studies focusing on the regional effects that climate and land use changes have on the magnitude and risks of floods. The research work pre-sented herein originates from the EU funded project EUROTAS, funded under the 4th FP – DG Research.

The study area is the Ali Efenti basin (Fig. 1), part of the Pinios River, located in central Greece. The basin suffers from frequent and hazardous storms and conse-quent flash floods, which the natural capacity of the river is inadequate to pass downstream for a large part of its length. This is mainly due to the topography of the river network, which varies from narrow passes to wide flood plains. The last decades, changes in land use, mainly deforestation, are exacerbating the flood problem.

For the assessment of the impact of climate change on the flow regime of the basin, two scenarios are applied. The first, HadCM2 is a transient, constructed by the Climatic Research Unit (CRU) of the University of East Anglia, with 2050 as the terminal year. The second scenario refers to time series constructed by PIK, based

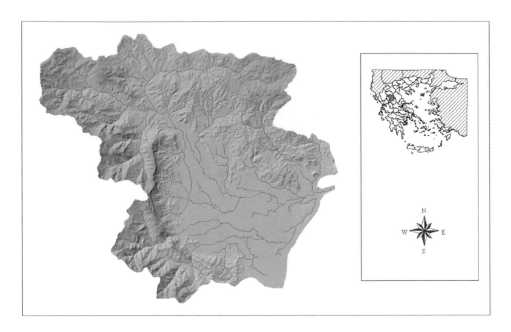

Figure 1. The study area and its location in mainland Greece.

on the outputs of the ECHAM GCM, with 2050 as the terminal year. For the study of the impact of land use change, a land use builder is used.

Study Area and Data Used

The Pinios river is located in the Thessaly district (central part of Greece). The total drainage area of the river is 9.450 km^2, with a varied topography from narrow gorges to wide flood plains. The Pinios catchment area consists of 15 sub basins drained by the main river and its 5 most important tributaries. The study focuses on the Ali Efenti sub basin (Fig. 1), for which reliable hydrometeorological time series of adequate length are available. The area of the catchment is 2.868,61 km^2. Some general characteristics of this basin are given in Table 1.

The climate is humid and temperate with substantial seasonal variations. The land cover of the basin (Fig. 2) consists of forest (57.12%), agricultural land (40.54%), urban land (1.67%) and wetlands (0.67).

Table 1. General characteristics of the study area.

Area (km^2)	2868.61
Mean Elevation (m)	539.73
Mean annual rainfall (mm)	933.3
Mean annual storm runoff coefficient	0.428
Mean annual flow (m^3/sec)	39.05

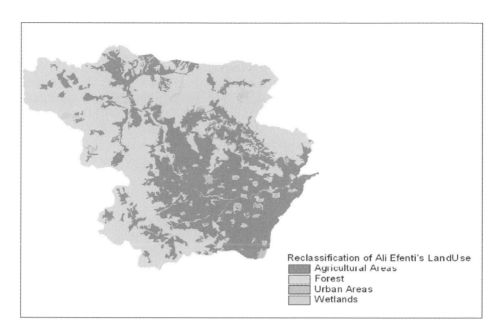

Figure 2. The land cover map of the study area.

Daily hydrometeorological data for a 33-year period (1960–1993) were acquired from 15 stations located within and near the study area. Those data refer to precipitation, temperature, wind velocity, relative humidity and sunshine duration values.

The basic input variable of the model, daily areal precipitation, was calculated by Thiessen method and by correcting for the elevation on the basis of a precipitation –elevation relationship. Daily minimum and maximum air temperature of the basin was calculated by associating the observations of min and max temperature of each station with its elevation and by establishing a relation between the temperature of the stations and the temperature at the mean elevation of the basin. Regarding the climate data, the requirements of the model are coarser and refer to average daily solar radiation for each month, average wind speed in month and average relative humidity in month. Mean daily runoff values at the outlet of the basin were available for the 1970–1993 period and were used for the calibration of the hydrological model. Spatial data used refer to a 20 x 20 m DEM of the basin, a digital land use map obtained after aggregation and reclassification of the original CORINE land use map and a digital soil map.

The Climate Change Scenarios

The sole information source on climate change comes from simulations done with General Circulations Models (GCMs). The climate change scenarios used in this study are the outputs of experiments conducted with two different GCMs. These models are the HadCM2, the second Hadley Centre coupled ocean-atmosphere

GCM (Mitchell et al., 1995; Johns et al., 1997) and the ECHAM4/OPYC3 GCM of the Max-Planck Institute for Meteorology (MPI) (Roeckner et al., 1996).

The HadCM2 data are grid based and expressed in the resolution 2.5^0 x 3.75^0 (latitude-longitude). Thus, only one grid covers the whole study area. The information was spatially downscaled into 10 grids of 0.5^0 x 0.5^0 resolution. Details on the downscaling can be found elsewhere (Mimikou et al., 1999a; Mimikou et al., 1999b). Changes refer to mean monthly values of precipitation (in % change) and temperature (in ^{0}C change). Using the 1961–1990 daily historical time series of precipitation and temperature, these changes were disaggregated into daily climatically changed time series.

While the aforementioned downscaling and disaggregation scheme is quite conventional, a more sophisticated was used for the downscaling of the outputs of the second GCM. The expanded downscaling (EDS) technique was developed by PIK specifically for the local and daily time scales that dominate most hydrological processes. The technique generated daily time series from 1860–2100 of precipitation and temperature for the study area by taking into account 30 year long time series observed in 36 meteorological stations within or near the study area. However, for comparison reasons in the study, the terminal year used was 2050.

The Land Use Change Scenarios

There is a strong linkage between land and water. Any man-induced changes on the landscape produce side effects on the water cycle by altering the boundaries in the soil profile that determines the partitioning of incoming water. For example soil surface serves as a division between flood flows and infiltration. Changes in land use may have significant effects on infiltration rates through the soil surface, on the water retention capacity of soils, on sub-surface transmissivity and thus on the production efficiency of rainfall (FAO, 1993).

A realistic set up of land use scenarios requires scenarios of future regional development. Land use decision-making is strongly influenced by socio-economic factors. As these particular future land use policies are complicated and beyond the scope of the current study, one hypothetical scenario target has been applied to the scenario builder. The scenario target refers to an increase of agricultural land and decrease of forest land (Table 2). Based on this definition of the scenario target, the land use builder delivered the modified land use map (Fig. 3), which was then used as input to the rainfall-runoff model, to deliver the modified flows.

Table 2. Scenario target: increase of agriculture, decrease of forest.

Land Use	Urban	Agriculture	Forest	Wetlands
Current (%)	1.7	40.5	57.0	0.7
Scenario (%)	1.7	**49.0**	**48.5**	0.7

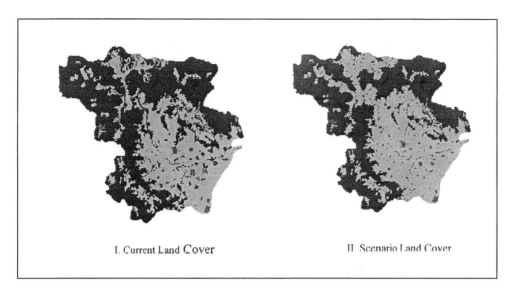

I. Current Land Cover   II. Scenario Land Cover

Figure 3. The current and modified land covers of the study area.

The Model

The Soil and Water assessment Tool (SWAT) is a physically based model used to simulate the hydrological cycle and its influence in the quantity and quality of the water and sediments (Arnold, J.G. et al., 1994). The model has the ability to describe all the physical activities taking place in a catchment and is thus considered to simulate the hydrological cycle in great detail. SWAT operates under a Geographical Information System (GIS), and more specifically under the ArcView platform, which gives more functionalities related with the spatial data analysis and management. The processes simulated by the model are Percolation, Surface Runoff, Erosion, Leaching, Evaporation, Plant growth and Agricultural management. The basic inputs can be classified into three categories: Anthropogenic (i.e. land use), Climatic (Precipitation, Solar radiation, Temperature etc.) and Watershed related inputs (i.e. catchment DEM, digital soil map, etc.). Outputs of the model refer to Flow, Sediments and Nutrients (Neitsch et al., 1999). However the current study focuses only in the flow results.

The model was calibrated against the 1970–1993 period of observed runoff. For the simulation, the model requires the first 6–7 years as a warming up period in order to account for the initial conditions. The monthly calibration was quite satisfactorily with a Nash number of 0.814 (Fig. 4). Daily calibration, as expected, had a lower Nash number, equal to 0.620 (Fig. 5).

Results

The methodology used to carry out the sensitivity study of the catchment is by running the SWAT model under climatically changed conditions and comparing the outputs to their baseline values (no climate change). The model simulates the flow

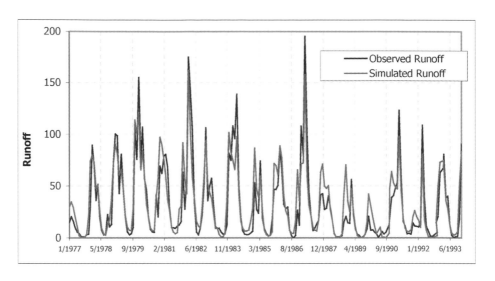

Figure 4. SWAT calibration in a monthly time step.

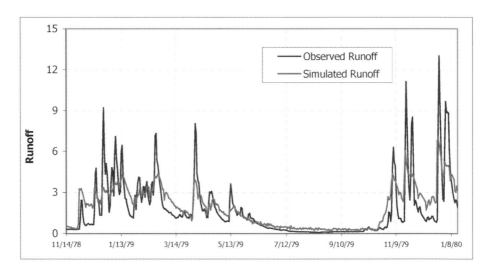

Figure 5. SWAT calibration in a daily time step.

as a function of precipitation. Thus, as expected, the changes in flow are according to the variation of the changes in precipitation, as given by the scenarios. Regarding floods, the Gumbel max distribution has been fitted to the annual daily maxima of the baseline and climatically changed flows. The floods with a 10, 20, 100 and 1000 year return period have been estimated and compared against each other (Table 3). The percentage changes indicate an increase of the flood magnitude for all return periods, for both scenarios (Fig. 6). Increases in flood magnitude in response to changes in climate have been addressed in other relevant studies as

Table 3. The historic and climatically changed 10, 20, 100 and 1000 year flows (m^3/s).

	RETURN PERIOD			
	10	**20**	**100**	**1000**
Historic	592.6	675.0	861.7	1125.9
HadCM2050 scenario	647.1	763.6	1027.3	1400.5
PIK scenario	672.6	838.6	1276.6	2025.0

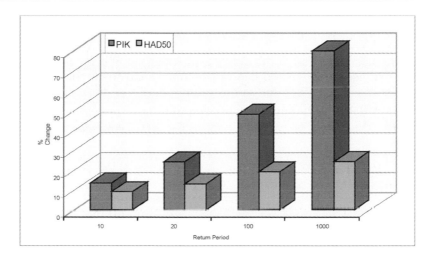

Figure 6. Percentage increase of the 10, 20, 100 and 1000 year flow for climatically changed conditions.

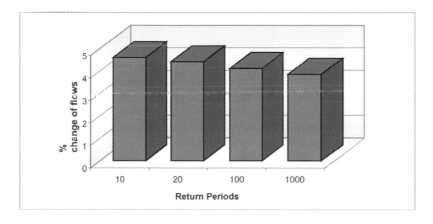

Figure 7. Percentage increase of the 10, 20, 100 and 1000 year flow after the reduction of forest land.

well (Knox, 1993). The PIK scenario gives clearly higher increases of the flood magnitude. Concerning the 1000-year flood, the increase can be as high as 79.85%. Regarding land use change scenarios, the Gumbel max distribution has been fitted to the flows under land use changed conditions. As can be seen in Fig. 7, an increase up to 5% is expected for the flows of different return periods.

Conclusions

The main conclusions drawn from this environmental impact assessment study can be summarised into the following points:

- Both climate change scenarios indicate a significant increase of the magnitude (and risk) of floods.
- The hydrologic response of the catchment under study is sensitive to land use change. Change of forestland into agricultural would result in increased runoff during wet months. Floods for all the return periods under examination are increasing up to 5%.

References

Arnold, J.G., Williams, J.R., Srinisavan, R., and King, K.W. (1994). "SWAT, Soil and Water Assessment Tool." *USDA, Agricultural Research Service, Temple, TX 76502.*

FAO, (1993). Land and Water integration and river basin management. *Proceedings in Rome*, Italy, 31 January – 2 February 1993.

Johns T.C., Carnell R.E., Crossley J.F., Gregory J.M., Mitchell J.F.B., Senior C.A., Tett S., and Wood R.A. (1997). The Second Hadley Center coupled ocean – atmosphere GCM: Model description, spinup and validation. *Climate Dynamics, 13, 103-134.*

Knox, J.C. (1993). Large increases in flood magnitude in response to modest changes in climate. *Nature, 361, 423-425.*

Mimikou M.A., Kanellopoulou S., and Baltas E. (1999a). "Human implication of changes in the hydrological regime due to climate change in Northern Greece." *Global Environmental Change*, 9, 139-156.

Mimikou M.A., Baltas E., Varanou E., and Pantazis K. (1999b). "Impacts of Climate Change on the Water Resources Quantity and Quality." *International Conference on Water, Environment, Ecology, Socio-economics, and Health Engineering.* Oct. 18-21, 1999, Seoul, Korea.

Mithchell J.F.B., Johns T.C., Gregory J.M. and Tett S. (1995). Climate response to increasing levels of greenhouse gases and sulphate aerosols. *Nature, Vol. 376, 501-504.*

Neitsch, S.L., Arnold, J.G., Williams, J.R. (1999). Soil and Water Assessment Tool. *User's manual,* Version 98.1.

Roeckner, E., Arpe, K., Bengtsson, L., Christoph, M., Claussen, M., Dümenil, L., Esch, M., Giorgetta, M., Schlese, U. and Schulzweida, U. (1996). The atmospheric general circulation model ECHAM-4: model description and simulation of present-day climate Max-Planck Institute for Meteorology, *Report No. 218*, Hamburg, Germany, 90 p.p.

Surface Water Hydrology, Singh, Al-Rashed & Sherif (eds)
© 2002 Swets & Zeitlinger, Lisse, ISBN 90 5809 363 8

Climate change impact on the river runoff: Regional study for the Central Asian region

Natalya Agaltseva and Ludmila Borovikova
Central Asian Research Hydrometeorological Institute
72 K. Makhsumova Str., 700052, Tashkent, Republic of Uzbekistan
e-mail: sanigmi@gimet.gov.uz

Abstract

The degree of impact of possible climate changes on the regime of mountain rivers of the Central Asia can be evaluated by sufficiently reliable mathematical models of the runoff formation in mountains. The basic mathematical model describes a complete cycle of the runoff formation, reflecting the main factors and processes: precipitation, dynamics of a snow cover, evaporation, contribution of melting and rain water to the catchment, glacial runoff, runoff transformation and losses in basin. The model complex consists of the model of snow cover formation in the mountains basin, model of glacial runoff and model of snowmelt and rainfall water inflow transformation in runoff. Calculations on regional climatic scenarios indicate that runoff volumes decrease. Longer-term assessments of the water recourses are more pessimistic, since, along with increasing evaporation, water resource inputs (snow and glaciers in the mountains) are continuously shrinking.

Introduction

The water resources of the Aral Sea Basin are jointly used by the Central Asian states. The river flow is concentrated in the two largest transboundary rivers: the Amudarya and Syrdarya Rivers, which run down from the mountains to the plains, cross the deserts and flow into the Aral Sea.

Uzbekistan is the major water consumer in the Aral Sea Basin. In accordance with interstate agreements, on average 43–52 km^2 of water per year as allotted for use by Uzbekistan from the boundary rivers. About 90% of river flow is formed beyond Uzbekistan's boundaries. Intensive irrigation has lead to the drying up of the Aral Sea. Salt marshes and salty, traveling sands are forming on the dried up bottom of the sea. Eolian transfer carries these salty sands to surrounding territories.

Under current conditions, water resource shortages in Uzbekistan, even a small but stable reduction of these resources presents a drastic problem. During the course of the last third of the 20th century, intensive irrigation from the flows of the Central Asian rivers caused the regrettable Aral Sea crisis: the drying-up of the Aral Sea level, a reduction of the delta's lakes system and drastic aggravation of the ecological situation in the Aral Sea Region. The anticipated climate change will cause additional adverse impact: increasing evaporation and salt migration: depleting groundwater reserves; reduction of humid landscapes; salinity growth in closed lakes; accelerated development of the water bodies' eutrophy. In addition, not a single climate scenario predicts increases in the flow of the Amudarya and Syrdarya Rivers; rather, considerable reduction in the flow is anticipated in the future. This will worsen the Aral Sea crisis.

The migration measures in response to climate change in the Aral Sea Region will also alleviate the Aral Sea crisis. Mitigation measures include adequate water delivery to the delta; formation of buffer protection zones of a chain of local water bodies; development of systems for regulating the water exchange in the lake system; phyto-reclamation; development of protected natural reserves; application of water-saving technologies in agriculture; further development of infrastructure for drinking water supply.

To mitigate the adverse impact of the water resources change, it will be necessary to establish reliable hydrometeorological monitoring in the flow formation zone and to use the available water resources carefully and effectively. Particular attention should be given to construction of reservoirs on mountain rivers, allowing for regulation of river regimes in accordance with the requirements of the water consumers.

Numerical Experiments

At present climatologists large pay great attention envisage to the problem of a climate change. The existing hypotheses of the possible climate change envisage the global warming. Global warming distorts the stable character of observation series. The climatic norms are changed. The distribution of a temperature regime and regime of moistening on the territory of Central Asia is determined by its location in the heart of continent, free access of cold air masses, huge mountain barrier in a southeast.

The precipitation is the major climatic characteristic, since it makes up a receipt part of water balance of a dry land. In Central Asia the natural conditions define deficit and variety in the distribution of precipitation.

Desert plain territory is considered to be extremely arid zone. As approaching to the mountain massif the amount of precipitation considerably increases, especially on western and south-west slopes opened for air flows carrying moisture.

The degree of impact of possible climate changes on the regime of mountain rivers of the Central Asia can be evaluated by sufficiently reliable mathematical models of the runoff formation in mountains. The basic mathematical model describes a complete cycle of the runoff formation, reflecting the main factors and processes: precipitation, dynamics of a snow cover, evaporation, contribution of melting and rain water to the catchment, glacial runoff, runoff transformation and losses in basin (Borovikova L.N., Denisov Ju.M., 1972).

The model complex consists of the model of snow cover formation in the mountains basin, model of glacial runoff and model of snowmelt and rainfall water inflow transformation in runoff.

The model of snow cover formation is based on representation of a field of precipitation norms by dependence them from height of area and characteristics macro-orography, and also hypothesis about temporary stability of relative coefficients of decomposition of this dependence in a series of Tailor. The accumulation of snow in basin are computed in model as result of balance of accumulation and loss of sediment precipitation in view of freezing water in snow thickness.

In the model glacial runoff the mathematical and physical-statistical models of processes of accumulation and melt of snow and describing annual cycle of a regime glaciation are accepted. In offered model of transformation of runoff mountains river basin is presented as two transforming reservoirs connected in parallel-successive order. The upper one series a analog of the upper aquifer with rapid inflow transformation. Lower reservoir describes transformation by deeper aquifer.

Important the question on accuracy is, from which the model allows to compute the hydrographs of rivers runoff for a vegetation period. The model complex was developed for an estimation and long-term forecasts of runoff of the mountains rivers. It is adapted for 18 hydrological objects of Central Asia, including inflows in large reservoirs and the estimations of quality of computation and forecast of the rivers runoff and inflows in reservoirs have shown rather high accuracy (efficiency – S/σ = 0,3 ÷ 0,6, «justification» of the forecasts 75 ÷ 94%).

The considered model complex was used for an estimation of influence of climate change on the rivers regime by numerical experiments. The meteorological scenarios using is artificially prescribed climate changes were worked out in different versions. For example, the temperatures norms of each decade increased during the whole vegetation period of during the separate months. The investigations of the «warming» effect showed, in particular the increase of temperature norms during spring-summer period (with other equal condition) causes the decrease of vegetation discharge value, while during the «warming» only in the summer period the vegetation discharge is increased negligibly.

As an example in the Table 1 the results of computation of vegetation and annual inflow water into Charvak Reservoir on the Chirchik River and Nurek Reservoir on the Vakhsh River, drainage areas them are 10 000 km² and 27 500 km² are given. In this case meteorological situations realizing of artificial changes of a climate, were designed in the following variants: the norm of decade temperatures of air increased on 2 °C during all year, or temperature increased only in warm (with 19 for 36 decades) or only in cold half-years (with 1 for 18 decades, and the numbering of decades begins with October 1). The variants with decreasing of temperatures on 2 °C, were similarly designed Besides the situations were designed, when during cold half-year of temperature are decreased, and in warm – increased on 2 °C. Last three variants in the table were designed in the assumption, that in each decade during all year two precipitation norms fall, and though it seems improbable to consider, that such situation is impossible there are no bases.

In Fig. 1 the combined hydrographs of inflow of water into Nurek of reservoir are given with various meteosituations. The analysis of results of computation the hy-

Table 1. Average vegetation and average annual discharges of water inflows in Charvak and Nurek of reservoir, designed with various meteorological situations.

№	Meteorological situations	Decade	Q_{veg} Charvak	Q_{veg} Nurek	Q_{annual} Charvak	Q_{annual} Nurek
1	$\overline{X},\overline{T}$	1 - 36	310,4	944,8	179,1	547,4
2	$\overline{X},\overline{T}$ +2°C	1 - 36	291,7	747,8	174,0	446,8
3	$\overline{X},\overline{T}$ -2°C	1 - 36	313,1	1197,4	177,5	676,1
4	$\overline{X},\overline{T}$ +2°C $\overline{X},\overline{T}$	1 - 18 19 - 36	286,6	932,8	171,3	543,3
5	$\overline{X},\overline{T}$ -2°C $\overline{X},\overline{T}$	1 - 18 19 - 36	325,0	955,4	183,4	550,7
6	$\overline{X},\overline{T}$ +2°C $\overline{X},\overline{T}$	19 - 36 1 - 18	314,2	1253,2	181,0	701,6
7	$\overline{X},\overline{T}$ -2°C $\overline{X},\overline{T}$	19 - 36 1 - 18	297,7	738,4	172,8	444,2
8	$2\times\overline{X},\overline{T}$	1 - 36	597,4	1270,0	326,4	712,9
9	$2\times\overline{X},\overline{T}$ -2°C	1 - 36	581,4	999,6	314,3	574,9
10	$2\times\overline{X},\overline{T}$ +2°C	1 - 36	577,4	1634,8	323,7	899,4

drograph has shown, that «warming» or «colding» only of winter period with precipitation equal by average of long-standing in each decade of appreciable influence on Q_{veg} and Q_{annual} have not rendered. «Warming» and «colding» during all year essentially had an effect on change of inflow in Nureks Reservoir. So Q_{veg} in the first case has decreased on 21%, and in second – has increased on the same size. «Warming» of spring-summer period with other equal conditions has caused increase of inflow in Nureks Reservoir, that is explained by presence powerful glacial in basin of the Vakhsh River and essential role glacial runoff (Q_{veg} has increased on 24%). In inflow in Charvaks Reservoir in this case essential changes have not taken place.

By increasing the average long-standing decade sums of precipitation twice Q_{veg} and Q_{annual} have increased approximately on 30%, and with increase of average long-standing temperature of air during all year on 2 °C, approximately on 40%.

Vulnerability of Surface Water

The assessment of vulnerability of water resources has been obtained by use of all designed climatic scenarios. Availability of the different scenario variants made it possible to take into account climatic scenarios and assess the most critical situations. Calculations were made for every basin-indicator zone for flow formation of the Syrdarya River (Chatkal, Pskem, Akhangaran, Ugam Kugart, Karadarya, Tentyaksay) and for the Amudarya River (Zeravshan, Vakhsh, Obikhingou, Vanch) with consequent integral evaluation.

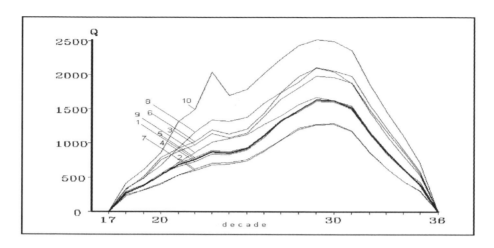

Figure 1. The hydrographs of inflow into Nurek Reservoir, designed with the various tasks of meteorological situations (variants 1-10).

Blanket of Snow

The blanket of show in the mountain is an important feeding source for the rivers of Uzbekistan. At present the contribution of show to total moisture delivery of the watershed surface amounts to 60-75%. Simulation modeling of the integrated snow reserves shows their gradual decrease in all spring months during the last decades.

The sustainable seasonal blanket of snow in the mountains is a level of 1400-1800 m. Melting of the seasonal blanket of snow, with exception of the densest accumulations of it in the negative forms of the relief, place during April-May.

Model calculations of snow reserves in the mountains under different climatic scenarios have demonstrated their gradual decrease due to growing aridity of the climate. Contribution of the snow is expected to decrease by 15-30%, especially for rivers which are snow-fed (Fig. 2). Changes in the snow border and a one-month shift in melting of snow are also expected.

Rain Runoff

The contribution of precipitation from rain might grow from 12-15% up to 25-35%, which negatively affect formation of snow reserves. This means that there would be proportionally more rain shower precipitation. This would result in rain flood, primarily on smaller water flows, expansion of soil erosion and flow turbidity. This foothill zones, especially in Fergana Valley, often experience heavy rains and showers. Given the expected changes in the climate, whole strips of foothills in this area could become intensive mudslides zones.

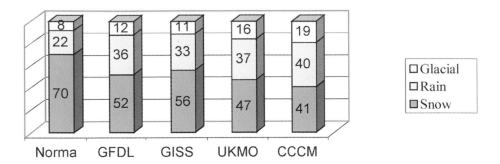

Figure 2. The correlation (%) of snow, rain and glacial components to the total annual water delivery to the Pskem River watershed for a base period under different climatic scenarios (the Syrdarya River basin, snow-rain type of river feeding).

The spring high water season would shift by one month in the annual course of river runoff fed by snow and rain (Fig. 3) and as a result, the regime of operation of hydro-technical facilities would change, too.

Glacial Runoff

There are glaciers in the basins of the Pskem River, Kashkadarya and Surkhandarya Rivers in Uzbekistan. Other glaciers that feed the rivers of Uzbekistan are situated beyond in territory.

Mountain glaciers are an important source of river feeding at the height of the summer season, when there is practically no water delivery from snow melting and

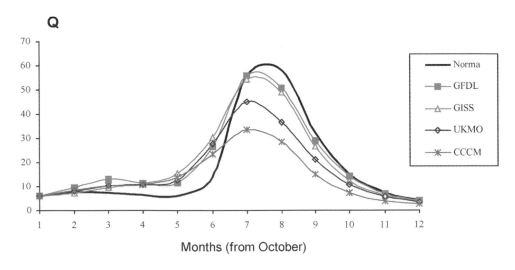

Figure 3. The Hydrograph of the Kugart River runoff for the base period under climatic scenarios (the Syrdarya River basin).

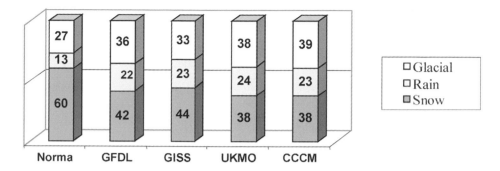

Figure 4. The correlation (%) of snow, rain and glacial contribution to the watershed of the Zeravshan River for the base period under different climatic scenarios (the Amudarya River basin, snow-glacial type of river feeding).

rains. Model calculations demonstrate that in the distant future, the glacier runoff would depend on the rates of reduction of mountain glaciation. At present, the annual glacial runoff of the rivers of the Syrdarya River basin amounts to 8-15%. Under different prognoses, increase in this flow of up to 20% is expected (Fig. 4). Contribution of glacial runoff to the rivers of the Amudarya River basin might grow 32-39% under the most "severe" climatic scenarios.

Prospective Assessment of Water Resources

An integrated runoff assessment was made for the Syrdarya and Amudarya Rivers based on numerical experiments with indicator-basins under different climatic scenarios.

During the cropping season, an increase is expected in evaporation from water surfaces of 15-20%. The most severe arid climate conditions in the watershed area were predicted under the CCCM model (average annual temperature increase by 6.5 °C, decline in annual precipitation rate by 11%). According to this model, if CO_2 concentration in the atmosphere is doubled, then the runoffs of the Syrdarya and Amudarya Rivers are expected to be reduced by 28 and 40%, respectively (Table 2).

The UKMO model also predicts unfavorable outputs where temperatures may increase by 5.2 °C, while annual precipitation increase by 6%. In this case, runoff could be reduced by 15-21%.

According to GFDL and GISS scenarios, average annual temperature in the catchment area would increase by 3-4 °C and average annual precipitation volume – by 10-15%. Under these scenarios, one could expect that no significant reduction in the Amudarya and Syrdarya runoff would occur.

Calculations of regional climatic scenarios by the year 2030 also indicate persistence of present runoff volumes accompanied by an increase in fluctuations from year. Longer-term assessments are more pessimistic, since, along with increasing

Table 2. Expected changes in water resources of the main rivers of the Aral Sea Basin under implementation of different climatic scenarios (% of the Base Norm).

Rivers	Basic Rate (km^3/year)	Climate scenarios				
		Regional, by the year 2030	GFDL	GISS	UKMO	CCCM
Syrdarya	37,9	+4	+1	-2	-15	-28
Amudarya	78,5	-3	0	-4	-21	-40

evaporation, water resource inputs (snow and glaciers in the mountains) are continuously shrinking.

Summary

Under the current conditions of water storage in Central Asia, even this small but continued reduction becomes a serious problem. The supposed flow reduction would be more pronounced during the vegetation period, thus being especially hazardous for irrigated farming and regional ecosystems (located on riverbanks and in deltas).

References

Borovikova L.N., Denisov Ju.M., Trofimova E.B. et al. (1972). "Mathematical simulation for mountainous rivers runoff process". Proc. SANIGMI, vol. 61(76).

Agaltseva N.A., Pak A.V. (1994). "The automated long-term forecasts of a runoff of the mountainous rivers of Central Asia". Proc. Institute of Cybernetics, vol. 150.

Surface Water Hydrology, Singh, Al-Rashed & Sherif (eds)
© *2002 Swets & Zeitlinger, Lisse, ISBN 90 5809 363 8*

The influence of ENSO on seasonal drought and excess rainfall in the Sultanate of Oman

Nazemosadat M. Jafar

Irrigation Department, Faculty of Agriculture, Shiraz University, Iran
e-mail: Jafar@hafez.shirazu.ac.ir

Abstract

The present study has examined the impact of El Niño-Southern Oscillation (ENSO) on the rainfall of 8 synoptic stations situated in various parts of Oman Sultanate. The available data till 1998 was used in computational analysis. Although a simple correlation analysis does not revealed significant coherence between the Southern Oscillation Index (SOI) and seasonal precipitation, the study has found that summer rainfall in Salalah (southern parts of the country) is influenced by intense ENSO events. For this region, the occurrence of warm ENSO phase was generally coincided with negative anomalies of the rainfall. The study has documented that, during warm events, the probability of the occurrence of flood and abundant summer rain is very low. The positive anomalies of the observed rainfall and flooding events were mostly harmonized with La Niña periods. Compared to warm periods, the median of the rainfall was found to be greater during cold ENSO phase.

Introduction

Sultanate of Oman with a population of 2.53 million (2000 estimate) and the area of 309,500 km^2 is situated over the southeast portion of the Arabian Peninsula (Figure 1). The climate of the country is one of the hottest in the world with limited rainfall and extensive humidity. The average annual temperature and rainfall is estimated to be around 28 °C and less than 100 millimeters, respectively (Microsoft, ENCARTA Interactive World Atlas, 2001). Water resources are limited and barren deserts occupy most parts of the country.

Oman falls naturally into three physical divisions consisting of a narrow coastal plain, ranges of mountains and hills, and an interior plateau. The coastal plain along the Gulf of Oman is known as Al Baatinah and is the country's principal agricultural region. Inland from the plain lies the Akhdar Mountains range. The highest peak in the range is Mount Sham, at an elevation of 3,026 meters. The coastal plain extending south along the Arabian Sea is largely barren.

The country has three distinct topographical features. The first consists of two, relatively fertile coastal strips; one in the north stretching from Muscat to the border with the United Arab Emirate, and the other surrounding the southern coastal town of Salalah. The second feature includes two mountainous regions; one in the north and the other bordering the Salalah plain in the south. The third feature is a sandy wasteland along the border with Saudi Arabia.

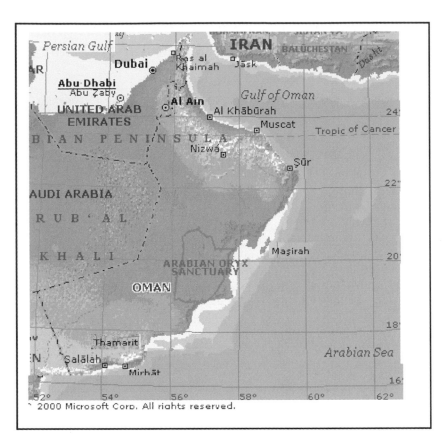

Figure 1. The geographical location of Oman Sultanate and the considered rainfall stations.

El Niño is the name given to the quasi-cyclic heating of the normally cool surface water of the equatorial Pacific Ocean, off the coast of Peru and Ecuador (Allan et al., 1996). The occurrence of intense El Niño is generally coincided with an above normal precipitation over western side of American continent. However, below normal rainfall and severe drought over Australia is expected during such events. The Southern Oscillation (SO) phenomenon has been described as a fluctuation in pressure field with opposed anomalies in the Australian region and southeastern Pacific Ocean, as first described by Walker (1924). This phenomenon is character-ized as a global change in atmospheric and oceanic conditions best observed over the Pacific Ocean and is associated with El Niño events. Ropelewski and Halpert (1996) have been explored the relationships between the SO Index (SOI) and pre-cipitation in global scale since early of the 20th century.

Since both the El Niño and Southern Oscillation phenomena are closely interlinked, the names are sometimes combined and the term ENSO is used to address both events. This terminology is adopted for the present study and is consistent with the

98

most research conducted by meteorologists and oceanographers (e.g. Dracup and Kahya, 1994; Ropelewski and Halpert, 1996). According to this adoption, El Niño refers to the warm (low) phase of the SO while La Niña refers to the cold (high) phase of this phenomenon.

The ENSO is a well-known teleconnection pattern that accounts for a large portion of rainfall extreme over various parts of the globe. A number of recent studies have confirmed that ENSO induce a significant impact on the variability of rainfall and water resources in America, Australia and many parts in Asia and Africa (e.g. Allan et al., 1996; Hastenrath et al., 1993; Ropelewski and Halpert, 1996; Nicholls, 1996; Drosdowsky, 1994 and 1995; Cordery and Opoku-Ankomah, 1994; Camberlin, 1995; Wu and Zhang, 1996; Nazemosadat and Cordery, 1997; Hodet, 1997). Nazemosadat and Cordery (2000a) have recently revealed that the autumn rainfall in Iran is negatively correlated with the Southern Oscillation Index (SOI). More (less) than usual autumn rainfall over nearly all parts of the country is, hence, associated with warm (cold) ENSO episodes. In winter, however, for a large portion of the country, more than normal rainfall is expected during La Niña events (Nazemosadat, 2001a, b; Nazemosadat and Cordery, 2000b). Moreover, in addition to rainfall, the surface climate of the Indian Ocean is also significantly influenced by ENSO events (Hastenrath et al., 1993; Nazemosadat, 1996). Since Oman is immediately adjacent to the Ocean, the meaningful associations between ENSO and rainfall of this country is expected.

The current study is hence prompted to investigate the geographical extend of the influence of ENSO over the region situated off the coast of the western Indian Ocean. The study has also focused to examine the possibility of the prediction of seasonal rainfall in Oman. Such an study is realized to be essential for efficient management of scarce water resources within the Sultanate. The SOI-rainfall relationships as well as the response of seasonal rainfall to the extreme El Niño and La Niña events are explored.

Data and Methods

Total monthly rainfalls for 8 key synoptic stations located over various parts of the country were gratefully supplied by the Directorate General of Civil Aviation and Meteorology in Oman (personal communication). Figure 1 illustrates the geographical location and Table 1 outlines duration, record length, total annual rainfall and the ratio of seasonal to annual precipitation for the considered stations. As indicated in this table, the record length of the data used varies from 56 years (in Salalah) to 18 years in Khasab and Thumrait. Although 30 years of continuous data is generally required for such analysis, the selected stations contain the highest amount of data. The total annual rainfalls vary from 40 mm in Thumrait to 369 mm in Siaq (Table 1). Annual rainfall at seed Airport, near Mascot the capital city of Oman is 88 mm (about 10% of average global precipitation). The Troup's (1965) Southern Oscillation Index (SOI) data, supplied by the Australian Bureau of Meteorology, is used as ENSO indicator.

Table 1. The period, record length, total annual rainfall and the ratio of seasonal rain to annual precipitation.

Station Name	Period	Record length	Total annual mm	Winter	Spring	Summer	Autumn
Salaleh	1943-98	56	110	0.09	0.26	0.52	0.13
Masirah	1956-98	43	70	0.37	0.38	0.11	0.15
Seed Airport	1975-98	24	88	0.57	0.17	0.05	0.21
Sur	1977-98	22	99	0.55	0.17	0.06	0.21
Siaq	1979-98	20	369	0.35	0.21	0.37	0.07
Majis	1980-98	19	120	0.64	0.07	0.03	0.26
Khasab	1981-98	18	219	0.69	0.04	0.00	0.26
Thumrait	1981-98	18	40	0.39	0.51	0.09	0.02

The simple correlation analysis was used for investigating the relationships between SOI and rainfall in seasonal scale. Moreover, the low (warm) and high (cold) phases of ENSO were also identified and the amounts of rainfall during these episodes were examined. The episodes that seasonal SOI were greater than +5 and less than −5, were considered as a La Niña (high or cold) and El Niño events (low or warm), respectively. For every station considered, the probability of the occurrence of positive and negative rainfall anomalies during these events was investigated. The median of seasonal rainfall during high and low phases was also computed and compared.

The time series of seasonal rainfalls were obtained by averaging the three month values of precipitation. Best results were obtained by defining winter, spring, summer and autumn from January to March, April to June, July to September and October to December, respectively. The same averaging procedure was performed to provide the seasonal time series of SOI data. Although missing values in the rainfall data were generally rare, they were excluded in computational analysis. The results presented here are therefore obtained from the available data.

Results and Discussion

Seasonal Distribution of Rainfall

The given values in Table 1 implies that winter rainfall is the major source of fresh water and makes up from about 9% (in Salalah) to 69% (in Khasab) of annual precipitation over different parts of the country. As a whole, it is estimated that this rainfall supplies about 46% of water resources within the country. For spring, the ratio is considerable in Thurmait (51%), Masirah (38%), Salalah (26%) and Siaq (21%). The ratio is, however, low for the other stations. It is estimated that spring rainfall that contributes about 22% of annual precipitation, is the second source of the country's water resources. In contrast to other stations, summer rain is substantial in Salalah and Siaq that supply about 52% and 37% of annual precipitation, respectively. For the other stations, the ratio is very small and negligible. As indicated in Table 1, from 2% (in Thumrait) to 26% (in Khasab) of total annual rainfall has been precipitated in autumn.

Correlation Analysis

Table 2 depicts the correlation coefficients between seasonal SOI and corresponding rainfall data for the period 1981–1998. For spring and summer, rainfall data of some stations contained a lots of zero values and the coefficients are hence not presented in this table. Since the data of this period was available for all stations, they are used as the basis for the evaluation and comparison of the effect of ENSO on the rainfall of Oman. The given results in Table 2 suggests that seasonal rainfall in considered station is not significantly (at 1% significance level) associated with SOI. The relatively strong correlation between winter SOI and rainfall data in Masirah is mostly associated with the abundant rainfall during intense El Niño 1983. After the data of this year was removed, the correlation coefficient has dropped to near zero. As depicted in this table, for the most of occasions, the sign of the correlation coefficients are negative that implies seasonal rainfall tend to be more (less) than usual for the episodes that SOI is positive (negative). In contrast to other stations, the SOI-rainfall relationships are positive in Salalah and Siaq during summer time. The positive relationship between these variables suggests that more (less) than usual of summer rain is generally coincided with positive (negative) SOI.

Nazemosadat (1996) has shown that, in autumn and winter, the atmospheric circulation over the Arabian Sea is significantly associated with ENSO. The poor correlation between SOI and rainfall in considered stations, particularly those situated over the coastal strip of the Arabian Sea is hence surprising and need further investigation.

Table 2. The correlation coefficients between SOI and rainfall for the period 1981–1998.

Station name	winter	spring	summer	autumn
Salaleh	-0.32	0.18	0.45	0.18
Misare	-0.56	0.15	-	-0.37
Seed Airport	-0.08	-0.20	-	-0.09
Sur	-0.47	0.02	-	-0.25
Siaq	0.04	-0.14	0.15	-0.42
Majis	-0.13	-0.17	-	-0.16
Khasab	-0.35	0.13	-	-0.31
Thurmait	0.03	-	-	-

Extreme ENSO Events and Seasonal Rainfall

For every individual station, the median of seasonal precipitation during warm and cold ENSO phases were computed and used as an index for the evaluation of the response of rainfall to these phases. The given results have indicated that summer rainfall in Salalah is the only sensitive incident to the extreme phases of ENSO. For the other stations, the impact of these phases on seasonal rainfall was generally found to be either negligible or difficult to assess due to the shortage of record length. It is noteworthy to recall that the coastal region near Salalah is a notable fertile area within the country and summer rainfall of this region consists of more than 50% of annual precipitation. The results of the present study are, therefore, important for seasonal prediction of rainfall and efficient management of water and land resources in this arable area.

For about 75% of observed events, the occurrence of warm ENSO phase was co-incided with the below normal of summer rainfall in Salalah. Figure 2 illustrates the concurrent fluctuation of SOI and the normalized anomalies of rainfall for the spells that SOI is less than –5. The anomaly values are multiplied by 10 to be more competitive with SOI data. This Figure implies that the positive anomalies of the rainfall during El Niño episodes is marginal and flooding events is, therefore, not expected during such episodes. In other words, although minimal above normal of summer rain is expected during warm periods (with probability about 0.25), the possibility of the occurrence of flood is near zero. On the other hand, the analysis has revealed that the most ever seen drought (in 1972) has occurred during warm ENSO phase.

The concurrent fluctuation of SOI and normalized anomalies of the rainfall during the spells that SOI was more than +5 is shown in Figure 3. Since the frequency of positive and negative rainfall anomalies are almost identical, the Figure implies that the occurrence of cold ENSO phase cannot be used as a reliable indicator for the prediction of drought or flood events. The results suggest that, although El Niño spells are generally associated with the shortage of summer rainfall in Salalah, during cold events, the occurrence of drought and excess rainfall is mostly independent from the fluctuations of SOI.

The magnitude of the median of summer rainfall during cold and warm episodes was examined and estimated as 17.0 and 19.2 (millimeter per month) respectively.

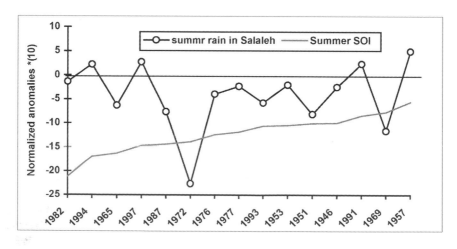

Figure 2. Concurrent fluctuations of SOI and summer rainfall in Salalah during warm ENSO phase. All data are sorted according to SOI values.

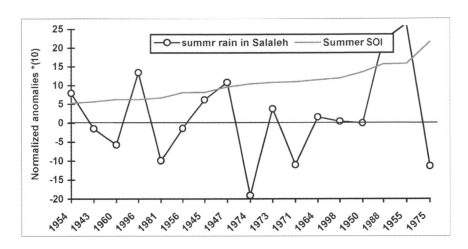

Figure 3. Concurrent fluctuations of SOI and summer rainfall in Salalah during cold ENSO phase. All data are sorted according to SOI values.

These figures suggest that, compared to warm events, the greater amount of summer rainfall (in Salalah) is expected during cold periods. The given results suggest that, in summer, the easterly wind near the tropical regions of the Indian Ocean is generally intensified (weaken) during cold (warm) spells. As these moisture-laden easterly winds strengthen during cold ENSO phase, bring about more rainfall over southern portion of the country. Less than usual rainfall is expected when the occurrence of warm phase weakens this easterly air-flow.

SOI and Extreme Rainfall Events

The associations of extreme drought and excess rainfall in Salalah with SOI were also investigated. It was found that positive anomalies of summer rain and flooding events are mostly associated with cold ENSO phase. As indicated in Figure 4, among fourteen of highest rainfall events (1/4 of all data), in 75% of cases, excess rainfall is concurrent with positive SOI values. As indicated in this Figure, above normal rainfall has not been occurred when SOI is less than –5. On the other hand, abundant rainfall and flooding event (such as 1945, 1974, 1988, 1995, 1996 and 1955) has occurred when SOI was greater than 5. The negative sign of SOI during summer 1963, with highest value of observed rainfall, is striking. The linkage between extreme summer droughts and ENSO was also investigated. The results do not exhibit coherence between these two phenomena.

Summary

A number of studies have indicated that El Niño Southern Oscillation (ENSO) phenomenon has significant influence on surface climate and water resources over various parts of the globe. However, the associations between the occurrence of seasonal rainfall in Oman Sultanate and this phenomenon have not yet been comprehensively explored. Such exploration is beneficial for seasonal prediction of

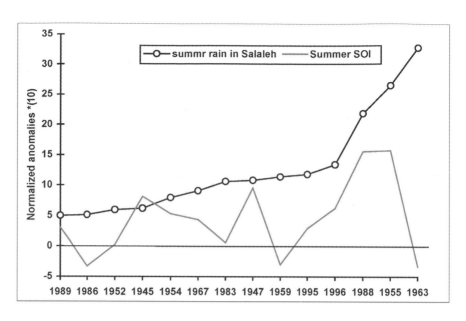

Figure 4. Concurrent fluctuations of SOI and summer rainfall in Salalah during
wet years.

rainfall and efficient managing of scarce available fresh water in this country and
Arabian Peninsula. Moreover, the results of the present study could be useful for
better understanding of the interactions between ENSO, Indian Ocean surface climate and atmospheric circulation over the Arabian Peninsula. Monthly rainfall data
of 8 key meteorological station in various parts of the country as well as the SOI
data were used to examine the impact of ENSO on seasonal rainfall in Oman. The
results of simple correlation analysis have revealed that the SOI-rainfall relationships are not significant in all stations. However, when the magnitude of rainfall
during low and high phases of ENSO events were assessed, the response of
summer rainfall in Salalah to these phases was found to be promising.

The occurrence of warm ENSO phase was generally founds to be coincided with
below normal of summer rainfall in Salalah. During such phase, flooding events
and profound abundant rainfall are not expected. Compared to warm spells, the
magnitude of the median of summer rainfall has slightly been observed greater during cold episodes. It was concluded that, in summer, the easterly wind near the
tropical regions of the Indian Ocean is generally intensified (weaken) during cold
(warm) spells. The study has found that positive anomalies of summer rain and
flooding events are mostly (75% of observed cases) associated with cold ENSO
phase.

References

Allan, R.J., Bread, G.S., Close, A. Herczeg, A.L., Jones, P.D. and Simpson, H.J.
1996. Mean sea level pressure Indices of the El Niño Southern Oscillation: Rele-

vance to stream discharge in south-eastern Australia, *CSIRO divisional report 96/1, Division of Water Resources*, 23 p.

Camberlin, P. 1995. June-September rainfall in north-eastern Africa and atmospheric signals over the tropics: A zonal perspective, *Int. J. Climatol.*, **15**, 773-783.

Cordery, I. and Opoku-Ankomah, Y. 1994. Temporal variation of relations between sea surface temperature and New South Wales rainfall, *Aust. Met. Mag.*, **43**, 73-80.

Dracup, J.A., and Kahya, E. 1994. The relationships between US streamflow and La Niña events, *Water. Resour. Res.*, **30**, 2133-2141.

Drosdowsky, W. 1994. The Southern Oscillation. In: Will it rain? The effects of Southern Oscillation and El Niño on Australia, Edited by Partridge, I.J. 2nd edition, 12-21.

Drosdowsky, W. 1995. Analogue (Non-linear) forecasts of the Southern Oscillation Index time series, *NOAA Experimental Long-lead Forecast Bulletin*, **4,** 28-31.

Hastenrath, S., Nicklis, A. and Greischar, L. 1993. Atmospheric-hydrospheric mechanisms of western equatorial Indian Ocean, *J. Geophys. Res.*, **98**, 20219-20235.

Hodet, G.C. de 1997. Seasonal climate summary southern hemisphere (spring 1996): characteristics of a weak positive phase of the Southern Oscillation persist. *Aust. Met. Mag.*, **46**.153-160.

Nazemosadat, M.J. 1996. The impact of oceanic and atmospheric indices on rainfall variability, PhD Thesis, University of New South Wales, Sydney, Australia.

Nazemosadat, M.J. and Cordery, I. 1997. The influence of geopotential heights on New South Wales rainfalls, *Meteorology and Atmospheric Physics*, **63**, 179-193.

Nazemosadat, M. J. and Cordery, I. 2000a. On the relationships between ENSO and autumn rainfall in Iran, Int. *J. Climatol*, **13**, 51-64.

Nazemosadat, M.J. and Cordery, I. 2000b. The impact of ENSO on winter rainfall in Iran, proceeding of the 26th National and 3rd International Hydrology and Water Resources Symposium, Inst. Engs. Australia, 20-23 November 2000, Perth. 538-543.

Nazemosadat, M.J. 2001a. Winter rainfall in Iran: ENSO and aloft wind interactions. Iranian Journal of science and technology. In press.

Nazemosadat, M.J. 2001b. Winter drought in Iran: Associations With ENSO, Drought Network News, Vol (13) No. 1, 10-13.

Nicholls, N. 1996. Improved statistical model approaches for seasonal climate forecasting, presented in: Managing with Climate Variability Conference, Canberra, November 1995.

Ropelewski, C.F. and Halpert, M.S. 1996. Quantifying Southern Oscillation-precipitation relationships, *J. Climate*, **9**, 1043-1059.

Troup, A.J. 1965. The Southern Oscillation', *Q. J. R. Meteorol. Soc.*, 91, 490-506.
Trenberth, K.E. 1997. The definition of El Niño, *Bull. Amer. Meteor. Soc.*, **78**, 2771-2777.

Walker, G.T. (1924). Corelation in seasonal variations of weather, Mem, India Meteorol, Dept, 24, 75-131.

Wu, G.-X. and Zhang, X.-H. 1996. Research in China n climate and its variability, *Theor. Apll. Clomatol*, 55, 3-17.

Surface Water Hydrology, Singh, Al-Rashed & Sherif (eds)
© *2002 Swets & Zeitlinger, Lisse, ISBN 90 5809 363 8*

Simulation of climate change effects on crop actual evapotranspiration over Sudan

Nadir Ahmed Elagib*
Division of Civil Engineering and Construction, University of Paisley
High Street, Paisley PA1 2BE, Scotland, UK
e-mail: elagib@excite.com

Abstract

The goal of this study is to illustrate the effect of possible future climate change on actual evapotranspiration and evapotranspiration deficit for three major crops grown in Sudan, namely sorghum, groundnut and millet. Using a water balance model, the simulation analysis was performed under different working scenarios of climate change, namely +1.5 °C change in temperature combined with 0 and ±20% changes in rainfall. These scenarios were, then, compared with the baseline conditions of average monthly values for the period 1961–1990. The simulation results indicate that climate change might have a significant effect on the evapotranspiration regimes and the subsequent intensity of agricultural drought over Sudan. In view of the observed magnitudes of impacts in the past few decades, such simulated changes if occur in the future would undoubtedly have significant implications for land and water resources, which are already highly exploited by increasing and sedentary human and animal populations.

Introduction

It is now increasingly recognized that the plausible climate change suggested by the scientific community will be manifested as alterations in regional hydrological cycles, with consequent impacts on the quantity and quality of regional water resources (Gleick, 1986, 1989). Changes in the timing and magnitudes of precipitation, surface runoff and soil moisture are, thus, anticipated. Such changes, whether perceived as surplus or deficit, are likely to have profound implications for water availability to supply agricultural needs. The amount of water stored in the soil, penetrating to depths beneath the root zone or running off into rivers and streams will directly change as a result of changes in precipitation and/or evapotranspiration (Rosenberg et al., 1989; McKenney and Rosenberg, 1993). Dry-land agriculture would be inflicted by alteration in the timing, duration and severity of periods of water stress. A large increase in potential evapotranspiration can lead to soil moisture deficit and vegetation desiccation (Rind et al., 1990). Hence, changes in evapotranspiration will be crucial for determining the amount of water for crop irrigation (Ramírez and Finnerty, 1998). Spatial and temporal distribution of potential evapotranspiration will determine what type of crops to grow, crop rotations to be followed and the timing of the initiation of the cropping season, particularly for rain-

---

* Address for correspondence: 53 Cranborne Avenue, Tolworth, Surbiton, KT6 7JP, UK

fed agricultural systems, to take advantage of rainfall needed to sustain crop growth (Lansigan et al., 2000).

Water availability is critical to the agriculture-based economy of Sudan. The traumatic, protracted conditions of drought, which inflicted Sudan during the last few decades, have had enormous impact on agricultural production (Larsson, 1996; Ayoub, 1999). While annual rainfall revealed a decrease of up to > 20% (Hulme, 1990), mean annual temperatures for the period 1961–1990 showed a significant warming of 0.22–0.40 °C per decade (Elagib and Mansell, 2000a). Decreasing duration of sunshine, global radiation and wind speeds have also been witnessed (Elagib and Mansell, 2000b). These combined trends resulted in signs of increasing rates of calculated potential evapotranspiration and intensifying aridity (Elagib and Mansell, 2000b). In view of these climatic trends and the dilemma of plausible climate change (Hare, 1993), it is a paramount task to understand how such changes could affect the different water balance components (WBCs) if proper water resources management is sought. Evapotranspiration is the second important component next to rainfall in the water balance. The analysis of this loss is necessary to many subject areas, such as hydrology, agronomy, forestry, and land-resources planning (Singh and Xu, 1997). This study endeavours to explore the magnitudes and directions of response the actual evapotranspiration in Sudan for sorghum, groundnut and millet may take under a changing climate.

Data and Methods

Data

Thirty-one sites spread over five different climatic zones are selected (Table 1 and Fig. 1), with 1961–1990 climatological normals for maximum and minimum temperatures, rainfall, actual sunshine duration, wind speed and global radiation. The exception is Zalingei where the normals were obtainable for the reference period of 1951–1980. These stations represent four main geographical regions, namely southern (4°–10° N), central (10°–16° N), northern (16°–22° N) and coastal (along the Red Sea) regions. The regions respectively encompass 5, 18, 6 and 2 stations. The central region can also be divided into three sub-regions, namely western, middle and eastern regions, which respectively comprise 8, 6 and 4 stations. While the climatic variables are recorded at all the current stations, solar radiation is measured at only 16 stations, which constitute the total number of stations recording solar radiation in Sudan. Alternatively, it has been estimated from sunshine duration and geographical parameters (Elagib et al., 1999a, 1999b; Elagib and Mansell, 2000c). Actual sunshine duration data are not available for only one meteorological station, namely Nyala. In this case, the data (normals) were simply assumed to be the average for the neighbouring stations of El Fasher, Zalingei and Gazala Gawazat.

Water balance model

In addition to the many advantages offered by water balance modelling, such as accuracy, flexibility and ease of use, Gleick (1986) argues that it is a promising method for evaluating the regional hydrological effects of global climatic change.

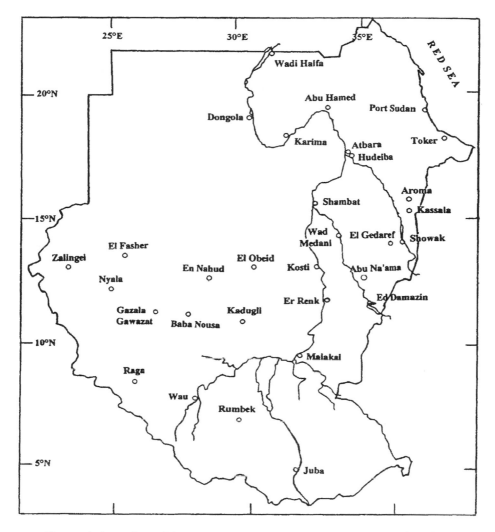

Figure 1. Location of the meteorological stations under consideration.

The hydrological process is an intricate interaction of land, plant and atmosphere, but its understanding can be approached via generating a water balance model (WBM), which accounts for all the necessary components over a selected time period. Under rain-fed conditions, these components may include rainfall *(P)*, runoff *(Q)*, deep percolation *(D)*, actual evapotranspiration *(AET)*, groundwater inflow *(GW)* and change in soil moisture storage *(ΔS)*. A simple water balance model is, then, of the form:

$$P - AET - Q - D + GW \pm \Delta S = 0 \qquad (1)$$

where all the parameters are in mm. Except the rainfall and runoff, detailed measurements of the other parameters are usually not available. Therefore, one must

109

look for an indirect way of quantifying these parameters with primary consideration of data availability and simplicity, reliability and applicability of the approach to all climates with minor modification. The monthly water balance approach is preferred because it is less data intensive and less sophisticated compared to those with shorter time-scale (Xu and Singh, 1998). Among the WBCs, *AET* is considered the most difficult one to measure since it is a function of climatological variables, soil moisture storage, canopy interception and vegetation growth rate (Kolka and Wolf, 1998).

Hoogmoed et al. (1991) indicated that upward flow of water is hardly expected to take place for the majority of soil conditions in the Sudano-Sahelian zone. Information available for Sudan does confirm this assertion. According to Shahin (1985), Musa (1986), Hulme (1986) and Hussein and El Daw (1989), the groundwater supplies in Sudan is quite deep. These conditions are unfavourable to vegetation, as plants hardly benefit from the deep water table. It is, hence, fair to assume that the term for groundwater (*GW*) in Eq. 1 is negligible.

The coarse-textured sands and gravels in much of the northern and western Sudan have high infiltration capacity (Oliver, 1969). Except in the event of very intense rainfall, these soils experience little runoff, and there is more opportunity for a larger proportion of rainfall to penetrate to deep layers. Overland flow can, thus, be assumed negligible under these soil conditions. In contrast, impermeable clay soils that cover extensive parts of the centre, east and south of Sudan crack when dry and expand, close the cracks and become sticky on wetting (Musa, 1986; Hussein and El Daw, 1989). The characteristics of rainfall are held responsible for reducing the infiltration capacity of the soil because large raindrops damage the soil by shattering and compacting it, thus sealing the surface pores by splashing of fines (Musa, 1986). Higher antecedent soil-moisture levels before heavy storms also result in lower infiltration capacity (Hulme, 1986), which restricts the availability of rainfall to plants and cause much of it to be lost as overland flow (Musa, 1986). Trampling of soil by the livestock in the dry season also helps to reduce the infiltration rate. These conditions cause rainwater to be dissipated by runoff while the remaining part that is held up near the surface is exposed to rapid evaporation (Oliver, 1965). However, Oliver (1969) reports that surface runoff during the wet season over half of the country occurs only for a few hours at a time and on a very limited occasions. Hence, it can be assumed that overland flow is eventually lost as evaporation and that deep percolation may be considered neglected in such clayey soils.

On the basis of the discussion outlined above, Eq. (1) can be reduced to:

$$P - AET - Q \pm \Delta S = 0 \qquad \text{(clay soils)} \qquad (2)$$

$$P - AET - D \pm \Delta S = 0 \qquad \text{(sand soils)} \qquad (3)$$

Table 1. Particulars of the meteorological stations.

Station	Station's Reference Number	Latitude (°N)	Longitude (°E)	Altitude (m)	Soil Type	Climatic Zone*
Southern region						
Juba	1	04.87	31.60	460	Clay	Dry sub-humid
Rumbek	2	06.80	29.70	420	Clay	Semi-arid
Wau	3	07.70	28.02	435	Clay	Dry sub-humid
Raga	4	08.47	25.68	545	Clay	Sub-humid
Malakal	5	09.55	31.65	390	Clay	Semi-arid
Western region						
Kadugli	6	11.00	29.72	500	Clay	Semi-arid
Baba Nousa	7	11.33	27.67	543	Sand	Semi-arid
Gazala Gawazat	8	11.47	26.45	480	Sand	Semi-arid
Nyala	9	12.07	24.88	655	Sand	Semi-arid
En Nahud	10	12.70	28.43	565	Sand	Arid
Zalingei	11	12.90	23.32	900	Sand	Semi-arid
El Obeid	12	13.17	30.23	570	Sand	Arid
El Fasher	13	13.63	25.33	730	Sand	Arid
Middle region						
Er Renk	14	11.75	32.78	380	Clay	Semi-arid
Ed Damazin	15	11.75	34.40	470	Clay	Semi-arid
Abu Na'ama	16	12.73	34.13	445	Clay	Semi-arid
Kosti	17	13.17	32.67	380	Clay	Arid
Wad Medani	18	14.40	33.48	405	Clay	Arid
Shambat	19	15.60	32.50	380	Clay	Arid
Eastern region						
El Gedaref	20	14.03	35.40	600	Clay	Semi-arid
Showak	21	14.22	35.85	380	Clay	Semi-arid
Kassala	22	15.47	36.40	500	Clay	Arid
Aroma	23	15.83	36.15	430	Clay	Arid
Northern region						
Hudeiba	24	17.57	33.93	350	Sand	Hyper-arid
Atbara	25	17.70	33.97	345	Sand	Hyper-arid
Karima	26	18.55	31.85	250	Sand	Hyper-arid
Dongola	27	19.17	30.48	225	Sand	Hyper-arid
Abu Hamed	28	19.53	33.33	315	Sand	Hyper-arid
Wadi Halfa	29	21.82	31.34	190	Sand	Hyper-arid
Coastal region						
Toker	30	18.43	37.73	020	Sand	Hyper-arid
Port Sudan	31	19.58	37.22	005	Sand	Hyper-arid

* Identified following the UNEP aridity index, i.e. the ratio of annual precipitation to potential evapotranspiration (UNEP, 1992; Hare, 1993).

A theoretical water-balance approach was first initiated by Thornthwaite and Mather (1957). The approach, henceforth referred to as the T-M model, requires minimal input data, such as temperature to calculate the potential evapotranspiration (*PET*), precipitation and soil water holding capacity. The water balance methodology illustrated by Thornthwaite and Mather and the different versions of it have been applied in many fields (e.g. Arnell, 1992; Kerkides et al., 1996; Feddema, 1998; Komuscu et al., 1998; Xiong and Guo, 1999; Leathers et al., 2000; Pimenta, 2000). Hydrologically justifiable modifications have been made to the WBM of Thonthwaite-Mather by many investigators (e.g. Alley, 1984; Kolka and Wolf, 1998; Bugmann and Cramer, 1998) to correct the errors stemming from the comparison of observed and estimated components. The refinement is mainly related to the soil moisture function. Generally, the mathematical structure of the T-M WBM (Alley, 1984; Bugmann and Cramer, 1998) can be outlined as follows:

Actual evapotranspiration, AET, for month i is calculated as

$$AET_i = PET_i \qquad \text{if } P_i \geq PET_i \qquad \text{(surplus condition)} \qquad (4)$$

$$AET_i = P_i + S_{i-1} - S_i \quad \text{if } P_i < PET_i \qquad \text{(deficit condition)} \qquad (5)$$

in which PET_i is the potential evapotranspiration for month i, P_i is the rainfall for month i, S_i is the soil moisture at the end of month i, and S_{i-1} is the soil moisture at the end of the preceding month.

$$S_i = \min\left[(P_i - PET_i + S_{i-1}), S_{max}\right] \qquad \text{if } P_i \geq PET_i \qquad (6)$$

$$S_i = S_{i-1}\exp\left[-\frac{(PET_i - P_i)}{S_{max}}\right] \qquad \text{if } P_i < PET_i \qquad (7)$$

where S_{max} is the maximum holding capacity of the soil in the root zone.

The change in soil moisture storage, ΔS, at the end of month i is

$$\Delta S_i = S_i - S_{i-1} \qquad (8)$$

The runoff for month i, Q_i, is

$$Q_i = P_i - AET_i - \Delta S_i \qquad (9)$$

In cases when $Q_i < 0$, runoff is set to 0. If runoff is negligible compared to deep percolation; then, Q_i may be replaced by D_i.

The moisture deficit, hereinafter called evapotranspiration deficit (ED), for month i is defined as

$$ED_i = PET_i - AET_i \qquad (10)$$

Evapotranspiration deficit could be considered as a practical index of agricultural drought. Thomas (2000) suggests that this term be taken to resemble closely the supplemental water needed, i.e. irrigation demand, for a given cropping system.

Most of the critiques of the T-M model concern the estimation of soil moisture availability and evapotranspiration. Traditionally, PET has been calculated using the temperature-based method of Thornthwaite (1948) because of its simplicity and use of easily accessible data. Lockwood (1999) extremely cautions the use of evapotranspiration estimates that are based solely on temperature, such as those obtained using the Thornthwaite method, in climate change studies. As a result of atmospheric warming, these methods give unrealistically large increases in *PET* values. It is believed that the FAO Penman-Monteith (Allen et al., 1994; Jensen et al., 1997) is more appropriate and would provide more reliable estimates of *PET* than the temperature-based method of Thornthwaite. This method has recently been recommended by the Food and Agriculture Organization (FAO) as the international standard (Allen et al., 1998).

Feddema (1998) states that the shortcoming of the T-M model is the inherent assumption that no moisture surplus condition is allowed unless the soil reservoir is filled, thus underestimating the runoff and overestimating evapotranspiration. In locations where rainfall intensity is high relative to soil permeability, however, a model that distinguishes between direct overland flow and infiltration would be more complex and impractical. This is owing to its requirement for significant increase in input data that are rarely available, characterization of soil infiltration properties and more detailed temporal and spatial characteristics. Wolock and McCabe (1999) also point out the difficulty and impracticability highlighted by Feddema regarding the inclusion of non-testable complexity in hydrological models.

The total amount of available water that a plant can extract from its root zone is calculated from the following formula (Allen et al., 1998):

$$S_{max} = 1000(\theta_{FC} - \theta_{PWP}) . D_r \qquad (11)$$

where S_{max} is the total available soil water in the root zone (mm); θ_{FC} is the water content at field capacity (m^3/m^3); θ_{PWP} is the water content at permanent wilting point (m^3/m^3); D_r is the rooting depth (m). The fraction of total available water that plant can extract from the root zone without experiencing water stress is

$$RAW = p . S_{max} \qquad (12)$$

where *RAW* is the readily available soil water in the root zone (mm); p is the fraction of S_{max} that can be depleted from the root zone before moisture stress occurs.

The use of a fixed S_{max} for the whole hydrological period is unrealistic for non-predominant plants (crops) because crops develop their roots until the maximum depth is reached. In this research, it is assumed that the root develops at a constant rate, following Agnew (1982). Also, *RAW* values have been used in stead of S_{max} to give consideration of water-stress conditions, which are expected to affect evapotranspiration under the arid conditions of Sudan. Since the crop growing season constitutes a part of the year, the rest of the year is considered as grass growing season and the corresponding *PET* values have been used to calculate the grass actual evapotranspiration.

Typical values for θ_{FC}, θ_{PWP} and D_r have been extracted from the FAO-56 (Allen et al., 1998) and the values adopted in this study for grass, sorghum, millet and groundnut growing in the clay and sand soils of the country are given in Tables 2 and 3. Although p is a function of evapotranspiration, a constant value is often used for the growing season rather than varying the value every month. A value of

Table 2. Soil water characteristics.

Soil Type	θ_{FC}	θ_{PWP}	$\theta_{FC} - \theta_{PWP}$*
Sand	0.07 – 0.17	0.02 – 0.07	0.05 – 0.11 (0.08)
Clay	0.32 – 0.40	0.20 – 0.24	0.12 – 0.20 (0.16)

* Figures between parentheses are those used in this study.

Table 3. Range of maximum effective rooting depth and water availability.

Crop	D_r* (m)	S_{max} (mm) Sand	S_{max} (mm) Clay	RAW (mm) Sand	RAW (mm) Clay
Grass	0.5 – 1.5 (1.0)	60	120	30	60
Sorghum	1.0 – 2.0 (2.0)	160	320	80	160
Millet	1.0 – 2.0 (2.0)	160	320	80	160
Groundnut	0.5 – 1.0 (1.0)	80	160	40	80

* Larger values are used for modelling soil water stress or for rain-fed conditions. Figures between parentheses are those used in this study.

0.50 for p is commonly used for many crops. The average depth of the predominant plants, such as grass, is usually of the order of 1 m (Milly, 1994).

To balance the net change in soil moisture from the beginning to the end of the period (12 months), a balancing routine is performed by setting the initial moisture content (S_1) to RAW, and the calculations are carried out until month 13; the P, PET and RAW for month 13 are those for month 1. Then, the calculation is repeated with the value of S_1 being replaced by S_{13} until the system converges, i.e. S_i values of two successive runs do not change significantly. Convergence normally occurred in the third run.

A few experiments have been performed herein to study the sensitivity of the monthly grass AET values from the T-M model to the RAW input. Two stations were selected from Table 1 to represent wet and dry climatic conditions, respectively Ed Damazin (semi-arid) and Kosti (arid), and the corresponding normal data have been used. The results demonstrate the following remarks:

- Under wet conditions, only the first and last few monthly values are those sensitive to the soil water capacity value. However, the effect of RAW value on the total seasonal or annual AET amount can, in general, be deemed minor.

- In the dry condition, no effect is expected since in this case the soil moisture storage is virtually nil and that the AET equals P.

- For crop water balance, thus, it is recommended that the hydrological year for running the WBM should commence in the month following the end of the crop growing season. This would minimize the effect of *RAW* value on the *AET* for the growing season of concern.

Crop evapotranspiration

The methodology published as FAO Irrigation and Drainage Paper No. 56 (Allen et al., 1998) has now become the international standard for determining the crop evapotranspiration (ET_c). It is calculated by multiplying the grass reference evapotranspiration (ET_o) by the crop coefficient (k_c), using the equation:

$$ET_c = k_c\, ET_o \qquad\qquad (13)$$

In this study, the single crop coefficient approach is adopted. The crop coefficient in Eq. 13 Integrates the characteristic effects that distinguish between the concerned crop and the reference crop (Allen et al., 1996). These characteristics are crop height, crop-soil surface resistance and crop-soil surface albedo. Crop information, such as the date of sowing and length of different growth stages within the season, is required in order to determine the values of k_c. The growth-stage period is divided into initial, development, mid-season and late-season stages. In the FAO-56 paper, the k_c curve is divided into the above four growth stages such that the initial and mid-season periods are characterized by horizontal lines, while the development and late-season periods are characterized by rising and falling lines, respectively.

General information on crop calendar in months for Sudan has been taken from different sources (Wilson, 1991; Sørbø, 1991; Haaland, 1991; Simpson and Simpson, 1991; Dickie, 1991; FAO, 1997). Because full details of the exact length of each crop stage in days were not available for this study, it has been decided to divide the local crop calendar (total days for the growing-season months) into stages by considering the stage length given in FAO-56 to be proportional to the local length of the stage. For example, if FAO-56 gives information on initial and total growth stages as 30 and 135 days, respectively, while the local information on total growth stage is 153 days, it is assumed that the local initial stage is $30 \times 153/135 = 34$ days. Table 4 shows the outcome of this approach.

For the development stage, k_c is taken as the average of coefficients for the initial (k_{cini}) and mid-season (k_{cmid}) stages, while that for the late (or end) season (k_{cend}) is taken as the average of k_{cmid} and k_{cend}. Since the water balance is calculated on monthly basis, it is assumed that one k_c value would be representative of the whole month. The k_c for a month shared between two different stages is also approximated to be the average of the k_c values for the previous month and the next stage. The month falling in the sloping range of k_c curve is given a k_c value representing the average of k_c values for the previous month and (1) the next stage in the case of development period or (2) the current stage in the case of end period. Whenever necessary, the graphical determination of k_{cini} was based on interval of wetting or irrigation of 14 days, as this is a standard watering interval in Sudan

115

Table 4. Length of crop growth stage.

Crop	Station's Reference Number	Initial	Develop-ment	Mid	Late	Total	Crop Calendar
				(days)			
Sorghum	1-5; 15	28	48	62	42	180	Apr-Sep
	6-14; 16-23	33	58	72	49	240	Jun-Dec
	24-31	33	58	74	50	215	Jul-Jan
Groundnut	1-5	34	44	44	28	150	May-Sep
	6-31	34	45	45	29	153	Jun-Oct
Millet	1-5; 15	26	38	71	45	180	Apr-Sep
	6-14; 16-31	31	45	84	53	214	Jun-Dec

(Ahmed et al., 1986; Hussein and El Daw, 1989; El-Tom and Osman, 1995). In this case, a range of k_{cini} values (Table 5) was obtained based on the range of ET_o and the type of soil (Table 1) for the region in question. An average value of k_{cini} is, then, assumed for the whole region. Table 6 gives the monthly k_c values used in this study for different crops and regions.

Climate change scenarios

Estimates of climate change scenarios are generally developed from general circulation models (Gleick, 1986), abbreviated as GCM. The assessments of the Intergovernmental Panel on Climate Change (IPCC) indicate that the global mean warming is likely to be within 1.5–4.5 °C by the year 2050 (Hare, 1993). However,

Table 5. Single crop coefficients (k_c).

Crop	Station's Reference Number	Initial	Mid	Late
Sorghum	1-5; 15	0.325-0.425 (0.38)	1.00-1.10 (1.05)	0.55
	7-13	0.175-0.19 (0.18)		
	6; 14; 16-23	0.30-0.40 (0.35)		
	24-31	0.15-0.18 (0.17)		
Groundnut	1-5	0.38-0.40 (0.39)	1.15	0.60
	7-13	0.175-0.19 (0.18)		
	6; 14-23	0.30-0.40 (0.35)		
	24-31	0.15-0.175 (0.16)		
Millet	1-5; 15	0.325-0.425 (0.38)	1.00	0.30
	7-13	0.175-0.19 (0.18)		
	6; 14; 16-23	0.30-0.40 (0.35)		
	24-31	0.20-0.30 (0.25)		

Note: Figure between parentheses is the average value used in the study.

Table 6. Monthly average crop coefficients (k_c).

Crop	Station's Reference Number	Apr	May	Jun	Jul	Aug	Sep	Oct	Nov	Dec	Jan
Sorghum	1-5; 15	0.38	0.72	0.89	1.05	0.80	0.68				
	7-13			0.18	0.62	0.84	1.05	1.05	0.93	0.74	
	6; 14; 16-23			0.35	0.70	0.89	1.05	1.05	0.93	0.74	
	24-31				0.17	0.61	0.83	1.05	1.05	0.93	0.74
Groundnut	1-5		0.39	0.77	0.96	1.15	0.88				
	7-13			0.18	0.67	0.91	1.15	0.88			
	6; 14-23			0.35	0.75	0.95	1.15	0.88			
	24-31			0.16	0.66	0.91	1.15	0.88			
Millet	1-5; 15	0.38	0.69	1.00	1.00	0.65	0.48				
	7-13			0.18	0.59	0.80	1.00	1.00	0.73	0.53	
	6; 14; 16-23			0.35	0.68	0.84	1.00	1.00	0.73	0.53	
	24-31				0.17	0.59	0.80	1.00	1.00	0.73	0.53

significant weaknesses of GCMs have been perceived by many users (Rind et al., 1990; Leavesley, 1994; Hulme, 1994; Wardlaw et al., 1996; Chattopadhyay and Hulme, 1997, 1997; Hulme and Viner, 1998). The difficulties associated with using GCMs include the coarse spatial and temporal resolution, the improper assessment of some variables, such as precipitation, evapotranspiration, runoff, soil moisture, etc., the time and expense required to operate such models and the variability of results produced by different models. These gaps existing between GCMs and hydrological models have been discussed elaborately by Xu (1999).

Given the above constraints of GCMs, some authors have examined the effects of hypothetical temperature rises within the range of +1 to +4.5 °C on evapotranspiration (e.g. Rosenberg et al., 1989; Abderrahman et al., 1991; McKenney and Rosenberg, 1993; Fennessey and Kirshen, 1994; Le Houérou, 1996). Others presented scenarios of potential evapotranspiration based on GCM scenarios of temperature rises (e.g. Palutikof et al., 1994; Zhang et al., 1996; Chattopadhyay and Hulme, 1997). Using drought indices as measures, the changes in regional moisture availability resulting from accepted ranges of temperature and precipitation scenarios have also been investigated (Rind et al., 1990; Palutikof et al., 1994; Le Houérou, 1996). Others assessed the probable consequences of changing climate on river basins by employing changing scenarios of one or more climate parameter(s), such as temperature, precipitation and potential evapotranspiration, in hydrological models. Examples of such studies are those generated by Měmec and Schaake (1982), Gleick, (1987), Mimikou et al. (1991), Holt and Jones (1996a, b), Roberts (1998), Najjar (1999) and Fowler (1999).

Temperature and rainfall scenarios derived from GCM experiments for the tropics (30° N to 30° S), with respect to the 1990 climate, have been presented by Hulme and Viner (1998). They showed that a 1.76 °C global warming might occur between late 2030s and 22nd century. Considerable regional differences exist in rainfall changes; drying and wetting scenarios have been estimated depending on the given region. This inconsistency of rainfall results has also been addressed by Hulme (1994) and Zinyowera et al. (1998). Rainfall is projected to decline by the year 2050 by about 10% in southern Africa and parts of the Horn of Africa, to range from a decrease of > 10% to an increase of ~10% and to increase by only 5% in Equatorial Africa. According to Hulme (1994), Ziyyowera et al. (1998) and Sircoulon et al. (1999), GCMs broadly agree that a moderate rise of temperature by 1–2 °C may take place in Africa by 2050, thus representing a warming rate of about 0.2 °C per decade. Roughly, the projected temperature increases are expected to increase potential evapotranspiration over Africa by 5–10% by the year 2050.

The sensitivity of rainfall over Sudan to global warming is by no means clear (Hulme, 1990). However, climatic records for the country during the recent decades suggest annual rainfall depletion of 0 to > 20% and an annual increase in temperature of 0.22° to 0.40 °C per decade during the 1961–1990. Although these rates are not interpreted as predictions, they reasonably lie within the predicted scenarios for the next few decades.

Based on the above review, and taking the normal period of 1961–1990 as a reference, this paper estimates the magnitude of the likelihood of future climate change on *AET* by adopting the general procedure described by Xu (1999) as follows:

1. Perturbation of +1.5 (ΔT) to the monthly historical mean temperature values is selected.
2. This warming scenario is imposed directly on the evapotranspiration model of FAO Penman-Monteith to generate potential evapotranspiration scenarios.
3. Rainfall scenarios (ΔP) of −20, 0 and +20% are applied to monthly normal values.
4. The response of the parameters to plausible scenarios of climate change is expressed as percent change from the average.

The above approach matches the idea of considering the full range of plausible scenarios for climate change response planning in Africa (Hulme, 1994). In this study, the response of *ED* to the climate change scenarios is expressed as change in mm from the average.

Results and Discussion

Sorghum

A 2 × 2° gridded map of the spatial distribution of total seasonal *AET* appears in Fig. 2. The *AET* isolines show a range of virtually zero mm in the driest part in the north to about 650 mm in the wettest part of the country (southwest). On regional basis, the *AET* range is 500–650 mm, 175–500 mm, 0–150 mm and 0–100 mm for the southern, central, northern and coastal regions, respectively. Under the scenario of 1.5 °C increase in temperature and no-change in rainfall amount (Fig. 3a), the change in total *AET* is surprisingly positive, though it does not exceed 4.0%, as

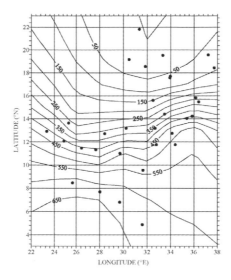

Figure 2. Total seasonal *AET* for sorghum (baseline).

in the extreme south. Such a scenario shows no effect at all on the *AET* rates over the northern half of the country. This can be explained by the fact that *AET* in these arid and hyper-arid areas is supply limited and does not exceed rainfall. In other words, without inducing a temperature rise, the baseline *PET* is already greater than rainfall.

The scenario of +1.5 °C and –20% change in *P* (Fig. 3b) would produce a corresponding change in *AET* ranging from 0% in the extreme southwest to as much as –20% in mid-central Sudan and northward. Fig. 3c shows the effect of the warm-wet scenario. The 20% increase in rainfall could offset the effect of 1.5 °C increase, and as a result, cause an increase in the *AET* of 3–21% from the south to the north of Sudan.

Under the present climatic conditions, southern Sudan experiences evapotranspiration deficit (*ED*) of 0–250 mm (Fig. 4). This range may increase by about 0–14 mm under warming-alone (Fig. 5a) and by about 30–80 mm under warming-drying (Fig. 5b). The warming-wetting scenario would reduce the deficit in this region by up to 30 mm (Fig. 5c). The moisture levels in the coastal region could become less compared to the present condition limit, which shows deficit of 650–750 mm, under the respective scenarios by 28–30 mm, 40–50 mm and 10–15 mm. Central Sudan exhibits the widest range of *ED* in the country under present conditions (150–825 mm). A 10–34 mm increase in deficit would be expected in case of a warming scenario. If this warming is compounded by a 20% reduction in rainfall amounts, agricultural drought would intensify by 45 mm in the west to over 100 mm in the areas of Abu Na'ama and Kosti (mid-central region).

On the other hand, if this warming is coupled with Δ*P* = +20%, moisture conditions would improve by 0 to 35 mm. At present, northern Sudan exhibits a deficit in the

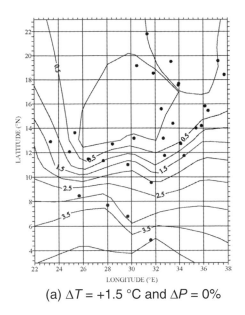

(a) $\Delta T = +1.5\ °C$ and $\Delta P = 0\%$

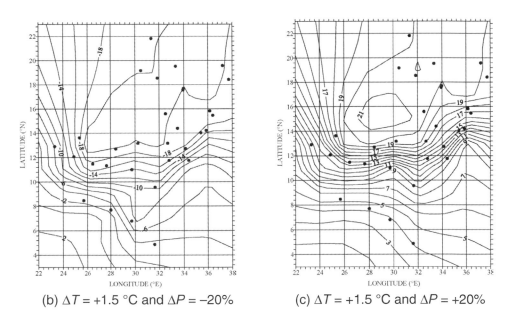

(b) $\Delta T = +1.5\ °C$ and $\Delta P = -20\%$ (c) $\Delta T = +1.5\ °C$ and $\Delta P = +20\%$

Figure 3. Change (%) in seasonal sorghum *AET* as a result of climate change scenarios.

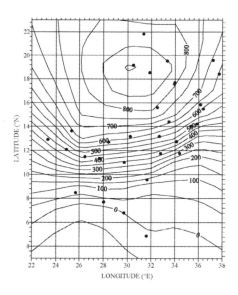

Figure 4. Total seasonal sorghum *ED* (baseline).

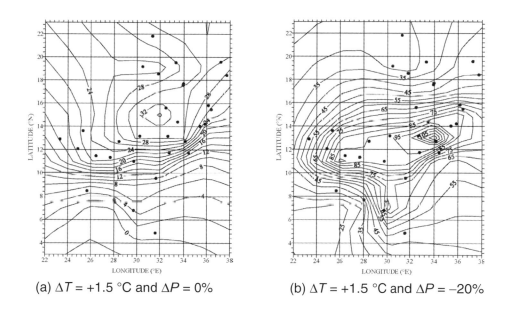

(a) Δ*T* = +1.5 °C and Δ*P* = 0%   (b) Δ*T* = +1.5 °C and Δ*P* = −20%

Figure 5. Change (mm) in seasonal sorghum *ED* under climate change scenarios.

range of 750–900 mm. A warming of 1.5°C may aggravate the deficit in this region by up to 30, 60 and over 25 mm for a ΔP of 0, −20 and +20%, respectively.

Groundnut

Fig. 6 shows the contours of the total baseline *AET* for groundnut. The hyper-arid zone of the country indicates *AET* of 0 to about 150 mm. The arid and semi-arid zones of central Sudan have values of less than 150 mm to above 500 mm. Over the semi-arid to sub-humid south, *AET* ranges from 500 to less than 600 mm. As such, it can be noticed that the spatial variability is highest in the central region and lowest in southern Sudan.

As shown in Fig. 7a, a warming of 1.5 °C would cause the *AET* to rise by between 2.0 and 3.6% in the region located south of 10° N. Within the 10°–16° N zone, the increase is from 0 to 2.4%. Northward, there is no effect of such an increase in temperature. Under the dry scenario (Fig. 7b), the gradient of the isolines becomes steeper as one moves to the southwest. A change of between −12 and 2.0% is likely to characterize the southern part of the country if a scenario of +1.5 °C combined with −20% change in *P* would occur (Fig. 7b). Further north, the change in *AET* would reach to a maximum of −20%. The wet scenario (Fig. 7c) indicates an increase in total *AET* of 3.0–8.0%, 6–20% and 20% for the south, centre and north/coast of Sudan, respectively.

The *ED* baseline displayed in Fig. 8 shows a range of 0–750 mm for the whole country. A southward shift of the isolines would occur under ΔT = +1.5 °C, thus exerting more deficit of 2–6 mm in the south, 6–22 mm in the centre and 22–24 mm in both the north and coastline (Fig. 9a). The worst scenario (Fig. 9b) would result in more deficit of up to > 100 mm, ~85 mm and 50 mm, as in the case of mid-eastern, southern and northern Sudan, respectively. A ΔP of +20% together with ΔT = +1.5 °C (Fig. 9c) could lead to less deficit in some areas but more deficit in others. An increase of ~15 mm in moisture levels may occur in southern Sudan, increasing to 45 mm in mid-central Sudan. The deficit, on the other hand, would intensify further by up to 20–25 mm in the northern and coastal regions.

Millet

Fig. 10 shows that the *AET* varies from 0 to 200 mm, from 150 to 500 mm and from 500 to 600 mm for the northern and coastal, central and southern regions, respectively. These ranges correspond to *ED* ranges of 650–850 mm, 200–750 mm and 0–200 mm, as illustrated in Fig. 11. The change in *AET* as a result of synchronized ΔT = +1.5 °C and ΔP = 0% increases from 0% in the extreme north to 10% in the extreme southwest (Fig. 12a), with a noticeable cluster of the contours in the area bound by 5°–13° N and 26°–30° E. This scenario could yield additional deficit of 0–10 mm in the south, 0–20 mm in mid/eastern Sudan and 10–30 mm north of 16° N, but there could be some improvement in certain western areas (Fig. 13a). In the case of ΔT = +1.5 °C accompanied by ΔP of −20%, a decrease in *AET* would occur over the whole country, ranging from 2% in the most humid areas to 20% in the driest locations (Fig. 12b).

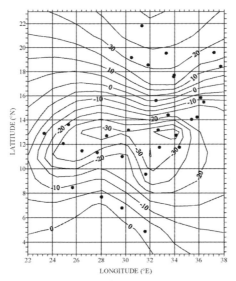

(c) $\Delta T = +1.5\ °C$ and $\Delta P = +20\%$

Figure 5. (*Continued*).

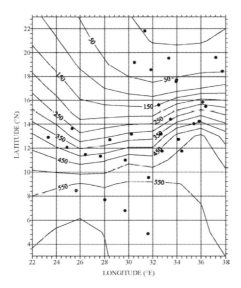

Figure 6. Total seasonal *AET* for groundnut (baseline).

123

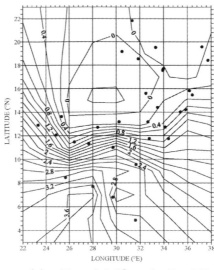

(a) $\Delta T = +1.5\ °C$ and $\Delta P = 0\%$

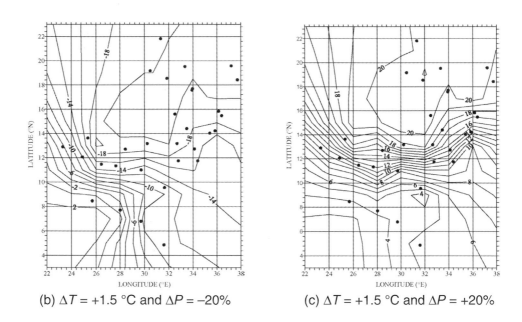

(b) $\Delta T = +1.5\ °C$ and $\Delta P = -20\%$ (c) $\Delta T = +1.5\ °C$ and $\Delta P = +20\%$

Figure 7. Change (%) in seasonal groundnut *AET* as a result of climate change scenarios.

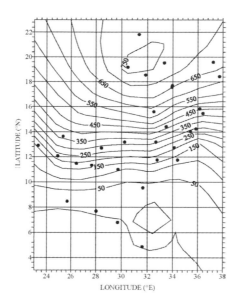

Figure 8. Total seasonal groundnut *ED* (baseline).

Such changes would result in a southward shift of the *ED* isolines, thus causing additional deficit, ranging from 20 to ~120 mm over the country (Fig. 13b). The change under this warm-dry scenario is quite complex. With both Δ*T* = +1.5 °C and Δ*P* = +20%, there would be more moisture for evapotranspiration, hence increasing the total *AET* by 4–11% in the south, by 5–20% in the centre and by 20% in the area north of 16° N (Fig. 12c). The effect of such changes would enhance the moisture conditions in southern, eastern, mid-central and most of western Sudan (Fig. 13c). But, the extreme western part of the country as well as the area north of 16° N would exhibit more moisture stress of up to 30 mm.

Overall

The data obtained for the total seasonal *AET* and *ED* have been presented as a function of latitude, as shown in Fig. 14 for *AET* and Fig. 15 for *ED*. Both parameters vary nonlinearly with latitude. Any change in the climate is shown to generate changes in the variability, as indicated by the values of the coefficient of variability (C_v) in Tables 7 and 8. In all the *AET* cases, the wet scenario results in least spatial variability while the dry scenario induce the highest spatial variability. The situation is reversed with respect to *ED*. For each crop, there appears no much difference in variability in both cases between the warming-alone and the present states, indicating that a 1.5 °C warming alone is not as decisive as when it is accompanied by a change in rainfall amount. Country-wide, the variability in *AET* is largest for groundnut and least for millet, while that in *ED* is least for sorghum but still largest for groundnut.

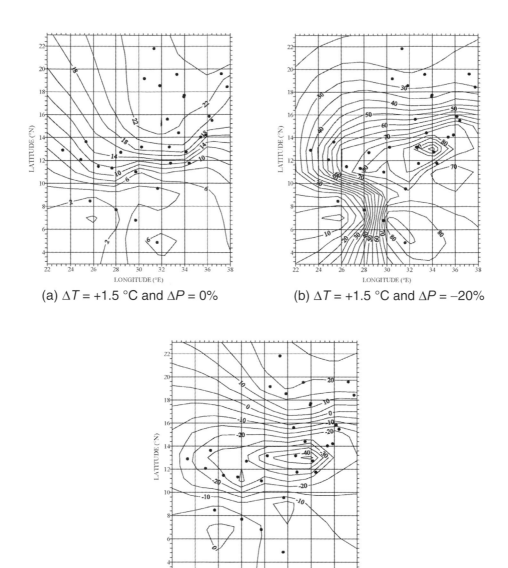

(a) $\Delta T = +1.5\ °C$ and $\Delta P = 0\%$

(b) $\Delta T = +1.5\ °C$ and $\Delta P = -20\%$

(c) $\Delta T = 1.5\ °C$ and $\Delta P = +20\%$

Figure 9. Change (mm) in seasonal groundnut *ED* under climate change scenarios.

126

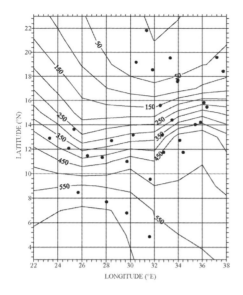

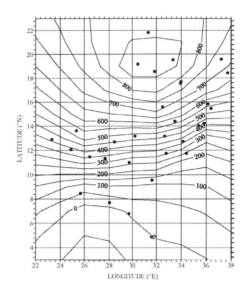

Figure 10. Total seasonal *AET* for millet (baseline).

Figure 11. Total seasonal millet *ED* (baseline).

Table 7. Climate change effect on spatial variability of total seasonal *AET*.

Crop	Coefficient of Variability (%)			
	Baseline	$\Delta T = +1.5\ °C$ $\Delta P = 0\%$	$\Delta T = +1.5\ °C$ $\Delta P = -20\%$	$\Delta T = +1.5\ °C$ $\Delta P = +20\%$
Sorghum	67.2	68.0	72.5	63.9
Groundnut	67.6	68.3	72.0	64.9
Millet	65.8	67.0	70.3	63.1

Table 8. Climate change effect on spatial variability of total seasonal *ED*.

Crop	Coefficient of Variability (%)			
	Baseline	$\Delta T = +1.5\ °C$ $\Delta P = 0\%$	$\Delta T = +1.5\ °C$ $\Delta P = -20\%$	$\Delta T = +1.5\ °C$ $\Delta P = +20\%$
Sorghum	55.2	54.5	48.4	55.2
Groundnut	76.6	75.3	62.6	83.3
Millet	59.7	59.3	51.8	63.6

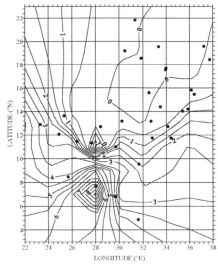

(a) ΔT = +1.5 °C and ΔP = 0%

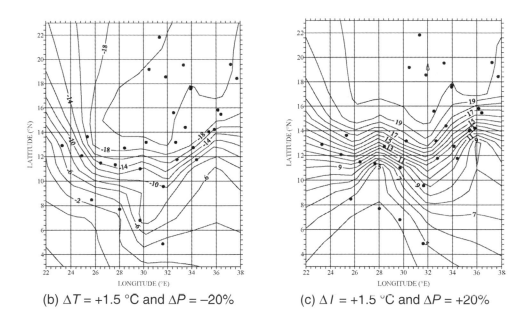

(b) ΔT = +1.5 °C and ΔP = −20%   (c) ΔI = +1.5 °C and ΔP = +20%

Figure 12. Change (%) in seasonal millet *AET* as a result of climate change scenarios.

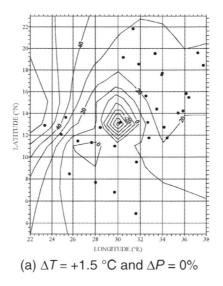

(a) $\Delta T = +1.5 \ ^{\circ}C$ and $\Delta P = 0\%$

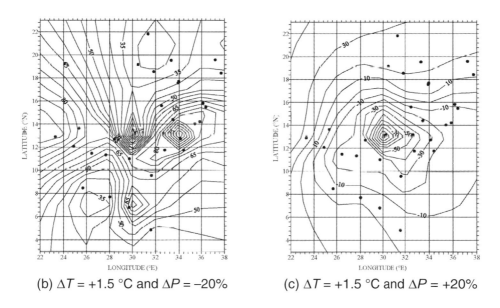

(b) $\Delta T = +1.5 \ ^{\circ}C$ and $\Delta P = -20\%$ (c) $\Delta T = +1.5 \ ^{\circ}C$ and $\Delta P = +20\%$

Figure 13. Change (mm) in seasonal millet *ED* under climate change scenarios.

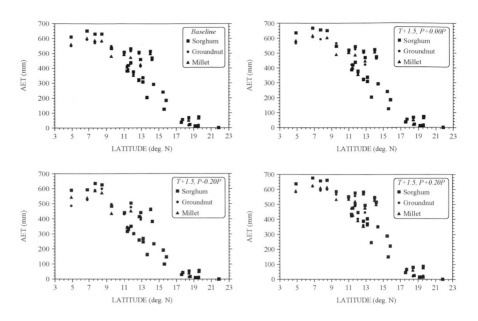

Figure 14. Total seasonal *AET* as a function of latitude.

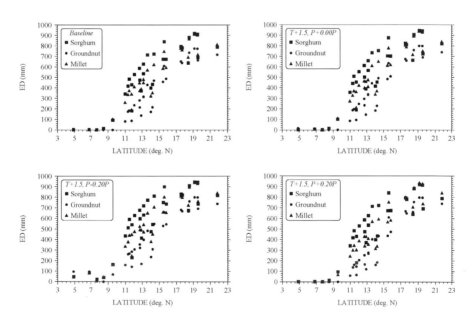

Figure 15. Total seasonal *ED* as a function of latitude.

Fig. 14 indicates that, in the zone located north of ~13.5° N, the three crops evapotranspire the same total amount of moisture even under a changing climate. On the other hand, the levels of *ED* for both the present and climate change scenarios (Fig. 15) indicate no observable difference in moisture stress between crops in the zone situated south of ~11.25° N. This indicates that, although the hydrological features in the wet part of the country varies more compared to the dry part from one crop to another, the variability in the level of agricultural drought is more pronounced in the latter region than in the former.

It should be spelled out, as many investigators did (e.g. Arnell, 1992; Chiew et al., 1995; Xu, 2000), that the magnitudes of changes presented in this study are not given as predictions of future changes. They are rather an attempt to explore the sensitivity of such hydrological regimes to possible climate change scenarios.

Conclusions

The results demonstrate spatial diversity of the moisture conditions over the country. Both elements vary non-linearly with latitude. These results suggest that rainfall has a major effect on the timing and magnitude of moisture levels in agricultural areas of Sudan. In other words, the water supply naturally available for plant growth and the demand for irrigation vary from site to site depending on the variations of rainfall. The total seasonal actual evapotranspiration for the crops considered in this study is estimated to lie in the range of 0–650 mm. On the other hand, evapotranspiration deficit estimates across the country range from 0 to 900 mm, 0 to 750 mm and 0 to 850 mm for sorghum, groundnut and millet, respectively. The severity and diversity of crop moisture stress in the country are most pronounced in the hyper-arid, arid and semi-arid regions, where there is already a strain in existing water resources and high population growth. Here, rainfall fails to recharge the soil moisture storage to subsequently meet the water requirements of plants (ET_c), thus causing agricultural drought.

Under an altered climate, the response of *AET* and *ED* are likely to depend much on the rainfall changes than to changes in temperature. A +1.5 °C alone may increase the total *AET* for sorghum, groundnut and millet by not more than 4.0, 3.6 and 9.0%, respectively. If this warming is associated with rainfall changes of ±20%, the *AET* was simulated to decline by up to 20% in the dry scenario and to increase by up to 20% in the wet scenario. The temperature rise combined with no-rainfall change or associated with rainfall changes would result in substantial changes in the moisture regimes over the country. The warm-dry scenario was simulated to compound the moisture deficit over the whole country by 30–115 mm, 10–100 mm and 30–110 mm in case of sorghum, groundnut and millet, respectively. In the warm-wet scenario, the country may experience three states of agricultural drought: no change, mitigation and exacerbation. Hence, the modification of the *ED* levels for the respective crops would take the ranges −35 to 30 mm, −45 to 25 mm and −100 to 30 mm. The hyper-arid region would not benefit from the 20% increase in the rainfall amounts, as it would still encounter more agricultural drought. Country-wise, any change in climate was simulated to result in changes in the variability of both *AET* and *ED* levels. In the case of *AET*, the spatial variability would increase, with both the warming-alone and the warming-drying scenarios generating higher variability whereas the wet scenario resulting in lower variability compared to present. An altered climate in the case of *ED* would augment the variability in the wet scenario but reduce it under the other two scenarios.

References

Abdelrrahman, W.A., Bader, T.A., Khan, A.U. and Ajward, M.H. (1991). "Weather modification impact on reference evapotranspiration, soil salinity and desertification in arid regions: a case study". *Journal of Arid Environments*, 20, 277-286.

Agnew, C.T. (1982). "Water availability and the development of rainfed agriculture in south-west Niger, West Africa". *Transactions of the Institute of British Geographers*, 7, 419-457.

Ahmed, S.E., Hussein, A.S.A. and Ahmed, A.A. (1986). "Water delivery programming with reference to Gezira Scheme". *Proceedings of the International Conference on Water Resources Needs and Planning in Drought Prone Areas*. Volume 2, Khartoum, Sudan, 843-859.

Allen, R.G., Smith, M., Pereira, L.S. and Perrier, A. (1994). "An update for the calculation of reference evapotranspiration". *ICID Bulletin*, 43(2), 35-92.

Allen, R.G., Smith, M., Pruitt, W.O. and Pereira, L.S. (1996). "Modification to the FAO crop coefficient approach". *Proceedings of the ASAE International Conference on Evapotranspiration and Irrigation Scheduling*, San Antonio, TX.

Allen, R.G., Pereira, L.S., Raes, D. and Smith, M. (1998). "Crop evapotranspiration – guidelines for computing crop water requirements". *FAO Irrigation and Drainage Paper 56*, Food and Agriculture Organization of the United Nations, Rome, Italy.

Alley, W.M. (1984). "On the treatment of evapotranspiration, soil moisture accounting, and aquifer recharge in monthly water balance models". *Water Resources Research*, 20(8), 1137-1149.

Arnell, N.W. (1992). "Factors controlling the effects of climate change on river flow regimes in a humid environment". *Journal of Hydrology*, 132, 321-342.

Ayoub, A.T. (1999). "Land degradation, rainfall variability and food production in the Sahelian zone of the Sudan". *Land Degradation and Development*, 10, 489-500.

Bugmann, H. and Cramer, W. (1998). "Improving the behaviour of forest gap models along drought gradients". *Forest Ecology and Management*, 103, 247-263.

Chattopadhyay, N. and Hulme, M. (1997). "Evaporation and potential evapotranspiration in India under conditions of recent and future climate change". *Agricultural and Forest Meteorology*, 87, 55-73.

Chiew, F.H.S., Whetton, P.H., McMahon, T.A. and Pittock, A.B. (1995). "Simulation of the impacts of climate change on runoff and soil moisture in Australian catchments". *Journal of Hydrology*, 167, 121-147.

Dickie, A. (1991). "Systems of agricultural production in southern Sudan". *The agriculture of the Sudan*, G.M. Craig, ed., Oxford University Press, Oxford, 280-307.

Elagib, N.A., Alvi, S.H. and Mansell, M.G. (1999a). "Day-length and extraterrestrial radiation for Sudan: a comparative study". *International Journal of Solar Energy*, 20(2), 93-109.

Elagib, N.A., Alvi, S.H. and Mansell, M.G. (1999b). "Correlationships between clearness index and relative sunshine duration for Sudan". *Renewable Energy*, 17(4), 473-498.

Elagib, N.A. and Mansell, M.G. (2000a). "Recent trends and anomalies in mean seasonal and annual temperatures over Sudan". *Journal of Arid Environments*, 45(3), 263-288.

Elagib, N.A. and Mansell, M.G. (2000b). "Climate impacts of environmental degradation in Sudan". *GeoJournal*, 50(4), 311-327.

Elagib, N.A. and Mansell, M.G. (2000c). "New approaches for estimating global solar radiation across Sudan". *Energy Conversion and Management*, 41(5), 419-434.

El-Tom, A.O. and Osman, N.E. (1995). "The informal channels of indigenous tenants in coping with the inefficient bureaucracy of the Gezira Scheme, Sudan". *GeoJournal*, 36(1), 79-85.

FAO (1997). Food and Agriculture Organization of the United Nations, http://www.foa.org/WAICENT/faoinfo/economic/giews/english/basedocs/sud/Sudtoc1e.htm

Feddema, J.J. (1998). "Estimated impacts of soil degradation on the African water balance and climate". *Climate Research*, 10, 127-141.

Fennessey, N.M. and Kirshen, P.H. (1992). "Evaporation and evapotranspiration under climate change in New England". *Journal of Water Resources Planning and Management,* 120(1), 48-69.

Fowler, A. (1999). "Potential climate change impacts on water resources in the Auckland region (New Zealand)". *Climate Research*, 11, 221-245.

Gleick, P.H. (1986). "Methods for evaluating the regional hydrologic impacts of global climatic changes". *Journal of Hydrology*, 88, 97-116.

Gleick, P.H. (1987). "The development and testing of a water balance model for climate impact assessment: modeling the Sacramento basin". *Water Resources Research*, 23(6), 1049-1061.

Gleick, P.H. (1989). "Climate change, hydrology, and water resources". *Reviews of Geophysics*, 27, 329-344.

Haaland, G. (1991). "Systems of agricultural production in western Sudan". In: *The agriculture of the Sudan*, G.M. Craig, ed., Oxford University Press, Oxford, 230-251.

Hare, F.K. (1993). "Climate variations, drought and desrtification". *WMO - No. 653*. World Meteorological Organization.

Holt, C.P. and Jones, J.A.A. (1996a). "Sensitivity analyses on the impact of global warming on water resources in Wales". In: *Regional hydrological response to climate change and global warming.* J.A.A. Jones, C.M., M.K. Woo and H.T. Kung, eds., Kluwer Academic Publishers, Netherlands, 317-335.

Holt, C.P. and Jones, J.A.A. (1996b). "Equilibrium and transient global warming scenario implications for water resources in Wales". *Water Resources Bulletin*, 32(4), 711-721.

Hoogmoed, W.B., Klaji, M.C. and Brouwer, J. (1991). "Infiltration, runoff and drainage in the Sudano-Sahelian zone". In: *Soil water balance in the Sudano-Sahelian zone.* M.V.K. Sivakumar, J.S. Wallace, C. Renard and C. Ciroux, eds., Proceedings of the International Workshop, Niamey, Niger, 18-23 February 1991, IAHS Publication No. 199, IASHS Press, Institute of Hydrology, Wallingford, UK, 85-98.

Hulme, M. (1986). "The adaptability of a rural water supply system to extreme rainfall anomalies in central Sudan". *Applied Geography*, 6, 89-105.

Hulme, M. (1990). "The changing rainfall resources of Sudan". *Transactions of the Institute of British Geographers*, 15, 21-34.

Hulme, M. (1994). "Regional climate change scenarios based on IPCC emissions projections with some illustrations for Africa". *Area*, 26(1), 33-44.

Hulme, M. and Viner, D. (1998). "Climate change scenario for the tropics". *Climatic Change*, 39, 145-176.

Hussein, A.S.A. and El Daw, A.K. (1989). "Evapotranspiration in Sudan Gezira Irrigation Scheme". *Journal of Irrigation and Drainage Engineering*, 115(6), 1018-1033.

Jensen, D.T., Hargreaves, G.H., Temesgen, B. and Allen, R.G. (1997). "Computation of ETo under nonideal conditions". *Journal of Irrigation and Drainage Engineering*, 123(5), 394-400.

Kerkides, P., Michalopoulou, H., Papaioannou, G. and Pollatou, R. (1996). "Water balance estimates over Greece". *Agriculture Water Management*, 32, 85-104.

Kolka, R.K. and Wolf, A.T. (1998). "Estimating actual evapotranspiration for forested sites: modification to the Thornthwaite model". *Southern Research Station Publications, Research Note SRS-6*, US Department of Agriculture, Forest Service, Southern Research Station, Asheville, N.C.

Komuscu, A.U., Erkan, A., and Oz, S. (1998). "Possible impacts of climate change on soil moisture availability in the south east Anatolya Development Project region (GAP): An analysis from an agricultural drought perspective". *Climatic Change*, 40, 519-545.

Lansigan, F.P., de los Santos, W.L. and Coladillaa, J.O. (2000). "Agronomic impacts of climate variability on rice production in the Philippines". *Agriculture, Ecosystems, and Environment*, 82, 129-137.

Larsson, H. (1996). "Relationships between rainfall and sorghum, millet and sesame in the Kassala Province, eastern Sudan". *Journal of Arid Environments*, 3, 211-223.

Leathers, D.J., Grundstein, A.J. and Ellis, A.W. (2000). "Growing season moisture deficits across the northeastern United States". *Climate Research*, 14, 43-55.

Leavesley, G.H. (1994). "Modelling the effect of climate change on water resources – a review". *Climatic Change*, 28, 159-177.

Le Houérou, H.N. (1996). "Climate change, drought and desertification". *Journal of Arid Environments*, 34, 133-185.

Lockwood, J.G. (1999). "Is potential evapotranspiration and its relationship with actual evapotranspiration sensitive to elevated atmospheric CO_2 levels?" *Climatic Change*, 41, 193-212.

McKenney, M.S. and Rosenberg, N.J. (1993). "Sensitivity of some potential evapotranspiration estimation methods to climate change". *Agricultural and Forest Meteorology*, 64, 81-110.

Měmec, J. and Schaake, J. (1982). "Sensitivity of water resource systems to climate variation". *Hydrological Sciences Journal*, 27(3), 327-343.

Milly, P.C.D. (1994). "Climate, soil water storage, and the average annual water balance". *Water Resources Research*, 30(7), 2143-2156.

Mimikou, M., Kouvopoulos, Y., Cavadias, G. and Vayianos, N. (1991). "Regional hydrological effects of climate change". *Journal of Hydrology*, 123, 119-146.

Musa, S.B. (1986). "Evaporation and soil moisture depletion in the Gedaref region of east-central Sudan". PhD Thesis, Department of Geography, University College, Swansea, U.K.

Najjar, R.G. (1999). "The water balance of the Susquehanna River basin and its response to climate change". *Journal of Hydrology*, 21, 7-19.

Oliver, J. (1965). "Evaporation losses and rainfall regime in central and northern Sudan". *Weather*, 20, 58-64.

Oliver, J. (1969). "Problems of determining evapotranspiration in the semi-arid tropics illustrated with reference to the Sudan". *Journal of Tropical Geography*, 28, 64-74.

Palutikof, J.P., Goodess, C.M. and Guo, X. (1994). "Climate change, potential evapotranspiration and moisture availability in the Mediterranean basin". *International Journal of Climatology*, 14, 853-869.

Pimenta, M.T. (2000). "Water balances using GIS". *Physics and Chemistry of the Earth (Part B)*, 25(7-8), 695-698.

Ramírez, J.A. and Finnerty, B. (1998). "CO_2 and temperature effects on evapotranspiration and irrigated agriculture". *Journal of Irrigation and Drainage Engineering*, 122(3), 155-163.

Rind, D., Goldberg, R., Hansen, J., Rosenzweig, C. and Ruedy, R. (1990). "Potential evapotranspiration and the likelihood of future drought". *Journal of Geophysical Research*, 95(D7), 9983-10004.

Roberts, G. (1998). "The effects of possible future climate change in evaporation losses from four contrasting UK water catchment areas". *Hydrological Processes*, 12, 727-739.

Rosenberg, N.J., McKenney, M.S. and Martin, P. (1989). "Evapotranspiration in a greenhouse-warmed world: a review and a simulation". *Agricultural and Forest Meteorology*, 47, 303-320.

Shahin, M. (1985). "Hydrology of the Nile basin". *Development in Water Science 21*, Elsevier, Amsterdam / Oxford / New York / Tokyo.

Simpson, I.G. and Simpson, M.C. (1991). "Systems of agricultural production in central Sudan and Khartoum Province". *The agriculture of the Sudan*, G.M. Craig, ed., Oxford University Press, Oxford, 252-279.

Singh, V.P. and Xu, C.Y. (1997). "Evaluation and generalization of 13 mass-transfer equations for determining free water evaporation". *Hydrological Processes*, 11, 311-323.

Sircoulon, J., Lebel, T. and Arnell, N.W. (1999). "Assessment of the impacts of climate variability and change on the hydrology of Africa". In: *Impacts of climate change and climate variability on hydrological regimes*, J.C. van Dam, ed., Cambridge University Press, Cambridge, 67-84.

Sørbø, G.M. (1991). "Systems of pastoral and agricultural production in eastern Sudan. In: *The agriculture of the Sudan*, G.M. Craig, ed., Oxford University Press, Oxford, 214-229.

Thomas, A. (2000). "Climatic changes in yield index and soil water deficit trends in China". *Agricultural and Forest Meteorology*, 102, 71-81.

Thornthwaite, C.W. (1948). "An approach toward a rational classification of climate". *The Geographical Review*, 38, 55-94.

Thornthwaite, C.W. and Mather, J.R. (1957). "Instructions and tables for computing potential evapotranspiration and water balance". *Publications in Climatology*, 10(3), Drexel Institute of Technology, Centerton, NJ.

UNEP (1992). "World atlas of desertification". Edward Arnold, London / New York / Melbourne / Auckland.

Wardlaw, R.B., Hulme, M. and Stuck, A.Y. (1996). "Modelling the impacts of climatic change on water resources". *Journal of the Chartered Institution of Water and Environmental Management*, 10, 355-364.

Wilson, R.T. (1991). "Systems of agricultural production in northern Sudan". In: *The agriculture of the Sudan*, G.M. Craig, ed., Oxford University Press, Oxford, 193-213.

Xiong, L. and Guo, S. (1999). "A two-parameter monthly water balance model and its application". *Journal of Hydrology*, 216, 111-123.

Wolock, D.M. and McCabe, G.J. (1999). "Explaining spatial variability in mean annual runoff in the conterminous United States". *Climate Research*, 11, 149-159.

Xu, C.-Y. (1999). "Climate change and hydrologic models: A review of existing gaps and recent research developments". *Water Resources Management*, 13(5), 369-382.

Xu, C.-Y. (2000). "Modelling the effects of climate change on water resources in central Sweden". *Water Resources Management*, 14, 177-189.

Xu, C.-Y. and Singh, V.P. (1998). "A review on monthly water balance models for water resources investigations". *Water Resources Management*, 12, 31-50.

Zhang, M., Geng, S., Ransom, M. and Ustin, S.L. (1996). "The effects of global warming on evapotranspiration and alfalfa production in California". http:// cstars.ucdavis.edu/papers/html/zhangetal1996/index.html

Zinyowera, M.C., Jallow, B.P., Maya, R.S. and Okoth-Ogendo, H.W.O. (1998). "Africa". In: *Regional impacts of climate change – an assessment of vulnerability*, T. Watson, M.C. Zinyowera and M.C. Moss ed., A special report of IPCC Working Group II of the IPCC, Cambridge University Press, Cambridge, 29-84.

Surface Water Hydrology, Singh, Al-Rashed & Sherif (eds)
© 2002 Swets & Zeitlinger, Lisse, ISBN 90 5809 363 8

Assessments of climate change impacts in semi-arid Northeast Brazil

José Nilson Beserra Campos, Ticiana Marinho de Carvalho Studart, Raimundo Oliveira de Sousa and Luíz Sérgio Vasconcelos Nascimento

CENARIDUS - Center of Hydrological and Environmental
Studies in Semi-Arid Regions
Department of Water Resources and Environmental Engineering
Universidade Federal do Ceará
P.O. Box 6018, Fortaleza, Ceará, Brazil, 60451-970
e-mail: nilson@ufc.br; ticiana@ufc.br

Abstract

The present paper analyses the global changes impacts in reservoir yield and efficiency under two different scenarios: (1) the scenario where precipitation and evaporation increases in the same ratio and the coefficient of surface run-off remains constant (e.g. annual inflow increases in the same ratio also) and (2) the scenario where precipitation and evaporation increases in the same ratio and the coefficient of surface run-off increases according to an polynomial rule. In evaluating the impacts of climate changes in the scenarios above described, there were used two different methodologies. First, the impact in reservoir's yield and efficiency were estimated using Monte Carlo simulation. In the second approach, the impact was estimated in an infinite reservoir. The annual inflows were considered, in both situations, having different variability pattern, translated by the coefficient of variation (CV = 0.2, 1.0 and 1.6).

Introduction

Projecting future climate change is highly problematic, once there are large uncertainties on future scenarios of population growth, economic development, life style choices, technological change and energy alternatives. One scenario often used for climate model studies employs rapid growth rates such that annual greenhouse gas emissions continue to accelerate. This is a useful scenario, once it provides a warning of the magnitude of climate change that may be possible if annual greenhouse gas emissions continue to increase.

There is a lot of uncertainties on how global change will affect the hydrologic cycle. An increase in the recycling rate of water is anticipated in response to higher global average temperatures. However, it is not expected to be spatially uniform.

The present study intends to evaluate the impacts of climate change in the availability of water in semi-arid Northeast of Brazil. The main focus is on the water yielded by surface reservoirs, which represents the main source for most portion of Brazil's Northeast, a very vulnerable region.

Methodology

A reservoir woks as a hydrologic transformation system. The water provided in a very irregular manner by nature is stored and released according to the demand. In that process, the inflows are partitioned in three portions: (1) the evaporation from the lake; (2) the reservoir's spills to downstream and (3) controlled volume, which is named "reservoir yield", and represents the system's availability. This third part represents the reservoir yield, or water availability, witch is focus of the present study.

The reservoir yield was estimated under two approaches. In the first approach, the yield was estimated solving the reservoir budget equation, in parametric formulation of Campos (1987, 1996), using Monte Carlo simulation. In the second, the yield was computed for an infinite reservoir according to an equation, obtained using also Monte Carlo simulation and the parametric formulation referred (Campos and Ibiapina, 1997; Campos, Ibiapina and Studart, 2001).

To introduce the climate change in the study, we analyzed two standards for hydrological modifications in a river basin located in Ceará State. For each standard, it were built scenarios for different values of precipitation, lake evaporation and reservoir inflows. The yield estimations were performed using both procedures: Monte Carlo Simulation and infinite reservoir yield equation.

The reservoir chosen for that study could be anyone. It was chosen Várzea do Boi Reservoir, object of prior studies by Campos et al. (2000) and Campos and Studart (2001), due its availability of data.

Standards assumed for hydrological changes

Due the uncertainties in the expected hydrological modifications in the river basin, we studied the impacts of climate change on water availability under two different standards of hydrological modifications in the basin. In each standard, it were built 10 scenarios grouped in an ensemble, named ensemble 1 and ensemble 2 respectively. In ensemble 1, precipitation, evaporation and inflows increases at the same ratio (K_X). In ensemble 2, precipitation and evaporation increases at the same ratio (K_X) and the coefficient of surface run-off increases according to Aguiar (1937) polynomial rule (K_R).

This is a kind of optimistic assumption, once the Aguiar's empirical rule is valid for a constant value for evapotranspiration. So, once an increase in evapotranspiration is expected as consequence of global change, it is also expected that the run-off increases a little less than the obtained by Aguiar's formulation. On the other hand, the first ensemble is a kind of pessimistic regarding to reservoir inflow, because some increase in run-off is expected due to the increase in precipitation.

The analysis were made considering annual inflows having different coefficient of variation: CV = 0.2 (characteristic of perennial river), CV = 1.0 and CV = 1.6 (very common values for intermittent rivers in semi-arid areas).

The computations to estimate the global change impacts on water availability were performed in the following steps:

1. For each ensemble, there were built 10 hydrological scenarios (K_X varying from 1.02 to 1.20) and computed the values of precipitation, evaporation and reservoir inflows.
2. For each scenario, there were estimated the reservoir yield and efficiency, according to two methodologies: a) using Monte Carlo simulation and b) using an empirical formulation for an infinite reservoir (Campos and Ibiapina, 1997; Campos, Studart and Ibiapina, 2001).

Yield estimation using Monte Carlo simulation

The reservoir water budget was solved using Campos (1987; 1996) formulation. The procedure consists in making a parameterization of the budget equation and computes the yield using Monte Carlo simulation. The water budget can be represented by Equations 1 and 2, as follows:

$$Z_{t+1} = Z_t + I_t - M - \left(\frac{1}{2}\right) \times (A_{t+1} + A_t) \times E - S_t \tag{1}$$

with

$$S_t = \max\left(Z_t + I_t - M \times \left(\frac{1}{2}\right) \times (A_{t+1} + A_t) \times E_v - K; 0\right) \tag{2}$$

where Z_{t+1} and Z_t – storage at the beginning of the (t+1)th and tth years, respectfully; I_t – inflow into the reservoir during the tth year; M – release from the reservoir during year the tth (M is assumed constant from year to year); A_{t+1} and A_t – lake area at the beginning of the (t+1)th and tth years; E_v – mean evaporation depth during the dry season (E_v is assumed constant from year to year); K – reservoir capacity and S_t – spill from the reservoir during the tth year.

The reservoir and lake morphologies are described, respectively, by $Z(h) = \alpha.h^3$ and $A(h) = 3.\alpha.h^2$. The α value can be estimated from a regression equation between the lake level (h) and the stored volume (Z). Considering Z(h) and A(h), Equation 1 can be rewritten as Equation 3:

$$Z_{t+1} = Z_t + I_t - 3\alpha^{1/3} - \left(\frac{Z_{t+1}^{2/3} + Z_t^{2/3}}{2}\right) E - S_t \tag{3}$$

The reservoir budget equation is reduced to the following parametric presentation:

$$f_m = \Phi\left(CV, G, f_k, f_e\right) \tag{4}$$

where G is the reservoir reliability (assumed 90%) in this study, f_K is the dimensionless capacity (K/μ), f_E is the dimensionless evaporation factor computed by $(3\alpha^{1/3}E_V)/\mu^{1/3}$ and f_M is the dimensionless release in steady state conditions estimated by $f_M = M/\mu$.

The Equation 4 is solved for synthetic inflows, for an gamma distribution function, using a coefficient of variation of annual inflows equal to 0.2, 1.0 and 1.6.

The yield from an infinite reservoir

To estimate the global impacts on an infinite reservoir yield (theoretically, the maximum yield from a basin controlled by a single reservoir), it was used the equation defined by Campos, Studart and Ibiapina (2001). Assuming reliability (G) of 90%, the authors obtained a general equation for reservoir's efficiency (η_M) on the steady state of the storage process (Equation 5).

$$\eta_M = 0,99 \times \exp\left(\frac{-f_E}{1,5031 - 1,7104 \times CV + 0,8555 \times CV^2 - 0,1528 \times CV^3}\right) \tag{5}$$

Only two input parameters are necessary: the coefficient of variation of annual inflows (CV) and the dimensionless factor of evaporation (f_E).

The infinite reservoir efficiency (η_M) and the dimensionless factor of release (f_M) are related by $\eta_M = 0.95\ f_M$ and the reservoir annual release (M) is given by $M = 1.05.\mu.\eta_M$.

Results

The results obtained were grouped and analyzed by hydrological modifications standards (ensembles) and by the methodology used in yield evaluation.

Ensemble 1 estimated by Monte Carlo simulations

The results (Tables 1 to 3 and Figures 1 and 2) show that, in these assumptions, there is a positive trend in water availability (M). The reservoir efficiency, however, follows an opposite tendency. Considering CV = 0.2, for instance, if the mean precipitation and evaporation increases 20%, the reservoir yield rises from 27.10 hm^3/year to 31.5 hm^3/year (approximately 1.2%). Regarding to reservoir's efficiency, it remains constant for a significance of 0.01.

Table 1. Values of annual yield (M) and reservoir efficiency (η) for a 90% reliability, computed by Monte Carlo simulations, in ensemble 1 (CV = 0.2).

K_X	μ (hm³/year)	σ (hm³/year)	f_E	M (hm³/year)	f_M	$\eta = 0.95\ f_M$
1.00	42.58	8.52	0.36	27.10	0.64	0.60
1.02	43.43	8.69	0.36	27.80	0.64	0.61
1.04	44.28	8.86	0.37	28.29	0.64	0.61
1.06	45.13	9.03	0.37	28.51	0.63	0.60
1.08	45.99	9.20	0.38	29.00	0.63	0.60
1.10	46.84	9.37	0.38	29.57	0.63	0.60
1.12	47.69	9.54	0.39	29.68	0.62	0.59
1.14	48.54	9.71	0.39	30.40	0.63	0.60
1.16	49.39	9.88	0.40	30.56	0.62	0.59
1.18	50.24	10.05	0.40	31.07	0.62	0.59
1.20	51.10	10.22	0.41	31.50	0.62	0.59

Table 2. Values of annual yield (M) and reservoir efficiency (η) for a 90% reliability, computed by Monte Carlo simulations, in ensemble 1 (CV = 1.0).

K_X	μ (hm³/year)	σ (hm³/year)	f_E	M (hm³/year)	f_M	$\eta = 0.95\ f_M$
1.00	42.58	42.58	0.36	11.15	0.26	0.25
1.02	43.43	43.43	0.36	11.39	0.26	0.25
1.04	44.28	44.28	0.37	11.29	0.25	0.24
1.06	45.13	45.13	0.37	11.54	0.26	0.24
1.08	45.99	45.99	0.38	11.61	0.25	0.24
1.10	46.84	46.84	0.38	11.76	0.25	0.24
1.12	47.69	47.69	0.39	11.75	0.25	0.23
1.14	48.54	48.54	0.39	11.39	0.23	0.22
1.16	49.39	49.39	0.40	11.94	0.24	0.23
1.18	50.24	50.24	0.40	12.17	0.24	0.23
1.20	51.10	51.10	0.41	12.21	0.24	0.23

Table 3. Values of annual yield (M) and reservoir efficiency (η) for a 90% reliability, computed by Monte Carlo simulations, in ensemble 1 (CV = 1.6).

K_X	μ (hm³/year)	σ (hm³/year)	f_E	M (hm³/year)	f_M	$\eta = 0.95\ f_M$
1.00	42.58	68.13	0.36	4.64	0.11	0.10
1.02	43.43	69.49	0.36	4.79	0.11	0.10
1.04	44.28	70.85	0.37	4.68	0.11	0.10
1.06	45.13	72.22	0.37	4.78	0.11	0.10
1.08	45.99	73.58	0.38	4.63	0.10	0.10
1.10	46.84	74.94	0.38	4.75	0.10	0.10
1.12	47.69	76.30	0.39	4.63	0.10	0.09
1.14	48.54	77.67	0.39	4.70	0.10	0.09
1.16	49.39	79.03	0.40	4.55	0.09	0.09
1.18	50.24	80.39	0.40	4.64	0.09	0.09
1.20	51.10	81.75	0.41	4.52	0.09	0.08

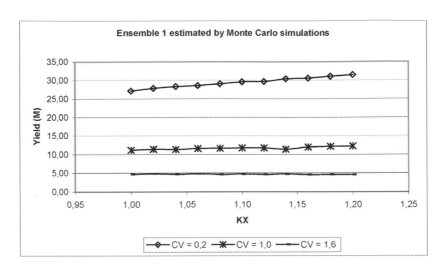

Figure 1. Values of annual yield (M) for a 90% reliability, computed by Monte Carlo simulations, in ensemble 1 (CV = 0.2 to 1.6).

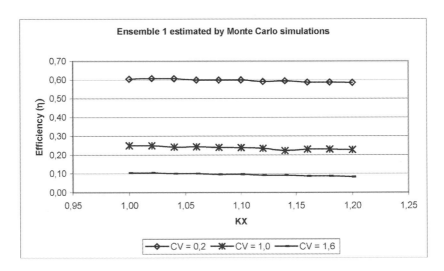

Figure 2. Values of reservoir efficiency (η) for a 90% reliability, computed by Monte Carlo simulations, in ensemble 1 (CV = 0.2 to 1.6).

Ensemble 1 with yield estimated by infinite reservoir equation

The analysis shows (Tables 4 to 6 and Figures 3 and 4) that there is a positive trend in reservoir yield. Notice that this yield represents the maximum controllable yield in the basin. So, regarding to maximum regulation capacity, in case of hydrologically balanced changes (ensemble 1), the impact of global change in the reservoir will be slightly positive. Despite of that, the reservoir efficiency decreases in all cases studied.

Table 4. Values of annual yield (M) and reservoir efficiency (η) with 90% of reliability for an infinite reservoir in ensemble 1 (CV = 0.2).

K_X	μ (hm³/year)	f_E	M (hm³/year)	$\eta = 0.95\ f_M$
1.00	42.58	0.36	32.74	0.73
1.02	43.43	0.36	33.26	0.73
1.04	44.28	0.37	33.78	0.73
1.06	45.13	0.37	34.29	0.72
1.08	45.99	0.38	34.80	0.72
1.10	46.84	0.38	35.31	0.72
1.12	47.69	0.39	35.81	0.72
1.14	48.54	0.39	36.31	0.71
1.16	49.39	0.40	36.81	0.71
1.18	50.24	0.40	37.30	0.71
1.20	51.10	0.41	37.79	0.70

Table 5. Values of annual yield (M) and reservoir efficiency (η) with 90% of reliability for an infinite reservoir in ensemble 1 (CV = 1.0).

K_X	μ (hm³/year)	f_E	M (hm³/year)	$\eta = 0.95\ f_M$
1.00	42.58	0.36	21.40	0.48
1.02	43.43	0.36	21.62	0.47
1.04	44.28	0.37	21.83	0.47
1.06	45.13	0.37	22.04	0.47
1.08	45.99	0.38	22.25	0.46
1.10	46.84	0.38	22.45	0.46
1.12	47.69	0.39	22.64	0.45
1.14	48.54	0.39	22.83	0.45
1.16	49.39	0.40	23.02	0.44
1.18	50.24	0.40	23.20	0.44
1.20	51.10	0.41	23.38	0.44

Table 6. Values of annual yield (M) and reservoir efficiency (η) with 90% of reliability for an infinite reservoir in ensemble 1 (CV = 1.6).

K_X	μ (hm³/year)	f_E	M (hm³/year)	$\eta = 0.95\ f_M$
1.00	42.58	0.36	14.90	0.33
1.02	43.43	0.36	14.98	0.33
1.04	44.28	0.37	15.06	0.32
1.06	45.13	0.37	15.13	0.32
1.08	45.99	0.38	15.20	0.31
1.10	46.84	0.38	15.26	0.31
1.12	47.69	0.39	15.32	0.31
1.14	48.54	0.39	15.38	0.30
1.16	49.39	0.40	15.43	0.30
1.18	50.24	0.40	15.49	0.29
1.20	51.10	0.41	15.53	0.29

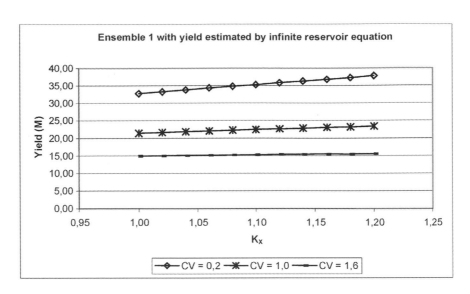

Figure 3. Values of annual yield (M) with 90% of reliability for an infinite reservoir in ensemble 1 (CV = 0.2 to 1.6).

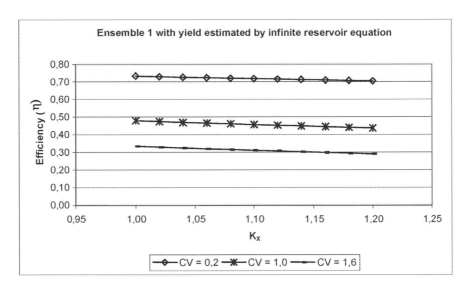

Figure 4. Values of reservoir efficiency (η) with 90% of reliability for an infinite reservoir in ensemble 1 (CV = 0.2 to 1.6).

Ensemble 2 with yield estimated by Monte Carlo simulations

Ensemble 2 embraces the scenarios where precipitation and evaporation increase in the same ratio (K_X) and run-off increases in a rate K_R, according to Aguiar's rule (Equation 6):

146

$$R(mm) = 28{,}53 \times H - 112{,}9 \times H^2 + 351{,}91 \times H^3 - 118{,}74 \times H^4 \qquad (6)$$

where H is the mean annual precipitation over the basin and R(mm) is the run-off in millimeters, for an unit basin. Using H = 0.52 as initial condition, one can find the reference run-off for an unit basin (R°mm) equal to 25.02 mm. The other values of Rmm (for each precipitation H) are evaluated by Equation 6. The K_R values are obtained dividing the correspondent Rmm by R°mm. The mean annual inflow (μ) in each scenario K_X is estimated by multiplying the mean annual inflow of the initial conditions (42.58 hm³/year) by the correspondent K_R. The results are showed in Tables 7, 8 and 9 and Figures 5 and 6.

Table 7. Values of annual yield (M) and reservoir efficiency (η) for a 90% reliability, computed by Monte Carlo simulations, in ensemble 2 (CV = 0.2).

K_X	H (m)	R (mm)	K_R	μ	σ	f_E	M	f_M	$\eta = 0.95\ f_M$
1.00	0.52	25.11	1.00	42.58	8.52	0.36	27.10	0.64	0.60
1.02	0.53	26.43	1.05	44.82	8.96	0.36	28.68	0.64	0.61
1.04	0.54	27.80	1.11	47.15	9.43	0.36	30.32	0.64	0.61
1.06	0.55	29.22	1.16	49.56	9.91	0.36	31.83	0.64	0.61
1.08	0.56	30.69	1.22	52.06	10.41	0.36	33.15	0.64	0.60
1.10	0.57	32.22	1.28	54.64	10.93	0.36	34.71	0.64	0.60
1.12	0.58	33.79	1.35	57.31	11.46	0.37	36.26	0.63	0.60
1.14	0.59	35.42	1.41	60.07	12.01	0.37	38.00	0.63	0.60
1.16	0.60	37.10	1.48	62.92	12.58	0.37	39.76	0.63	0.60
1.18	0.61	38.83	1.55	65.85	13.17	0.37	39.80	0.60	0.57
1.20	0.62	40.61	1.62	68.88	13.78	0.37	43.85	0.64	0.60

Table 8. Values of annual yield (M) and reservoir efficiency (η) for a 90% reliability, computed by Monte Carlo simulations, in ensemble 2 (CV = 1.0).

K_X	H (m)	R (mm)	K_R	μ	σ	f_F	M	f_M	$\eta = 0.95\ f_M$
1.00	0.52	25.11	1.00	42.58	42.58	0.36	11.15	0.26	0.25
1.02	0.53	26.43	1.05	44.82	44.82	0.36	11.68	0.26	0.25
1.04	0.54	27.80	1.11	47.15	47.15	0.36	12.34	0.26	0.25
1.06	0.55	29.22	1.16	49.56	49.56	0.36	13.12	0.26	0.25
1.08	0.56	30.69	1.22	52.06	52.06	0.36	13.66	0.26	0.25
1.10	0.57	32.22	1.28	54.64	54.64	0.36	14.28	0.26	0.25
1.12	0.58	33.79	1.35	57.31	57.31	0.37	14.73	0.26	0.24
1.14	0.59	35.42	1.41	60.07	60.07	0.37	15.42	0.26	0.24
1.16	0.60	37.10	1.48	62.92	62.92	0.37	16.19	0.26	0.24
1.18	0.61	38.83	1.55	65.85	65.85	0.37	16.90	0.26	0.24
1.20	0.62	40.61	1.62	68.88	68.88	0.37	17.71	0.26	0.24

Table 9. Values of annual yield (M) and reservoir efficiency (η) for a 90% reliability, computed by Monte Carlo simulations, in ensemble 2 (CV = 1.6).

K_X	H (m)	R (mm)	K_R	μ	σ	f_E	M	f_M	$\eta = 0.95\,f_M$
1.00	0.52	25.11	1.00	42.58	68.13	0.36	4.63	0.11	0.10
1.02	0.53	26.43	1.05	44.82	71.72	0.36	4.96	0.11	0.11
1.04	0.54	27.80	1.11	47.15	75.44	0.36	5.22	0.11	0.11
1.06	0.55	29.22	1.16	49.56	79.30	0.36	4.44	0.09	0.09
1.08	0.56	30.69	1.22	52.06	83.29	0.36	5.70	0.11	0.10
1.10	0.57	32.22	1.28	54.64	87.42	0.36	6.04	0.11	0.10
1.12	0.58	33.79	1.35	57.31	91.70	0.37	6.03	0.11	0.10
1.14	0.59	35.42	1.41	60.07	96.11	0.37	6.32	0.11	0.10
1.16	0.60	37.10	1.48	62.92	100.67	0.37	6.65	0.11	0.10
1.18	0.61	38.83	1.55	65.85	105.36	0.37	7.02	0.11	0.10
1.20	0.62	40.61	1.62	68.88	110.21	0.37	7.29	0.11	0.10

As can be saw in Tables 7, 8 and 9, there is a positive trend in water availability for K_X – more accentuated for smaller CV's. Considering CV = 0.2, for instance, when precipitation increases 4%, the reservoir yield increases 62%. That is, in the global change assumptions of ensemble 2, it is expected a reasonable gain in water availability. The reservoir efficiency, however, remain practically inalterable.

Ensemble 2 with yield estimated by infinite reservoir equation

The analysis showed that there is a positive trend in reservoir yield, independently of the CV assumed. Regarding to maximum capacity of regulation, in case of hydrologically balanced changes (ensemble 2), the impact of global change in the

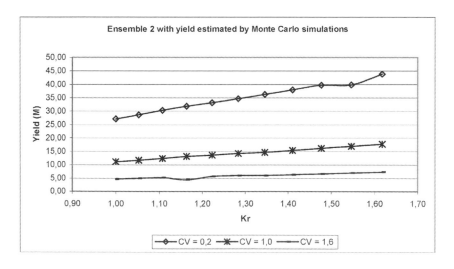

Figure 5. Values of reservoir yield (M) with 90% of reliability using Monte Carlo simulation reservoir in ensemble 2 (CV = 0.2 to 1.6).

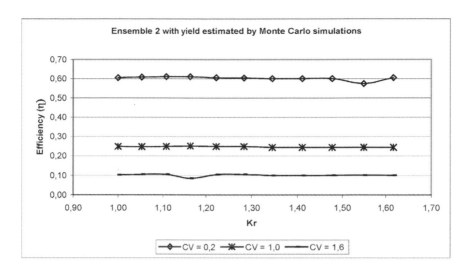

Figure 6. Values of reservoir efficiency (η) with 90% of reliability using Monte Carlo simulation reservoir in ensemble 2 (CV = 0.2 to 1.6).

reservoir will be positive regarding to water availability. Regarding the efficiency, otherwise, it remains inalterable (Tables 10, 11 and 12 and Figures 7 and 8).

Conclusions

The results of this research reached the following conclusions:

For the pessimistic scenario of hydrologically balanced changes called "ensemble 1" (precipitation, evaporation and inflows increasing at the same ratio) there are a positive trend in reservoir yield and a negative trend in reservoir efficiency in the

Table 10. Values of annual yield (M) and reservoir efficiency (η) for a 90% reliability for an infinite reservoir in ensemble 2 (CV = 0.2).

K_X	K_R	μ (hm³/year)	f_E	M (hm³/year)	$η = 0.95 f_M$
1.00	1.00	42.58	0.36	32.74	0.73
1.02	1.05	44.82	0.36	34.44	0.73
1.04	1.11	47.15	0.36	36.20	0.73
1.06	1.16	49.56	0.36	38.02	0.73
1.08	1.22	52.06	0.36	39.91	0.73
1.10	1.28	54.64	0.36	41.86	0.73
1.12	1.35	57.31	0.37	43.88	0.73
1.14	1.41	60.07	0.37	45.96	0.73
1.16	1.48	62.92	0.37	48.11	0.73
1.18	1.55	65.85	0.37	50.33	0.73
1.20	1.62	68.88	0.37	52.61	0.73

Table 11. Values of annual yield (M) and reservoir efficiency (η) for a 90% reliability for an infinite reservoir in ensemble 2 (CV = 1.0).

K_X	K_R	μ (hm³/year)	f_E	M (hm³/year)	$\eta = 0.95\ f_M$
1.00	1.00	42.58	0.36	21.40	0.48
1.02	1.05	44.82	0.36	22.48	0.48
1.04	1.11	47.15	0.36	23.61	0.48
1.06	1.16	49.56	0.36	24.77	0.48
1.08	1.22	52.06	0.36	25.97	0.48
1.10	1.28	54.64	0.36	27.22	0.47
1.12	1.35	57.31	0.37	28.50	0.47
1.14	1.41	60.07	0.37	29.83	0.47
1.16	1.48	62.92	0.37	31.20	0.47
1.18	1.55	65.85	0.37	32.61	0.47
1.20	1.62	68.88	0.37	34.06	0.47

Table 12. Values of annual yield (M) and reservoir efficiency (η) for a 90% reliability for an infinite reservoir in ensemble 2 (CV = 1.6).

K_X	K_R	μ (hm³/year)	f_E	M (hm³/year)	$\eta = 0.95\ f_M$
1.00	1.00	42.58	0.36	14.90	0.33
1.02	1.05	44.82	0.36	15.64	0.33
1.04	1.11	47.15	0.36	16.41	0.33
1.06	1.16	49.56	0.36	17.20	0.33
1.08	1.22	52.06	0.36	18.02	0.33
1.10	1.28	54.64	0.36	18.87	0.33
1.12	1.35	57.31	0.37	19.74	0.33
1.14	1.41	60.07	0.37	20.65	0.33
1.16	1.48	62.92	0.37	21.58	0.33
1.18	1.55	65.85	0.37	22.54	0.33
1.20	1.62	68.88	0.37	23.53	0.33

present river basin topology. This behavior is also observed in the assumption of an infinite reservoir. The negative trend in reservoir efficiency is more accentuated for larger values of coefficient of variation of annual inflows.

For the optimistic scenario of hydrologically balanced changes called "ensemble 2" (precipitation and evaporation increasing at the same ratio, and inflows increasing according Aguiar's polynomial rule) there is a positive trend in the reservoir yield for the present river basin topology as well as for the totally controlled basin. In both situations, no gains, or losses, are predicted in reservoir efficiency.

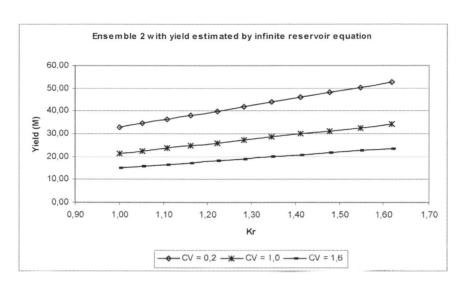

Figure 7. Values of reservoir yield (M) with 90% of reliability for an infinite reservoir in ensemble 2 (CV = 0.2 to 1.6).

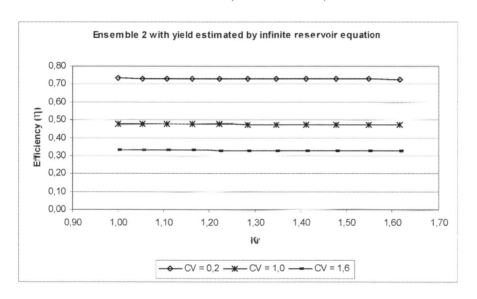

Figure 8. Values of reservoir efficiency (η) with 90% of reliability for an infinite reservoir in ensemble 2 (CV = 0.2 to 1.6).

Acknowledgments

The authors would like to thank CNPq and FUNCEME for supporting the research.

References

Aguiar, F.G. (1978). *Estudo Hidrométrico do Nordeste Brasileiro*. Departamento Nacional de Obras Contra as Secas. Boletim Técnico. 36 n. 2 jul./dez. Reimpressão.

Campos, J.N.B. (1987). *A Procedure for Reservoir Sizing on Intermittent Rivers Under High Evaporation Rate*. Fort Collins, Colorado State University. PhD thesis.

Campos, J.N.B., Vieira Neto, J. and Queiroz, E.A. (2000). Impacto Cumulativo da pequena açudagem: um estudo de caso do Açude Várzea do Boi, em Tauá, Ce. *V Simpósio de Recursos Hídricos do Nordeste*, Natal, RN.

Campos, J.N.B. (1996). *Dimensionamento de Reservatórios: O Método do Diagrama Triangular de Regularização,* Edições UFC,. Fortaleza.

Campos, J.N.B. and Ibiapina, N.G. (1997). Uma Equação para a Máxima Capacidade de Regularização em um Reservatório. In. *XII Simpósio Brasileiro de Recursos Hídricos*, Vitória, ES. Associação Brasileira de Recursos Hídricos.

Campos, J.N.B., Studart, T.M.C. and Ney G. Ibiapina (2001). Computing the Yield from an Infinite Reservoir. In: *American Geophysical Union 21th Annual Hydrology Days*, Colorado State University, Colorado, USA, April.

Campos, J.N.B. and Studart, T.M.C. (2001). Variabilidades Climáticas e Tendências Hidrológicas em Climas Semi-Áridos. In: *III Encuentro das Aguas*, Santiago, Chile.

Hoff, H. (2001). Climate Change and Water Avaiability. Lozán, J.L., Graβl H. and Hupter, P. In: *Climate ot the 21st Century: Changes ans Risk.* Wissenschaftliche Auswertungen, Hamburg.

Surface Water Hydrology, Singh, Al-Rashed & Sherif (eds)
© *2002 Swets & Zeitlinger, Lisse, ISBN 90 5809 363 8*

Sampling uncertainty of climate change impacts on water resources: Case study using data from Iran and England

Najmur R. Nawaz and Adebayo J. Adeloye
Department of Civil & Offshore Engineering, Heriot-Watt University
Edinburgh, EH14 4AS, United Kingdom
e-mail: civnrn@eudoramail.com

Abstract

It is now generally accepted that anthropogenic greenhouse gas emissions are largely responsible for rapid global warming. Moreover, according to the United Nations Intergovernmental Panel on Climate Change (IPCC, 2001), it is likely that human-induced climate change will persist for many centuries to come. It is therefore inevitable that water resources, in arid and semi-arid regions in particular, will be subject to increased stress. Consequently, numerous studies have been carried out to assess the impact of climate change on water resource systems such as reservoirs. Although these studies employed different schemes for baseline climate perturbation, and different models for simulating the resulting catchment rainfall-runoff response, the impacts assessment is based on the single realisation of the perturbed and baseline climate. As a result, it has been difficult to estimate the sampling variability of the assessed impacts or to attach confidence limits to them. In the current study, Monte Carlo simulation experiments were performed to assess the sampling uncertainties in impacts on water resources of projected climate changes. The investigation employed data from basins in Iran, and catchments in Yorkshire, England, both of which incorporate water supply reservoirs. The approach enabled the population of "impacts" to be obtained, which allowed risks of having different levels of impacts to be quantified.

Introduction

Many studies have been carried out over the last decade or so to assess the impact of climate change on water resource systems (see www.pacinst.org). Impact assessment often involves three distinct stages: The first stage is to construct catchment-scale GCM-based climate change scenarios and use these to perturb baseline (current) climate to obtain future climate. This is then followed by forcing a catchment response model with both the current and future climate to obtain the corresponding runoff records. Finally, the hydrological data series are then input into a water resource simulation model to obtain possible impacts. Because of uncertainties introduced at successive stages of the assessment, it is important that the assessed impacts are viewed with caution (Hulme et al., 1999).

Such uncertainties arise because of errors in global climate modelling and in "downscaling" climate to the catchment scale, the imprecision in hydrological and

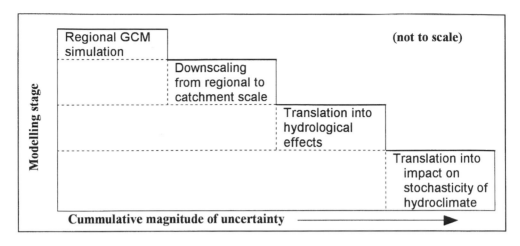

Figure 1. Schematic of the levels of uncertainty in successive stages of climate impact assessment (note the incremental uncertainties are not necessarily equal).

water resource systems modelling, and the limitation caused by using only single records for the impacts assessment (sampling uncertainty). These uncertainties are summarised in Fig. 1. It is possible that if the problem is posed in a sensitivity analysis context rather than prediction, which is commonplace (Wood et al., 1997) then climate modelling errors and possibly downscaling errors are largely removed. Similarly, where both the baseline and future hydrology are based on the hypothesised rainfall-runoff scheme, the uncertainty due to this should also cancel out, leaving only the sampling uncertainty of the hydroclimate.

In this work, extended Monte Carlo simulation experiments are used to assess the sampling uncertainties in impacts on water resources of projected climate changes. The approach enabled the "population of impacts" to be obtained as opposed to the single realisation of impact possible with traditional methods of assessment. The use of the methodology is demonstrated by its application to water resource systems in Iran (a semi-arid climate) and England (a temperate climate).

Methodology

The impact assessment methodology, based on recommended protocol (Carter et al., 1994; 1999), is summarised in Fig. 2. Fig. 2 shows that the traditional approach to impacts assessment utilises only single streamflow records (both baseline and future). An extension of this methodology adopted in this study employs a Monte Carlo simulation technique to characterise the sampling uncertainties of the assessed water resources impacts. It can also be seen in Fig. 2 that once relevant data have been collected, a hydrological model needs to be calibrated and validated.

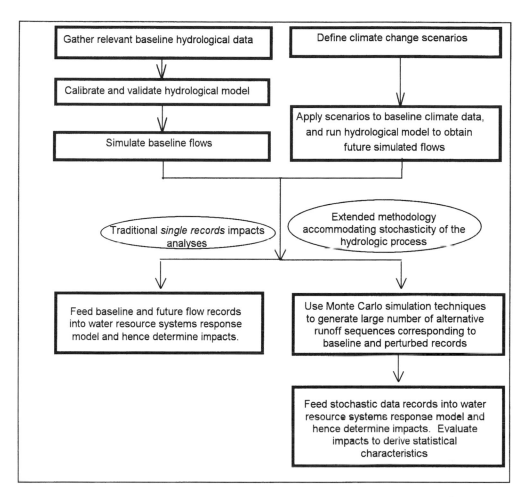

Figure 2. Deterministic and stochastic methodologies for climate change water resources impacts assessment (note the stochastic methodology enables sampling uncertainty of the impacts to be characterised).

Hydrologic Modelling

The hydrologic modelling to determine rainfall-runoff response for the Iranian catchments used the monthly conceptual water balance model of Xu and Halldin (1997). For the English catchments, the daily water balance model MODHYDROLOG (Chiew and McMahon, 1994) was employed as it was decided to investigate this system in greater detail.

The model of Xu is a single store model (Xu et al., 1996), which, in its basic form, has three parameters controlling respectively, actual evapotranspiration (AE), slow and fast runoff (see Fig. 3). The model has been successfully calibrated for many catchments from ten countries with satisfactory results (Vandewiele and Ni-Lar-

155

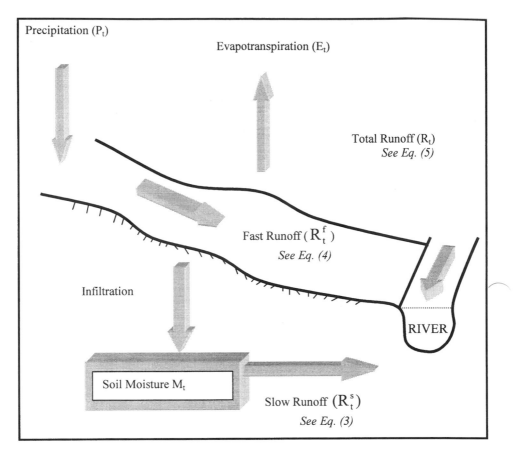

Figure 3. Schematic of Xu's conceptual monthly-water balance model (adopted from Xu et al., 1996, with modifications).

Win, 1998). An extended version of the model can simulate snow accumulation and melting and has six parameters. Together with monthly runoff, both models can accept as inputs different combinations of monthly precipitation, potential evapotranspiration (PE), temperature and humidity. In general, time series data of monthly PE are preferred but where these are unavailable, the PE is estimated internally from temperature and/or humidity using empirical relationships. The snow module is also driven by the temperature. Monthly runoff and other water balance components are the outputs. Table 1 summarises the equations of the basic model without the incorporation of the snow module.

MODHYDROLOG is a conceptual daily rainfall-runoff model structured around five moisture stores as shown in Fig. 4. All five stores are inter-related by catchment processes as shown in Fig. 4 and formulated in Table 2. The model requires daily precipitation and PE as input and simulates groundwater recharge in addition to runoff. The model has 19 parameters which simulate soil moisture and surface water movement. Two parameters can be fixed, leading to only 17 parameters for optimisation. Reungoat (2000) showed that even when the parameters requiring optimisation are reduced to seven, the model performs adequately. MODHYDROLOG has been extensively tested in arid and temperate climates (Chiew and McMahon, 1994; Reungoat, 2000) and used in a number of climate impacts investigations.

Table 1. Main equations of Xu's conceptual monthly water balance model – without snow module (Vandewiele and Xu, 1991).

Potential evapotranspiration:	$E_t = a_4(\Omega_t^+)^2$	(1)
Actual evapotranspiration:	$E_t' = \min\left\{W_t\left(1-e^{-a_1E_t}\right), E_t\right\}$	(2)
Slow runoff:	$R_t^s = a_2\left(M_{t-1}^+\right)$	(3)
Fast runoff:	$R_t^f = a_3\left(M_{t-1}^+\right)P_t^a$	(4)
Total runoff:	$R_t = R_t^s + R_t^f$	(5)
Complete water balance equation:	$M_t = M_{t-1} + P_t - E_t' - R_t$	(6)

Where $W_t = P_t + M_{t-1}^+$ is the available water; $M_{t-1}^+ = \max(M_{t-1}, 0)$ is the available storage; $P_t^a = P_t - E_t\left(1-e^{P_t/\max(E_t,1)}\right)$ is the active rainfall; P_t and Ω_t are monthly precipitation and air temperature, respectively; E_m and Ω_m are long-term average potential evapotranspiration and air temperature, respectively; a_1 parameter is a function of soil permeability; a_2 and a_3 are the storage constant and fast runoff parameters, respectively.

Table 2. The main equations of MODHYDROLOG (Chiew and McMahon, 1994).

Infiltration:	$I_{ft} = \min\left\{(a_{coeff}\ e^{-a_{sq}M/a_{smsc}}),(P-a_{rinsc})\right\}$	(7)
Overland flow:	$R_{over} = P - a_{rinsc}I_{ft}$	(8)
Depression flow:	$R_{dep} = e^{-a_{md}a_{dsc}/R}(a_{dsc} - a_{ads}M_{dep})$	(9)
Surface runoff:	$R_{sur} = R_{over} - R_{dep}$	(10)
Interflow:	$R_{int} = a_{sub}(M/a_{smsc})I_{ft}$	(11)
Groundwater recharge:	$R_{gw} = a_{crak}(M/a_{smsc})(I_{ft} - R_{int})$	(12)
Soil moisture flow:	$R_m = I_{ft} - R_{int} - R_{gw}$	(13)
Actual evapotranspiration:	$E' = \min\{a_{em}(M/a_{smsc}), E\}$	(14)
Deep seepage:	$R_{seep} = a_{vcond}(M_{gw} - a_{dlev})$	(15)
Baseflow/river recharge:	R_{bf} or $R_r = a_{rk1}M_{gw}^+ + a_{rk2}\left(1-e^{-a_{rk3}M_{gw}^+}\right)$	(16)

Where M is the soil moisture store; P, E, R are precipitation, potential evapotranspiration and runoff, respectively; M, M_{dep} and M_{gw} are respectively the soil moisture store, depression store and groundwater store; a_{coeff}, a_{md}, a_{dsc} etc. are model parameters defined in Table 3.

Table 3. 19 MODHYDROLOG parameters.

Parameter	Description
a_{rinsc}	Interception storage capacity
a_{coeff}	Infiltration loss parameter
a_{sq}	Exponent in infiltration capacity equation
a_{ads}	Fraction of total catchment area with depressions
a_{dsc}	Depression storage capacity
a_{rmd}	Exponent in depression flow equation
a_{sub}	Constant of proportionality in interflow calculation
a_{crack}	Constant of proportionality in groundwater recharge calculation
a_{smsc}	Soil moisture store capacity
a_{em}	Maximum vegetation controlled rate of evapotranspiration
a_{locate}	Parameter fixing origin in cycle of seasonal fluctuation of COEFF, CRAK and SUB
a_{seas}	Parameter fixing amplitude in seasonal fluctuation of COEFF, CRAK and SUB
a_{power}	Routing exponent
a_{co}	Routing coefficient
a_{rk1}	Constant of proportionality in linear component of stream-aquifer flow equation
a_{rk2}	Constant of proportionality in exponential component of stream-aquifer flow equation
a_{rk3}	Exponent in exponential component of stream-aquifer flow equation
a_{vcond}	Constant of proportionality in deep seepage equation
a_{dlev}	Deep seepage parameter

Reservoir Planning Analysis

Reservoir analysis was achieved using the modified sequent peak algorithm (SPA) (Lele, 1987; Adeloye and Montaseri, 1998) which is an extension of the basic SPA (Thomas and Burden, 1963). The basic SPA estimates the failure-free capacity of an initially full single reservoir as the maximum of all the sequential deficits. Modifications to the basic SPA were first made by Lele (1987), and later by Adeloye and Montaseri (1998). Modifications made by Lele (1987) enabled both surface net evaporation losses and different reliability levels to be considered explicitly using an iterative procedure. The recent modification carried out by Adeloye and Montaseri (1998) allowed the SPA to be applicable to multiple reservoir systems.

Monte Carlo Experiments

The generation of alternative runoff data utilised a parametric, multivariate annual lag-one autoregressive model, AR(1), which is also known as the Markov model. This was then followed by disaggregation to monthly flows using a Valencia-Schaake (VS) (Valencia and Schaake, 1973) scheme.

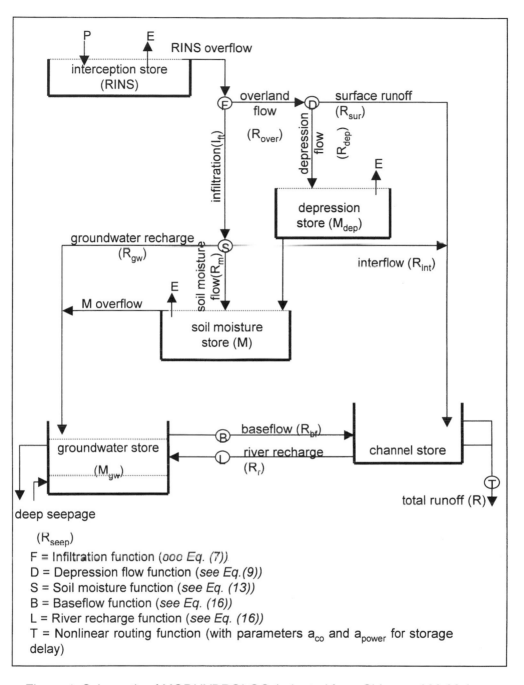

Figure 4. Schematic of MODHYDROLOG (adopted from Chiew and McMahon, 1994, with modifications).

The adopted multivariate stochastic data generation procedure ensured that any inter-site correlation between the flows is preserved. It was decided to ensure that the generated paired baseline-perturbed streamflow records, when analysed, provided the change in reservoir yield for that realisation pair. The way in which this was achieved was to consider the future at each reservoir site as an additional fictitious site.

Case Study

Catchments and Data

Iranian Catchments
The first reservoir system, located in the semi-arid Urmia region of northwest Iran, comprises three catchments: Baranduz, Shahr, and Nazlu. These catchments are located between 37°6′ and 38°0′ northern latitude, and 44°19′ and 45°5′ eastern longitude as shown in Fig. 5. The catchments are situated in the mountainous areas of the region where the minimum altitude is approximately 1300 meters above sea level. The mean annual temperature over the baseline period for the region is about 9.0 °C with summer temperatures reaching as high as 29 °C. During winter, however, temperatures average −2 °C and there are on average 125 freezing days annually in the region (MWP, 1995). Consequently, snowmelt plays a major role in the hydrology of the catchments.

The Iranian catchments have a total area of 2751 km^2 that is drained by a number of tributaries of the main drainage channels of the Baranduz, Shahr, and Nazlu rivers (see Fig. 5). At present, without the presence of any dams, water from the Iranian catchments is simply lost via the three major rivers to the saline Urmia Lake. Consequently, three reservoir systems are proposed by the Iranian Ministry for Water and Power to control and regulate the rivers at the location of reservoirs (see Fig. 5). A simplified schematic of the Iranian system is shown in Fig. 6.

English Catchments
The English reservoir system is located in Yorkshire, England and consists of three direct catchments namely Hebden, Luddenden, and Ogden. These catchments, which are considerably smaller than the Iranian catchments, are located between 53°41′ and 53°50′ northern latitude, and 1°53′ and 2°8′ western longitude as shown in Fig. 7. The catchments are upland in character with minimum and maximum altitudes of 250 and 510 meters above sea level, respectively. The mean annual temperature over the baseline period for the region is about 8.2 °C.

The English system comprises a total of eleven inter-linked reservoirs and is therefore a fairly complex system as shown in Fig. 7. Consequently, a five-reservoir (i.e. Gorple, Widdop, Walshaw Dean, Luddenden and Ogden) simplified configuration of the system (see Fig. 8) was considered in the analysis. Some relevant catchment characteristics for both systems are shown in Table 4.

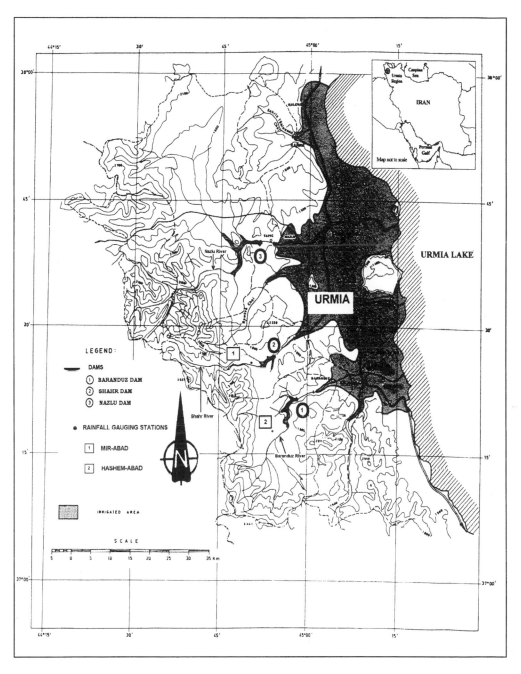

Figure 5. Detailed map of the Urmia reservoir systems (adopted from MWP, 1995, with modifications).

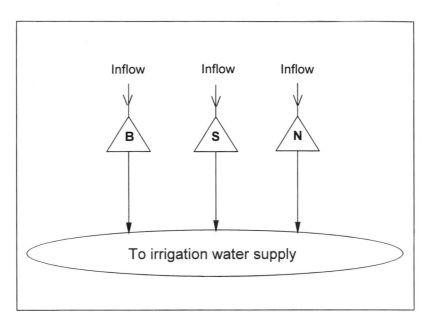

Figure 6. Simplified schematic of the Iranian multiple reservoir system
(B: Baranduz, S: Shahr, N: Nazlu).

Table 4. Characteristics of the catchments analysed.

System	Name	Area (km^2)	Mean runoff (mm)	Mean rainfall (mm)	Mean PE (mm)	Mean E$_o$ (mm)
Iranian	Baranduz	618	461.9	420.5	528.3	1251.2
	Shahr	418	411.0	420.5	517.7	1225.5
	Nazlu	1715	234.4	440.0	509.6	1206.4
	Group total	2751	312.3	432.7	515.0	1219.4
English	Gorple	8.02	989.4	1469.9	545.0	735.3
	Widdop	9.00	1025.1	1413.4	545.0	735.3
	W. Dean	9.41	985.4	1397.1	545.0	735.3
	Luddenden	6.46	824.5	1112.9	545.0	735.3
	Ogden	5.39	828.9	1081.4	545.0	735.3
	Group total	38.28	946.3	1323.8	545.0	735.3

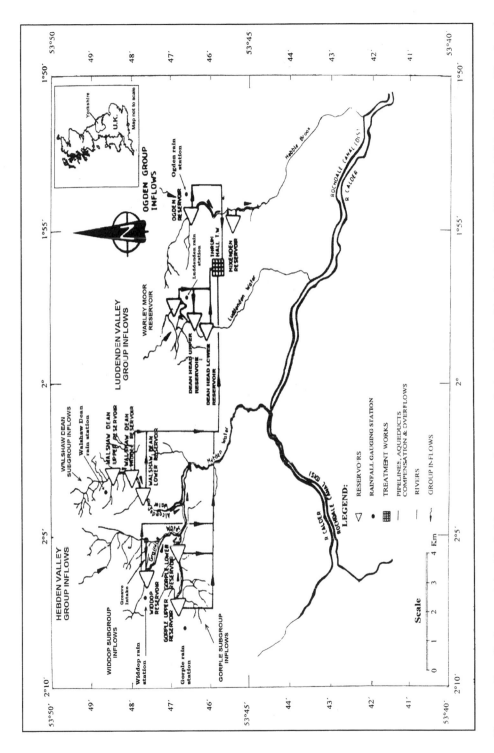

Figure 7. Detailed map of the Yorkshire reservoir system (adopted from Yorkshire Water RRDY, 1991, with modifications).

163

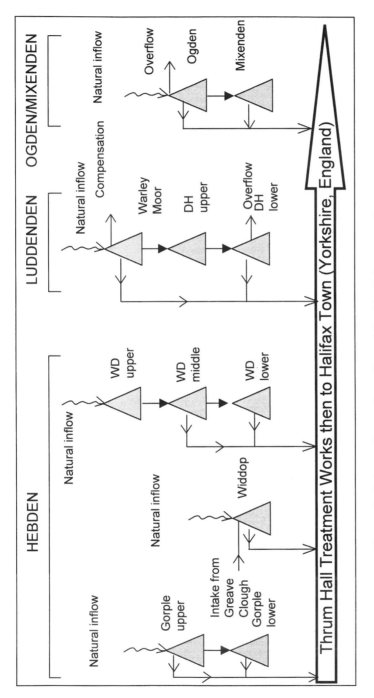

Figure 8. Schematic configuration of the English reservoir systems.

Hydroclimatological Data

Monthly precipitation records at two gauging stations (Mir-Abad and Hashem-Abad stations shown in Fig. 5) for the baseline period were available for the Iranian catchments. Also available for the Urmia region were the monthly baseline average temperature data. These data will be required as input to the monthly rainfall-runoff model of Xu and are used to include the effects of snowpack. Mean monthly open-water evaporation data (E_0) derived from Class A Pan evaporation measurements are provided by the Iranian Ministry of Water and Power (MWP, 1995). Along with precipitation data, E_0 data will be used to include the effects of reservoir net surface-fluxes on reservoir yield. Additionally, baseline monthly inflow data for each of the Iranian sites were also available.

Daily baseline (1961–1990) data of precipitation, potential evapotranspiration (PE), E_0 and runoff were available for the English catchments. Additionally, minimum and maximum daily temperature, and observed daily number of sunshine hours required for stochastic weather generation (described later) were also available.

Climate Change Scenarios

Iranian Scenarios

A simple climate model called MAGICC (Model for the Assessment of Greenhouse Gas Induced Climate Change) and the SCENGEN (SCENario GENerator) regional climate change scenario generator were used to generate precipitation and temperature scenarios for the Iranian catchments.

MAGICC comprises a gas model and upwelling diffusion-energy balance (UD/EB) climate model and an ice melt & thermal expansion (IM/TE) model. The gas models convert each of the main greenhouse gases into atmospheric concentrations and uses this to calculate radiation forcing. The UB/EB climate model calculates global mean temperature response to a given radiation forcing and the IM/TE model computes sea-level change.

SCENGEN is able to generate global and regional scenarios of climate change using results from MAGICC and a wide range of GCMs (Hulme et al., 1995). It achieves this by scaling output from MAGICC to agree with output from a particular GCM by (i) determining the change in regional climate (temperature, precipitation etc.), per degree of global warming as simulated by a GCM (known as the climate sensitivity), and (ii) by applying this change to the global mean temperature simulated by MAGICC. The scaling factor (S_f) can be expressed in mathematical terms as $S_f = \Delta T_{year} / \Delta T_{2x}$, where ΔT_{2x} is the climate sensitivity for the particular GCM and ΔT_{year} is the global temperature change for the year simulated by MAGICC.

Using output from the so-called Simple Climate Models (SCMs) such as MAGICC to obtain regional climate change scenarios has proved quite popular (e.g. IPCC, 1990, 1992; CCIRG, 1996) over recent years. This is because this technique allows regional climate change scenarios (that are in broad agreement with GCM output) to be determined in a significantly shorter time frame than a GCM. Consequently, the SCM can be forced with numerous greenhouse gas emissions scenarios to obtain a wide range of regional climate change scenarios.

SCENGEN was used together with results from MAGICC and HadCM2 GCM and the resulting scenario was named IRHAD, connoting Iranian scenario based on a Hadley Centre GCM.

Perturbation of baseline climate at the Iranian sites was based on a simple perturbation scheme that directly applies the mean differences in climate derived from the GCM simulation experiment (i.e. twelve mean monthly factors) to observational baseline data directly. Table 5 contains the precipitation and temperature scenarios for the Iranian sites.

English Scenarios
Scenarios for the English catchments used output from the following three GCMs:

(i)   CGCM1 – Canadian Centre for Climate Modelling and Analysis first Generation Atmosphere-Ocean GCM;
(ii)   CSIRO1 (CSIRO-Mk2b) – Australian Commonwealth Scientific and Industrial Research Organisation first Generation first generation Atmosphere-Ocean GCM;
(iii)   HadCM3 – UK Hadley Centre for Climate Prediction and Research third Generation Atmosphere-Ocean GCM.

Scenarios representative of the climate in the 2020s, 2050s and 2080s compared to the 1961–1990 baseline were used. This resulted in a total of nine climate change scenarios of radiation, temperature and precipitation. The GCM temperature and radiation data were used to construct PE scenarios using a simple energy-balance approach. Since GCM output is at a relatively coarse spatial scale, it was downscaled to the English catchments' scale using linear interpolation.

Table 5. IRHAD rainfall and temperature scenarios for the Iranian sites – based on HadCM2-2020s.

Month	Change in rainfall from baseline in mm	Change in rainfall (% of baseline mean)	Rise in temperature from baseline (^{o}C)
Jan	-1.1	-3	0.7
Feb	-1.5	-4	0.8
Mar	-1.6	-3	1.0
Apr	-1.0	-1	0.8
May	-1.9	-3	0.8
Jun	-1.2	-4	1.1
Jul	-0.5	-7	1.2
Aug	-0.1	-3	1.0
Sep	-0.3	-6	1.3
Oct	0.8	3	1.0
Nov	0.6	1	0.5
Dec	-0.7	-2	0.5

The complete range of climate change scenarios (applied to all the English sites) expressed as percentage change from the baseline are presented in Figs 9–11. The corresponding absolute monthly changes in temperature, precipitation and PE from the baseline are provided in Tables 6–8.

As mentioned before, perturbation of baseline climate at the Iranian sites was based on a 'simple perturbation' approach. A critical limitation of this technique is that the temporal structure (such as length of dry periods and interannual variability) of the perturbed records will be the same as the historic record. To overcome this limitation, the LARS stochastic weather generator (LARS-WG) (Racsko et al., 1991) was employed for the English system. LARS-WG has been applied in impacts studies by Semenev and Porter (1994), and Semenev and Barrow (1997). The stochastic weather generator was calibrated using baseline climate data and the parameters, including the variability parameter, were then perturbed using the climate change scenarios (e.g. see Wilks, 1992; Buishand and Beckmann, 2000). A new sequence of daily climatic variables representing the future were thus produced.

The application of LARS-WG first involved generating baseline hydroclimate (i.e. minimum and maximum daily temperature, daily precipitation and solar radiation) using the historical baseline records. Given that observational radiation data were unavailable at the sites, the model was supplied with observed sunshine hours data instead. These data are used internally by LARS-WG to obtain estimates of radiation.

After determining the baseline radiation time-series data, these were converted to PE using temperature data along with the Bowen ratio method (see Shaw, 1994). It was necessary to simulate baseline PE because this PE rather than the observed PE would be used in subsequent analyses. This is to allow modelling errors in simulated baseline and future PE to cancel each other out.

After simulating the baseline hydroclimate, climate scenarios were used by LARS-WG to generate future hydroclimate. The use of the weather generator enabled variability to be introduced in to the future time-series.

Results and Discussion

Hydrologic Modelling

In the hydrologic modelling, the snow process was considered for Iran and the snow module was driven by temperature. MODHYDROLOG (snow process not incorporated) was used for modelling the runoff in England. The monthly water balance model used for runoff simulation in the Iranian catchments was calibrated over 1970–1980, with validation being over 1981–1985. The English models were calibrated over 1962–1975 with validation taking place over the period 1976–1990. Model performance was excellent during calibration with an R^2 range of 0.94–0.97 for Iran and 0.96–1.00 for the English catchments. Performance over validation,

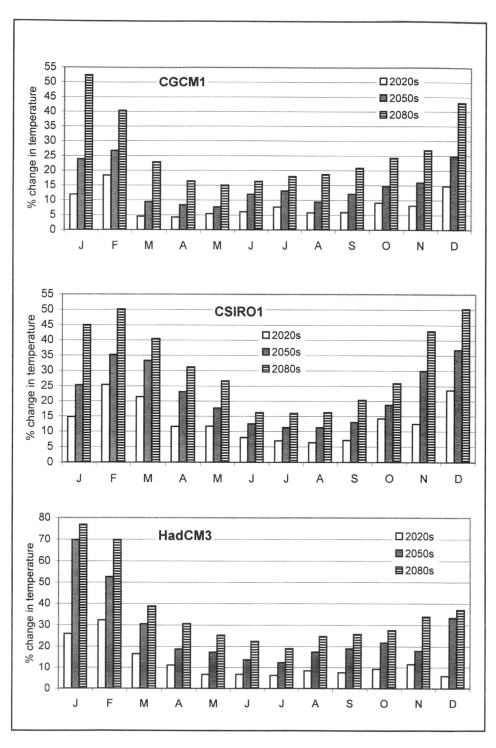

Figure 9. Percentage change in temperature (from baseline) at English sites.

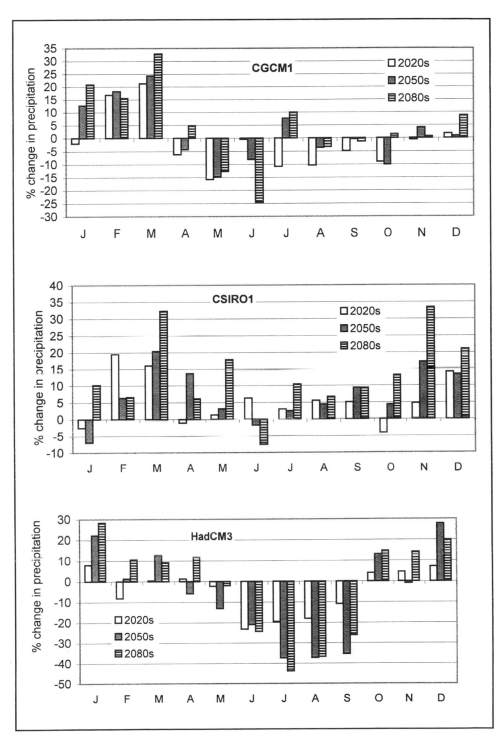

Figure 10. Percentage change (from baseline) in precipitation at English sites.

169

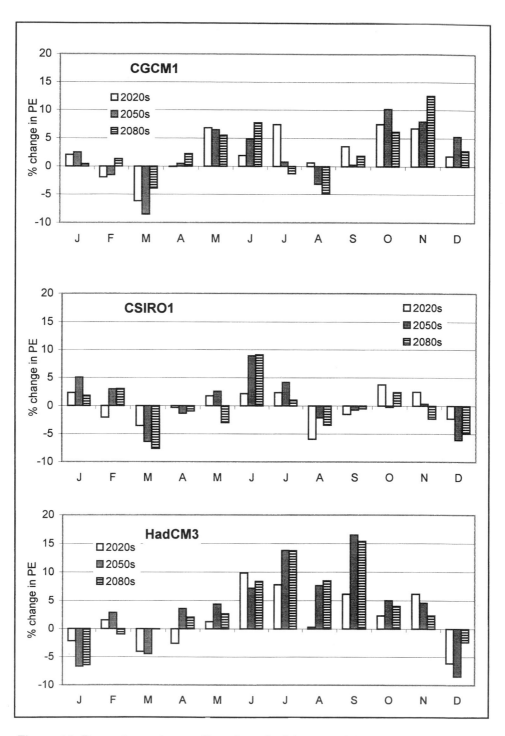

Figure 11. Percentage change (from baseline) in potential evapotranspiration at English sites.

Table 6. Absolute changes (from baseline) in temperature (^{o}C) at English sites.

Scenario	Jan	Feb	Mar	Apr	May	Jun	Jul	Aug	Sep	Oct	Nov	Dec
C2	0.3	0.5	0.2	0.3	0.5	0.8	1.1	0.8	0.7	0.9	0.4	0.5
C5	0.6	0.7	0.4	0.5	0.8	1.6	1.9	1.4	1.5	1.4	0.9	0.9
C8	1.4	1.1	1.0	1.1	1.5	2.1	2.6	2.7	2.6	2.3	1.5	1.5
A2	0.4	0.7	0.9	0.8	1.2	1.0	1.0	0.9	0.9	1.3	0.7	0.9
A5	0.7	0.9	1.4	1.5	1.8	1.6	1.7	1.6	1.6	1.8	1.6	1.3
A8	1.2	1.3	1.7	2.0	2.6	2.1	2.3	2.3	2.5	2.4	2.3	1.8
H2	0.7	0.8	0.7	0.7	0.7	0.9	0.9	1.2	0.9	0.9	0.6	0.2
H5	1.9	1.4	1.3	1.2	1.7	1.8	1.8	2.5	2.3	2.0	1.0	1.2
H8	2.1	1.8	1.7	2.0	2.5	2.9	2.8	3.5	3.2	2.6	1.8	1.3

C: CGCM1, A: CSIRO1, H: HadCM3, 2: 2020s, 5: 2050s, 8: 2080s

Table 7. Absolute changes in precipitation (mm) at English sites.

Scenario	Jan	Feb	Mar	Apr	May	Jun	Jul	Aug	Sep	Oct	Nov	Dec
C2	-5.0	19.0	33.3	2.1	-20.5	-4.6	-14.9	-6.8	-12.3	-21.1	-7.5	5.5
C5	16.8	21.8	42.2	-2.1	-13.5	-13.9	9.1	8.0	-4.0	-29.2	2.9	10.3
C8	33.2	17.0	46.9	6.9	-8.9	-25.7	11.3	5.5	-1.4	-7.5	-6.5	7.3
A2	-2.0	27.1	28.9	3.3	7.4	7.0	9.1	18.6	16.3	-17.8	6.6	33.4
A5	-14.7	6.5	35.7	17.7	7.4	-5.2	4.8	17.4	12.1	-2.5	14.2	33.5
A8	18.0	7.3	52.7	17.5	30.2	-19.1	11.1	18.0	8.2	7.1	58.0	43.0
H2	11.5	-14.4	1.4	3.3	0.7	-30.0	-26.6	-21.6	-11.9	-7.4	8.6	16.1
H5	37.8	1.8	22.3	-8.6	-20.2	-29.8	-44.8	-59.8	-60.3	5.8	-4.8	50.3
H8	48.1	13.7	10.9	6.0	-9.4	-32.6	-53.2	-55.6	-46.7	12.9	18.6	36.3

C: CGCM1, A: CSIRO1, H: HadCM3, 2: 2020s, 5: 2050s, 8: 2080s

Table 8. Absolute changes in potential evapotranspiration (mm) at English sites.

Scenario	Jan	Feb	Mar	Apr	May	Jun	Jul	Aug	Sep	Oct	Nov	Dec
C2	0.3	-0.3	-2.1	0.0	5.4	1.6	6.3	0.5	1.8	2.3	1.2	0.2
C5	0.3	-0.2	-2.9	0.3	5.2	4.0	0.7	-2.4	0.2	3.2	1.4	0.7
C8	0.1	0.2	-1.3	1.1	4.4	6.4	-1.1	-3.6	1.0	1.9	2.2	0.3
A2	0.3	-0.3	-1.2	-0.2	1.4	1.8	2.0	-4.4	-0.8	1.2	0.4	-0.3
A5	0.7	0.5	-2.1	-0.7	2.1	7.3	3.5	-1.6	-0.4	-0.1	0.1	-0.8
A8	0.2	0.5	-2.5	-0.5	-2.4	7.5	0.9	-2.5	-0.2	0.8	-0.4	-0.6
H2	-0.3	0.2	-1.4	-1.3	1.0	8.1	6.6	0.2	3.2	0.7	1.1	-0.8
H5	-0.9	0.4	-1.5	1.8	3.5	5.9	11.6	5.7	8.6	1.6	0.8	-1.1
H8	-0.8	-0.2	0.0	1.0	2.1	6.9	11.6	6.3	8.0	1.2	0.4	-0.3

C: CGCM1, A: CSIRO1, H: HadCM3, M: GCM simulated change, 2: 2020s, 5: 2050s, 8: 2080

though not as good as that over calibration, was generally very good with R^2 ranges between 0.70–0.95 for the Iranian catchments and 0.87–1.00 for the English catchments. Figs 12–15 compare the observed and simulated runoff for the Iranian and English sites during calibration and validation. The fits are generally good.

Runoff Sensitivity to Climate Change

Fig. 16 summarises the percentage changes (from baseline) in mean monthly runoff expected by the 2020s for the Iranian system. For lack of space, results are only presented for the aggregated (global) systems.

The changes in runoff can be traced back to the precipitation changes. For example, the largest reduction in runoff (21%) occurs in the month of July. This arises due to the largest reduction in precipitation during this month (7%). This confirms the dominance of precipitation on runoff (Gleick, 2000).

Changes in mean monthly runoff for the English aggregated system resulting from nine climate change scenarios are provided in Fig. 17. In common with the above results for Iran, the results indicate that changes in runoff follow a similar pattern to precipitation changes presented earlier (see Figure 10). For instance, the CSIRO1 scenario predicted increases in precipitation throughout most of the year with the maximum increases expected in March and November. Similarly, the CSIRO1 scenario results in an increase in runoff during most of the year with maximum increases expected in March and November (see Fig. 17).

The scenarios based on the CGCM1 and HadCM3 GCMs are showing a tendency for reduced runoff during the summer. The largest reduction in runoff results from HadCM3 in September during the 2050s (66% = 38 mm) whilst the largest increase of 54% (59 mm) is expected under the CSIRO1 in November during the 2080s. Such large reductions will no doubt lead to significant reductions in future reservoir yield.

Climate Change Impacts on Reservoir Yield and its Sampling Uncertainty

The yield assessment for the Iranian system involved determining 500 yield changes (from baseline) for a storage capacity of 233×10^6 m^3 which is about 27% MAF. For the English case study, it was decided that the 500 runoff replicates giving the yields should be doubled to reflect the detailed analysis in this study. 1000 yield changes for the English system were evaluated for the existing storage capacity of 11.13×10^6 m^3 (30% MAF) which is the capacity of the five-reservoir system. Figs. 18 and 19 show the empirical distribution of the yield changes based on the traditional single records approach as well as the Monte Carlo approach, for the Iranian and English systems, respectively. Information extracted from these figures is presented in Table 9. Based on the traditional approach, the IRHAD scenario for Iran results in reduced yield at both the 100% and 98% reliability by the 2020s. Because IRHAD had predicted consistently lower precipitation which in turn

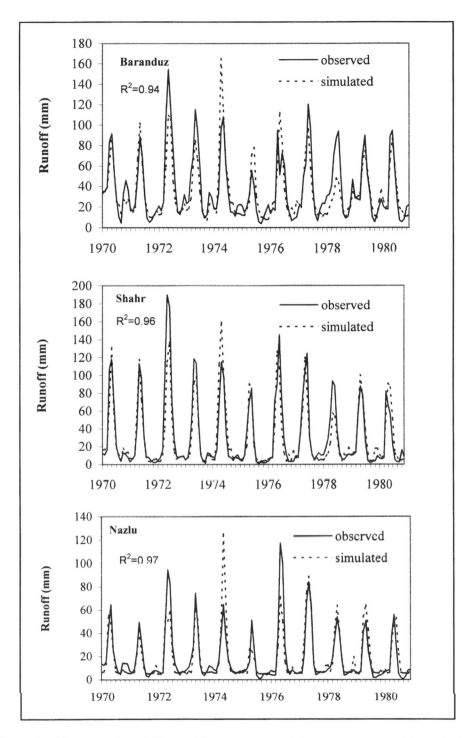

Figure 12. Hydrographs of Xu model monthly simulated and observed flows in Iran over calibration period.

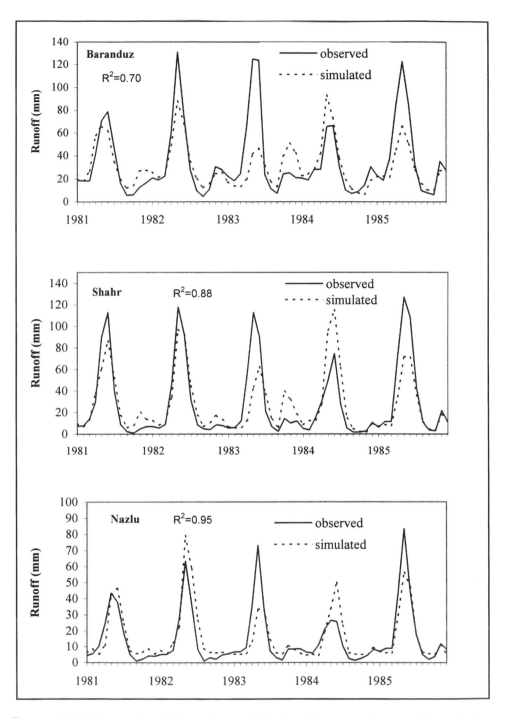

Figure 13. Hydrographs of Xu model monthly simulated and observed flows in Iran over validation period.

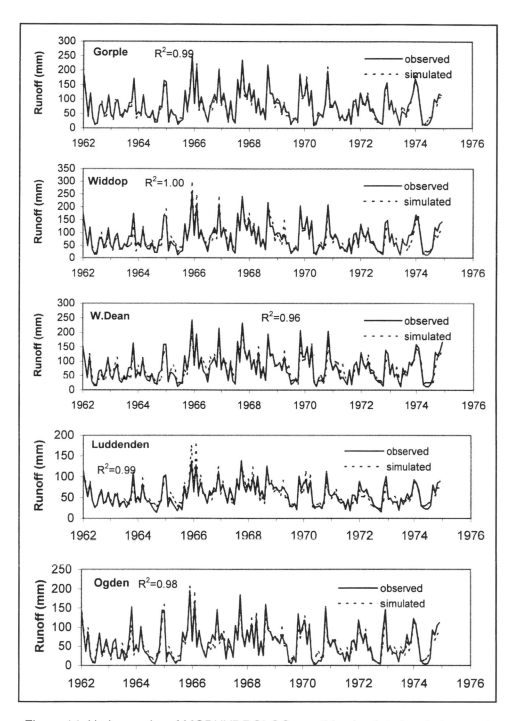

Figure 14. Hydrographs of MODHYDROLOG monthly simulated and observed
flows in England over calibration period.

175

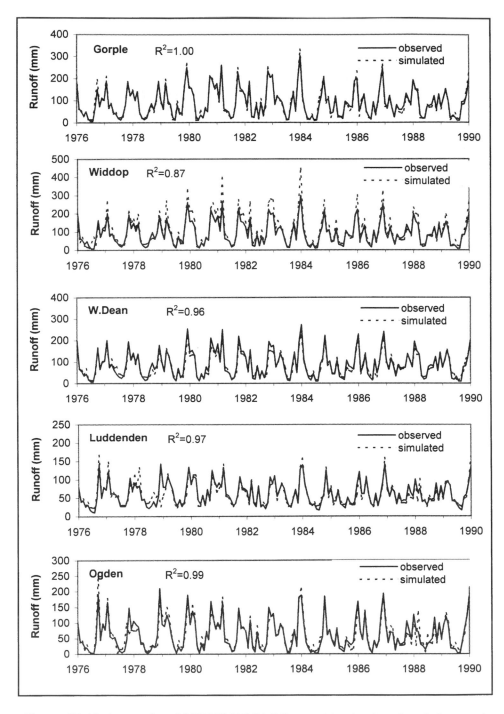

Figure 15. Hydrographs of MODHYDROLOG monthly simulated and observed flows in England over validation period.

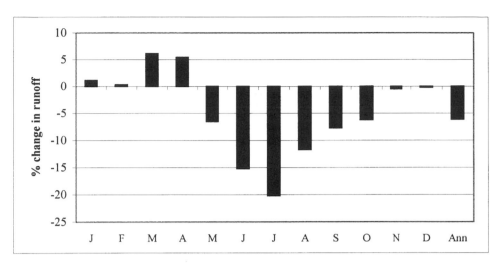

Figure 16. Percentage change in runoff at Iranian catchments resulting from IR-HAD scenario (based on HadCM2-2020s).

translated to consistently lower runoff, this observed behaviour of yield change is not unexpected. While the predicted changes in yield given in Table 9 are modest on the basis of the traditional single records approach, this is untrue when the sampling variability of the input records is taken into account. For example, in Table 9, it can be seen that yield changes vary between −1.9% and −10.8% at the no-failure risk level (100% reliability).

It can be seen from Table 9, that for the English system, at the 100% reliability level and based on the traditional approach, yield changes vary from −1.8% to +2.8% for the CGCM1 scenarios. However, use of the extended Monte Carlo approach leads to variations in yield changes from about −15% to +19%. The change in the median is from −2.7% to +0.7%. The results indicate a reduction in yield by the 2020s and a subsequent increase by the 2050s and 2080s. This pattern of change is consistent with the runoff scenarios presented in Fig. 17. It may be recalled that the CGCM1 scenarios resulted in summer runoff reduction by the 2020s whilst small increases in runoff were expected throughout most of the year by the 2050s and 2080s.

The changes in yield resulting from the CSIRO1 scenarios are more extreme than the CGCM1 based changes. For these scenarios, results based on the traditional approach indicate that the reservoirs will be able to provide more water in the future. The increased reservoir yield is expected to amount to 10.5%, 9.5% and 10.9% by the 2020s, 2050s and 2080s, respectively. Based on the Monte Carlo approach, the yield changes range from −3% (2020s) to a massive +28% (2080s). As with yield changes resulting from CGCM1, the CSIRO1 yield changes can be traced back to runoff changes that were summarised in Fig. 17. It might be recalled that the CSIRO1 scenarios resulted in increased runoff throughout the year in the future.

177

Table 9. Percentage change in yield (from baseline) for the Iranian and English aggregated reservoir systems resulting from climate change scenarios.

System	Reliability (%)	Scenario	Single Record Method	Monte Carlo Method			
				Mean	Minimum	Maximum	Median
Iranian	100	IRHAD	-3.4	-3.4	-10.8	-1.9	-3.5
	98	IRHAD	-3.5	-3.7	-8.0	-2.3	-3.9
English	100	C2	-1.8	-2.6	-14.9	8.7	-2.7
		C5	2.0	0.9	-7.8	18.7	0.7
		C8	2.8	0.6	-9.1	13.2	0.5
		A2	10.5	5.9	-3.0	21.1	5.8
		A5	9.5	5.5	-4.7	19.9	5.3
		A8	**10.9**	9.7	-0.3	**28.3**	9.4
		H2	-1.2	-3.7	-12.5	7.4	-4.0
		H5	**-8.2**	-10.9	**-19.2**	5.7	-11.1
		H8	-3.7	-6.9	-17.8	8.5	-7.1
	98	C2	-3.0	-3.4	-7.5	2.1	-3.5
		C5	-0.5	0.3	-4.3	11.3	0.1
		C8	0.5	0.7	-6.1	8.5	0.6
		A2	5.5	5.0	1.2	12.7	4.9
		A5	6.4	5.4	-0.1	15.3	5.2
		A8	**11.1**	9.1	2.5	**18.5**	8.9
		H2	-4.2	-4.6	-8.6	4.0	-4.8
		H5	**-10.5**	-10.7	**-14.1**	-0.6	-10.9
		H8	-6.1	-6.7	-11.2	5.1	-6.8

Bold font indicates extreme changes in yield for the English system
C: CGCM1, A: CSIRO1, H: HadCM3, 2: 2020s, 5: 2050s, 8: 2080s

In common with the yield changes observed for the Iranian system, the English results for the HadCM3 scenarios indicate reductions in yield. Based on the traditional approach, the reductions range from −1.2% (2020s), −8.2% (2050s) and −3.7% (2080s). The Monte Carlo approach leads to more variability, and the changes range from a minimum of −19% (2050s) to a maximum of +8.5% (2080s).

Highlighted in Table 9 (in bold font) are the most severe changes in yield for the English system that can be expected to occur based on the traditional and Monte

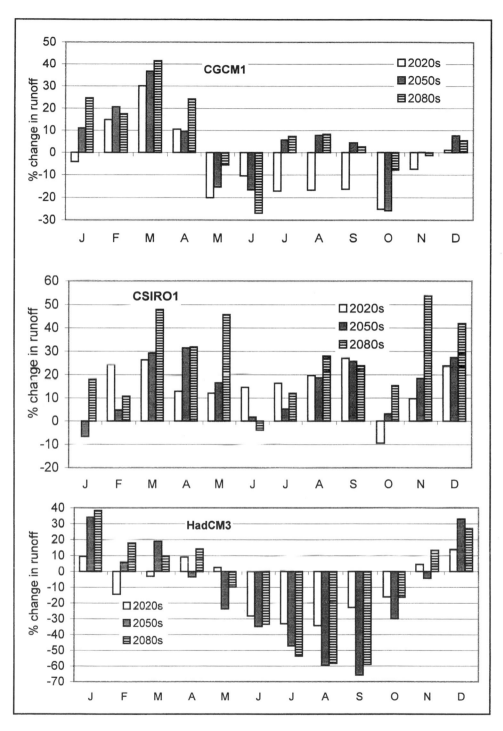

Figure 17. Percentage change in mean monthly runoff at English sites.

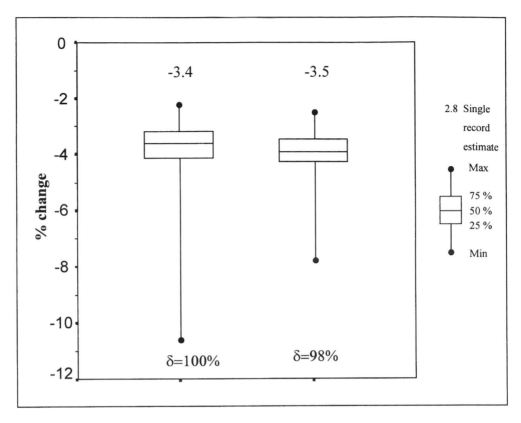

Figure 18. Empirical box plots of yield changes for Iranian aggregated system for the HadCM2-based IRHAD (δ = time-based reliability).

Carlo approach. At the 100% reliability level, and based on the traditional approach, the largest rise in yield results from the CSIRO1 2080s (A8) scenario (10.9%). On the other hand, the largest reduction results from HadCM3 2050s (H5) scenario (8.2%). Based on the Monte Carlo results, these changes are +28.3% (A8) and −19.2% (H5). The current yield from the English aggregated system is 60.9 Megalitres/day (Mld). Based on the traditional approach and according to all scenarios, the future yield will vary between 55.9 Mld and 67.5 Mld. However, based on the Monte Carlo approach, future yield is likely to show much greater variations − between 49.2 Mld and 78.1 Mld. The mean yield changes (of 1000 yield estimates) vary from −10.9% to −3.7% for the HadCM3 scenarios.

Conclusions

Two important conclusions can be drawn from this study as follows:

1. A Monte Carlo simulation approach to climate change water resources impacts assessments is more comprehensive than the traditional 'single' records approach. This is because Monte Carlo simulation experiments enable the 'popu-

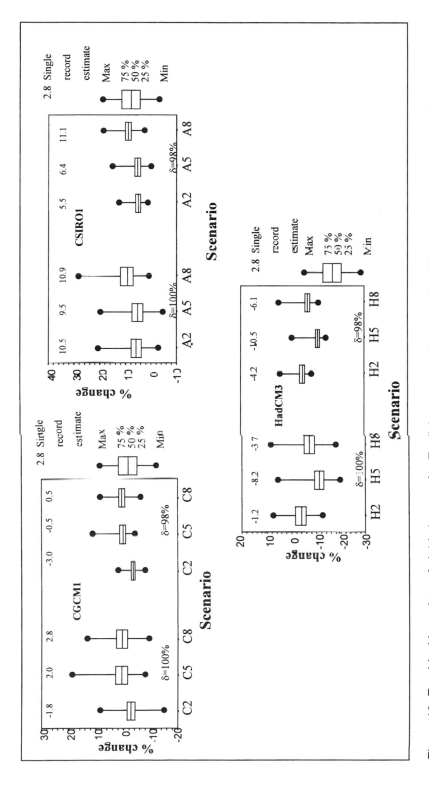

Figure 19. Empirical box plots of yield changes for English aggregated system for different climate scenarios (δ = time-based reliability; C = CGCM1; A = CSIRO1; H = HadCM3; 2 = 2020s; 5 = 2050s; 8 = 2080s).

181

lation of impacts' to be obtained as opposed to the single realisation of impact possible with traditional methods of assessment. These impacts can then be subject to standard statistical analysis to determine the probability attached to a specific 'impact' for instance.

2. Climate change water resources impacts are highly uncertain because of differences in GCM projections. While all the three GCMs agree on the likely change and direction of future temperature in the English catchments, projections of precipitation changes often vary from one GCM to another. This is a major problem for water resources impacts assessment since precipitation often has a much bigger impact on runoff than evaporation. Of the climate change scenarios used in the English case study, those based on the UK HadCM3 GCM indicate drier future conditions whilst wetter conditions are predicted by the Australian CSIRO1 GCM for the same catchments in England. Consequently, use of scenarios based on different GCMs had led to opposite impacts on the same water resources system. The impacts reported in this research should therefore be viewed as a likely range of projections rather than predictions.

Acknowledgements

We would like to thank Dr Jason Ball of Yorkshire Water Services Ltd., UK for providing relevant data. Thanks are also due to Dr Chong-Yu Xu of Uppsala University, Sweden, the British Atmospheric Data Centre (BADC), UK Climate Impacts Programme (UKCIP), Climate Research Unit (CRU) at the University of East Anglia, UK, Canadian Climate Centre, and the Australian Commonwealth Scientific & Industrial Research Organisation (CSIRO). The first author would also like to gratefully acknowledge the financial support, in the form of a scholarship, from Heriot-Watt University.

References

Adeloye, A.J. and Montaseri, M. (1998). "Adaptation of a single reservoir technique for multiple reservoir storage-yield-reliability analysis". In: *Water: a looming crisis? Proc. Int. Conf. on world water resources at the beginning of the 21st Century, UNESCO*, edited by Zebidi, H., Paris, 349-355.

Buishand, T.A. and Beckmann, B.R. (2000). "Development of daily precipitation scenarios at KNMI". Paper presented at the ECLAT-2 Blue Workshop (EW-3): Climate scenarios for water related and coastal impact, a concerted action towards the improved understanding and application of results from climate model experiments in European climate change impact research, May 10-12, 2000.

Carter, T.R., Parry, M.L., Nishioka, S. and Harasawa, H. (1994). "Technical guidelines for assessing climate change impacts and adaptations". Intergovernmental Panel on Climate Change, University College London, Centre for Global Environmental Research, Tsukuba.

Carter, T.R., Hulme, M. and Lal, M. (1999). "Guidelines on the use of scenario data for climate impact and adaptation assessment", version 1. Intergovernmental Panel on Climate Change, Task Group on Scenarios for Climate Impact Assessment.

CCIRG (Climate Change Impacts Review Group) (1996). *Review of the potential effects of climate change in the United Kingdom*, Second Report. Department of Environment, London, HMSO.

Chiew, F.H.S. and McMahon, T.A. (1994). "Application of the daily rainfall-runoff model MODHYDROLOG to 28 Australian catchments". *J. Hydrology*, 153(1-4), 383-416.

Gleick, P.H. (2000). *The World's Water 2000-2001*. Island Press, Washington D.C.

Hulme, M. (1996). *The 1996 CCIRG scenario of changing climate and sea level for the United Kingdom*, Climate Impacts LINK technical note 7. Climatic Research Unit, University of East Anglia, Norwich.

Hulme, M., Raper, S.C.B. and Wigley, T.M.L. (1995). "An integrated framework to address climate change (ESCAPE) and further developments of the global and regional climate modules (MAGICC)". *Energy Policy*, 23(4/5), 347-355.

Hulme, M., Mitchell, J., Ingram, W., Lowe, J., Johns, T.C., New, M. and Viner, D. (1999). "Climate change scenarios for global impacts studies". *Glob. Environ. Change*, 9, S3-S19.

IPCC (1990). *Climate change: The IPCC Scientific Assessment*, edited by Houghton, J., Jenkins, G. and Ephraums, J. Cambridge University Press, Cambridge.

IPCC (1992). *Climate change 1992: The Supplementary Report to the IPCC Scientific Assessment*, edited by Houghton, J.T., Callander, B.A. and Varney, S.K. Cambridge University Press.

IPCC (2001). *Climate change 2001: The scientific basis, contribution of working group I to the Third Assessment Report of the Intergovernmental Panel on Climate Change (IPCC)*, edited by Houghton, J.T., Ding, Y., Griggs, D.J., Noguer, M., van der Linden, P.J. and Xiaosu, D. Cambridge University Press.

Lele, S.M. (1987). "Improved algorithms for reservoir capacity calculation incorporating storage-dependent losses and reliability norm". *Water Resour. Res.*, 23(10), 1819-1823.

MWP (Ministry of Water and Power, Iran) (1995). *Reservoir systems project at Urmia Region* (translated from Persian).

Racsko, P., Szeidl, L. and Semenov, M. (1991). "A serial approach to local stochastic weather models". *Ecol. Modelling*, 57, 27-41.

Reungoat, A. (2000). "Rainfall-runoff modelling with MODHYDROLOG", unpublished MSc Thesis, Heriot-Watt University, Edinburgh, UK.

Semenov, M.A. and Barrow, E.M. (1997). "Use of a stochastic weather generator in the development of climate change scenarios". *Climatic Change*, 35(4), 397-414.

Semenov, M.A. and Porter, J.R. (1994). "The implications and importance of non-linear responses in modelling of growth and development of wheat". In: *Predictability and non-linear modelling in natural sciences and economics,* edited by Grasman, J. and van Straten, G., Wageningen.

Shaw, E.M. (1994). *Hydrology in practice*. Chapman and Hall, London.

Thomas, H.A. and Burden R.P. (1963). *Operations research in water quality management*. Harvard University Press, Cambridge, Mass.

Valencia, D. and Schaake, J.C. (1973). "Disaggregation processes in stochastic hydrology". *Water Resour. Res.*, 9(3), 580-585.

Vandeweile, G.L. and C-Y. Xu (1991). "Regionalisation of physically-based water balance models in Belgium, application to ungauged catchments". *Water Resour. Managt*, 5, 199-208.

Vandewiele, G.L. and Ni-Lar-Win (1998). "Monthly water balance models for 55 basins in 10 countries", *Hydrological Sci. J.*, 43(5).

Wilks, D.S. (1992). "Adapting stochastic weather generation algorithms for climate change studies". *Climatic Change*, 22, 67-84.

Wood, A.W., Lettenmaier, D.P. and Palmer, R.N. (1997). "Assessing climate change implications for water reservoirs planning". *Climate Change*, 37, 203-228.

Xu, C-Y., Seibert, J. and Halldin, S. (1996). "Regional water balance modelling in the Nopex area: Development and application of monthly water balance models", *J. Hydrology*, 180(1-4), 211-236.

Yorkshire Water RRDY (1991). *Re-assessment of resource design yield (RRDY) project 3: Calder area*, Technical report, 1, prepared by W.S. Atkins Ltd, England.

Surface Water Hydrology, Singh, Al-Rashed & Sherif (eds)
© 2002 Swets & Zeitlinger, Lisse, ISBN 90 5809 363 8

Climate conditions, moisture and lakes of south-eastern Kazakhstan in the late Pleistocene and Holocene

Aleksey V. Babkin
State Hydrological Institute
23 Second Line, St. Petersburg, 199053 Russia
e-mail: abav@mail.ru

Abstract

The region of south-eastern Kazakhstan is an arid and semi-arid area of internal flow which embraces the catchment areas of several large closed lakes. In geo-logical past the areas of the lakes were significantly variable. Climate and moisture conditions of lakes existence in present and past epochs were modeled with using the method of joint solution of heat and water balance equations of water surface and landscapes of the basin. The nomograph of dependence of area of the lakes from precipitation and area of glaciation in the basin was computed. It was shown that in the time of lake transgressions of late the Pleistocene and Holocene the desert and semi-desert landscapes of south-eastern Kazakhstan were changed for the dry steppe.

Introduction

The land surface of the Earth is subdivided into areas of internal (endorheic) and external (exorheic) flow. The areas of external flow occupy around 80% of land surface. As a rule, these territories have favorable humidity conditions. Precipitation is sufficient to form the surface run off. Rivers of these areas discharge their water into the World ocean.

The endorheic areas because of the large distance of their location from oceans, forms of relief of continents or peculiarities of atmosphere circulation are mainly arid and semi-arid territories. There are low norms of precipitation and poor river systems here. Very often the rivers form their flow in mountain areas. On the plains the transit river water is lost for evaporation and taken away for irrigation. The surface run off is accumulated in the enclosed water bodies, seas or lakes not connected with ocean. Moisture conditions of internal flow areas are reflected in general by levels of enclosed water bodies.

The lack of water resources impedes the development of economy of many countries situated within the endorheic areas. Among all continents the largest areas of internal flow 12,3 million km^2 are located in Asia. The Asian endorheic regions are the Central internal flow area, the Arabian Peninsular, the Tar desert and the Central Anatolian area with the Dead Sea basin (Alyushinskaya, 1974).

During the geological past the relief, climate, moisture conditions and flow systems of the Earth were constantly transformed. The large enclosed water bodies appeared and dried up. But the direct data about changes of surface flow systems and humidity of different regions is not too much. Their estimations permit to reconstruct climatic conditions only for periods of late Pleistocene and Holocene.

The aim of this work is the reconstruction of climatic and moisture conditions of lakes existence of the south-eastern Kazakhstan, large region of Asian Central internal flow area in the late Pleistocene and Holocene. The approach based on the solution of heat and water balance equations of water surface of lakes and land of their catchment areas may be applicable for estimation of climate and moisture conditions of other endorheic areas in the present and past epochs.

Lakes of the South-Eastern Kazakhstan

A few large lakes are located in the vast intermountain depression of south-eastern Kazakhstan. These lakes are divided into two groups by not high watershed inside the Balkhash and Ala-Kul catchment. In the early 1950s, the largest of these lakes, Balkhash and Ala-Kul had the water surface area of 17,030 and 2,455 km^2, respectively. They were close to each other in their levels of 340 m above m.s.l.

Geological and geomorphologic data are testimony to a considerable variability of their areas and levels in the past. By analyzing of their hollows' structure Kurdyukov (1952) distinguished revealing traces of two transgression shore lines, 10 and 60 m above the modern level, outlined by isohypses 350 and 400 m above m.s.l. (Fig. 1). These shore lines were related by him and afterwards specified by Khrustalev and Thernousov (1983) to the periods of the Wurmian glaciation and the Holocene optimum.

During the Wurmian maximum transgression the territory of the four modern lakes Balkhash, Ala-Kul, Sasyk-Kul and Uyaly was occupied by one huge water body Balkhash–Ala-Kul. Its area was more then 100,000 km^2, which was five times as much as the total area of the modern lakes.

During the less considerable Holocene optimum transgression the two isolated lakes existed: Balkhash of 35,400 and Sasyk–Ala-Kul of 6,250 km^2 in water surface area.

According to Shnitnikov (1976) the area of Balkhash–Ala-Kul drainage basin is equal to 501,000 km^2. Approximately 0,33 part of the basin area is occupied by mountains. There is some glaciation in the mountain part of the Balkhash–Ala-Kul basin, which modern area doesn't exceed 2,000 km^2.

Glaciation of any part of endorheic drainage basin causes the increase of level and area of lake because run off from glacier is always larger then run off from any other iceless surfaces. The glacier in the basin of endorheic lake is an independent factor which influences to the lake area and level.

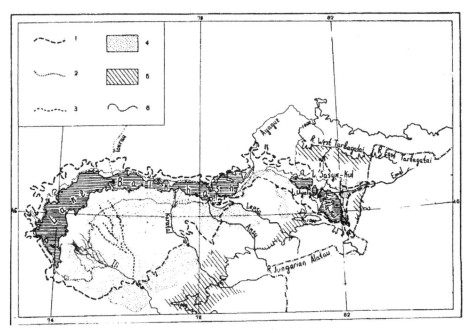

1 – *Wurmian Glasiation lakes water surface, 2 – Holocene optimum lakes water surface, 3 – Dried river–beds of River Ili, 4 – Sanded territories, 5 – Loesses, 6 – 1000 m elevation.*

Figure 1. Paleo lake basins of south-eastern Kazakhstan (Kurdyukov, 1952).

Climate and Moisture Conditions of the Balkhash–Ala-Kul Basin

The present study deals with reconstruction of climatic and moisture conditions of the Balkhash–Ala-Kul basin by the method of joint solution of its heat and water balance equations. This method allows expressing the lake surface area through the areas of different landscapes of the rest part of the basin and their climatic characteristics.

The water body drainage basin is subdivided into three landscape types successively alternating each other: glacier, iceless land and lake water surface. According to Kutzbach (1980) the joint equation of the heat and water balances for the endoheic basin may be presented as follows:

$$LP = \frac{1}{1+B_w}R_w \cdot a_w + \frac{1}{1+B_b}R_b \cdot a_b + LE_g \cdot a_g \qquad (1)$$

where: P is the mean annual basin–average precipitation; L is the evaluated as 0,077 W m^{-2} mm^{-1} annum latent heat of evaporation; R_w, R_b, B_w, B_b are annual values of radiation balance and Bowen ratio of the lake and the iceless basin respectively; E_g is the evaporation from glacier; a_w, a_b, a_g are parts of the lakes area, iceless land area and glacier area from total area of the Balkhash–Ala-Kul basin so that

187

$$a_w + a_b + a_g = 1 \qquad (2)$$

The works devoted to modern radiation, heat and water balances of the largest lakes of the world are generalized by Adamenko (1985). The Balkhash lake radiation balance R_w and Bowen ratio B_w are equal to 99,5 W m^{-2} and 0,3 respectively. The radiation balance and the Bowen ratio of the land, R_b and B_b, written out from charts, composed by Budyko (1971), are equal to 50 W m^{-2} and 2,5. These values of climatic characteristics presuppose evaporation from lake surface and basin 1000 and 185 mm annum^{-1} respectively.

Krenke (1982) showed, that the mean annual evaporation from glacier is equal to 0.

To calculate by the Formula (1) it is necessary to take into account relationship between the glacier area and the climatic characteristics of lake and iceless basin. Let us assume that with increasing of glaciation area the values of climatic characteristics of lake and iceless land should approach step by step to climatic characteristics of glacier. For the approximate estimation it is possible to suppose the linear dependences:

$$R_w = R'_w - a_g(R'_w - R_g) \qquad (3)$$

$$R_b = R'_b - a_g(R'_b - R_g) \qquad (4)$$

$$B_b = B'_b - a_g(B'_b - B_g) \qquad (5)$$

where R'_w, R'_b are the radiation balance of the lakes and iceless land when glacier is absent ($a_g = 0$), B'_b is the Bowen ratio of land when glacier is absent. R_g, B_g are annual values of radiation balance and Bowen ratio of glacier, assumed equal to 0. The lake Bowen ratio B_w assumes independent from glacier constant.

The other relationship, which should take into account, is dependence of iceless land evaporation from precipitation. Main properties of this dependence were considered by Budyko (1971). He showed that this relationship is satisfactory described by the formulas of Shrieber and Oldekop. This dependence may be expressed by the iceless land Bowen ratio. For modern value of glaciation ($a_g = 0,003$) the iceless land Bowen ratio presents as follows:

$$B_b = \frac{R_b - LE_b}{LE_b} \qquad (6)$$

where: E_b is evaporation from land.

The Bowen ratio of the iceless land is calculated with account of relationship between evaporation and precipitation expressed by the Oldekop formula:

$$E_b = k\bar{E} \, th\frac{P}{\bar{E}} \qquad (7)$$

where: $\overline{E}$ is potential rate of evaporation from iceless land, $k = 0,87$ is the correctional coefficient which produces the same accordance between the Bowen ratio of the land and precipitation as Formula (1) for modern values of climatic characteristics and landscape areas of the Balkhas–Ala-Kul basin.

Fig. 2 illustrates the dependence of evaporation estimated by the Shrieber formula (curve 1), Oldekop formula (curve 2) and by Equation (7) (curve 3) from precipitation. If the moisture is insufficient the estimates of evaporation by the Equation (7) are smaller then calculated by the Shrieber and Oldecop formulas. When the moisture is sufficient or abundant the Equation (7) gives the intermediary values than estimated by that two formulas.

Fig. 3 illustrates the dependence of the lakes area of the Balkhash–Ala-Kul basin from glaciation area and precipitation. The isolines of the equal lakes area are shown by dotted lines: the modern lakes area ($a_w = 0,041$), the Holocene optimum lakes area ($a_w = 0,083$), the Wurmian glaciation lake area ($a_w = 0,204$).

Now when $a_w = 0,041$ and $a_g = 0,003$ the mean precipitation in the Balkhash–Ala-Kul basin is equal to 220 mm annum^{-1}.

Presupposing that during the Holocene climatic optimum $a_g \rightarrow 0$, for this period mean precipitation is equal to 330 mm annum^{-1}. During the Holocene climatic optimum in the Balkhash–Ala-Kul region precipitation exceeded its modern value by 110 mm annum^{-1}. This magnitude of precipitation increase consents with the data of Budyko et al. (1991), where it was noted that during the Holocene climatic optimum precipitation exceeded its modern values by 100–150 mm annum^{-1}.

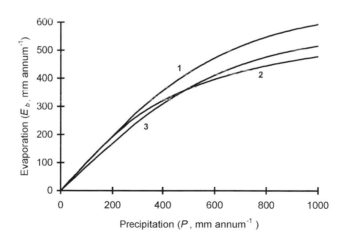

Figure 2. Possible dependence between evaporation from land and precipitation onto the Balkhash–Ala-Kul drainage basin. Estimates made by formulas:
1 – Oldekop, 2 – Shrieber, 3 – this paper Formula 7.

189

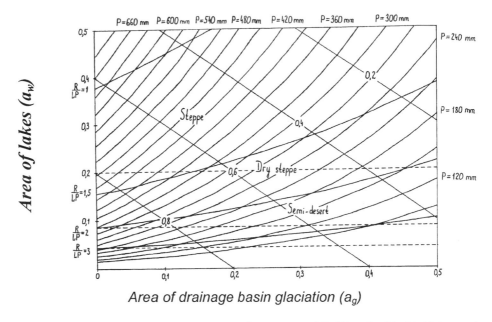

Figure 3. Relationship between water surface area of Balkhash–Ala-Kul lakes, glaciation area and mean annual precipitation.

Paleoclimatical reconstructions of Borzenkova (1992) pointed out that during the Wurmian glaciation mean precipitation exceeded its modern value by 50 mm annum^{-1}. It means (Fig. 3) that during existence of the huge Balkhash–Ala-Kul water body the sufficient area of glaciation was 20–25% of the drainage basin area. This glaciation area was smaller than mountain area and really was covered by glacier.

In order to estimate the scale of climatic changes needed by the lakes' area of the Balkhash–Ala-Kul to reach one or another value one should calculate the radiation dry index of its basin $R_b(LP)^{-1}$ by dividing of Equation (1) right and left parts to R_b.

Fig. 3 shows values of radiation dry index of iceless land determined by Budyko (1971). Now the radiation dry index is close to its value on the boundary of the desert and semi-desert natural zones. This value consents with physic-geographical description of the Balkhash–Ala-Kul basin of Shnitnikov (1976).

During the Holocene optimum transgression radiation dry index was approximately equal to its value on the semi-desert and steppe zones boundary. During the Wurmian glaciation the maximal transgression was when the dry steppe dominated at the iceless land.

Conclusion

Modeling of climate conditions of the south-eastern Kazakhstan lakes existence showed that in the time of the Wurmian glaciation the precipitation exceeded its modern value by 50 mm annum^{-1} while glaciers covered up to 25% of basin area.

During the Hololcene climatic optimum precipitation exceeded its modern value by 110 mm annum^{-1}. If climatic conditions of Holocene optimum are considered as analog of future climate (Budyko et al., 1991), it may be supposed that the region of the south-eastern Kazakhstan will be less arid.

A considerable increase of water surface area of Balkhash–Ala-Kul lakes could occur due to insignificant changes of climatic conditions. So the huge late Pleistocene Balkhash–Ala-Kul water body could exist under climatic conditions when desert-semi-desert landscape in its basin was changed by the dry steppe.

References

Adamenko, V.N. (1985). *Climate and lakes*. Gidrometeoizdat, Leningrad. 263 pp. (in Russian).

Alyushinskaya N.M. (1974). "Water balance of Asia". In: *World water balance and water resources of the Earth*. Gidrometeoizdat, Leningrad. 227–241 (in Russian).

Borzenkova, I.I. (1992). *Climatic change in the Cenozoic*. Gidrometeoizdat, St. Petersburg. 248 pp. (in Russian).

Budyko, M.I. (1971). *Climate and life*. Gidrometeoizdat, Leningrad. 491 pp. (in Russian).

Budyko, M.I. *et al*. (1991). *Forthcoming climate change*. Gidrometeoizdat, Leningrad. 271 pp. (in Russian).

Khrustalev, Iu.P., Tchernousov, S.Ya. (1983). "Main steps of lake Balkhash development in Holocene". *Papers of the USSR Academy of Science*. Vol. 291, No. 6, 1468–1491 (in Russian).

Krenke A.N. (1982). *Mass-exchange in glacier systems over the territory of the USSR*. GIMIZ, Moscow. 288 pp. (in Russian).

Kurdyukov, K.V. (1952). "Paleo lake basins of the south-eastern Kazakhstan and the climatic factors of their temporal existence". *Proc. of the USSR Academy of Science*. Ser. geographical, No. 2, 11–24 (in Russian).

Kutzbach, I.E. (1980). "Estimates of past climate at paleolake Chad, North Africa, based on a hydrological and energy balance model". *Quaternary Research*, Vol. 14, No. 2, 213–223.

Shnitnikov, A.V. (1976). "Lakes of Middle Asia (historical variability and modern state)". In: *Large lakes of Middle Asia and some ways of their managment*. Nauka, Leningrad. 15–133 (in Russian).

Section 3: Rainfall

Surface Water Hydrology, Singh, Al-Rashed & Sherif (eds)
© *2002 Swets & Zeitlinger, Lisse, ISBN 90 5809 363 8*

Multifractals and the study of extreme precipitation events: A case study from semi-arid and humid regions in Portugal

M. Isabel P. de Lima
IMAR – Institute of Marine Research, Coimbra Interdisciplinary Centre
Department of Forestry, Agrarian Technical School of Coimbra
Polytechnic Institute of Coimbra, Bencanta, 3040-316 Coimbra, Portugal
e-mail: iplima@mail.esac.pt

Daniel Schertzer
Laboratoire de Modelisation en Mecanique
Pierre & Marie Curie University, 4 Place Jussieu, Paris Cedex 05, France
e-mail: schertze@ccr.jussieu.fr

Shaun Lovejoy
Physics Department, McGill University
3600 University st., Montréal, Québec, H3A 2T8 Canada
e-mail: lovejoy@physics.mcgill.ca

João L.M.P. de Lima
IMAR – Institute of Marine Research, Coimbra Interdisciplinary Centre
Department of Civil Engineering, Faculty of Science and Technology –
Campus 2, University of Coimbra, 3030-290 Coimbra, Portugal
e-mail: plima@dec.uc.pt

Abstract

Precipitation is a highly non-linear hydrologic process that exhibits wide variability over a broad range of time and space scales. This variability involves a large dynamic range, which in certain cases leads to catastrophic events, both related to drought and flood situations. It is thus important to analyze precipitation with methods that have the potential to assess the full range of precipitation fluctuations.

Multifractal theory can play an important role in such studies. Moreover, multifractal methods are innovative for the analysis of extremes, offering practical tools for assessing correctly their probability of occurrence.

This work presents results of the multifractal analysis of precipitation time series from semi-arid and humid regions in Portugal, concentrating on the statistics of extreme events. The data were recorded daily over a period of more than 50 years.

Introduction

Precipitation is the hydrological process having perhaps the greatest impact on everyday life. Because precipitation is the driving agent of many other processes, its temporal and spatial variability are important issues in many studies and areas of research (e.g. hydrology, hydraulics, agronomy, soil pollution, water resources). This highly non-linear hydrologic process exhibits wide variability over a broad range of time and space scales. Such variability involves a large dynamic range, which in certain cases leads to catastrophic events, both related to drought and flood situations.

The study of precipitation, and the requirements of hydrological and engineering applications, have led to many different approaches to analyzing and modeling temporal precipitation, and to the development of various design procedures. However, there are some problems related to the study of precipitation because the strongly irregular fluctuations of this process are difficult to capture instrumentally (because of technical limitations of the measuring devices) and to handle mathematically (many approaches fail to grasp the extreme variability of precipitation). It is often necessary to determine precipitation values with very low probability of being exceeded, in particular for the design of major hydraulic structures (e.g. spillways on large dams) or the management of hydrologic systems. This is very important because of the associated high risk to lives and property. It is thus important to analyze precipitation with methods that have the potential to assess the full range of precipitation fluctuations.

Multifractal theory can play an important role in such studies. The multifractal analysis of precipitation is based on the invariance of properties across scales, and it takes into account the persistence of the variability of the process over a range of scales. Multifractal methods are innovative for the analysis of extremes, offering practical tools for assessing correctly their probability of occurrence. One of the most remarkable predictions of multifractals corresponds to the existence of heavy tails (i.e., power-law tails) in the precipitation intensity probability distributions.

This work presents results of the multifractal analysis of precipitation time series from both semi-arid and humid regions in Portugal, concentrating on the statistics of extreme events. The data were recorded daily over a period of more than 50 years.

Other multifractal precipitation studies are reported by e.g. Ladoy et al. (1991, 1993), Tessier et al. (1992, 1993, 1994, 1996), Hubert (1992, 1995), Hubert and Carbonnel (1993), Hubert et al. (1993), Olsson and Niemczynowicz (1994, 1996), Lima and Bogardi (1995), Olsson (1995, 1996), Svensson et al. (1996), Burlando and Rosso (1996), Over and Gupta (1996), Koutsoyiannis and Pachakis (1996), Harris et al. (1996), Onof et al. (1996), Menabde et al. (1997), Bendjoudi et al. (1997), Lima (1998, 1999), Lima and Grasman (1999), Jothityangkoon et al. (2000).

Some Aspects of Multifractal Theory

Multifractal theory (Hentschel and Procaccia, 1983; Grassberger, 1983; Schertzer and Lovejoy, 1983), which has evolved from fractal theory (Mandelbrot, 1977; 1982), is based on the recognition that the type of variability of processes and systems exists for a range of scales. It is thus a scaling theory, where the term scaling (or scale-invariance) is used to indicate that certain features of a dynamic system are independent of scale. Scaling theories apply to processes and systems without a characteristic scale. Scaling is expected to hold from some large (outer or upper) scale down to a small (inner or lower) scale. This behavior leads to a class of scaling rules (power laws) characterized by scaling exponents. This allows the relationship of variability between different scales to be quantified. Statistical properties of scale-invariant systems at different scales (i.e., on large and small scales) are related by a scale-changing operation that involves only scale ratios. Scaling theories are developed in a non-dimensional framework, because one looks for features that are independent of the physical size of the object of study.

One can use standard spectral methods and analysis to test for scale-invariance. The most familiar consequence of scaling is the power-law behavior that is expected in the energy (power) spectra of scaling processes (e.g. Mandelbrot, 1982; Schertzer and Lovejoy, 1985, 1987; Ladoy et al., 1991; Lovejoy and Schertzer, 1995):

$$E(\omega) \approx \omega^{-\beta} \qquad (1)$$

where ω is the wave-number, $E(\omega)$ is the energy, and β is the spectral exponent. For temporal processes, the wave-number can be approximated by $\omega \sim 1/\tau$, with τ being the magnitude of any time interval. Thus, in this application, ω is a frequency. The energy (power) spectrum is obtained from the Fourier transform of the autocorrelation function, which is a second order moment (see e.g. Press et al., 1989; Haslings and Sugihara, 1993).

One way to investigate the multifractal temporal structure of the precipitation process is to study the (multiple) scaling of the probability distributions of the non-dimensional precipitation intensity, ε_γ (e.g. Schertzer and Lovejoy, 1987). The precipitation intensity threshold level is evaluated with the order of singularity γ of the intensities $\varepsilon_\lambda \sim \lambda^\gamma$ (e.g. Frisch and Parisi, 1985; Halsey et al., 1986; Schertzer and Lovejoy, 1987), where λ is called a scale ratio (it is the ratio between the largest scale of interest and the scale of homogeneity). The scaling of the probability distributions is given by the exponent function $c(\gamma)$:

$$\Pr(\varepsilon_\lambda \geq \lambda^\gamma) \approx \lambda^{-c(\gamma)} \qquad (2)$$

In literature, the function $c(\gamma)$ is called the codimension function (see e.g. Lavallée et al., 1991). Eq. (2) holds for proportionality constants varying slowly with λ and depending weakly on γ (e.g. Schertzer and Lovejoy, 1989; Lovejoy and Schertzer,

1991). This statistical characterization of multifractals arises directly from multiplicative cascade processes (see e.g. Schertzer and Lovejoy, 1987, 1988, 1989, and Lovejoy and Schertzer, 1990).

Another (equivalent) way to investigate the multifractal temporal structure of precipitation is with the statistical moments of the precipitation intensity (Schertzer and Lovejoy, 1987). The scaling of the moments of the precipitation intensity is described by the exponent function $K(q)$. The notion of moment can be generalized to any real value q. The moments scaling function $K(q)$ satisfies

$$\langle \varepsilon_\lambda^q \rangle \approx \lambda^{K(q)} \tag{3}$$

where $<\varepsilon_\lambda^q>$ is the (ensemble) average q^{th} moment of the precipitation on a scale specified by λ.

The two multifractal scaling exponent functions $c(\gamma)$ and $K(q)$ are related through the formulas

$$K(q) = \max_\gamma \{q\gamma - c(\gamma)\} \quad \text{and} \quad c(\gamma) = \max_q \{\gamma q - K(q)\} \tag{4}$$

which are Legendre transform pairs that establish a one-to-one correspondence between orders of singularity γ and statistical moments q (e.g. Frisch and Parisi, 1985). Both scaling exponent functions $c(\gamma)$ and $K(q)$ are (theoretically) non-linear increasing functions (concave functions). However, two distinct statistical mechanisms lead to discontinuities in the empirical multifractal scaling exponent functions, which are associated with special linear forms of these functions. These mechanisms are divergence of moments (above a certain critical order) and finite sampling.

The divergence of moments results from the multifractal singular behavior at the small-scale limit (i.e., $\varepsilon_\lambda \to \infty$ as $\lambda \to \infty$): $<\varepsilon_\lambda^q> = \lambda^{K(q)} \to \infty$, for all moments $q > q_D$ (it is $K(q) > 0$ for $q > 1$). Empirical moments, which are averages of empirical values, are always finite: the divergence of moments means in this case that the empirical moments increase without limit as the sample size increases. This statistical behavior occurs because the sum of independent contributions is determined by the largest of the contributions (i.e., rare events will have dominant contributions).

There is equivalence between the divergence of moments (for $q > q_D$) and the algebraic fall-off of the probability distribution for extreme events (e.g. Feller, 1971; Mandelbrot, 1974; Schertzer and Lovejoy, 1985; Lovejoy and Schertzer, 1985). The tail of the probability law determines the relative frequency of extreme behavior. The slope of this tail is the critical order for divergence of statistical moments, q_D:

$$\Pr(\varepsilon_\lambda > s) \approx s^{-q_D} \tag{5}$$

where s is a sufficiently large intensity-threshold. The smaller the exponent q_D, the more extreme is the fluctuation of the process.

The Scale-Invariant and Multifractal Behaviour of the Temporal Structure of Precipitation: Examples from Portugal

The Precipitation Data from Ponte de Lima and Tavira

The precipitation data used in the scale-invariant analysis were recorded at two locations in Portugal: Ponte de Lima and Tavira. Figure 1 shows the location of the measuring stations in Portugal: Ponte de Lima is located in the north and Tavira in the south of Mainland Portugal. The two measuring stations are separated by approximately 520 km. In the south of Mainland Portugal the climate can be classified as Mediterranean and semi-arid. In the north, where Ponte de Lima is located, the climate is much more humid and cooler, due to the influence of the Atlantic Ocean. The coordinates of these measuring stations are approximately 41°46' N and 8°36' W, for Ponte de Lima, and 37°07' N and 7°39' W, for Tavira. The altitude of the measuring sites is, respectively, 15 m and 25 m above mean sea level.

The precipitation measuring devices are of the 20-14-G type (according to the classification by Sevruk and Klemm, 1989); they have horizontal openings of 200 cm^2 at 1.5 m above the ground surface. The gauges were observed daily. The resolution of the measurements is 0.1 mm of precipitation. Trace precipitation of less than 0.1 mm is disregarded and such days are considered dry (zero-precipitation days).

The precipitation recorded in Ponte de Lima and Tavira, for the years 1941 to 1994, is shown ine Figure 2. Figure 2(a) shows the annual precipitation, Figure 2(b) shows the monthly precipitation and Figure 2(c) shows the monthly average precipitation. Table 1 contains some descriptive statistics of precipitation at the two locations. The precipitation recorded in Ponte de Lima and Tavira illustrates the temporal and spatial variability of precipitation in Portugal. Both locations exhibit a marked seasonal distribution of precipitation during the year. In Ponte de Lima, of the 648 months that constitute this sample, 6 months had less than 0.1 mm (roughly 0.9% of the sample). In Tavira, in the same period, there were 121 months with less than 0.1 mm of precipitation (roughly 18.7% of the sample).

Scaling and Multifractal Analysis of the Daily Precipitation Data

This section presents results of the investigation of scaling and multifractal behavior in the temporal structure of precipitation from Ponte de Lima and Tavira, in Portugal (see previous section). The data were recorded from 1941 to 1994. In the different analyses, given below, the statistics are accumulated for the 54 years covered by the data.

To non-dimensionalize time measurements, one assumes that the duration of the longest period of interest is equal to 1. If this period has a duration T, then the

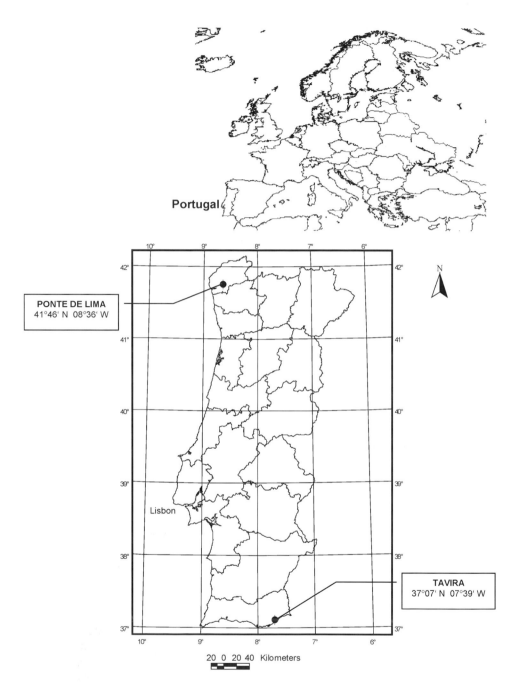

Figure 1. Map of Mainland Portugal localizing the two meteorological stations used in this study (Ponte de Lima and Tavira). On top, positioning of Portugal in Europe.

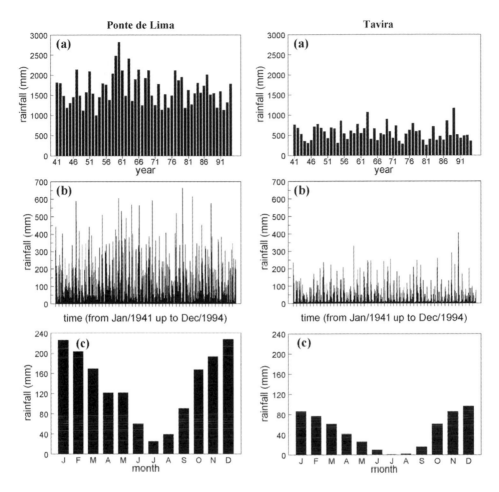

Figure 2. Precipitation (in mm) in Ponte de Lima and Tavira, Portugal, for the years 1941 to 1994: (a) annual precipitation; (b) monthly precipitation; and (c) monthly average precipitation.

magnitude of any time interval τ should be divided by T. Any time scale corresponding to τ can then be characterized by a scale ratio λ, with $1/\lambda = \tau/T$. In practical applications it is common that the scale ratio λ is decreased gradually by a factor of 2. To non-dimensionalize the precipitation intensity on a time scale of resolution λ, the intensity can be divided by the ensemble average intensity of the process. For precipitation this means the climatological average precipitation. Nevertheless, in practice, one generally uses the average intensity of the sample, which corresponds to the largest scale of interest ($\lambda = 1$). Let the (average) precipitation intensity for $\lambda = 1$ be $<R_1>$, where the angular brackets $< >$ mean (ensemble) average. So if the precipitation intensity on a time interval λ^{-1} is R_λ, the corresponding non-dimensional intensity is $\varepsilon_\lambda = R_\lambda / <R_1>$, hence $<\varepsilon_\lambda> = 1$.

Table 1. Descriptive statistics of precipitation data from Ponte de Lima and Tavira (Portugal), for the period 1941-1994.

Type of precipitation data (1941-1994)	Measuring stations	
	Ponte de Lima	Tavira
Mean annual precipitation	1643.6 mm	563.1 mm
Coef. of variation (annual prec.)	0.235	0.347
Precipitation in wettest year	2820.7 mm	1165.8 mm
Precipitation in driest year	994.4 mm	250.2 mm
Average monthly precipitation	137.0 mm	46.9 mm
Coef. of variation (monthly prec.)	0.901	1.279
Precipitation in wettest month	664.1 mm	405.2 mm
Precipitation in driest months	0 mm*	0 mm*
Precipitation in wettest day	137.2 mm	186.0 mm

* Daily precipitation below 0.1 mm is disregarded.

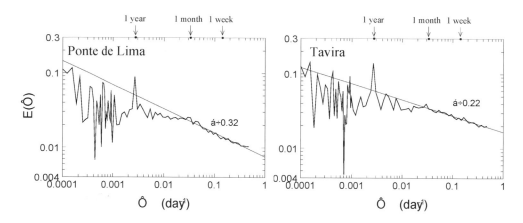

Figure 3. Energy spectra obtained for daily precipitation from Ponte de Lima and Tavira, Portugal, from 1941 to 1994.

The energy spectra obtained for daily precipitation from Ponte de Lima and Tavira are plotted in Figure 3. The spectra have been smoothed for high frequencies. The spectra exhibit power-law behavior, confirming the presence of scale invariance in the temporal structure of precipitation over a wide range of scales. The power-law behavior of both spectra extends from 1 day up to about one-and-a-half months. The spectral exponent β, in Eq. (1), is estimated as 0.32, for the data from Ponte de Lima, and 0.22, for the data from Tavira. Parameters β were estimated from the absolute values of the slopes of the regression lines fitted to the right-hand side scaling regions of the spectra, plotted in log-log axis. The lower magnitude of β found for the data from Tavira indicates the higher variability of precipitation at this location, compared to the behavior of precipitation from Ponte de Lima. The spectral peaks at $\omega \approx 0.0027$ day^{-1} observed in both spectra correspond to the annual cycle frequency.

The scaling of the statistical moments can be tested with log-log plots of the average q^{th} moment of the precipitation intensity ε_λ, observed on scales of different levels of resolution λ, against the scale ratio λ. For the data from Ponte de Lima and Tavira, Figure 4 shows these log-log plots for time scales from 1 day ($\lambda = 256$) up to 8.5 months ($\lambda = 1$). Figure 4(a) shows moments larger than 1 and Figure 4(b) moments smaller than 1. The moments q plotted in Figure 4 are indicated in the legend.

The power-law behavior observed in the moments plots from Ponte de Lima and Tavira confirms the presence of scale invariance in the temporal structure of precipitation over a large range of scales. The scaling range seems to extend from

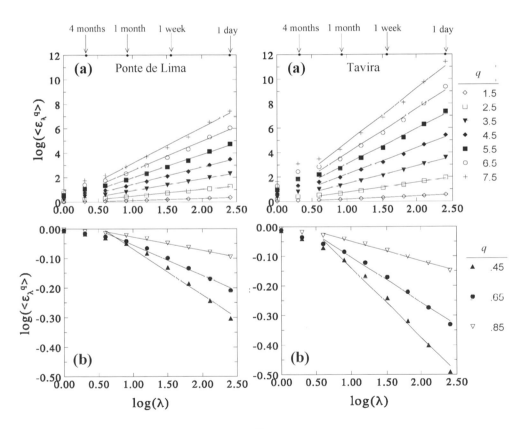

Figure 4. Log-log plots of the average q^{th} moments of the precipitation intensity ε_λ on scales between 1 day ($\lambda = 256$) and 8.5 months ($\lambda = 1$), against the scale ratio λ: (a) for moments larger than 1; and (b) for moments smaller than 1. The plots are for the data from Ponte de Lima and Tavira, Portugal.

one day up to about two months. Moreover, the plots in Figure 4 show the multi-fractality of precipitation through the different scaling exponents associated with the various moments q.

The empirical scaling functions $K(q)$, in Eq. (3), which describe the statistics of precipitation from Ponte de Lima and Tavira, are obtained from the regression lines fitted to the moments plots in Figure 4, over the relevant range of moments q of precipitation intensity. These functions are shown in Figure 5.

The different behavior observed in the scaling of the moments, for the data from Ponte de Lima and Tavira, indicates the higher non-linear variability present in the temporal structure of precipitation from Tavira. The two empirical $K(q)$ functions in Figure 5 have both non-linear and linear sections. The non-linear sections are concave increasing functions. The linear sections in these functions correspond to a special type of statistical behavior that can be explained by divergence of moments above a certain critical order (i.e., for all moments $q > q_D$). This divergence of moments, which would only be observed for infinite sample sizes, is empirically associated with discontinuities in the empirical multifractal scaling functions. In Figure 5, arrows indicate the estimates of the critical moment q_D obtained from the analysis of the precipitation intensity probability distribution functions (see, below, Figure 6), which agree with the estimates of q_D obtained from the empirical $K(q)$ functions.

The lower critical order moment, q_D, observed for the data from Tavira, shows empirically that precipitation at this location is characterized by stronger irregular fluctuations. The slopes of the linear sections of the moments scaling functions, for moments $q > q_D$, give estimates of the largest orders of precipitation singularity γ_{max} that are present in the two finite samples: 0.75, for Tavira, and 0.52, for Ponte de Lima.

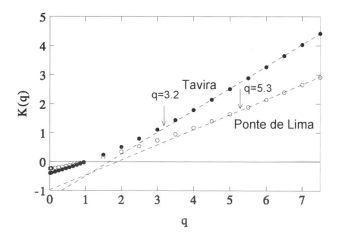

Figure 5. Empirical moments scaling function (dotted lines), determined with daily precipitation from Ponte de Lima and Tavira (Portugal) recorded from 1941 to 1994, for the range of scales from 1 day up to 64 days.

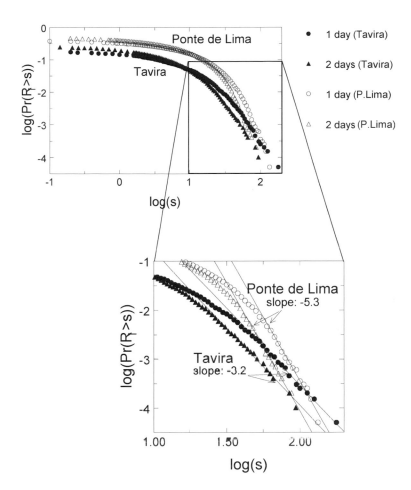

Figure 6. Empirical probability distributions of precipitation on time scales of 1 and 2 days, for the data from Ponte de Lima and Tavira, Portugal. Precipitation intensities are expressed in mm/day.

The study of the empirical functions $K(q)$ near $q = 0$ yields estimates of the codimension $c(\gamma_{min}) = -K(0)$ of 0.40, for the data from Tavira, and 0.24, for the data from Ponte de Lima. These values are related to a minimum non-zero observation of the precipitation process, of singularity γ_{min}. Being the fractal dimension of the geometric structure that is the "support" of the precipitation process $D = 1 - c(\gamma_{min})$, estimates of D are 0.60, for Tavira (in the semi-arid region), and 0.76, for Ponte de Lima (in the humid region). These results are consistent with the fact that, within the time domain, precipitation occurrences in Tavira is sparser than in Ponte de Lima, which can be related to recurrent droughts in that region, associated with a desertification process.

Figure 6 shows examples of probability distribution functions of precipitation on different scales (1 and 2 days), for the data from Ponte de Lima and Tavira. These functions exhibit algebraic tails (heavy tails), which indicates divergence of statistical moments for moments $q > q_D$. The critical order moment q_D, in Eq. (5), is given by the absolute value of the slope of the algebraic tail. The absolute values of the slopes of the regression lines fitted to the tails of the empirical probability distributions are 5.3, for the data from Ponte de Lima, and 3.2, for the data from Tavira. For the two data sets, the values of the moments q_D, estimated in this way, agree with the behavior of the empirical functions $K(q)$ (see Figure 5). The discontinuities in these functions are observed at roughly the same critical moments.

The different values obtained for the critical moment q_D (5.3 and 3.2, respectively, for Ponte de Lima and Tavira), suggest that there are differences in the variability of precipitation, depending on the climate. The lower critical order for divergence of moments, obtained for the data from the semi-arid region, indicates stronger fluctuations in precipitation intensity and the presence of larger singularities (i.e., intensities) in the process. The probability of occurrence of certain high-intensity precipitation events is greater for Tavira than for Ponte de Lima, whereas for low-intensity events the opposite is observed.

Concluding Remarks

In this study, analyses of daily precipitation data from two measuring stations in Portugal, one in a semi-arid region and the other in a humid region, show that the temporal structure of precipitation exhibits multifractal properties across a wide range of scales. Scaling was observed to hold from 1 day up to approximately 2 months. Other studies have shown that the scaling and the multifractal behavior of precipitation extend down to scales of the order of minutes (e.g. Olsson, 1995; Lima, 1998, 1999; Lima and Grasman, 1999).

Results show that scale invariant and multifractal analyses are able to quantify differences in precipitation variability. In this case study, the data from the semi-arid region exhibited higher variability than the data from the humid region. For the two data sets, the precipitation intensity probability distributions showed heavy tails (i.e., power-law tails). The behavior of the tail of the probability distributions is important for studying extreme values. Its slope gives empirically the critical order for divergence of moments. The absolute values of the slopes of the heavy tails of the precipitation intensity probability distributions were approximately 3 for the data from the semi-arid region, and about 5 for data from the humid climate.

Comparison of the probability distributions of precipitation at the two locations in Portugal shows that the probability of exceeding certain very large precipitation events is greater in the semi-arid region than in the humid region, although the annual precipitation in the humid region is much higher. Moreover, heavy tails indicate that the probability of exceeding certain precipitation events is much greater than the probability predicted by more "traditional" models (see e.g. Ladoy et al., 1991). These facts should be taken into account in engineering studies and design procedures. Also, extrapolation of hydrologic behavior from areas of different climatic conditions will most certainly fail, especially, which concerns low-probability extreme events.

Analysis of the dependency of the multifractal behavior on climatological and geographical factors (i.e., on precipitation-generating mechanisms) is essential and should be investigated further. It requires a systematic study of precipitation from different origins.

Acknowledgments

This study was funded by the Foundation for Science and Technology (Research Project FCT – PRAXIS/C/ECM/12018/1998) of the Portuguese Ministry of Science and Technology, Lisbon, Portugal. Maria de Fátima E.S. Coelho, from the Portuguese Meteorological Institute, in Lisbon, provided the precipitation data, within the referred research project.

References

Bendjoudi, H., Hubert, P., Schertzer, D., and Lovejoy, S. (1997). "Multifractal point of view on rainfall intensity-duration-frequency curves", *C. R. Acad. Sci. Paris*, Série II, Fasc. A, 325 (5), 323-326.

Burlando, P., and Rosso, R. (1996). "Scaling and multiscaling models of depth-duration-frequency curves for storm precipitation". *Journal of Hydrology*, 187(1-2), 45-64.

Feller, W. (1971). *An introduction to probability theory and its applications*. John Wiley & Sons, New York, USA, vol. II (second edition).

Frisch, U., and Parisi, G. (1985). "Fully developed turbulence and intermittency". In: Ghil, M., Benzi, R., and Parisi, G. (eds.) *Turbulence and predicability in geophysical fluid dynamics and climate dynamics*, Proc. of International School of Physics "Enrico Fermi", Course LXXXVIII, Italian Physical Society, North-Holland, Amsterdam.

Grassberger, P. (1983). "Generalized dimensions of strange attractors", *Physics Reviews Letters*, 97A(6), 227-230.

Halocy, T.C., Jensen, M.H., Kadanoff, L.P., Procaccia, I., and Shraiman, B.I. (1986). "Fractal measures and their singularities: The characterization of strange sets", *Physics Reviews Letters*, 33A(2), 1141-1151.

Harris, D., Menabde, M., Seed, A., and Austin, G. (1996). "Multifractal characterization of rain fields with a strong orographic influence", *Journal of Geophysical Research*, 101(D21), 26405-26414.

Hastings, H.M., and Sugihara, G. (1993). *Fractals: A user's guide for the natural sciences*, Oxford University Press, Oxford, U.K.

Hentschel, H.G.E., and Procaccia, I. (1983). "The infinite number of generalized dimensions of fractals and strange attractors", *Physica*, 8D, 435-444.

Hubert, P. (1992). "Analyse multifractale de champs temporels d'intensité des précipitations", *Rencontres hydrologiques Franco-Roumaines*, UNESCO, Paris, 3-6 September, 1991, SC/92/WS/48, 379-386.

Hubert, P. (1995). "Fractals et multifractals appliqués à l'étude de la variabilité temporelle des précipitations". In: Feddes, R.A. (ed.), *Space and time scale variability and interdependencies in hydrological processes*, Cambridge University Press, U.K., 175-181.

Hubert, P., and Carbonnel, J.-P. (1993). "Analyse multifractale et précipitations extremes". *Proceedings of the Sixth Scientific Assembly of the International Association of Meteorology and Atmospheric Physics and Fourth Scientific Assembly of the International Association of Hydrological Sciences*, Yokohama, Japan, July 11-23, 1993.

Hubert, P., Tessier, Y., Lovejoy, S., Schertzer, D., Schmitt, F., Ladoy, P., Carbonnel, J.P., Violette, S., and Desurosne, I. (1993). "Multifractals and extreme rainfall events", *Geophysical Research Letters*, 20(10), 931-934.

Jothityangkoon, C., Sivapalan, M., and Viney, N.R. (2000). "Tests of a space-time model of daily rainfall in southwestern Australia based on nonhomogeneous random cascades", *Water Resources Research*, 36(1), 267-284.

Koutsoyiannis, D., and Pachakis, D. (1996). "Deterministic chaos versus stochasticity in analysis and modeling of point rainfall series". *Journal of Geophysiscal Research*, 101(D21), 26441-26451.

Ladoy, P., Lovejoy, S., and Schertzer, D. (1991). "Extreme variability of climatological data: scaling and intermittency". In: Schertzer, D., and Lovejoy, S. (eds.), *Non-linear variability in Geophysics: scaling and fractals*, Kluwer Academic Publishers, The Netherlands, 241-250.

Ladoy, P., Schmitt, F., Schertzer, D., and Lovejoy, S. (1993). "Variabilité temporelle multifractale des observations pluviométriques à Nîmes", *C.R. Acad. Sci. Paris*, Série II, 317, 775-782.

Lavallée, D., Schertzer, D., and Lovejoy, S. (1991). "On the determination of the codimension function". In: Schertzer, D., and Lovejoy, S. (eds.), *Non-linear variability in Geophysics: scaling and fractals*, Kluwer Academic Publishers, The Netherlands, 99-110.

Lima, M.I.P. de (1998). *Multifractals and the temporal structure of rainfall*. Ph.D. Thesis, Wageningen Agricultural University, Wageningen, The Netherlands.

Lima, M.I.P. de (1999). "A fractal and multifractal study of hourly precipitation time-series from The Netherlands". In: V.P. Singh, II Won Seo and J.H. Sonu (eds.), *Hydrologic Modeling*, Water Resources Publications, LLC, USA, 15-28.

Lima, M.I.P. de, and Bogardi, J.J. (1995). "Multifractals in hydrological studies: the analysis of rainfall time series", *Proc. of the UNESCO International Conference Statistical and Bayesian Methods in Hydrological Sciences*, September 11-13, 1995, UNESCO, Paris, France, Vol. II.

Lima, M.I.P. de and Grasman, J. (1999). "Multifractal analysis of 15-min and daily rainfall from a semi-arid region in Portugal". *Journal of Hydrology*, Elsevier, 220: 1-11.

Lovejoy, S., and Schertzer, D. (1985). "Generalized scale invariance in the atmosphere and fractal models of rain". *Water Resources Research*, 21(8), 1233-1250.

Lovejoy, S., and Schertzer, D. (1990). "Multifractals, universality classes and satellite and radar measurements of cloud and rain fields", *Journal of Geophysical Research*, 95(D3), 2021-2034.

Lovejoy, S., and Schertzer, D. (1991). "Multifractal analysis techniques and the rain and cloud fields from 10^{-3} to 10^6 m". In: Schertzer, D., and Lovejoy, S. (eds.), *Non-linear variability in Geophysics: scaling and fractals*, Kluwer Academic Publishers, The Netherlands, 111-144.

Lovejoy, S., and Schertzer, D. (1995). "Multifractals and rain". In: Kunzewicz, Z.W. (ed.), *New Uncertainty Concepts in Hydrology and Water Resources*, UNESCO series in Water Sciences, Cambridge University Press, New York, 62-103.

Mandelbrot, B. (1974). "Intermittent turbulence in self-similar cascades: divergence of high moments and dimension of the carrier". *Journal of Fluid Mechanics*, 62, part II: 331-358.

Mandelbrot, B. (1977). *Fractals: form, chance and dimension*, Freeman, San Francisco.

Mandelbrot, B. (1982). *The fractal geometry of nature*, Freeman, San Francisco.

Menabde, M., Harris, D., Seed, A., Austin, G., and Stow, D. (1997). "Multiscaling properties of rainfall and bounded random cascades". *Water Resources Research*, 33(12), 2823-2830.

Olsson, J. (1995). "Limits and characteristics of the multifractal behaviour of a high-resolution rainfall time series". *Nonlinear Processes in Geophysics*, 2(1), 23-29.

Olsson, J. (1996). "Validity and applicability of a scale-independent, multifractal relationship for rainfall". *Journal of Atmospheric Research*, 42, 53-65.

Olsson, J., and Niemczynowicz, J. (1994). "Multifractal relations in rainfall data. In: Kettunen, J., Granlund, K., Paasonen-Kivekäs, M., and Sirviö, H. (eds.), *Spatial and temporal variability and interdependencies among hydrological processes*,

Proc. of Nordic Seminar, September 14-16, 1994, Kirkkonummi, Finland. NHP Report No. 36, 110-119.

Olsson, J., and Niemczynowicz, J. (1996). "Multifractal analysis of daily spatial rainfall distributions". *Journal of Hydrology*, 187(1-2), 29-43.

Onof, C., Northrop, P., Wheater, H.S., and Isham, V. (1996). "Spatiotemporal storm structure and scaling property analysis for modelling". *Journal of Geophysical Research*, 101(D21), 26415-26425.

Over, T. M., and Gupta, V. K. (1996). "A space-time theory of mesoscale rainfall using random cascades". *Journal of Geophysical Research*, 101(D21), 26319-26331.

Press, W.H., Flannery, B.P., Teukolsky, S.A., and Vetterling, W.T. (1989). *Numerical recipes, The art of scientific computing* (Fortran version). Cambridge University Press, Cambridge, UK.

Schertzer, D., and Lovejoy, S. (1983). "Elliptical turbulence in the atmosphere, *Proc. of Fourth Symposium on Turbulent Shear Flows*, Karlshule, West Germany, 11.1-11.8.

Schertzer, D., and Lovejoy, S. (1985). "The dimension and intermittency of atmospheric dynamics". In: Bradbury, L.J.S., Durst, F., Launder, B.E., Schmidt, F. W., and Whitelaw, J.H. (eds.), *Turbulent shear flow 4*. Selected papers from the Fourth International Symposium on "Turbulent Shear Flows", Springer-Verlag, Berlin, 7-33.

Schertzer, D., and Lovejoy, S. (1987). "Physical modeling and analysis of rain and clouds by anisotropic scaling multiplicative processes", *Journal of Geophysical Research*, 92(D8), 9693-9714.

Schertzer, D., and Lovejoy, S. (1988). "Multifractal simulations and analysis of clouds by multiplicative processes". *Atmospheric Research*, 21, 337-361.

Schertzer, D., and Lovejoy, S. (1989). "Nonlinear variability in Geophysics: multifractal simulations and analysis". In: Pietronero, L. (ed.), *Fractals' physical origin and properties*, Plenum Press, New York, 49-79.

Sevruk, B., and Klemm, S. (1989). "Types of standard precipitation gauges". In: Sevruk, B. (ed.), *Precipitation measurements*. Proceedings of WMO/IAHS/ETH International Workshop on "Precipitation Measurements", December 3-7, 1989, St. Moritz, Switzerland, 227-232.

Svensson, C., Olsson, J., and Berndtsson, R. (1996). "Multifractal properties of daily rainfall in two different climates", *Water Resources Research*, 32(8), 2463-2472.

Tessier, Y., Lovejoy, S., and Schertzer, D. (1992). "Universal multifractals: theory and observations for rain and clouds". *Proc. of 11th International Conference on Clouds and Precipitation*, Montreal, Canada, 2, 1098-1101.

Tessier, Y., Lovejoy, S., and Schertzer, D. (1993). "Universal multifractals: theory and observations for rain and clouds". *Journal of Applied Meteorology*, 32(2), 223-250.

Tessier, Y., Lovejoy, S., and Schertzer, D. (1994). "Multifractal analysis and simulation of the global meteorological network". *Journal of Applied Meteorology*, 33(12), 1572-1586.

Tessier, Y., Lovejoy, S., Hubert, P., Schertzer, D., and Pecknold, S. (1996). "Multifractal analysis and modeling of rainfall and river flows and scaling, causal transfer functions". *Journal of Geophysical Research*, 101(D21), 26427-26440.

Surface Water Hydrology, Singh, Al-Rashed & Sherif (eds)
© *2002 Swets & Zeitlinger, Lisse, ISBN 90 5809 363 8*

Modelling regional rainfall

Rajib M. El-Geroushi and Abdulwahab M. Bubtaina

Civil Engineering Department, University of Garyounis
P.O.Box 13073, Benghazi, Libya
e-mail: bubtaina@yahoo.com

Abstract

The quantity of rainfall is considered as the most important factor in the design of any hydraulics structure or in the study of water resources of catchments area or region. The most difficult problem in this field is how to determine the quantity of rainfall of desired Average Return Period for the study area; this problem increases in the areas where no records are available or in shortage of records. For the importance of this hydrological factor, an empirical model of regional rainfall has been created. The model has been done by using statistical analysis of records of 59 climatic station cover all area of Libya. The model is done using the methodology of Flood-Index (surveying dept. USA). This model makes very possible to determine the annual quantity of rainfall for any period and to any region in Libya.

Introduction

Many hydrological processes are so complex that they can be interpreted and explained only in a probabilistic sense. Hydrologic events appear as uncertainties of nature and are the result of an underlying process with random or stochastic components. The information to investigate these processes is contained in records of hydrologic observations. Methods of statistical analysis provide ways to reduce and summarize observed data, to present information in precise and meaningful form, to determine the underlying characteristics of the observed phenomena, and to make predictions concerning future behavior. In dealing with single historical sequences of hydrologic variables, the predictive value of fitted distributions is limited because the records are generally short and the sampling errors corresponding large. Additional and more reliable information often can be obtained within a homogeneous region by correlating dependent hydrologic variables with other causative or physically related factors. In such ways, hydrologic characteristic within the region can be summarized, and estimates of statistical parameters can be derived from general regional relationships. Typical example is the prediction of rainfall depths and frequencies in ungauged or incompletely gauged areas from characteristics of well-guaged sites in the same area. Definition of the regional boundaries depends on the parameters or variables to be estimated. In some cases, significant generalizations can be made overlarge physiographic regions (mean annual precipitation in the United States, for example). The index-flood method proposed by the U.S. Geological survey is an example of summarizing regional characteristics successfully.

In some regions of our country (Libya), the historical records of the rainfall are not available or are not longe enough, for that an attempt has been made to construct a model of regional rainfall based on the same methodology adopted in Index-flood method, to solve the problem of missing records of rainfall, where the proposed model can find the rainfall depth for any frequency (return period) for any region or area in Libya regardless, if the historical records are available or not.

This model has been done through the following steps:

Statistical analysis for the records, where found out all the required statistical parameters. The statistical analysis include:

The size test of the records (data).
The quality test of the records.
The homogeneity test of records.

Plotting the point frequency curves, which include:

Point frequency analysis.
Probability distribution of the records.

Determination the rainfall value corresponding to the average return period (frequency) oftheaverage annual rainfall.

Determination of rainfall values corresponding to some frequencies.

Determination the form of regional rainfall model.

The Statistical Analysis

The generalization of any statistical model (empirical model) is provided by the homogeneity of the region from where the records are taken. This homogeneity can be proved statistically by finding out the correlation factor between different point (station) in the region. The homogeneity of the region used in proposed regional model has been measured for the all 59 climatic stations (points) distributed over the country (Libya), for that a software has been used to measure all the necessary statistical parameter such as: size of sample, average, mode, variance, standard deviation, standard error, min. value, max. value, range, lower quartile, upper quartile, Skewness, Kurtosis ... etc. These parameters make the predicting of future behavior of the phenomena is possible by understanding their physical meaning, in addition, that these parameters are necessary to test the data used in the statistical analysis. After finding out the statistical parameters, the data used to establish the proposed model subjected to some tests to assure the validity of the model. For that three important tests have been made on the data used for the proposed model.

Size Test of the Data

The reliability of statistical parameters depends mainly on the size of the data, this size, for natural phenomena depends on the frequency of this phenomena which depends on the climatic and natural characteristic of the region.

In semi-arid regions, the required data (records) to be used for statistical analysis should not be less than 40 years, otherwise extending of the short data "S" is possible if a long data "L" is available. This technique is based on finding ratio of some statistical parameters such as: average "X", skewness coefficient C_s, variance coefficient C_v or standard deviation.

$$\frac{C_{SS}}{C_{SL}} \quad , \quad \frac{C_{VS}}{C_{VL}} \quad , \quad \frac{X_S}{X_L} \tag{1}$$

where

C_{SS} : Skewness coefficient for short reading (data).
C_{SL} : Skewness coefficient for long reading (data).
C_{VS} : Variance coefficient for short reading (data).
C_{VL} : Variance of coefficient of long reading (data).
X_S : Average of short reading (data).
X_L : Average of long reading (data).

If the ratio in Eq. 1 does not exceed $10 \cong 15\%$, the short reading consider as a reasonable for the analysis, otherwise, the short readings should be extended providing that, the correlation factor between the records of large and short readings is not less than 75%.

In the proposed model, found that, it was necessary to extend the records of some stations to the longest records (58 years), where the regression analysis has shown high correlation factor ($83 \cong 99\%$).

Quality Test of the Data

Reliability of a regional model depends on the reality and quality of data used in the model, the quality of the data is confirmed by study the homogeneity and consistency of the data. In rainfall the quality of the records is determined by the double mass curve. In the proposed model, the records used have no missed readings, and where have some missed records, double mass curve is drown to find these missed records.

The Homogeneity of Records (Time-series)

Not only the reliability and reality of the data is required for the data used in an empirical model but also the consistency of the data is very required to assure the succeed of the model. The consistency of the data means that the origin of the data is same, for the rainfall data will not be consistence if there is an artificial rain record used in the model. In the proposed model, the consistence data are used.

Plotting the Point Frequency Curves

The rainfall frequency curves represent the relation between the frequency (average return period) for each recorded value. The average return period is the average spent time between two values exceeding or equal a certain value, practically if a hydraulic system is designed for a frequency of 1000 years (for example) of some value of a natural phenomena, that means the probability of this value to happen is as small as 0.001 during any year of 1000 years. The frequency curve is drawn for all the 59 stations using Gumbel's distribution. To predict the rainfall values for the design frequency, probability distribution of the records has been drawn.

The method followed requires determination of some values of the phenomena out of available records (100-500 and 1000 years), this process is known as a hydraulic forecasting which can be done mathematically or statistically.

Statistically, this forecasting depends on the probability theory and probability distribution. The probability distributions are good tool to identify some natural variable. For each natural phenomena there is a suitable distribution by which the predicted values can be assumed accurate and precise, the Gumbel distribution is fit for the hydrological phenomena.

In the proposed model, Gumbel distribution has been used to determine rainfall values of frequencies out of records, where this distribution is drawn for the 59 stations. Gumbel distribution or extreme values theory taken in consideration in the drawings only the max. and min. values.

Mathematically Gumbel distribution can be put in very simple form as following:

$$X = u + \left(\frac{1}{a}\right) \ln T \qquad (2)$$

where

$$u = \overline{X} - \left(\frac{0.57772}{a}\right) \qquad (3)$$

and

$$\frac{1}{a} = 0.7797 \sigma_s \qquad (4)$$

where

X : Value of the phenomena after T of years.
$\overline{X}$: Main value of the phenomena's value.

$\dfrac{1}{a}$: The mode of the phenomena value.

σ_s : The standard deviation of the phenomena value.

Graphical technique can be used to fit the Gumbel extreme value distribution to the observed data, the rainfall data are ranked according to magnitude and frequency coverage return period is estimated by using the equation

$$T = \frac{n+1}{m} \tag{5}$$

where

n is the number of years of records.
m is the order number in the rank.
T is the average returning period.

For the proposed model, the ranked data and the computed average return period (frequency) and plotted using a software. Fitting a straight line thorough the plotted data completes the analysis. The straight line can be conveniently extrapolated as desired.

Determination the Rainfall Value Correspondent to the Average Return Period (Frequency) of the Average Annual Rainfall

The average annual value of the phenomena according to the Gumbel distribution is the correspondent value of time T = 2.33 year which found from the frequancy factor K, where

$$K = -0.7797\left(0.5772 + \ln\ln\frac{T}{T-1}\right) \tag{6}$$

The frequency factor is the number of standard deviation above and below the main to attain the probability point of interest. For two-parameters distribution it varies only with probability, or the equivalent in frequency analysis, the average return period, T.

From Equation (6), K = 0 at T = 2.33 year. Hence, in the frequency analysis using Gumbel distribution, the return period of the phenomena is 2.33 year. Determination this value can be done through the Eq. (2) by substituting T = 2.33 after finding the average mode, and the standard deviation for each point (station) or graphically as used in the proposed model.

Determination of Rainfall Value for Some Frequency (Average Return Period)

From the frequency curves, rainfall values are determined for frequencies of 1, 10, 50, 100, 500 and 1000 years using Eq. (2) or graphically by projecting a vertical line from each frequency value to be intersected with the curve, and then the value of correspondent rainfall can be found.

Model of Regional Rainfall

The final form of the model has been found using the following steps:

Finding the main annual rainfall values for each frequency by dividing the rainfall value resulted from Paragraph 4 by the rainfall values resulted from Paragraph 3.

Calculating ratios of main annual rainfall by computing the median for each frequency selected in Paragraph 4.

Plotting the selected frequencies T versus the main ratio of the rainfall on probability paper.

Table 1 shows the computation of the first and second steps of the model of regional rainfall procedure. Fig. 1 shows the graphical form of the proposal model. It shows a very strong correlation (95%) and very good consistency.

Application of the Model

Assume that there is a plan to construct a hydraulic structure in ungauged area, or the records one short for statistical analysis. The model can give initial suggestion on the amount of rainfall expected on some selected return period (frequency). In this case, main annual rainfall can be determined (200 mm for example), and say, the required frequency is 100 years from the Fig. 1, ratio of annual rainfall can be found (3.3), and the amount of rainfall in mm for 100 years average return period is equal to $200 \times 3.3 = 6.60$ mm, or mathematically P_{100}.

Table 1. Ratio of main annual rainfall for the selected frequencies.

1000	500	100	50	20	10	1	Frequency / Station	Serial No.
5.0	4.5	3.5	3	2.43	1.95	0.5	Jofra	1
3.3	3.1	2.4	2.2	1.8	1.6	0.6	Sidi Masri	2
4.6	4.2	3.3	2.9	2.3	1.9	0.5	Martoba	3
11.4	10.1	7.4	6.3	4.6	3.4	-	Sabha	4
5.4	5.0	3.8	3.3	2.6	2.1	0.4	Gambot	5
2.5	2.4	2.0	1.8	1.5	1.4	0.8	Guba	6
4.2	3.9	3	2.6	2.1	1.7	0.5	Magroon	7
9.5	8.5	6.3	5.4	4	2.9	-	Jaloo	8
2.7	2.5	2.0	1.9	1.6	1.4	0.8	Kaser Libya	9
2.4	2.2	1.9	1.7	1.5	1.4	0.8	Susa	10
4.2	3.3	2.6	2.3	1.9	1.6	0.6	Tamymi	11

Table 1. (continued).

4.1	2.7	2.9	2.6	2.1	1.8	0.6	Hnya	12
2.6	2.4	1.9	1.7	1.4	1.2	0.6	Taknes	13
2.9	2.7	2.2	2.0	1.7	1.5	0.7	Safsaf	14
3.4	3.1	2.5	2.2	1.9	1.6	0.7	Ain mara	15
2.2	2.1	1.8	1.6	1.4	1.3	0.8	Abrak	16
4.3	3.9	3.1	2.7	2.2	1.8	0.6	Amsyat	17
3.4	3.2	2.5	2.3	1.9	1.6	0.7	Tobrok	18
3.4	3.1	2.5	2.2	1.9	1.6	0.7	Tokra	19
4.2	3.8	3	2.6	2.1	1.8	0.6	Ras Hilal	20
2.5	2.3	1.9	1.8	1.5	1.35	0.8	Aum Ruzm	21
10.3	9.3	6.8	5.7	4.3	3.3	-	Kofra	22
4	3.7	2.9	2.5	2.1	1.7	0.6	Benyna	23
4.3	3.9	13.04	2.7	2.2	1.8	0.6	Derna	24
2.4	2.2	1.9	1.7	1.5	1.4	0.8	Gaminas	25
6.1	4.6	4.2	3.6	2.8	2.4	0.3	Gdames	26
2.14	2.02	1.7	1.6	1.4	1.3	0.84	Dryana	27
4.6	4.3	3.2	2.8	2.3	1.9	0.6	Gazala	28
3.6	3.3	2.6	2.3	1.95	1.6	0.66	Tamina	29
2.3	2.2	1.9	1.7	1.5	1.32	0.84	Zoytena	30
18	16	11.7	9.7	7	5.13	-	Tazrbo	31
3.33	3.05	2.4	2.2	1.8	1.6	0.7	Myna Benghazi	32
3.5	3.2	2.5	2.2	1.8	1.5	0.53	Ajdabya	33
4.4	4	3.12	2.7	2.2	1.84	0.56	Bardya	34
2.4	2.2	1.9	1.7	1.5	1.33	0.8	Mosrata	35
3.7	3.4	2.7	2.4	2.0	1.7	0.6	Tripoli	36
3.9	3.5	2.8	2.4	2.0	1.7	0.6	Marawa	37
2.9	2.7	2.3	?	1 7	1.5	0.7	Tarhona	38
4.3	3.9	3.03	2.7	2.2	1.76	0.54	Ben Walid	39
3.76	3.43	2.7	2.45	2	1.7	0.6	Foyhat	40
7.0	6.4	4.6	4.04	3.14	2.6	0.18	Hoon	41
13.6	12.1	8.86	7.43	5.43	4.1	-	Jagbob	42
3.52	3.2	2.5	2.3	1.9	1.6	0.7	Tripoli airport	43
3.8	3.5	2.8	2.4	2.0	1.7	0.65	Slanta	44
4.5	4.1	3.15	2.8	2.2	1.85	0.54	Sorman	45
4.8	4.4	3.3	2.94	2.33	1.9	0.44	Mazra Tahrir	46
5.6	5.03	3.85	3.4	2.6	2.2	0.4	Bouargoob	47
3.3	3.03	2.42	2.2	1.9	1.6	0.7	Misa	48
5.6	5.3	4.04	3.43	2.73	2.12	0.3	Naser base	49
4.1	3.8	3	2.6	2.1	1.76	0.52	Azyzya	50
6.1	5.5	4.14	3.6	2.8	2.2	0.4	Naloot	51
3.2	2.91	.24	2.1	1.73	1.54	0.64	Zwara	52
-	4.4	3.4	3.0	2.3	1.9	0.5	Garyan	53
6.0	3.7	4.3	3.7	2.8	2.2	0.57	Faydya	54
3.33	3.16	2.5	2.3	2.0	1.66	0.66	Byada	55
3.14	2.85	2.3	2.0	1.17	1.43	0.57	Tolmetha	56
4.0	3.8	2.92	2.48	1.88	1.8	0.48	Benghazi	57
2.45	2.15	1.78	1.56	1.27	1.17	0.88	Bayda	58
2.92	2.83	2.25	2.08	1.7	1.42	0.66	Shahat	59
4.6	4.25	3.3	2.9	2.3	1.9	0.55	Median	

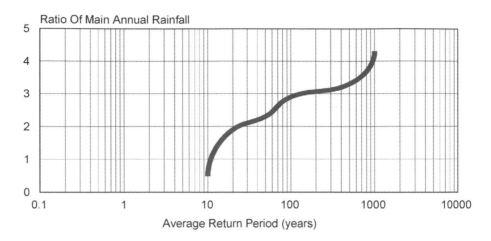

Figure 1. Model of regional rainfall.

Conclusion

The model of regional rainfall is an empirical formula established using the methodology of index-flood method which depends on the frequency analysis which is done using extreme-value theory (Gumbel distribution) the analysis of region which based on study of statistical parameter of the records of each time-series which subjected to some statistical test to prove their consistency and homogeneity which make very important factors for any regional model to be generalized.

The proposed model can be used to determine the rainfall depth for any frequency to any region or area in Libya by knowing the main annual rainfall with no need to the historical records in the study area or region.

References

Haan, Charles T. (1977) Statistical methods in hydrology. 1st ed., Iowa State University.

Linsley, Rayk et al. (1975) Hydrology for Engineers, 2nd ed. McGraw-Hill Kogakusha, Ltd.

Schutz, E.F. (1973) Problems in applied hydrology. Water Resources Publications, Fort Collins, Colorado 80521, USA.

Sokolov, A.A. et al. (1976) Flood flow computation methods complied from world experience. The Unesco Press.

Surface Water Hydrology, Singh, Al-Rashed & Sherif (eds)
© 2002 Swets & Zeitlinger, Lisse, ISBN 90 5809 363 8

Probabilistic occurrence of the average autumn rainfall in Iran related to ENSO events

M.J. Nazemosadat,
Irrigation Department, Agriculture Faculty, Shiraz University, Iran
e-mail: jafar@hafez.shirazu.ac.ir

Abstract

The average rainfall is considered as an essential threshold for the evaluation of climatological condition in a particular year. The influence of the El Niño-Southern Oscillation (ENSO) phenomenon on the probabilistic occurrence of average rainfall (POAR) was explored for the period 1951–1995. The study has been carried out for autumnal rainfall in both seasonal and monthly time-scales. It was found the POARs have increased (decreased) during warm (cold) episodes for the most parts of Iran. However, for the eastern parts of the Caspian Sea coasts, warm (cold) incidents induce a significant decline (improvement) in POAR values. Successful seed germination, early growth of crops, less autumnal air pollution, more than usual water storage in dam reservoirs and increase in soil water content are expected during intense warm episodes.

Introduction

The Islamic republic of Iran (Figure 1) has an area of 1,648,000 Km^2 and a population of 65 millions (1995 estimate). As a whole, the country has arid or semiarid climates with average annual rainfall about 250 mm. Two mountain ridges (Alborz and Zagros), which run northwest-northeast and northwest-southeast respectively, play an influential role in determining the amount and distribution of rainfall over the country (Figure 2). The peaks of Alborz and Zagros attain the heights of 5671 and around 4000 meters, respectively (Brawer, 1988).

As is shown in Figure 2, two uninhabited great deserts called Dasht-e Lut and Dasht-e Kavir that occupy around one-sixth of the total area of the country are stretched over the central parts of Iran. In spite of sever dryness condition over these deserts and their nearby areas, some regions including the Zagros and Alborz highlands as well as the coastal strip of the Caspian Sea usually receive moderate precipitation. Rainfall generally occurs from October to March with extreme events being most prominent during January and February. Summer rainfall is also considerable over the Caspian Sea shores. Although total annual rainfall over the northern sides of Alborz range may reach 1800 mm, for central and eastern deserts, it may decline to only 50mm (Khalili, 1992).

The occurrence of droughts and floods are common and the severity and hardships of these natural disasters have frequently hit the rural regions as well as the urban societies. Drought limits the cultivation of dry farming crops and affects the productivity of irrigated lands. Moreover, due to massive overgrazing, large-scale soil ero-

Figure 1. The geographical location of Iran and the position of rainfall stations whose data were analyzed in this study.

sion is expected during dry spells. The occurrence of recent severe drought (1998–2001) cost the nation billions of dollars damages. During this period, dam reservoirs within the country dropped to their lowest records and water supply for agricultural and drinking proposes faces serious dilemma in most parts of the country including Tehran (the country's capital city).

Excess rainfall has also repeatedly caused devastating floods in different regions. According to the report provided by the Iranian Meteorological Organization (IMO), in March 1998, flood event in southwestern districts cost the lives of 6 people and millions dollars of damages. Due to the devastating flood in August 2001, more than 250 people have died (in Golestan Province southeastern parts of the Caspian Sea coast) and billions dollars of damages were reported in residential compounds, livestock's, roads and infrastructures. It is estimated that atmospheric and climatic incidents (i.e floods, droughts and lightning) account for about 97% of the cost of natural disaster in most parts of the country.

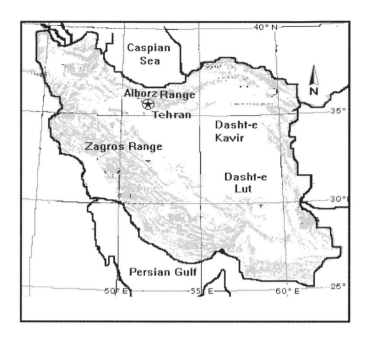

Figure 2. Relief map of Iran. The Alborz and Zagros Ranges and the two main deserts (Dasht-e Lut and Dasht-e Kavir) are shown.

It has been recently documented that the rainfall variability and the occurrence of drought and flood events in Iran are influenced by the El Niño and Southern Oscillation (ENSO) phenomena (Nazemosadat, 1999; Nazemosadat and Cordery, 2000a, b; Nazemosadat, 2001a, b). A brief description of the phenomena, its extreme phases and their impact on global climate are addressed in a number of studies (Allan et al., 1995 and 1996; Hastenrath et al., 1993; Lough, 1992 and 1997; Nicholls, 1996; Drosdowsky, 1994 and 1995; Cordery and Opoku-Ankomah, 1994; Frederiksen and Balgovind, 1994; Camberlin, 1995; Beltrando and Camberlin, 1993; Wu and Zhang, 1996; Nazemosadat and Cordery, 1997; Hodel, 1997; Dracup and Kahya, 1994; Ropelewski and Halpert, 1996; Kahya and Dracup, 1993).

Nazemosadat and Cordery (2000a) have been shown that autumn rainfall in Iran is negatively correlated with the Southern Oscillation Index (SOI). The relationships were found to be strong and consistent over some regions comprising the southern foothills of the Alborz Mountains, northwestern districts and central areas (Figure 2). For these regions, more than usual rainfall is, generally, coincided with El Niño (warm or low phase) episodes. On the other hands, lower than usual rainfall is also expected during La Niña (cold or high phase) spells. In winter, however, for most parts of the country, more than normal rainfall is expected during La Niña events (Nazemosadat and Cordery, 2000b; Nazemosadat, 2001a, b). The influence of ENSO on rainfall in Iran was generally found to be stronger and more persistent during autumn than in winter.

The above-mentioned studies about Iran have, generally, focused on the total amount rather than probabilistic occurrence of rainfall during various phases of ENSO events. With the consideration that the average rainfall is an essential threshold for the evaluation of climatological condition in a particular year, the present study has focused to measure the probability of the occurrence of this threshold during extreme phases of ENSO. It is noteworthy to notice that the rainfall data over most parts of Iran (southern and eastern regions in particular) do not follow the normal distribution. Due to this constraint, the magnitude of medians (that deliver probability concept) are generally less than corresponding average values of rainfall. The ratio of average to median autumn rainfall was found to be around 1.20 for most of the stations and may reach to near 2 in Bandarlengeh, Bandarabas, Zahedan, Zabol and Chahbahar. The average rather than median values of the rainfalls are, therefore, considered as a minimum requirement for successful seed germination, grain productivity and fair availability of water resources.

The term of "probabilistic occurrence of average rainfall (POAR)" was hence defined to examine the occurrence probability of average rainfall during various ENSO phases. During the periods that POAR is high, more than average rainfall is expected and the grain-growers usually, encounters lower than usual risk. On the other hand, higher risk and lower production yield are expected for the episodes that POARs are less than usual. The magnitude of POAR is hence important for agriculturists, soil erosion researchers, forest and water resources specialists, builders, economists, tourism managers and most of the prominent decision-makers.

Since autumn rainfall is significantly associated with ENSO events (Nazemosadat and Cordery, 2000a), the alternation of this event from one state (e.g. warm phase) to another (e.g. cold phase) could crucially influence the POAR magnitude. Therefore, the comparative degree of seed germination, productivity of crops, livestock products, soil erosion, tourist population, economic condition, air pollution and general health condition in Iran could be associated with the ENSO phases and corresponding POAR values.

Because autumn rainfall contributes a significant portion of Iranian water resources, the present study has focused on the assessment of POAR during this particular season. The aims of the study are prompted as:

1. Computing and mapping of POAR values for the circumstances that whole available rainfall data are included in the analysis. All the analyses were carried out in both monthly and seasonal time-scales.
2. Delineating the spatial distribution of POARs during the extreme episodes of ENSO events.
3. Discussing the causes of the discrepancies between the produced maps.

Data and Methods

Total monthly rainfalls for 40 key synoptic stations located over various parts of the country are the basic data used in this analysis. These data were derived from the yearly Weather-books published by the Iranian Meteorological Organization (IMO) for the period 1951–1995. Figure 1 illustrates the geographical location of the rainfall stations and Figure 3 shows the ratio of autumn rainfall to annual precipitation. Total annual rainfalls vary from 62 mm in Yazd and Zabol to 1779 mm in Bandaranzali. The rainfall amounts in eastern and southern parts are generally less than the precipitation over western and northern parts of the country. The Troup's (1965) Southern Oscillation Index (SOI) data, which have been used as the ENSO indicator, were supplied by the Australian Bureau of Meteorology.

Autumn rainfall constitutes from about 18% (in Bam) to 47% (in Bandaranzali) of annual precipitation over different parts of the country. For the coastal strip of the Caspian Sea and the western half of Iran, the ratio of autumn rainfall to annual precipitation is larger than the eastern half. The ratio is also relatively high for the areas near the northern part of the Persian Gulf.

The time series of autumn rainfall were obtained by averaging the three monthly values of precipitation. Best results were obtained by defining autumn from October to December. The same averaging procedure was performed to provide the seasonal time series of Southern Oscillation Index (SOI) data. Although missing values in the rainfall data were generally rare, a variety of missing data were found for some rainfall stations such as Chahbahar and Bandarlengeh (Figure 1). Missing values were excluded in computational analysis and the results presented here were obtained from the available data.

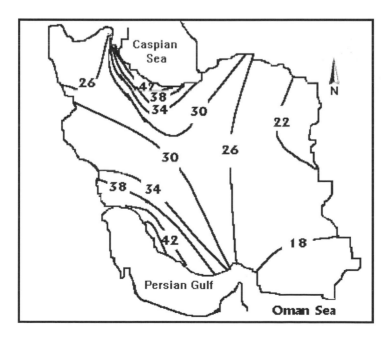

Figure 3. The spatial distribution of the ratio of autumn rainfall to annual precipitation.

Trenberth (1997) has provided listings of El Nñio and La Niña events after 1950 as defined by SSTs in the Niño 3.4 and exceeding ± 0.4 °C threshold. The present study has used these lists for the detection of the warm and cold phases of the SO phenomenon. The identified warm and cold phases were, commonly, found to correspond with the episodes that autumn SOI were greater than +5 and less than (-5), respectively. After the identification of extreme episodes of ENSO events, rainfall time series for these episodes were constructed.

The time series of the rainfall during November are considered as the representative of monthly data sets. For every station, three separate lists of November and seasonal time series of rainfall data were provided. The first list contained the whole available data the other two lists comprised rainfall data of the years coincided with warm and cold phases of the ENSO events, respectively. The monthly and seasonal data sets were then ranked for each individual month. Long-term average of monthly and seasonal data as well as the mean values of the rainfall during cold and warm episodes were then computed for every station.

Since the considered rainfall time series did not fit with a unique distribution and a variety of zero values were observed in monthly data, Weibull formula (Mutreja, 1990) was adapted for the computation of the probability. In this formulation, when the rainfall time series are ranked from 1 to n in descending order, the POAR is defined as:

$$POAR = 100(m/(n+1))$$

where m is either the rank of average value or the rank of nearest (larger) value to the average in the ranked data set.

The magnitudes of average rainfalls were found to be widely different among the stations. For reducing the effects of such hetrogeneties in station records, the POARs were computed for each individual data set.

Results and Discussion

POAR Values for November

The spatial distribution of November POAR for the whole available data and warm episodes are depicted in Figures 4 and 5, respectively. Comparing these two Figures suggests that the occurrence of warm ENSO events increases the magnitude of POAR over the southern parts of Iran more than northern districts. The probability also shows around 15% increase for the stations situated over western coasts of the Caspian Sea. However, for Gorgan, located over the eastern corner of the coasts, warm incidents induce a significant decline in November POAR. It is worthwhile to note that, for all of the considered stations with the exception of Gorgan and Abadan, average November rainfall during El Niño events were found to be more than long-term averages.

Figure 6 illustrates November POAR for La Niña episodes. Compared with Figures

226

4 and 5, Figure 6 implies that the occurrence of cold events decreases the probability over most parts of Iran. The only exception was found for Gorgan where significant increase in POAR is coincided with cold phase periods. For 9 recorded La Niña events from 1950–1995, November rainfall in Gorgan were more than the

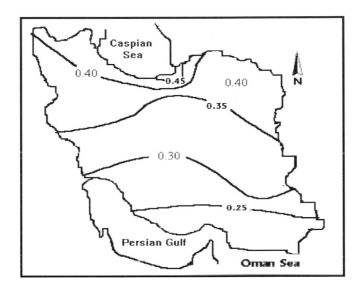

Figure 4. The spatial distribution of POAR (in percent) for November when the whole available records are considered.

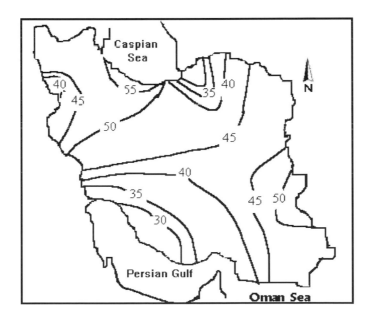

Figure 5. Same as Figure 4 but data during warm episodes are considered.

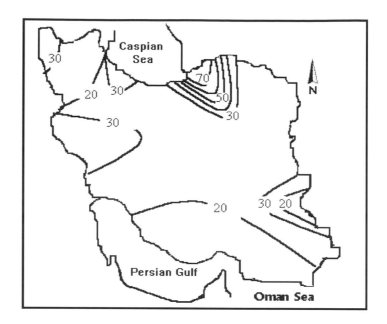

Figure 6. Same as Figure 4 but data during cold episodes are considered.

long-term average for 7 events (POAR = 78%). The exceptions were found during 1970 and 1971 for which the rainfall were below average. The recent devastating flood (1999) in the eastern portion of the Caspian Sea coasts (including Gorgan and nearby regions) could also be attributed to the cold phase of ENSO event.

A sea-saw fluctuation on the POAR values is evident over the eastern and western margin of the Caspian Sea shores. This suggests that the source region and the mechanism of rain formation over these two distinct areas are different. The results imply that the occurrence of warm episodes increases the frequency of moisture-laden airmasses (originated over the Black Sea and nearby waters) toward the northern and northwestern districts of Iran. Although the western portion of the Caspian Sea coasts benefit from such an air motion, the impact of these moisture laden airmasses on Gorgan autumn rainfall is, however, less significant during warm events. The pressure system over the Caspian Sea (rather than Mediterranean Sea regions) probably plays the most influential role in the fluctuation of autumn rainfall in Gorgan and the nearby areas. The occurrence of La Niña periods could probably direct the moisturized (by the Caspian Sea) southeasterly winds toward eastern coasts more than usual. Rodionov (1994) has not been noted such an influential role of ENSO event on the surface climate of the Caspian Sea.

POAR Values for Autumn

The spatial distributions of autumn POAR for the whole collected data are mapped in Figure 7. The figure implies that this index is less than 50% for the most parts of the country. Maximum (minimum) POARs were observed in Oromieh and shahroud

(Iranshahr and Chahbahar) for which the index was 50% and 51% (25% and 27%), respectively. This compared to Figures 4, 7 and 10, the POARs are, however, slightly larger for most of the considered stations.

The geographical distribution of autumn POARs during warm episodes are shown in Figure 8. As indicated in this figure, the occurrence of El Niño has meaningfully increased the probability for most of the considered stations. For about 30% of sta-

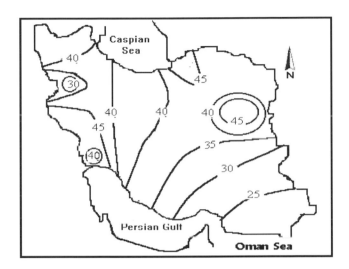

Figure 7. The spatial distribution of POAR (in percent) for autumn when the whole available records are considered.

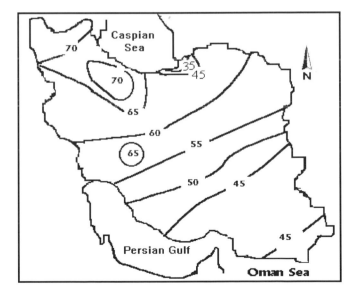

Figure 8. Same as Figure 7 but data during warm episodes are considered.

tions that are mainly situated over the northwestern portion of the country, western margin of the Caspian Sea coasts and southern foothill of the Alborz mountains, the POAR values were found to be around 70%. The probabilities were observed more (less) than 50% for 28 (12) stations generally situated over the western (eastern) half of Iran.

The results suggest that, in autumn, the occurrence of warm ENSO events induce a significant improvement in POAR values and Iranian water resources. Therefore, successful seed germination and the early growth of crops, less air pollution, more than usual water storage in dam reservoirs (Mohsenipour) and soil layers are expected to happen during intense warm episodes.

In contrast to warm events, the incidence of cold phase represses the autumn POARs for most parts of the country. The only station for which cold phase has caused a significant increase in POAR is Gorgan (POAR = 80%). The average monthly rainfall of this station has also increased during cold events. Except for three stations of Abadan, Boushehr and Khoramabad for which POARs were around 50%, for the other stations the index was generally found to be less than 35%. For Abadan, Boushehr, located over the western portion of the Persian Gulf, more than usual rainfall is expected during cold events. However, for the other stations, La Niña events, generally causes less than normal rainfall. In autumn, a meaningful depletion in national water resources is, therefore, expected during cold events.

Conclusion

It has been already shown that, in Iran, ENSO accounts for a large portion of the variance in autumn rainfall. The Probabilistic Occurrence of Average Rainfall (POAR) associated with ENSO extremes was assessed in this study. In November and when the whole rainfall records are considered, the greatest (lowest) POARs were associated to the coastal strips of the Caspian Sea (southeastern districts). For the most parts of Iran, the occurrence of El Niño (La Niña) events has caused a significant increase (decrease) in November POARs. During warm epochs, the November POARs were found to be about 50 for the most parts of the northern and northwestern parts of the Iran. In contrast to the most parts of the country, the occurrence of cold (warm) episodes in November has increased (decresed) POARs over the eastern side of the Caspian Sea coasts. For the stations situated over the northwestern portion of the country, the western margin of the Caspian Sea coasts and the southern foothill of the Alborz mountains, the autumn POARs during warm episodes were found to be around 70%. On the other hand, with the exception of the eastern portion of the Caspian Sea coasts and the northwestern parts of the Persian Gulf shores, cold ENSO events have generally caused a considerable decline in autumn POARs.

References

Allan, R.J., Lindesay, J. and Reason, C. 1995. Multidecadal variability in the climate system over the Indian Ocean, *J. Climate*, **8**, 1853-1873.

Allan, R.J., Bread, G.S., Close, A. Herczeg, A.L., Jones, P.D. and Simpson, H.J., 1996. Mean sea level pressure Indices of the El Niño Southern Oscillation: Relevance to stream discharge in south-eastern Australia, *CSIRO divisional report 96 / 1, Division of Water Resources*, 23 p.

Camberlin, P. 1995. June-September rainfall in north-eastern Africa and atmospheric signals over the tropics: A zonal perspective, *Int. J. Climatol.*, **15**, 773-783.

Cordery, I. and Opoku-Ankomah, Y. 1994. Temporal variation of relations between sea surface temperature and New South Wales rainfall, *Aust. Met. Mag.*, **43**, 73-80.

Beltrando, G. and Camberlin, P. 1993. Interannual variability of rainfall in the eastern Horn of Africa and indicators of atmospheric circulation, *Int. J. Climatol*, **13**, 533-546.

Brawer, M. 1988. Atlas of the Middle East, Macmillan Publishing Company, New York, 140 p.

Drosdowsky, W. 1994. The Southern Oscillation. In: Will it rain? The effects of Southern Oscillation and El Niño on Australia, Edited by Partridge, I.J. 2nd edition, 12-21.

Drosdowsky, W. 1995. Analogue (Non-linear) forecasts of the Southern Oscillation Index time series, *NOAA Experimental Long-lead Forecast Bulletin*, **4**, 28-31.

Dracup, J.A. and Kahya, E. 1994. The relationships between US streamflow and La Niña events, *Water. Resour. Res.*, **30**, 2133-2141.

Frederiksen, C.S. and Balgovind, R.C. 1994. The influence of the Indian Ocean / Indonesian SST gradient on the Australian winter rainfall and circulation in an atmospheric GCM, *Q. J. R. Meteorol. Soc.*, **120**, 923-952.

Hastenrath, S., Nicklis, A. and Greischar, L. 1993. Atmospheric-hydrospheric mechanisms of western equatorial Indian Ocean, *J. Geophys. Res.*, **98**, 20219-20235.

Hodet, G.C. de 1997. Seasonal climate summary southern hemisphere (spring 1996): Characteristics of a weak positive phase of the Southern Oscillation persist. *Aust. Met. Mag.*, **46**.153-160.

Kahya, E. and Dracup, J.A. 1993. US streamflow patterns in relation to the El Niño Southern Oscillation, *Water. Resour. Res.*, **29**, 2491-2503.

Khalili, A. 1992. Fundamental study of Iranian water resources, climatological aspects: Parts 1 and 2, Jamab Consultant Reports, Iranian Ministry of Energy, 892 p. (in Persian).

Lough, J.M. 1992. Variations of sea surface temperature off north-eastern Australia and associations with rainfall in Queensland: 1956-1987, *Int. J. Climatol.*, **12**, 765-782.

Lough, J.M. 1997. Regional indices of climate variation: Temperature and rainfall in Qeensland, Australia, *Int. J. Climatol.*, **17**, 55-66.

Mutreja, K.N. 1990. Applied Hydrology, Tata McGraw Hill Publishing Company, New Delhi, 959 p.

Nazemosadat, M.J. and Cordery, I. 1997. The influence of geopotential heights on New South Wales rainfalls, *Meteorology and Atmospheric Physics*, **63**, 179-193.

Nazemosadat, M.J. 1999. The impact of ENSO on autumnal rainfall in Iran, *Drought News Network,* **11**, 15-19.

Nazemosadat, M.J. and Cordery, I. 2000a. On the relationships between ENSO and autumn rainfall in Iran. *Int. J. climatol*, 13, 51-64.

Nazemosadat, M.J. and Cordery, I. 2000b. The impact of ENSO on winter rainfall in Iran, proceeding of the 26th National and 3rd International Hydrology and Water Resources Symposium, Inst. Engs. Australia, 20-23 November 2000, Perth. 538-543.

Nazemosadat, M.J. 2001a. Winter rainfall in Iran: ENSO and aloft wind interactions, *Iranian Journal of Science and Technology.* In press.

Nazemosadat, M.J. 2001b. Winter drought in Iran: Associations With ENSO, *Drought Network News,* Vol (13), No.1, 10-13.

Nicholls, N. 1996. Improved statistical model approaches for seasonal climate forecasting, presented in: Managing with Climate Variability Conference, Canberra, November 1995.

Rodionov, S.N. 1994. Global and regional climate interaction: The Caspian Sea experiment, Kluwer Academic Publisher, Boston, USA, 241 p.

Ropelewski, C.F. and Halpert, M.S. 1996. Quantifying Southern Oscillation-precipitation relationships, *J. Climate*, **9**, 1043-1059.

Troup, A.J. 1965. 'The Southern Oscillation', *Q. J. R. Meteorol. Soc.*, 91, 490-506. Trenberth, K.E., (1997). The definition of El Niño, *Bull. Amer. Meteor. Soc.*, **78**, 2771-2777.

Wu, G.-X. and Zhang, X.-H. 1996. Research in China n climate and its variability, *Theor. Apll. Climatol*, 55, 3-17.

Section 4: Evaporation and evapotranspiration

Surface Water Hydrology, Singh, Al-Rashed & Sherif (eds)
© *2002 Swets & Zeitlinger, Lisse, ISBN 90 5809 363 8*

Study of surface evaporation in a semi-arid region

S. Poongothai and S. Thayumanavan

Centre for Water Resources, Department of Civil Engineering
Anna University, Chennai-600025, India
e-mail: spoong86@yahoo.com

V. Rangapathy

Department of Civil Engineering, Annamalai University
Annamalainagar, Chidambaram-608001, India

Abstract

Water resources management plays a significant role in the development of any country. Many areas of the world are arid and semi-arid. The water loss due to evaporation and evapotranspiration is the major problem in these regions. The problem caused by the loss of water stored in lakes and reservoirs for irrigation and domestic use by evaporation during the summer months in these regions is enormous and perennial. In India the estimated area of arid, semiarid and long dry spell regions is about 200,000 km^2 (Ramdas, 1966). The annual water loss due to evaporation with estimated depth of 3 m is of the order of $1.32*10^{12}$ gallons (Verma, 1996) in India. Therefore estimation of evaporation and evapotranspiration are required in the design of reservoirs, irrigation systems, scheduling and frequency of irrigation, water balance and simulation studies and watershed planning and management. In this paper, Pan and Penman methods of estimating surface evaporation at Annamalainagar, TamilNadu, India, were studied and results were compared and presented. The study area is a semi-arid region, plain delta and lightly vegetated with mainly paddy crops. This study undertakes to analyse the methods of estimating evaporation so that suitable values of evaporation can be applied at the time of water management and reservoir operation practices. The regression and correlation analysis was carried out between the Pan and Penman evaporation values. Also Penman evaporation was related to evapotranspiration.

Introduction

Evaporation and evapotranspiration are the most important and the most complicated phases of the hydrologic cycle. These phases redistribute the heat energy between surfaces and atmosphere. Evaporation is the process by which water from liquid or solid state passes into the vapour state and is diffused into atmosphere.

For the occurrence of evaporation it is necessary to have (i) a supply of water, (ii) a source of heat and (iii) vapour pressure deficit, $(e_s - e_a)$. By Dalton's law (1802),

$$E = c (e_s - e_a) \tag{1}$$

where E is the evaporation, e_s is the saturated vapour pressure of water corresponding to water temperature and e_a is the actual vapour pressure of air above the free surface and c is a coefficient to account for the other factors which affect evaporation (Jayarami Reddy, 1994).

The methods of estimating evaporation may be divided into five categories. They are (i) Empirical equation method, (ii) Water balance method, (iii) Energy balance method, (iv) Mass transfer method and (v) Combined energy balance and Mass transfer method (Penman's method). Out of these methods, the Penman's method is most popular because (i) it requires only usually available meteorological data and (ii) it applies both the theories of Energy balance and Mass transfer methods.
The study area, Annamalainagar is a Township adjoining Chidambaram, lying in latitude 11°24′ N and longitude 79°44′ E at M.S.L. + 5.79 m. The region is essentially coastal as the Bay of Bengal is just 5 km as the crow flies. The area is a semi arid, plain delta and lightly vegetated with mainly paddy crops.

The Indian Meteorological Observatory situated at Annamalainagar is about 240 km south of Chennai. The observations recorded by this Observatory in local mean time: 7.00 h, 8.30 h and 14.00 h were taken for the purpose of study.

Estimation of Evaporation

Methods Employed

This study has attempted to estimate the evaporation at Annamalainagar using Standard Pan (Evaporimeter) and Penman methods (Jayarami Reddy, 1994).

Data Base

The following meteorological data are collected for four years 1990, 1991, 1992 and 1993:

(i)   The daily maximum and minimum temperature in °Celsius.
(ii) Humidity in (%).
(iii) Wind velocity in km/h.
(iv) Bright sunshine hours.
(v) Pan evaporation in mm.

Daily values of Penman evaporation are obtained for all the four years by running a simple program in *FORTRAN*. These values are estimated for different values of albedo, mean temperature and mean vapour pressure with same Pan coefficient.

The monthly, weekly and daily evaporation values for Penman method are obtained and compared with Pan evaporation values. This study has concentrated more on albedo = 0.10, which is more suitable for Annamalainagar. For this particular data, the results are regressed and correlated.

From the results it is noted that the Penman evaporation is very high in almost all the months even though it tends to equal or go below the Pan evaporation values around May, June and July. There is consistent overestimation of evaporation values by Penman method. Hence it is inferred that the evaporation values estimated by Penman method may be more nearly evapotranspiration values.

Input for the Program

The following data are input:

(i)   Mean temperature (t). Temperature in $^{\circ}K = t\,^{\circ}C + 273^{\circ}$.
(ii)   Atmospheric pressure P = 752mm of Hg.
(iii)   Wind velocity, m/s.
(iv)   Saturation vapour pressure (e_s) corresponding to 't'.
(v)   Mean vapour pressure (e_a) form the equation h = 100 e_a/e_s where h is humidity.
(vi)   Evaporation E_a from free surface (cm/day) is calculated from the relation ship of e_a, e_s and wind velocity.
(vii)   The radiation (cal/cm^2/day) received at the top of atmosphere with reference to the latitude and month of the year.
(viii)   The maximum possible bright sunshine hours with reference to the latitude and time of year.

A sample input data pertaining to January 1990 is presented in Table 1.

The above-mentioned inputs are applied to determine the Penman evaporation for all the days of 1990-1993. The results are tabulated and compared with the observed Pan evaporation values.

Results and Discussion

A comparison between the daily Pan and Penman evaporation values for the days of January 1990 with albedo = 0.05, pan coefficient = 1.0, temperature factor = 1.0 and vapour pressure factor = 1.0 was made and presented in Table 2. The above details were prepared monthly for all the four years 1990-1993. A sample of monthly evaporation values computed for the year 1990 with albedo = 0.10 was presented graphically in Fig. 1.

The trend or lack of fit, was observed in daily evaporation values in all the four years with albedo = 0.05 and Pan coefficient = 1.0. Actually Penman evaporation is very high in almost all the months. In fact a Pan coefficient of about 0.8 is usually applied for the Pan values because Pan values themselves are liable to be overestimated (Chow et al., 1988). But in this study, Pan coefficient is not reduced and kept as 1.0 itself because the Pan evaporation values may go down further, considering on this line Penman values are further over estimating the evaporation values.

Table 1. Meteorological data, January 1990.

Date	Pan evapn. mm	Mean temp. °C	Mean Vap.pr. mm of Hg	Wind velo. m/s	Sun shine hours h	Temp. in ° K	satur. Vap.pr. mm. of Hg	Radiant in Cal/cm²/d	Po.Br. Sunsh. Hrs.
D	PE	t	EA	Vel	SA	TK	ES	QO	PBS
1	2.20	24.75	24.81	0.78	7.70	297.75	31.21	745.99	11.53
2	3.20	25.00	26.60	1.33	8.70	298.00	31.67	746.94	11.53
3	2.60	24.50	24.75	1.42	9.90	297.50	30.75	747.88	11.53
4	3.00	23.50	22.30	1.61	7.80	296.50	28.96	748.83	11.53
5	3.60	24.00	24.91	3.19	7.80	297.00	29.83	749.78	11.53
6	1.40	23.00	26.12	2.11	0.00	296.00	28.09	750.73	11.53
7	0.60	23.75	24.10	2.69	0.00	296.75	29.39	751.67	11.53
8	2.60	24.75	26.22	2.44	2.10	297.75	31.21	752.62	11.53
9	3.40	24.25	25.44	2.28	5.00	297.25	30.29	753.57	11.53
10	1.60	23.50	24.04	1.50	3.20	296.50	28.96	754.51	11.53
11	4.00	23.25	24.24	0.97	10.40	296.25	28.52	755.46	11.53
12	4.00	23.00	21.35	1.08	10.00	296.00	28.09	756.41	11.53
13	2.60	21.50	18.85	0.28	9.70	294.50	25.65	757.36	11.53
14	3.20	21.50	17.19	0.42	7.90	294.50	25.65	759.48	11.53
15	2.90	22.50	18.13	0.36	10.50	295.50	27.26	761.60	11.53
16	3.20	22.25	20.26	0.47	10.10	295.25	26.84	763.73	11.53
17	3.00	22.75	20.89	0.83	10.00	295.75	27.67	765.85	11.53
18	3.80	22.50	19.90	0.81	10.00	295.50	27.26	767.97	11.53
19	3.30	21.25	18.43	0.67	10.50	294.25	25.25	770.10	11.53
20	3.00	20.75	14.45	0.72	10.00	293.75	24.49	772.22	11.53
21	4.40	23.10	15.97	0.53	11.00	296.10	28.26	774.34	11.53
22	3.20	22.60	20.57	0.42	10.80	295.60	27.42	776.46	11.53
23	3.30	23.50	22.01	0.33	10.70	296.50	28.96	778.59	11.53
24	3.80	23.40	22.16	0.78	10.80	296.40	28.78	780.71	11.53
25	3.30	23.10	21.34	0.97	10.40	296.10	28.26	782.83	11.53
26	3.00	22.25	21.07	0.61	11.00	295.25	26.84	784.96	11.53
27	3.60	23.00	20.51	0.47	10.40	296.00	28.09	787.08	11.53
28	4.00	24.00	21.48	0.97	10.50	297.00	29.83	789.20	11.53
29	4.00	23.75	21.75	1.25	9.00	296.75	29.39	791.33	11.53
30	4.40	24.00	22.97	0.86	9.90	297.00	29.83	793.45	11.53
31	3.40	23.65	21.62	1.19	10.30	296.65	29.22	795.57	11.53

Table 2. Calculation of Penman evaporation, January 1990.

Date	Pan evaporation mm	Penman evaporation mm	Difference mm
1	2.20	5.75310	-3.55310
2	3.20	6.30070	-3.10070
3	2.60	6.65440	-4.05440
4	3.00	5.80030	-2.80030
5	3.60	6.16910	-2.56910
6	1.40	2.84830	-1.44830
7	0.60	3.35830	-2.75830
8	2.60	4.07110	-1.47110
9	3.40	5.05350	-1.65350
1	1.60	4.19580	-2.59580
11	4.00	6.57280	-2.57280
12	4.00	6.35430	-2.35430
13	2.60	5.74070	-3.14070
14	3.20	5.20930	-2.00930
15	2.90	6.11880	-3.21880
16	3.20	6.14500	2.94500
17	3.00	6.32660	-3.32660
18	3.80	6.27120	-2.47120
19	3.30	6.16510	-2.86510
20	3.00	5.81890	-2.81890
21	4.40	6.37630	-1.97630
22	3.20	6.54460	-3.34460
23	3.30	6.70510	-3.40510
24	3.80	6.87590	-3.07590
25	3.30	6.73320	-3.43320
26	3.00	6.72840	-3.72840
27	3.60	6.57950	-2.97950
28	4.00	6.98510	-2.98510
29	4.00	6.53650	-2.53650
30	4.40	6.83730	-2.43730
31	3.40	6.99780	-3.59780
	97.6	184.8279	-87.2279

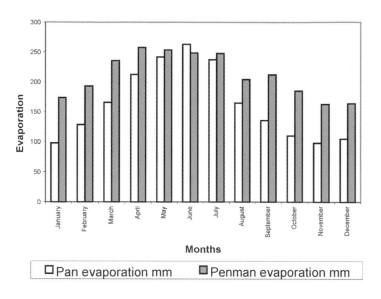

Figure 1. Monthly evaporation with Albedo = 0.10, 1990.

Trials with Changed Parameters

One of the reasons for the overestimation by the Penman method may be due to the adoption of some improper parameter values. Hence it was decided to change the value of albedo (Al.), mean temperature and vapour pressure gradient.

This study has evaluated Penman evaporation with the following four different sets of values with Pan coefficient = 1.0.

(i)   Albedo = 0.05, TF = 1.0, VPF = 1.0;
(ii)   Albedo = 0.10, TF = 1.0, VPF = 1.0;
(iii)   Albedo = 0.10, TF = 0.9, VPF = 1.0;
(iv)   Albedo = 0.10, TF = 1.0, VPF = 1.1.

Where TF = Temperature factor & VPF = Vapour pressure factor.

Albedo

As the landscape of Annamalainagar is only thinly vegetated, a value of 0.10 was felt sufficient however a value of more than 0.25 is recommended for densely vege-tated area (Chow et al.,1988). The daily Penman evaporation values for four years were calculated for the set (ii). The results show a reduction in the estimated Pen-man values by 10% compared to the results obtained with albedo = 0.05 (see Table 3).

Table 3. Monthly evaporation averaged over four years, 1990-1993.

Months	Pan evaporation mm	Penman evaporation mm			
		Al. = 0.05	Al. = 0.10	Al. = 0.10 TF = 0.90	Al. = 0.10 VPF = 1.1
January	110.25	190.15	178.91	178.55	178.91
February	126.15	206.19	194.14	193.73	195.53
March	168.10	252.90	238.76	242.71	240.72
April	197.61	272.13	257.85	257.49	259.84
May	243.40	294.90	280.54	279.82	281.52
June	254.98	265.01	252.77	252.17	250.43
July	247.68	253.44	241.90	241.37	238.27
August	195.75	240.67	229.07	228.57	226.08
September	153.43	232.35	220.37	219.83	219.45
October	104.83	194.94	184.62	184.19	183.00
November	84.30	167.85	158.61	158.27	156.15
December	97.20	170.41	160.88	160.55	157.52

Mean Temperature

It is not necessary to use mean temperature in the computation of Penman evaporation because maximum temperature may exist for a shorter duration whereas minimum temperature may prolong for a longer duration (or vice-versa). Hence it was decided to reduce the mean temperature by 10%, that is a temperature factor of 0.90 is used for calculating the next set of results (set (iii)), a reduction factor of 0.90 was chosen as 10% can be accepted as a normal deviation in hydrological data. The results were obtained for albedo = 0.10 and temperature factor = 0.90. A comparison of values shows only a very marginal improvement in the results (see Table 3).

Vapour Pressure Gradient

According to Dalton's law of evaporation, E is proportional to vapour pressure gradient, hence it was decided to reduce the value of vapour pressure gradient and observe the resulting solutions. A 10% increase was applied to e_a and the results were observed. A perusal of observed results does not show any improvement in Penman evaporation values.

Thus the changes in values of albedo, mean temperature and vapour pressure have altered the results as the theory of evaporation would expect, but the improvement is only very marginal (see Table 3). For ready reference the monthly mean of Pan and Penman evaporation values for set (ii) data, January 1990-93 is graphically presented in the form of bar chart in Fig. 2.

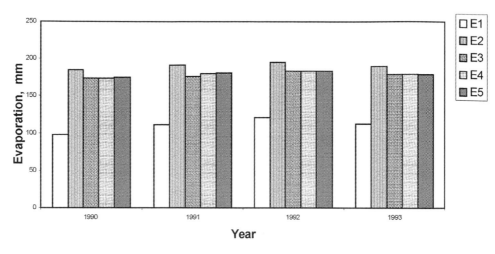

E1 = Pan Evaporation
E2 = Penman Evaporation (Albedo = 0.05)
E3 = Penman Evaporation (Albedo = 0.10)
E4 = Penman Evaporation (Albedo = 0.10, Temperature factor = 0.90)
E5 = Penman Evaporation (Albedo = 0.10, Vapour pressure factor = 1.10)

Figure 2. Monthly mean evaporation values, January 1990-1993.

Regression and Correlation Analysis

Monthly Evaporation

The monthly evaporation values for the months January to December were obtained and their values for each set of data were presented in Table 3.

The set (ii) data were *linearly regressed* and the results were furnished in Fig. 3. A straight-line y = 0.60x + 117.54, where x represents Pan evaporation and y represents Penman evaporation, was obtained. The correlation coefficient was found to be 0.92, which is highly satisfactory. Also the above equation was subjected to *chi-square test for goodness of fit* and its calculated value was found to be 12.17, which is less than the tabulated value of chi-square for eleven degrees of freedom at 5% level of significance.

Weekly, Daily and Annual Evaporation

Similar procedure was adopted for weekly, daily and annual evaporation values. The results were tabulated in Table 4. The annual evaporation values for both methods were presented for all the four sets of data in Table 5. The table shows how the Penman method is highly overestimating the evaporation values.

Regression and Correlation Analysis

Monthly Evaporation

The monthly evaporation values for the months January to December were obtained and their values for each set of data were presented in Table 3.

The set (ii) data were *linearly regressed* and the results were furnished in Fig. 3. A straight-line y = 0.60x + 117.54, where x represents Pan evaporation and y represents Penman evaporation, was obtained. The correlation coefficient was found to be 0.92, which is highly satisfactory. Also the above equation was subjected to *chi-square test for goodness of fit* and its calculated value was found to be 12.17, which is less than the tabulated value of chi-square for eleven degrees of freedom at 5% level of significance.

Weekly, Daily and Annual Evaporation

Similar procedure was adopted for weekly, daily and annual evaporation values. The results were tabulated in Table 4. The annual evaporation values for both methods were presented for all the four sets of data in Table 5. The table shows how the Penman method is highly overestimating the evaporation values.

From the Figs. 3 to 5 and Table 4, it is inferred that there is a consistent relationship between Pan and Penman evaporation values. The straight line fit is very satisfactory and the difference between the Pan and Penman evaporation values are evident and pronounced.

Table 4. Regression and correlation analysis of set (ii) data.

Evaporation values	Equation for the fitted straight line	Correlation co-efficient	Remarks
Monthly	y = 0.60x + 117.54	0.92	Refer Fig. 3, averaged over four years
Weekly	y = 0.59x + 27.47	0.91	Refer Fig. 4, averaged over a week for four years
Daily	y = 0.59x + 3.92	0.91	Refer Fig. 5, averaged over a week for four years
Annual	y = 0.19x + 216.33	—	Results can be improved by applying more number of values, Refer Table 5

Table 5. Annual evaporation, 1990-1993.

Year	Pan evapora-tion mm	Penman evaporation mm	Difference mm	Remarks
1990	1957.20	2675.96	-718.76	Albedo = 0.05
1991	2029.30	2772.72	-743.42	
1992	2029.90	2748.67	-718.77	
1993	1918.24	2766.34	-848.10	
1990	1957.20	2534.78	-577.58	Albedo = 0.10
1991	2029.30	2625.84	-596.54	
1992	2029.90	2606.82	-576.92	
1993	1918.24	2622.19	-703.95	
1990	1957.20	2528.83	-571.63	Albedo = 0.10
1991	2029.30	2624.76	-595.46	Temp. Factor 0.9
1992	2029.90	2601.10	-571.20	
1993	1918.24	2616.24	-698.00	
1990	1957.20	2521.20	-564.00	Albedo = 0.10
1991	2029.30	2626.19	-596.89	Vapour pressure Factor = 1.10
1992	2029.90	2589.71	-559.81	
1993	1918.24	2603.34	-685.10	

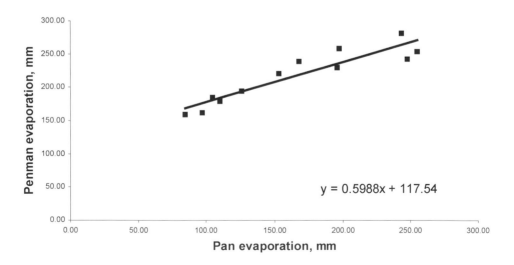

$y = 0.5988x + 117.54$

Figure 3. Regression of monthly evaporation values averaged over four years, 1990-1993 (Albedo = 0.10).

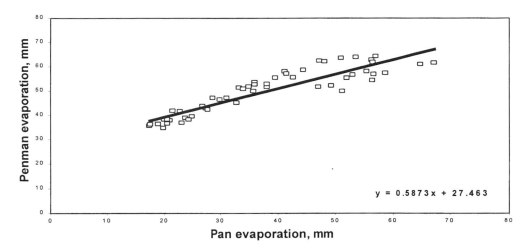

Figure 4. Regression of weekly evaporation values averaged over four years, 1990-1993 (Albedo = 0.10).

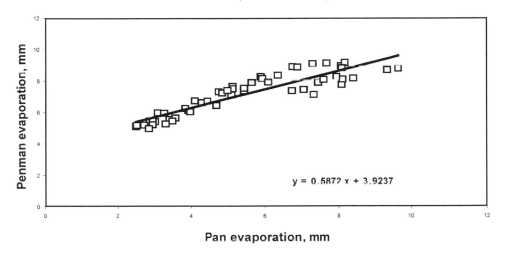

Figure 5. Regression of daily evaporation averaged over a week for four
years, 1990-1993 (Albedo = 0.10).

Evaporation and Evapotranspiration

The relationship between the evapotranspiration (ET) and Pan evaporation E_p is usually expressed as $ET = KE_p$, where K is the crop factor and is found to vary according to the stage of growth of the crop.

The reason for the vast difference between the Pan and Penman evaporation values may be as follows. Pan evaporation is measured from a Pan with stagnated water surface and which the temperature at water surface may influence and its

245

corresponding saturated vapour pressure. But in Penman's, the mean air temperature and its corresponding saturated vapour pressure are adopted. Thus it happens that a water surface evaporation is being compared with estimated evaporation from a land surface, in other words, the Penman evaporation perhaps is actually evaluating an evapotranspiration than a pure evaporation. To support this discussion, the observed values of ET and Pan evaporation at Annamalainagar, September 1977, were presented in Table 6, when paddy IR20 was being cultivated. The difference between these two values is found to be similar to the difference be-

Table 6. Comparison of Pan evaporation & evapotranspiration, Sep. 1977.

Day	Pan evaporation mm	Evapotranspiration mm	Difference mm
1	4.50	7.00	-2.50
2	4.90	6.90	-2.00
3	6.40	8.40	-2.00
4	4.80	5.80	-1.00
5	4.80	6.30	-1.50
6	4.10	5.10	-1.00
7	4.00	5.50	-1.50
8	5.00	7.00	-2.00
9	5.40	7.20	-1.80
10	5.10	6.90	-1.80
11	5.60	7.30	-1.70
12	5.50	9.70	-4.20
13	6.20	7.40	-1.20
14	5.80	6.60	-0.80
15	4.90	6.50	-1.60
16	3.20	4.90	-1.70
17	3.40	5.30	-1.90
18	4.00	7.30	-3.30
19	5.10	6.10	-1.00
20	3.60	6.00	-2.40
21	3.80	7.40	-3.60
22	3.20	5.80	-2.60
23	3.80	7.40	-3.60
24	4.00	5.70	-1.70
25	4.00	5.90	-1.90
26	3.80	6.50	-2.70
27	3.50	7.30	-3.80
28	2.40	4.90	-2.50
29	2.30	5.20	-2.90
30	1.90	4.60	-2.70

tween Pan and Penman evaporation values observed from 1990 to 1993 (Refer Tables 2 and 6).

The consistent overestimation by the Penman method leads to the inference that the evaporation values estimated by Penman method are more nearly ET values. This inference can be accepted as the modified Penman evaporation method. Thus the straight-line equations obtained by regression analysis can be used to determine ET values from the given Pan evaporation values.

Conclusions

The following conclusions can be made on the basis of this study.

(i)   At Annamalainagar, the evaporation values estimated by Penman method are consistently higher than evaporation values measured with Pan. Therefore, the applicaion of Penman equation at Annamalainagar should be done with caution.
(ii)   For small individual changes in mean temperature or mean vapour pressure, the Penman method is not sensitive. However when albedo value was taken as 0.10 instead of 0.05, the results were improved.
(iii)   There is a clear linear correlation between Pan evaporation and penman evaporation. This correlation may enable to predict evapotranspiration values if a proper value for albedo is used in Penman equation.

Future work will involve the study of effect of ignoring advected energy and stored energy in the Penman evaporation equation and their effect on the estimation of evaporation values.

References

Chow, V.T., Maidament, D.R., and Mays, L.W. (1988). "Applied Hydrology". McGraw-Hill Book Company, New York.

Jayarami Reddy, P. (1994). "A text book of Hydrology". Laxmi Publications, New Delhi.

Jayarami Reddy, P. (1987). "Stochastic Hydrology". Laxmi Publications, New Delhi.

Ramdas, L.A. (1996). International symposium on water evaporation control, UNESCO South Asia Science Corporation Office, New Delhi.

Varma, C.V.J. (1996). "Manual on evaporation and its restriction from free water surfaces". Central Board of irrigation and power, Oxford & IBH publishing Co. pvt. Ltd., New Delhi.

Surface Water Hydrology, Singh, Al-Rashed & Sherif (eds)
© 2002 Swets & Zeitlinger, Lisse, ISBN 90 5809 363 8

Quantifying volumetric evaporation in a semi-arid region of Iran using the modified Sequent Peak Algorithm reservoir planning model

M. Montaseri
Department of Irrigation Engineering, Urmia University, Iran

A.J. Adeloye
Department of Civil and Offshore Engineering, Heriot-Watt University
Edinburgh EH14 4AS, UK
e-mail: a.j.adeloye@hw.ac.uk

Abstract

Open water evaporation from reservoir surfaces constitutes a major demand on available water resources in arid and semi-arid regions of the world, which if not considered during planning of such reservoirs can result in significant under-sizing of facilities. However, despite this, only few planning analysis techniques accommodate the effect of evaporation, because its consideration complicates the analysis and because evaporation measurements are often unavailable. It will therefore be useful if simple rules for evaporation loss adjustment can be provided for use in such circumstances. In this work, reservoir simulation studies based on a modified form of the sequent peak algorithm were carried out to quantify reservoir surface evaporation losses during reservoir critical drawdown periods. The studies used weather and flow data from three basins in north-western Iran and the results were used to develop a tentative rule for adjusting capacity estimates for evaporation in the region.

Introduction

The main causes of water losses from dam impoundment include seepage, infiltration and evaporation but while seepage and infiltration can be largely controlled by appropriate choice of construction materials and methods, evaporation loss is not so easy to control. This inevitable evaporation loss from reservoir surfaces can be significant in arid and semi-arid regions. For example, Gan et al. (1991) used data from Australia and they found that the storage capacity for satisfying a given yield increases when reservoir surface evaporation losses are included in the analysis. Fennessey (1995) analysed a number of Massachusetts, USA, systems and found that for a given reservoir capacity, the deliverable yield decreases by about 6% when evaporation losses are considered compared with when they are ignored. It is therefore important that evaporation should be considered during the planning analysis of water resources systems made up of dams, so that the resulting reservoir capacity estimate will be adequate to meet both the useful consumptive demands and the inevitable evaporation losses.

However, while it is important to allow for surface evaporation in dam reservoir analysis, most analysts often fail to do this for four main reasons:

(a) evaporation depends on the exposed reservoir surface area which in turn depends on the amount of water in storage, both of which vary temporally during analysis;
(b) there are very few methods of reservoir planning analysis which can explicitly accommodate net evaporation;
(c) those methods which can accommodate evaporation are often much more complex to implement than those that do not; and
(d) measurements of open water evaporation are scarce and significant uncertainties are associated with existing expressions for predicting evaporation from climatic data.

Because of the above problems, it has become the usual practice for most analysts to add an arbitrary allowance to the without-evaporation storage capacity estimate, in the hope that such will be sufficient to meet the inevitable evaporation losses. However, while this practice is expedient, there is often no guidance on what this allowance should be or on how the allowance should vary with the yield, storage capacity and reliability. Answers to these questions have been provided in this work.

Using streamflow, evaporation and rainfall data from three sites in the semi-arid climate of Iran, reservoir capacity and volumetric evaporation losses were assessed using a modified form of the Sequent Peak Algorithm (SPA) reservoir planning model. The evaporation losses were then expressed as ratio of the reservoir capacity estimate. The results were then used to prepare a graphical guide for correcting for evaporation in reservoir analysis, which is more objective than using arbitrary adjustments. Such information should prove useful in evaporation-data-scarce regions or during preliminary reservoir planning analyses.

In the next Section, details of the modified SPA are presented, highlighting its advantages over other reservoir planning techniques (see McMahon and Mein, 1986), particularly because it could be used to plan to different reliability and vulnerability targets. Subsequent Sections describe the data used, results obtained and the main conclusions and recommendations of the investigation.

Reservoir Storage-Yield-Performance Planning Using the SPA

Reservoir planning is concerned with determining the storage required in a reservoir for meeting a given yield with an acceptable level of performance. Performance measures often considered include the reliability, resilience, vulnerability, and sustainability (Loucks, 1997). The time-based reliability is the proportion of total time for which a reservoir system meets the demand, and can be expressed as:

$$R = 100(T - f)/T \qquad (1)$$

where T is the total number of time periods in the simulation, f is the number of periods during which there is insufficient water to meet the demand, and R is the time-based reliability (%). The time-based reliability is simple to determine and use; however, it neither gives any idea about the magnitude of the water shortage during failure nor the ability of the reservoir to recover from failure.

The resilience is the probability that a reservoir would recover from failure and can be estimated using $\beta = f_c /f$, where β is the resilience; and f_c is the total number of continuous sequences of failures. For example, a reservoir having four failure periods, with each failure period followed by at least a period of normal operation is more resilient (resilience = 1.0) than another reservoir whose failure periods follow one another without an intervening period of successful performance (resilience = 0.25).

The vulnerability measures the magnitude of water shortage and can be taken as the mean of the maximum shortfalls occurring in each of the continuous failure periods:

$$\eta = \frac{\sum_{k=1}^{f_c} max.(sh_k)}{Df_c} \tag{2}$$

where η is the vulnerability expressed as ratio of the demand D; and max. (sh_k) is the maximum shortage in continuous failure sequence k. The complement of Eq. (2), i.e. $1 - \eta$, has some resemblance to the volumetric reliability commonly used in reservoir performance assessment (see Nawaz and Adeloye, 1999). The volumetric reliability is the total water supplied divided by the total water demanded over the simulation period T and, if the shortage per failure period is fixed for example, it can be readily shown that the two indices are related by:

$$R_v = 1 - \eta(1 - 0.01R) \tag{3}$$

where R_v is the volumetric reliability and R is the time-based reliability defined in Eq. (1). Indeed, if consideration of volumetric reliability is limited to the failure periods within T, then the volumetric reliability is exactly equal to the complement of the vulnerability, $1 - \eta$. This result follows immediately from Eq. (3), since over the failure periods, the time-based reliability R is zero. In this work, subsequent uses of the volumetric reliability will be taken as $(1 - \eta)$.

Finally, there has recently been an increasing tendency to use sustainability criteria for reservoir planning assessment (see e.g. Takeuchi et al., 1998). However, most of the available sustainability criteria – e.g. fairness, reversibility, consensus (see Simonovic et al., 1997) – have yet to find widespread use in quantitative water resources analysis precisely because they are largely subjective and hence cannot be easily expressed in mathematical terms. However, one plausible criterion of sustainability as far as reservoir planning is concerned would be a figure of merit FM (Simonovic, 1998) that integrates the effects of reliability, resilience and volumetric reliability. Apart from serving as a sustainability criterion, such a FM will also remove the complexity of assessing the performance of reservoir systems, caused by the numerous possible trade-offs between the reliability, resilience and vulnerability, as demonstrated by Moy et al. (1986). The sustainability index introduced by Loucks (1997) is one such FM and is defined as:

$$\varphi = 0.01R(\beta)(1-\eta) \tag{4}$$

where φ is the sustainability. Other FM's have been used in the literature, such as the drought risk index (Zongxue et al., 1998); however, the sustainability index in Eq. (4) is plausible given its multiplicative nature which ensures that $0 \le \varphi \le 1$, the usual range for all other reservoir performance indices. The multiplicative form also ensures that more weight is given to the lowest of the three component indices R, β and $(1 - \eta)$ defining φ. For example, if any of the indices is zero, then whatever values are taken by the other indices become irrelevant since their product will still be zero. Furthermore, as noted by Loucks (1997), changing the index whose value is originally high will have the least effect on the sustainability index. Thus since most reservoirs are often designed for high time-based and volumetric reliabilities, then attempting to further improve on these two indices will produce the least effect on the sustainability index. In such circumstances, better results will be obtained by improving the resilience.

There are very few reservoir-planning techniques that can design to specified targets of the reliability, vulnerability and resilience. One of such few techniques is the SPA, a critical period approach utilising a mass balance equation in reservoir sequential deficits.

The Basic SPA

In the original, basic form of the SPA, i.e. when there are no failures, the required reservoir capacity of an initially full reservoir for meeting the demand is obtained as the maximum of all the sequential deficits as follows (Loucks et al., 1981):

$$K_t = \begin{cases} K_{t-1} + D_t + E_t - I_t; & \text{if} > 0.0 \\ 0.0; & \text{otherwise} \end{cases} \tag{5a}$$

$$K_a^* = \max(K_t); \quad t = 1,2,3\ldots T, +1, \ldots, 2T \tag{5b}$$

where K_{t-1} is the volumetric sequential deficit at the beginning of period t; K_t is the corresponding volumetric deficit at the end of t; D_t is the volumetric demand during period t; E_t is the volumetric net evaporation during t; I_t is the volumetric inflow during period t; K_a^* is the exact capacity estimate; and T is the total number of periods in the data record. Note that because the reservoir is assumed to be initially full, then $K_o = 0.0$. It should also be noted that although the formulation in Eq. 5 requires two cycles of the data record, the second cycle is unnecessary if the reservoir refills at the end of the first cycle, i.e., if $K_T = 0.0$. Finally, because E_t depends on the exposed surface area of the reservoir, which in turn depends on the storage, E_t cannot be explicitly included in the basic SPA and is thus ignored. The resulting capacity is therefore an approximation, denoted by K'_a.

Accommodating Evaporation in the SPA to Determine Exact Capacity

Evaporation depends on the non-linear area-storage function at a dam site; however, such non-linear relationships are sometimes approximated by linear functions of the type (Dandy et al., 1997):

$$A_t = aS_t + b \tag{6}$$

where A_t is the reservoir exposed surface area at the beginning of t ($\times 10^6$ m^2), S_t is the corresponding storage ($\times 10^6$ m^3), and a and b are empirical constants obtainable by regression fitting of the area-storage data at the site. With Eq. 6 established, the volumetric evaporation loss ($\times 10^6$ m^3) during t can then be approximated by the mean over the interval [t, t+1] as:

$$E_t = en_t(A_t + A_{t+1})/2 \tag{7}$$

where A_{t+1} is the surface area at end of t; and en_t is the depth of net evaporation, i.e. evaporation minus rainfall (m) during t. Using Eq. 6, Eq. 7 can be expressed in terms of storage as:

$$E_t = en_t(b + 0.5a(S_t + S_{t+1})) \tag{8}$$

Lele (1987) first reported a modification of the SPA to accommodate evaporation using Eq. 7. A modification of Lele's approach reported by Adeloye et al. (2001) was used in this study. The incorporation of evaporation by Adeloye et al. (2001) takes place in a number of iterative steps. First, a forward pass is made using the basic SPA to determine K'_a and identify the critical period. (The critical period is uniquely identified in the SPA as that between K'_a and the first occurrence of zero in the K_t series, moving backward from K'_a). Then the approximate storage states (S_t) are estimated by subtracting K_t from K'_a; the S_t's are then used to obtain the surface evaporation E_t's with Eq. 8. The estimates of the E_t are then included in the SPA and a new estimate of K'_a is obtained; this process is repeated until the capacity settles down, usually after four or five iterations. The new algorithm employs a convergence criterion given by Eq. 9:

$$\left| \frac{(K_a - K'_a)}{K_a} \right| \leq 0.0001 \tag{9}$$

where K_a and K'_a are the estimates of the active storage capacity in any two successive iterations respectively.

253

It is rarely the case that a reservoir system will be designed to be failure-free. Usually, some limited failures are allowed as long as the resulting performance of the reservoir is satisfactory. Performance norms that can be specified as inputs in the modified SPA are the reliability (see Eq. 1) and vulnerability (see Eq. 2); the resilience and sustainability are output performance indices. In the study, the water shortage was fixed for all periods, making the vulnerability equal to this fixed shortfall quantity. As noted by Adeloye et al. (2001), fixing the vulnerability in this way can help to cure the "misbehaviour" of reservoir capacity statistics once reported by Pretto et al. (1997).

To incorporate both the reliability and vulnerability norms, the number of failure periods, f, must first be specified using:

$$f = T(1 - 0.01R) \tag{10}$$

where R is the reliability (%) defined in Eq. 1. Then the release in each of the f failure periods has to be set to the full demand less the vulnerability. This reduced release will henceforth be referred to as the drought draft. Given that the critical period is a period of extreme low flows, it is logical to assume that the failures will occur during the critical period. The critical period is however liable to shift when evaporation is being considered. Thus, the iteration to reduce the release over f failure periods starts by assuming that the first failure period coincides with the end of the critical period as identified using the modified SPA for the no-failure case, and incorporating evaporation. The basic SPA without evaporation is then run with the demand in this first failure period, $f = 1$, set to the drought draft while the demand in all other periods is set to the full demand. This will lead to the estimation of the approximate storage K_a' with a single failure period. Evaporation is then added as described previously and the exact storage capacity K_a^* for a single failure period is then determined.

The iteration then moves to $f = 2$, i.e. with two failure periods in the simulation. One of these failure periods has been identified and was used in the previous iteration; a second one, which is distinct from the first, has now to be identified. The way this is achieved is to examine the end of the new critical period following convergence at $f = 1$: if this end period is not the same as the one used in the first iteration, then it is taken as the second failure period; otherwise the second failure period is taken as the period just preceding it. By doing this, it will be ensured that no single period is used more than once as a failure period, and that ultimately the total number of independent failure periods considered in the iteration is equal to that corresponding to the given reliability as estimated using Eq. 10. This process is then continued until releases for f independent failure periods have been adjusted. The resulting capacity is that required for the reliability and vulnerability.

Given an appropriate operating policy, the modified SPA can also be applied for planning multiple reservoir systems. This has been described in Nawaz et al.

(1999) for a system of parallel reservoirs using the Space Rule (Oliveira and Loucks, 1997).

Application

River Basins

The above methodology was applied to three river basins – Baranduz, Shahr and Nazlu – in N-W Iran. These basins are located between $37°6'$ and $38°0'$ northern latitudes, and $44°19'$ and $45°5'$ eastern longitudes, with a minimum altitude of 1300 m above sea level (Fig. 1). More than 96% of the annual rainfall occurs during the winter season from October to July, with approximately 50% of this falling as snow.

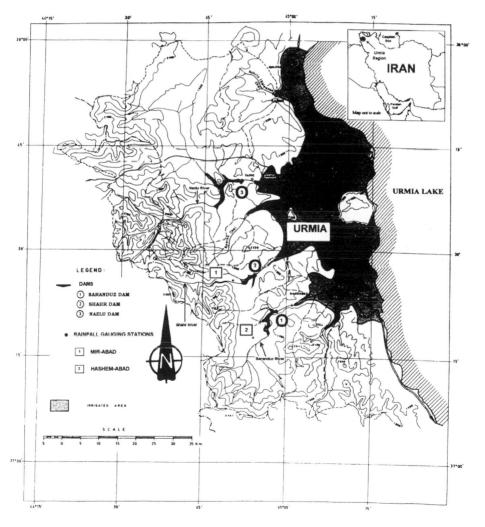

Figure 1. Map showing the location of the proposed reservoir sites in Urmia region, Iran.

Summer temperatures can reach 29 ^{0}C leading to excessive evaporative losses from reservoir surfaces.

The three basins and their planned reservoirs are very strategic for the water resources situation in the region, as they will provide municipal water for Urmia city and irrigation water for Urmia irrigation area. The management of the rivers, including gauging and control of abstraction, is done by the Iranian Ministry of Water and Power. The total irrigated area will be 66000 ha, with the annual demands from each of the Baranduz, Shahr and Nazlu reservoirs being estimated as 66%, 76%, and 80% of the respective mean annual runoff (Ministry of Water & Power, Iran, 1995). In addition, a quarter of the total withdrawal from the planned reservoir on Shahr River will be used for Urmia city water supply. Relevant basin characteristics at the respective reservoir sites are shown in Table 1.

Data Availability

Monthly runoff data records are available at each of the sites for 44 years (1950-1993). Annual summary statistics for these runoff records are shown in Table 1. In general, the coefficients of variation (CV) of the annual runoff in Table 1 are moderate, which would indicate that both within-year and over-year storage behaviours would be significant at the sites (Montaseri and Adeloye, 1999). With this being the case, it will be important that the reservoir analysis uses monthly time series data of runoff so as not to under-estimate reservoir total storage capacity.

Open water pan evaporation and rainfall measurements are also available and these are summarised in Table 2. These are required for characterising the reservoir surface net evaporation loss (see Eqs 6–8). It is quite clear from Table 2 that, in general, the monthly mean rainfall is lower than the corresponding open water evaporation, implying that a net outflow flux of water from the reservoir surfaces will occur in most months of the year.

Finally, area-storage data for the sites obtained for the Iranian government by consultants working on the Urmia project were also available. These area-storage data were used to fit equations of the type in Eq. 6 for converting net evaporation depths into volumetric units as described in a previous section. The resulting regression equations are A = 1.13 + 0.041S (Baranduz); A = 0.9 + 0.046S (Shahr) and A = 1.69 + 0.032S (Nazlu). Fig. 2 shows plots of the area-storage data and their fitted linear approximations, which show that such linear approximations to the area-storage relationship, are reasonable.

Table 1. Characteristics of the river basins.

| Site | Reservoir | Area km^2 | Annual flow | | | |
			Mean (10^6 m^3)	CV	Skewness	lag-1 correlation
1	Baranduz	618	272.7	0.34	1.21	0.34
2	Shahr	418	170.3	0.38	1.04	0.38
3	Nazlu	1715	390.3	0.41	1.03	0.41

Table 2. Mean monthly rainfall (mm) and open water evaporation (mm) at the sites.

Month	Baranduz		Shahr		Nazlu	
	Evap	Rain	Evap	Rain	Evap	Rain
Jan.	21.0	25.7	21.3	37.7	20.3	27.7
Feb.	34.4	25.7	34.5	39.6	33.8	31.7
Mar.	62.6	44.7	59.7	58.5	58.7	55.9
Apr.	93.5	57.6	92.4	69.9	91.4	78.1
May.	134.8	52.0	130.1	58.7	127.6	77.6
Jun.	173.6	17.2	167.5	25.8	165.6	35.1
Jul.	198.1	2.8	193.5	6.1	191.2	8.5
Aug.	181.5	2.7	187.2	3.5	184.8	3.1
Sep.	155.2	1.4	150.3	3.3	148.2	8.4
Oct.	109.1	20.7	104.2	29.4	102.0	29.4
Nov.	61.5	32.7	59.0	38.7	57.8	42.4
Dec.	25.9	35.1	25.8	47.5	25.0	34.5

Results and Discussion

Monte Carlo Experiments

Rather than employing the single historic flow data records at the sites, 1000 alternative inflow runoff sequences were generated using stochastic models calibrated with the historic records. The stochastic modelling took account of the uncertainties in the historic parameter estimates, using the approach recommended by Stedinger and Taylor (1982). Prior to using the replicated data in the reservoir planning analysis, the adequacy of the stochastic data generation models was checked by comparing various attributes of the generated and observed data. These include the mean, variance, skewness, lag-1 serial correlation coefficient, and annual runoff frequency curves. The adequacy tests were satisfactory. Details of the stochastic modelling and the results of the adequacy tests are provided by Montaseri (1999).

Stochastic modelling of the rainfall and evaporation data, which determine the net evaporation from the reservoir surface, was not carried out; rather the mean monthly values of these variables as shown in Table 2 were used throughout. As shown by Fennessey (1995), there is little difference between using mean monthly evaporation and the time series of monthly evaporation data when assessing the impact of reservoir evaporation on storage-yield relationship. Consequently, ignoring the stochasticity in the monthly evaporation and rainfall data, whilst helping to significantly reduce the required computational effort, should have little impact on the outcome of the study.

Relevant Reservoir Planning Characteristics

The mean of the 1000 capacity estimates are shown in Table 3, for annual demands varying from 20 to 80% of the mean runoff. For the 98% reliability, the shortfall (or vulnerability) during failures was fixed at 30% of the full demand. In

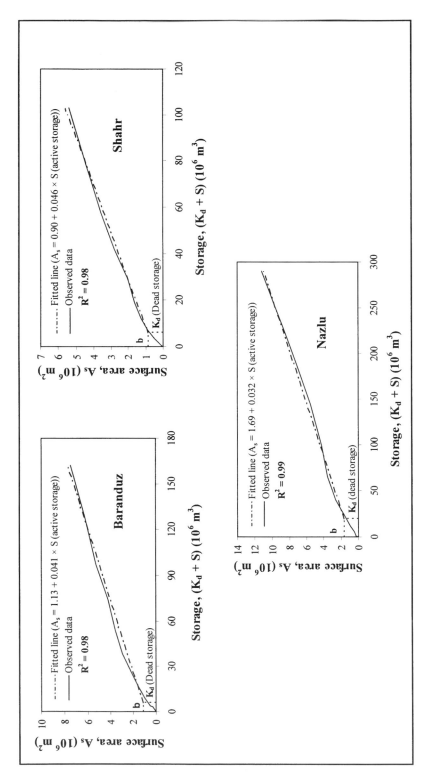

Figure 2. Reservoir surface area-storage relationship for the Iranian sites.

general, the expected behaviour in the storage-demand relationship was obtained, in which the storage increases as the demand ratio increases. Although the results in Table 3 have not been plotted as classical storage-demand functions (see Nawaz and Adeloye, 1999), it is nonetheless evident in Table 3 that the elasticity of the storage, with respect to the demand, is much greater at high demands than at low demands. In other words, at low demand ratios, increasing the demand merely causes the storage to increase marginally whereas as the demand ratio becomes high and approaches unity, significant increase in storage will be required to meet a very modest increase in the demand ratio. This was why Nawaz and Adeloye (1999) suggested that it might not be economical to attempt to enhance reservoir yield by increasing storage when the demand ratio is already high. In such situations, developing alternative sources, such as groundwater and river abstraction schemes, might offer a much more effective solution.

The mean critical periods in months are shown in Fig. 3. This information is relevant because the evaporation loss will be estimated over the critical period; so the longer the critical period, the larger is the volume of evaporation loss expected to be. As seen in Fig. 3, the mean critical period is largely constant over a wide range of demands until the 60% annual demand level when significant increases in the critical period with increasing demand occur. Since total evaporation loss depends on both the reservoir storage and length of the critical period then, based on the results in Fig. 3, evaporation is likely to be disproportionately much higher at high demands than at low demand ratios.

The relationship between the reliability, resilience, demand and vulnerability is shown in Fig.4. According to Fig. 4, a reservoir is more resilient at higher vulnerability for a given demand ratio. Similarly, high demand systems exhibit lesser resilience than low demand systems, albeit slightly. Indeed, as noted by Montaseri and Adeloye (1999), the change in slope of the resilience-reliability-vulnerability-demand curves actually occurred at the region of transition from within-year to over-year storages. These results are consistent with those obtained by Moy et al. (1986), who investigated the trade-offs between reservoir performance criteria and concluded that an increase in vulnerability can result from either a more reliable or a more resilient reservoir situation.

Fig. 5 shows, for a fixed time-based reliability of 98%, the relationship between the sustainability index, the demand ratio and the vulnerability. Plots for the other time-based reliability targets considered in the study are similar and are provided by Montaseri (1999). As remarked in an earlier Section, the sustainability index is a

Table 3. Mean capacity estimates ($\times 10^6$ m^3) at the sites (R = 98%, vulnerability = 30%).

Site	Demand Ratio (%)						
	20	30	40	50	60	70	80
Baranduz	15.10	29.80	46.56	65.19	87.00	123.63	217.92
Shahr	8.17	16.49	27.08	41.61	65.33	103.88	173.09
Nazlu	30.83	53.21	79.39	114.05	171.20	270.08	458.67

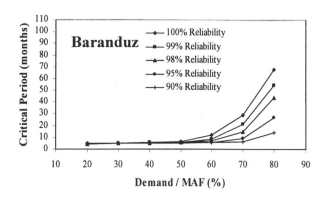

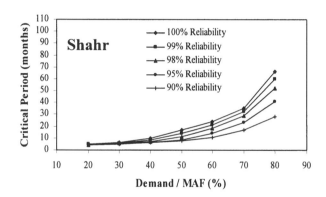

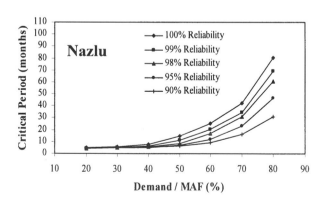

Figure 3. Variation of mean critical periods (months) with demand and reliability at the sites.

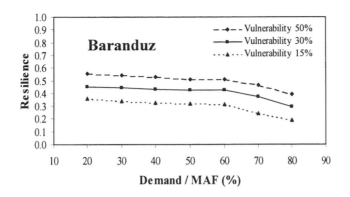

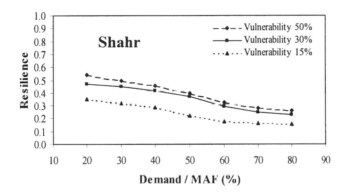

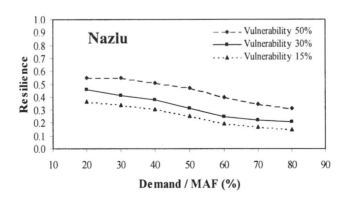

Figure 4. Variation of mean resilience with demand and vulnerability at the sites.

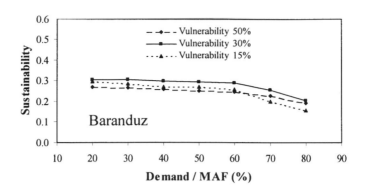

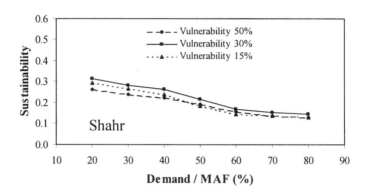

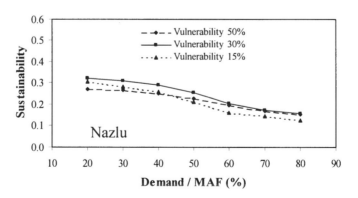

Figure 5. Variation of sustainability index with demand and vulnerability for the three sites. (R = 98%; resilience are as presented in Figure 4.)

relatively recent introduction in the water resources literature when compared with the reliability, resilience and vulnerability, which have been used in characterising reservoir performance for some time. Consequently, published reservoir planning studies incorporating quantitative measures of sustainability are few or, to the best of our knowledge, non-existent. The results in Fig. 5 show that reservoir sustainability will decrease with increasing demand for any given vulnerability, although the marginal reduction is higher at higher demands than at low demands. The time-based reliability used for deriving Fig. 5 is 98% and the volumetric reliability ranged from 50% (vulnerability = 50%) to 85% (vulnerability = 15%). However, despite the high values of these two indices, the sustainability index rarely exceeds 0.3, a value more compatible with the resilience (see Fig. 4). Since both the time-based and volumetric reliabilities are already high, any further increase in these two performance indices will, as noted by Loucks (1997), have little or no influence on the sustainability index. The most effective way of increasing the sustainability therefore is by improving the resilience. An increased resilience is best achieved by decreasing the demand ratio for a given capacity (see Moy et al., 1986).

Another conclusion, albeit tentative, from Fig. 5 is that a low shortfall (or vulnerability) during failure does not necessarily imply a high sustainability. Indeed, the sustainability in Fig. 5 seems to be peaking at the 30% vulnerability level.

Evaporation Losses

Table 4 contains, for the 98% time-based reliability, the mean volumetric evaporation losses from the reservoirs over their respective critical periods. These results show that the volumetric evaporation loss increases as the demand (and hence reservoir capacity) increases, as expected. The estimated evaporation losses also decrease as the reliability decreases, which is also expected given that the capacity decreases as reliability decreases. Further, these behaviours of volumetric evaporation are also plausible because the critical period over which it is calculated lengthens as the demand increases (Fig. 2).

However, the marginal increase in evaporation losses is not uniform and while for example an increase in demand by 10% from 20% to 30% causes the evaporation loss to increase by 40%, the same change in demand from 70% to 80% almost quadruples the volumetric evaporation loss. This is a consequence of the flattening of a typical storage-demand function in the region of high demand ratios as observed previously (see also Nawaz and Adeloye, 1999). In such regions of high demand ratios, the required capacity (and its surface) area will very large which, combined with the very long critical periods, will ensure that volumetric evaporation loss is very high.

Rule for Evaporation Loss Adjustment During Planning

To enhance the usefulness of the evaporation loss results for reservoir planning, the volumetric evaporation losses have been expressed as ratios of the capacity and these evaporation ratios are shown in Fig. 6. The evaporation ratio versus demand ratio curves are concave in shape, with a minimum of about 0.05, occurring in the 40%–50% demand region. Further, the evaporation ratio increases as the reliability increases; however, the impact of the reliability is more pronounced at high demands possibly because of the response of storage and the critical periods to the demand as noted previously.

Comparing Figs. 6a–c, it would be observed that, in general the evaporation ratio at all sites is within 3% of one another. This modest inter-site variation in the evaporation ratios means that average values over the three sites can be taken as being representative of the evaporation loss ratios for a given reliability. These average ratios are plotted against the demand ratios in Fig. 6d. The solid thick line in Fig. 6d represents the average evaporation ratios over all the five time-based reliabilities, which can be used in situations where reliability consideration is not of concern.

To use the results in Fig. 6d to correct for evaporation, the estimate of the capacity without evaporation will first be obtained using the basic SPA or another suitable technique. Then entering Fig. 6d with the demand ratio, an estimate of the evaporation ratio will be read off the appropriate reliability curve or, where the reliability is not being considered, the thick curve in the Figure. Multiplying this evaporation ratio by the estimated capacity will give an estimate of the volumetric evaporation loss, which can then be added to the capacity estimate to give an estimate of the exact reservoir capacity.

Conclusion

Using data from three river basins in Iran, a tentative rule has been developed which can be used to correct reservoir capacity estimates for evaporation losses. The graphical rule expresses evaporation in terms of the demand and reliability, both of which are known *a priori* during reservoir planning. The rule is therefore easy to use and does not require knowledge of the actual evaporation rate at the location of interest. This will aid reservoir planning in regions where evaporation data for sequential analysis are unavailable. Further improvement in the rule will be possible by considering other sites in other regions of the world, thereby covering a wider range of climates and evaporation regimes. If this is carried out, then a simple expression relating the evaporation ratio to the demand and reliability can be developed. The development of such a regression equation is the focus of further work.

Table 4. Mean net evaporation losses ($\times 10^6$ m^3) for the sites (R = 98%, vulnerability = 30%).

Site	Demand Ratio (%)						
	20	30	40	50	60	70	80
Baranduz	1.60	2.00	2.47	2.93	3.83	8.04	32.69
Shahr	1.14	1.47	1.98	3.16	5.88	11.32	29.77
Nazlu	2.13	2.50	3.02	4.45	9.07	20.26	58.25

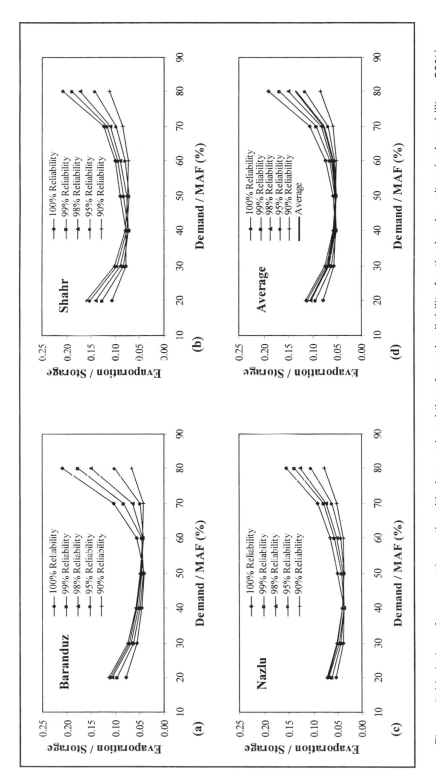

Figure 6. Variation of evaporation ratio with demand and time-based reliability for the Iranian sites (vulnerability = 30%).

References

Adeloye, A.J., and Montaseri, M. (1998). "Adaptation of a single reservoir technique for multiple reservoir storage-yield-reliability analysis". Proc. Int. Conf. (H. Zebidi, ed.), UNESCO, Paris, 349-355.

Adeloye, A.J., Montaseri, M., and Garmann, C. (2001). "Curing the misbehaviour of reservoir capacity statistics by controlling shortfall during failures using the modified sequent peak algorithm". Wat. Resour. Res., 37(1), 73-82.

Dandy, G.C., Connarty, M.C., and Loucks, D.P. (1997). "Comparisons of methods of yield assessment of multiple reservoir systems". J. Wat. Resour. Plan. Magt., ASCE, 123(6), 350-357.

Fennessy, N.M. (1995). "Sensitivity of reservoir-yield estimates to model time step surface-moisture fluxes". J. Wat. Resour. Plan. Managt., ASCE, 121(4), 310-317.

Gan, K.C., McMahon, T.A., and O'Neil, I.C. (1991). "Sensitivity of reservoir sizing to evaporation estimates". J. Irr. Drain. Engrg., ASCE, 177(3), 324-335.

Hashimoto, T., Stedinger, J.R., and Loucks, D.P. (1982). "Reliability, resiliency and vulnerability criteria for water resource system performance evaluation". Wat. Resour. Res., 18(1), 14-20.

Lele, S.M. (1987). "Improved algorithms for reservoir capacity calculation incorporating storage-dependent losses and reliability norm". Wat. Resour. Res., 23(10), 1819-1823.

Ministry of Water & Power, Iran (1995). "Reservoir systems project at Urmia Region". (Translated from Persian).

Montaseri, M. (1999). "Stochastic investigation of the planning characteristics of within-year and over-year reservoir systems". PhD thesis, Heriot-Watt University, Edinburgh, UK.

Montaseri, M., and Adeloye, A.J. (1999). "Critical period of reservoir systems for planning purposes". J. Hydrol., 224(3-4), 115-136.

Moy, W.-S., Cohon, J.L., and ReVelle, C.S. (1986). "A programming model for analysis of reliability, resilience, and vulnerability of a water supply reservoir". Wat. Resour. Res., 22(4), 489-498.

Loucks, D.P. (1997). "Quantifying trends in system sustainability". Hydro. Sci. J., 42(4), 513-530.

Nawaz, N.R., and Adeloye, A.J. (1999). "Evaluation of monthly runoff estimated by a rainfall-runoff regression model for reservoir yield assessment". Hydrol. Sci. J., 44(1), 113-134.

Nawaz, N.R., Adeloye, A.J., and Montaseri, M. (1999). "The Impact of climate change on storage-yield curves for multi-reservoir systems". Nordic Hydrol., 30(2), 129-146.

Oliveira, R., and Loucks, D.P. (1997). "Operating rules for multiple reservoir systems". Wat. Resour. Res., 33(4), 839-852.

Simonovic, S.P. (1998). Sustainability criteria for possible use in reservoir analysis. In: (K. Takeuchi et al., eds) Sustainable reservoir development and management, IAHS Publ. No. 251, 55-58.

Simonovic, S.P., Burn, D.H., and Lence, B.J. (1997). "Practical sustainability criteria for decision-making". Int. J. Sust. Dev. and World Ecology, 4(4), 231-244.

Takeuchi, K., Hamlin, M., Kundzewicz, Z.W., Rosbjerg, D., and Simonovic, S.P. (eds) (1998). "Sustainable reservoir development and management". IAHS Publ. No. 251.

Zongxue, X., Jinno, K., Kawamura, A., Takesaki, S., and Ito, K. (1998). "Performance risk analysis for Fukuoka water supply system". Wat. Resour. Managt., 12, 13-30.

Surface Water Hydrology, Singh, Al-Rashed & Sherif (eds)
© 2002 Swets & Zeitlinger, Lisse, ISBN 90 5809 363 8

Evaluation of evaporation and heat and moisture monitoring during post-irrigation drying

Chao He
Program for Doctor's Degree, Graduate School of Engineering
Fukui University, 3-9-1 Bunkyo, Fukui 910-8507, Japan
e-mail: chao@anc.anc-d.fukui-u.ac.jp

Teruyuki Fukuhara
Department of Architecture and Civil Engineering, Faculty of Engineering
Fukui University, 3-9-1 Bunkyo, Fukui 910-8507, Japan
e-mail: teruyuki@anc.anc-d.fukui-u.ac.jp

Yasuhide Takano
Department of Civil Engineering, Faculty of Science and Engineering
Kinki University, 3-4-1 Kowakae, Higashi-Osaka, Osaka 577-8502, Japan
e-mail: takano@civileng.kindai.ac.jp

Abstract

In order to better understand post-irrigation drying phenomena, measurements of evaporation flux and heat and moisture transfer monitoring in soil under bare surface were carried out in the United Arab Emirates. The evaporation flux was directly measured with a new type of evaporation pan and an electric balance. This paper describes a simple model to analyze the evaporation flux resulting from heat, liquid water and water vapor movement that occur in the near-surface soil layer during the post-irrigation drying phase. The validity of the proposed evaporation model is examined by comparing numerical with experimental results. It is concluded that the proposed evaporation model and the present monitoring system are adequate to evaluate the water budget of post-irrigation in arid regions.

Introduction

Arid regions are potentially attractive places for plant production because they are rich in solar energy needed to facilitate photosynthesis. Greening in arid regions could become an important key to solving global environmental problems such as the food crisis and global warming. The sustainability of agriculture in arid regions depends critically on economizing irrigation water, i.e. on the optimal usage and management of groundwater. To accomplish this matter, it is essential to establish the experimental techniques to evaluate evaporation flux from soils after watering and to monitor the heat and moisture transfer in soils.

In order to better understand post-irrigation drying phenomena of soil under bare surface, micro-meteorological observations and heat and moisture transfer measurements in soil under bare surface conditions were made continuously in the

United Arab Emirates (U.A.E.) starting in 1996 (Takano et al., 1999). The vertical profile of water vapor density during post-irrigation drying was obtained and it was shown that the increase in the thickness of the surface dry layer after watering can be monitored from water vapor density profile. In these observations, however, the measurement of evaporation flux was not conducted, so that knowledge of the evaporation was not sufficient.

On the other hand, the authors pointed out that the water vapor density in soil pores reaches a maximum at the interface between the dry surface layer and the unsaturated layer beneath it and that the internal evaporation rate can be calculated by the Fickian diffusion equation with acceptable accuracy, assuming a linear vapor density profile in the dry layer (Fukuhara et al., 1994a). Moreover, to estimate the soil surface evaporation by $\alpha - \beta$ method, the relationship between the volumetric water content of the soil surface layer and α and β were obtained in combination a "soil surface layer energy model" with indoor column experiments (Futagami et al., 1997).

Base on such our research background, additional experiments were carried out in U.A.E. in 2000 to examine the practical use of the proposed evaporation model, using the data from a new type of evaporation pan and from the heat and moisture transfer measurements. This paper describes the validity of the proposed evaporation model and monitoring system and presents the water budget of post-irrigation.

Outline of experiment and measurement techniques

Post-irrigation drying experiments were carried out on an experimental site at Hamuraniyah Agriculture Research Station, in Ras Al Khaima Emirate, U.A.E. in September 2000. The soil at the experimental site is classified as sandy loam with a mean grain size of 0.08 mm.

The measurement system is comprised of long-wave radiation and albedo meters, anemometers, thermo-hygrometers, Time Domain Reflectometry (TDR) moisture probes and evaporation pans (see Figure 1 and Photograph 1). Air temperature and relative humidity were measured with thermo-hygrometers, placed at 6 different heights ranging from 0.02 m to 1.0 m above the soil surface. Soil temperature and soil pore relative humidity profiles were obtained with soil thermo-hygrometers, inserted at 8 different depths ranging from 0.01 m to 0.5 m below the soil surface. Water vapor density was calculated from temperature and relative humidity. Volumetric water content profile was measured with 8 TDR moisture probes, set at different depths ranging from 0.01 m to 0.2 m below the soil surface.

The evaporation pan is made of stainless steel and its inside diameter and depth are 0.25 m and 0.15 m, respectively. Two pans were prepared for the evaporation experiments and were filled with the soil of the experimental site. Level of the pans was adjusted as shown in Photograph 1 until the soil surface in the pan and the surrounding soil surface become the same level. The evaporation flux was calculated from the time decrement of the pan weight measured by an electric balance with the minimum reading of 0.1 g.

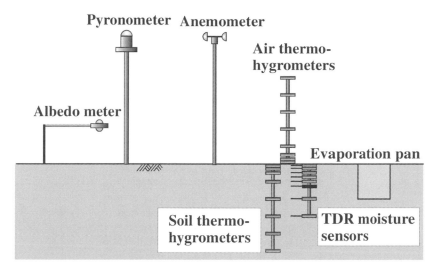

Figure 1. Experimental apparatus.

Photograph 1. Appearance of experimental site.

Watering on the bare soil (3 m x 3 m) was performed two times. First irrigation was started from 11:00 September 7, and the volume and temperature of sprinkled water were 4.5×10^{-2} m^3 (watering flux: 1.28×10^{-2} kg/m^{2}s) and 34.9 degrees centigrade, respectively. The second irrigation was started from 11:00 September 10. The major difference between two irrigations was the volume of sprinkled water. The water of 9.0×10^{-2} m^3 was uniformly sprinkled at a rate of 2.09×10^{-2} kg/m^{2}s in the second one.

271

Results of heat and moisture transfer monitoring

Meteorological conditions

Figures 2 (a) and (b) show the time variations of global radiation and long-wave radiation from sky, and of air temperature and relative humidity at a height of 1.0 m from 11:00 September 6 to 11:00 September 15. Same diurnal patterns were repeated through the experimental period.

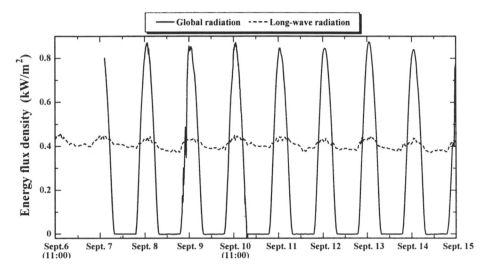

(a) Time variations of global radiation and long-wave radiation from sky

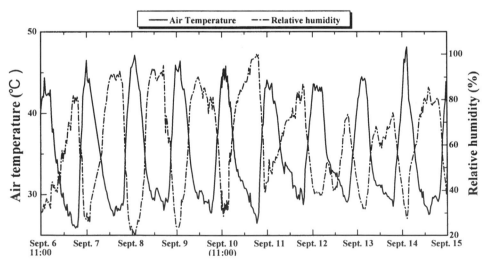

(b) Time variations of air temperature and relative humidity

Figure 2. Time variations of meteorological date.

Volumetric water content

Figure 3 shows the time variations of the volumetric water contents at depths of 0.01, 0.02, 0.04, 0.06, 0.08, 0.1, 0.15 and 0.2 m from 11:00 September 6 to 11:00 September 15. Immediately after watering (expressed with an allow in Figure 3), a sharp increase of the volumetric water content was observed near the soil surface.

It can be seen that the sprinkled water infiltrated downward to about 0.04 m below the soil surface for the first irrigation and about 0.08 m below the soil surface for the second one.

The volumetric water contents near the soil surface decreased as a consequence of the evaporation and downward infiltration with the passage of time. For the first irrigation, the volumetric water contents near the soil surface returned to original moisture level in about 3 days after watering, while significant diurnal variations of the volumetric water contents were admitted at least until 5 days after watering for the second one. This difference in the drying of the soil surface layer is caused by the volume of the sprinkled water.

Soil temperature

Figure 4 shows the time variations of the soil temperatures at depths of 0.01, 0.03, 0.05, 0.15, 0.3 and 0.5 m from 11:00 September 6 to 11:00 September 15. Immediately after watering, a rapid decrease of the soil temperature was observed near the soil surface as a result of sensible heat transfer to the cooler sprinkled water and because of the latent heat loss associated with evaporation. From the next day of the watering, diurnal variations and time delays associated with heat conduction became evident near the soil surface but the temperature variation at a depth of 0.5 m was no longer significant.

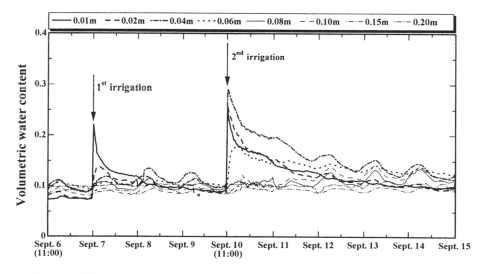

Figure 3. Time variations of volumetric water contents at different depths.

273

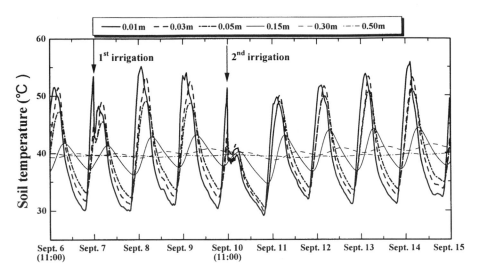

Figure 4. Time variations of soil temperatures at different depths.

Water vapor density profile

Figures 5 (a) and (b) present the time variations of the water vapor density profiles between a height of 0.5 m and a depth of 0.5 m from 6:00 September 10 to 6:00 September 11 (the first one day after the second irrigation) and from 6:00 September 14 to 6:00 September 15 (4 days after watering), respectively.

As Figure 5 (a) indicates, when the soil surface was wet due to the watering, the maximum water vapor density was appeared at the soil surface in the daytime (from 12:00 to 15:00). Moreover the difference between the atmospheric water vapor density at a height of 0.02 m and the soil pore water vapor density at a depth of 0.01 m was larger during the daytime (for example, 12:00 or 15:00) than during the nighttime (from 21:00 to 6:00). This fact suggests that the evaporation rate during daytime becomes higher than that during nighttime for soil surface evaporation. We shall discuss this point later.

Subsequently, when the soil surface layer was dry because of evaporation, the maximum water vapor density was appeared below the soil surface in the daytime (see Figure 5 (b), from 12:00 to 18:00). This suggests a shift from soil surface evaporation to internal evaporation. Since the position of the maximum vapor density corresponds to the interface between the dry surface layer and the unsaturated layer beneath it (Fukuhara et al., 1994a), it is known that an upward movement of the water vapor originates at the interface and that eventually the water vapor is released to the atmosphere.

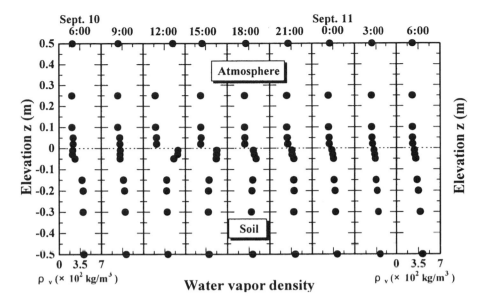

(a) 6:00 a.m., Septemper 10 ~ 6:00 a.m., Septemper 11

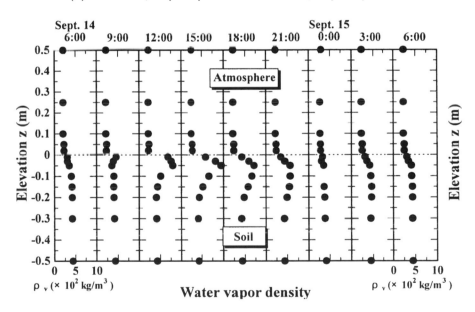

(b) 6:00 a.m., Septemper 14 ~ 6:00 a.m., Septemper 15

Figure 5. Time variations of water vapor density profiles
above and blow soil surface.

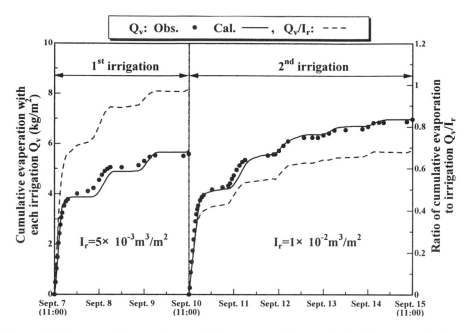

Figure 6. Time variations of cumulative evaporation and irrigation water budget.

Results of evaporation and evaluation of evaporation model

Evaporation and irrigation water budget

Figure 6 shows the time variations of cumulative evaporation, Q_v, per unit area after watering for the first and second irrigation. The ratio of Q_v to the watering discharge, I_v, per unit area, i.e. Q_v/I_v is also shown as a broken line in the same figure. For both irrigations, the time increment of Q_v was biggest within several hours just after watering, associated with the soil surface evaporation. Moreover the evaporation rate was higher during daytime than during nighttime, regardless of soil surface evaporation and internal one. The evaporation flux became small with the passage of time because of the decrease in the volumetric water content, i.e. drying of the soil surface layer. The diurnal variation of Q_v, however, was still significant even five days after watering. From our visual observation, it was confirmed that the soil surface was completely dry in about 4 hours and 7 hours after watering for first irrigation and for the second one, respectively. Therefore this period may belong to the soil surface evaporation stage and then the internal evaporation stage follows to it.

As for Q_v/I_v of the first irrigation, it was seen that about 80% of the sprinkled water was lost within one day after watering. Eventually, most of sprinkled water was lost in 3 days after watering. On the other hand, about 50% of the sprinkled water was lost within one day after watering for the second irrigation and about 70% was lost for 5 days after watering.

Evaluation of evaporation

The evaporation mass flux density for soil surface evaporation period can be calculated by the following $\alpha - \beta$ method:

$$M_v = \beta \; \gamma_m (\alpha \; \rho_{vs} - \rho_{va})$$ (1)

where, M_v: evaporation mass flux density (kg/m^{2}s), β: coefficient associated with area fraction of evaporation and resistance of water vapor movement in soil layer, γ_m: coefficient of water vapor transfer (m/s), α: relative humidity of soil surface, ρ_{vs}: water vapor density at soil surface (kg/m^3), ρ_{va}: water vapor density of atmosphere (kg/m^3). The values of α and β equal to 1 for the soil surface evaporation period (Futagami et al., 1997). Note that γ_m depends on wind velocity.

On the other hand, when the evaporation takes place at the interface, M_v can be expressed according to the following Fickian law:

$$M_v = -\xi \; \eta \; (\varepsilon - \theta) \; D_{atm} \frac{d\rho_v}{d\,z}\bigg|_{z=0}$$ (2)

where, ξ: correction coefficient for liquid-vapor interfacial area (3-D interfacial area / 2-D apparent area), η: tortosity (= 0.67), ε: porosity, θ: volumetric water content (m^3/m^3), D_{atm}: diffusivity of water vapor in atmosphere (m^2/s) and z: vertical coordinate (m). The suffix, $z = 0$ means soil surface. The value of ξ ranges between 2 and 3 (Fukuhara et al., 1994b) and turns out to be the order of $(dT/dz)_a/dT/dz$ (T: temperature, suffix a refers to air in soil pores) defined by Philip and de Vries (1957). Note that D_{atm} is a function of temperature.

Since all parameters except γ_m and ξ in Eqs. (1) and (2) are measured through this experiment, these two unknown parameters, γ_m and ξ can be calculated reversibly using Eqs. (1) and (2), respectively. The representative value of γ_m was 7.6 x 10^{-3} m/s and ξ was 2.1. A solid line in Figure 6 represents the calculated cumulative evaporation and the calculated result agreed well with the observations throughout the experimental period.

Conclusions

In order to better understand post-irrigation drying phenomena and to evaluate the water budget of irrigation, heat and moisture transfer measurements in soil under bare surface were carried out in the United Arab Emirates in September 2000. The time variation of evaporation flux after watering was measured by a new type of evaporation pan and an electric balance. A simple model (Fickian type diffusion theory) was proposed to analyze internal evaporation flux. The $\alpha - \beta$ method was applied to soil surface evaporation. The validity of the presented model was shown by a good agreement between calculated and experimental results.

The following conclusions can be deduced from the present study.

1. The evaporation pan is useful for a field experiment.
2. The evaporation has a diurnal variation because the evaporation flux is larger during daytime than during nighttime, regardless of soil surface and internal evaporation.
3. The proposed evaporation model and heat and moisture monitoring system are adequate to estimate the water budget of post-irrigation in arid regions. Especially, a vertical profile of water vapor density above and below soil surface is one of important informational sources to monitor the drying of surface soil layers.

References

Fukuhara, T., Sato, K. and Baba, T. (1994a). "Movement of water vapor in sand column and mechanism of evaporation". *J. Hydroscience and Hydraulic Engineering*, JSCE 12(1), 47-55.

Fukuhara, T., Sato, K. and Imai, T. (1994b). "Interaction between evaporation, movement of water vapor and heat transfer in sand layer under constant meteorological conditions". *J. Hydraulic, Coastal and Environmental Engineering*, JSCE No. 503/II-29, 29-38 (in Japanese).

Futagami, S., Takano, Y., Fukuhara, T. and Sato, K. (1997). "Heat and moisture transfer between sand surface and atmosphere by surface layer model – Consideration of thickness of sand layer by $\alpha-\beta$ method". *Annual J. Hydraulic Engineering*, JSCE 41, 37-42 (in Japanese).

Philip, J.R. and deVries, D.A. (1957). "Moisture movement in porous materials under temperature gradients". *Trans. Amer. Geophys. Union*, 38, 222 - 232.

Takano, Y., Fukuhara, T. and Sato, K. (1999). "Considerations of seasonal micrometeorological characteristic in the United Arab Emirates and evaporation-drying after watering the soil by the "soil thermo-hygrometer" method". *J. Japan Society of Hydrology and Water Resources*, JSHWR 12(4), 327-337 (in Japanese).

Surface Water Hydrology, Singh, Al-Rashed & Sherif (eds)
© *2002 Swets & Zeitlinger, Lisse, ISBN 90 5809 363 8*

The concept of virtual lysimeters to measure groundwater recharge and evapotranspiration

Ferdindnand J. Kastanek, Genia Hauer and Willibald Loiskandl
Institute of Hydraulics and Rural Water Management
University of Agricultural Sciences Vienna (BOKU)
Muthgasse 18, A1190 Vienna, Austria
e-mail: kastanek@mail.boku.ac.at

Abstract

Virtual lysimeters are devices similar to traditional lysimeters to investigate the hydrological (or substantial) balance of a soil body but without separation from its environment. Improvements of measuring equipment facilitate the application of this simple concept to record the soil water balance in the field. The components of the hydrological cycle are determined by applying soil physical principles on soil water movement and storage and measured soil physical parameters like soil water content and soil matric potential. First experiences with this concept at field scale at the experimental field of the University of Agricultural Sciences Vienna proved the usefulness of this method but there are some suggestions for improvements. Virtual lysimeters have the main advantage to maintain the natural boundary conditions, to be cheap (no expansive constructions), not to obstruct root development and no side wall effect of water seepage may occur. There exist some problems to measure matric potential close to the soil surface. Moreover restricted usage of tensiometers and gypsum blocks are not solved satisfactory till today. It is recommended to calibrate TDR or FDR sensors in the field, soil temperature should be measured and all sensors should be installed from the soil surface. When soil conditions allow it, sensors should be installed at least to a depth of two meters. It is still a problem to measure deep seepage (groundwater recharge), but analyzing different flow conditions allows to estimate its amount from measured data by use of the gradient of the total soil water potential and changes of the soil water content.

Introduction

Soil water retention and soil water movement are physical processes, which may be described by physical means. Their knowledge is of fundamental importance to understand the relationship between soil and water. In practice, it is possible to measure the evolved soil physical parameters in the field and to analyze their changes with time and depth. This has been done since a long time to describe soil water balance, but now new sensors are available, which give more accurate results and are easy to use. Kastanek (1995) and Loiskandl et al. (1998) found some drawbacks of traditional lysimeters for the determination of actual evapotranspiration and they suggested to study the soil water balance directly in the field. This concept was called virtual lysimeter.

Soil Water Retention

There are two properties to characterize soil water retention: the soil water content and the soil water potential. Their mutual dependence is the well known soil water characteristics, which is of outstanding importance in soil physics. It is common practice to express soil water content either as mass of water per unit mass of dry soil (water mass content) or as volume of water per unit bulk volume of soil (volumetric water content) dried to a constant mass at 105°C (SSSA, 1987). Standard methods to take and dry disturbed or undisturbed soil samples are not well suited for field research, but they are not meaningless for calibration and testing indirect methods like neutron meters, Time Domain Reflectometers (TDR) or Frequency Domain Resonance (FDR) methods.

At the experimental field of the Institute of Hydraulics and Rural Water Management (IHLW) of the University of Agricultural Sciences, Vienna (BOKU) near Vienna experiments are conducted to test different methods to measure soil water content in the field. Figure 1 shows the arrangement of different sensors:

2 TDR sensors (Trime and Trase), 3 replicates.
2 FDR sensor (Diviner, 3 replicates, and Enviroscan, 2 replicates).
1 Neutron Probe, 3 replicates.
(1 FDR Sensor – Gopehr, no replication).
Set of tensiometers for soil water tension.

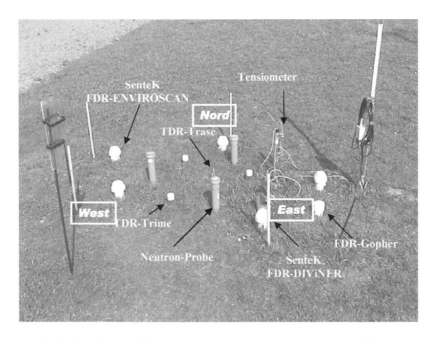

Figure 1. Field site of joint FAO-IAEA-IHLW project to compare different methods to measure soil water content (Cepuder et al., 2001).

In Figure 2a, 2b and 2c distributions of measured water contents with depth are plotted for 12 June, 2001. Additional actual soil water contents have been determined simultaneously with core samples. These reference data are also presented in Figure 2a, 2b and 2c. Calculations of soil water contents from measured parameters specific to the device used, were performed with the equations proposed by the manufacturers. At first glance the different methods seems to give quite diverging results. But these differences also may be caused by spatial inhomogeneities of the soil. The comparison of the three replicates of the sample core meas-

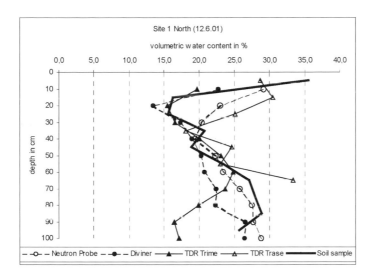

Figure 2a. Comparison of different soil water content sensors, group North.

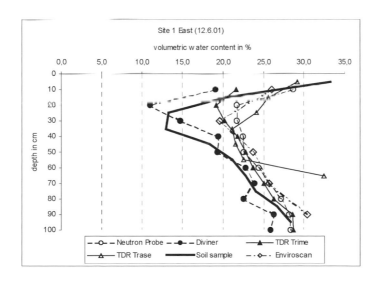

Figure 2b. Comparison of different soil water content sensors, group East.

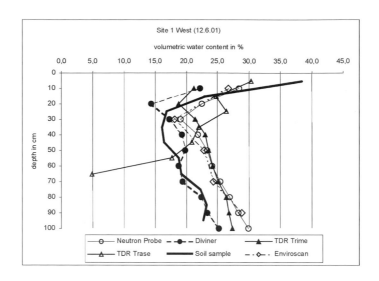

Figure 2c. Comparison of different soil water content sensors, group West.

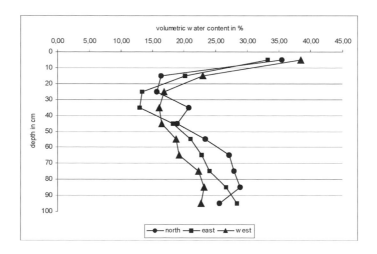

Figure 2d. Comparison of volumetric water content at the three different locations (West, East and North).

urements (Figure 2d) shows similar deviations. With the exception of the Trase sensors all other sensors show a similar tendency, hence deviations may be explained partly by spatial differences in soil properties. But nevertheless the sensors should be calibrated for actual field conditions.

Force fields acting on soil water are less obvious than water contents but even more important to explain soil water movement. Work caused by these forces is accumulated or emitted as potential energy. Energy difference may cause potentially movement, which may be visualized as forces. Therefore force (a vector) originating from a force field may be explained by spatial differences of potential energy contents, its gradient (a scalar). There are a lot of different forces acting on soil water expressed by its individual potential, but (scalar) potentials may be easily added and expressed as total soil water potential. The concept of soil water potential is the source to explain and estimate soil water movement.

According to the definition of ISSS (1976) the total soil water potential of the constituent water in soil at temperature T_0 is defined as the amount of useful work per unit quantity of pure water that must be done by means of externally applied forces to transfer reversibly and isothermally an infinitesimal amount of water from the standard state to the soil liquid phase at the point under consideration (ISSS, 1976, Jury et al., 1991). The reference quantity to is not given; it may be the unit of weight (J/N = m ≡ head), unit of volume ($J/m^3 = N/m^2$ = Pa) or unit of mass (J/kg).

The potential energy content is not an absolute value, to cause movement only differences in the energy status are relevant. The potential energy has to be defined relative to standard systems. According to the ISSS (1976) the standard system is defined as a infinitesimal small cup filled with pure water at atmospheric pressure at a well defined but arbitrary elevation. The total potential of soil water may be divided into three components: the gravitational potential, the tensiometer pressure potential and the osmotic potential. For most flow processes in the field the osmotic potential may be neglected. The gravitational potential takes into account the amount of energy, which is needed to rise or which is gained if lowering a quantity of water. The difference in energy level due to the gravitational potential corresponds to the difference in height.

Since the pressure within groundwater – the hydrostatic pressure – is equal or greater than atmospheric pressure the tensiometer pressure potential is always greater than zero if the atmospheric pressure is taken as reference pressure. The hydrostatic pressure within the groundwater has always a positive value. The hydrostatic pressure is measured with a piezometric tube.

The tensiometer pressure potential encompasses all effects on soil water other than gravity and solute interactions (Jury et al., 1991). The effects should now be restricted to the influence of binding to soil solids and to capillarity of a rigid matric – the matric potential. The pressure of soil water, which is held by capillarity or by adsorptive forces, is less than zero. The matric potential is measured with tensiometers and blocks. Whereas the hydrostatic pressure is found below the groundwater level where the soil is water saturated, the matric potential occurs in the unsaturated (capillary) zone above the groundwater level. Since the pressure within the zone above the groundwater level is less than zero, no water can seep out of this zone into holes, ditches etc, where the pressure is equal to atmospheric pressure. The installation of a plate somewhere in the unsaturated zone and to wait if water is collected there gives no insight into deep percolation or groundwater recharge.

In groundwater hydraulics the sum of gravitational and hydrostatic potential is called hydraulic potential. Similar the hydraulic potential should be extended to comprise the sum of gravitational and tensiometer pressure or hydrostatic potential. This makes sense because the hydrostatic potential and the matric potential exclude each other.

The direction of flow is given by the gradient of the total potential, which in our case is equal to the hydraulic potential. Differences of the tensiometer pressure potential alone are meaningless if neglecting the gravitational potential. Water flows in the direction of decreasing total potential.

In the field it is possible to measure soil water content and the matric potential at the same time at the same point. By repeated systematic measurements the functional relationship between soil water content and matric potential – the soil water characteristic – may be found. Due to hysteresis this relationship is not unique, but nevertheless it is of fundamental importance of all soil water behavior. The soil water characteristics may be determined in the labor with the pressure plate method, the hanging water column method or with the equilibrium over salt solutions, all methods are very time consuming.

Soil Water Movement

Flow of water in saturated soils is governed by Darcy's law. In 1907 Buckingham modified Darcy's law for unsaturated soil. Table 1 shows the difference and their common features.

The well known Laplace differential equation for saturated soil is a special case of the Richards equation for unsaturated soil. Each real world flow problem is charac-

Table1. Flow equations for saturated and unsaturated soil.

DARCY for water saturated soil		*BUCKINGHAM* for unsaturated soil
$v = -k\,i = -k\,\partial H/\partial z$	Flow-Equation (in z-direction)	$v = -k(n_l)\,\partial H/\partial z$
$\partial v/\partial z = 0$	Water-Conservation-Equation (only flow in z-direction)	$\partial v/\partial z + \partial n_l v/\partial t = 0$
$\partial^2 H/\partial z^2 = 0$ (LAPLACE Differential-Equation)	Flow-Equation + Water-Conservation-Equation (in z-direction)	$\partial n_l/\partial t = \partial[k(n_l)\,\partial H/\partial z]/\partial z$ (RICHARDS-Equation)

v .... flow-rate (amount of water per unit of time and unit of area)
$\partial H/\partial z, i...$ gradient of hydraulic potential

k_0...saturated hydraulic conductivity H .. gravitational potential + hydrostatic potential (hydraulic potential)	$k(n_l)$ ... unsaturated hydraulic conducticity H ... gravitational potential + matric potential (hydraulic potential)

terized by definite boundary conditions which are unique for every flow problem. Solutions of Richards or Laplace equation have to comply with the corresponding boundary conditions. This hold also for field devices to record the soil water balance.

Soil Water Balance

With traditional lysimeters the components of the soil water balance of a separated soil body of known dimensions for a given time interval are measured:

$$P + I - R = ET + D + DW$$

P precipitation	R surface runoff	D deep seepage
I irrigation	ET actual evapotranspiration	DW change of stored water

Usually irrigation is included in the precipitation term and surface runoff may be neglected. The actual evapotranspiration is the only term which can not be measured directly. Rearranging for this term leads to the lysimeter equation:

$$ET = P - D - DW \text{ (Figure 3).}$$

This equation gives the theoretical background for traditional lysimeters to estimate actual evapotranspiration in the field. But it can be shown, that the flow pattern at field scale is difficult to be realized with the traditional lysimeter. In practice the deep seepage measured with traditional lysimeters is different from real field conditions.

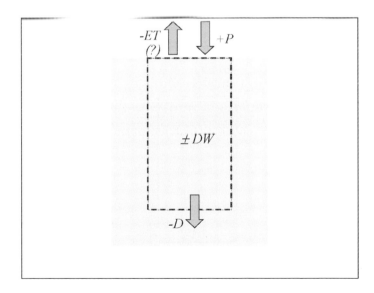

Figure 3. Water balance of a lysimeter.

Virtual Lysimeter

The general lysimeter principle is the estimation of the only component of the soil water balance, which can not be measured directly, the actual evapotranspiration (Figure 4).

Traditional lysimeters are completely or partly separated from the ambient soil body. At the lower boundary water may seep freely into the atmosphere (Figure 4). Since the soil water pressure at this level is equal to the atmospheric pressure a groundwater level is created artificially at the lower boundary for this kind of lysimeters. Moreover water movement into the lysimeter from the ambient soil is cut off.

Where deep groundwater levels exist the soil water pressure is negative in the ambient soil at the level of the bottom of a lysimeter. It is obvious that the flow problem within the lysimeter is therefore governed by other boundary conditions than in the ambient soil (Loiskandl et al.,1998). This was the reason to apply negative pressure at the bottom of a lysimeter (Figure 5). This pressure was set arbitrary to e.g. –0.3 bar. But nevertheless there is no reason why the soil water pressure should be –0.3 bar in the ambient soil at the level of the lysimeter bottom. Moreover the amount of deep seepage D depends on the applied negative pressure. Hence the measured outflow is also arbitrary.

It became evident that the boundary conditions of the flow problem within the lysimeter do not correspond to the flow problem in the ambient soil. But this system is easily improved by adjusting the applied pressure according to the real pressure in the ambient soil (Figure 6). For this purpose the pressure at the lower boundary of the lysimeter (the bottom) as well as in the ambient soil at the same level have to be measured (with tensiometers or blocks) and compared. Then the applied pres-

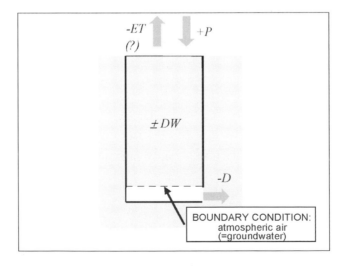

Figure 4. Principe of a traditional gravitational lysimeter.

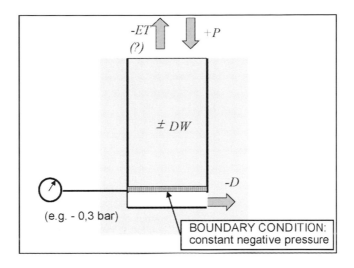

Figure 5. Principe of negative pressure controlled lysimeter with constant negative pressure.

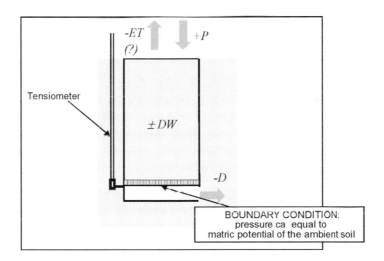

Figure 6. Principe of adjusted negative pressure controlled lysimeter.

sure has to be electronically adjusted to the value measured in the ambient soil. With some time lag it is possible to get the same boundary condition for the flow problem within the lysimeter as in the ambient soil. This type of lysimeter seems very promising but it is expensive and has some other drawbacks, e, g. restricted root development close to the lysimeter wall, problem of undisturbed soil within the lysimeter.

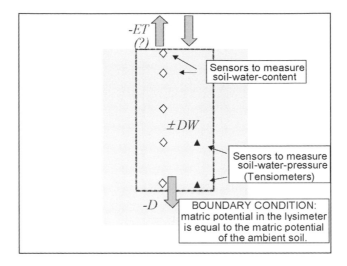

Figure 7. Principe of virtual lysimeter.

Similar to traditional lysimeters virtual lysimeters depend on the knowledge of the soil water balance, but they are not separated from the ambient soil. With virtual lysimeters the components of the soil water balance equation are measured directly in the field (Figure 7) and the water movement is derived from the soil water transport equation. Precipitation is measured in the traditional way with rain gauges. Changes of the stored water in the soil are estimated by analyzing changes of soil water profile contents with time. Sensors for the soil water contents are inserted into the soil from the soil surface with only little disturbance of the soil profile, there is no need to dig ditches or holes. Additional tensiometers are inserted to measure the matric potential. At first glance it does not seem necessary to use tensiometers to estimate the soil water balance, but the knowledge of the gradient of the total soil water potential is required to calculate the direction of water flow. This is the key point of virtual lysimeters because deep seepage is estimated by using the transport equation at the lower boundary. Consequently its knowledge is indispensable to estimate deep seepage.

Field Experiments

In 1999 our first virtual lysimeter was installed at the experimental field of the Institute of Hydraulics and Rural Water Management (IHLW) of the University of Agricultural Sciences Vienna (BOKU) to gain some experiences for further improvements. Figure 8 shows the measured values of water content and water pressure in four levels below soil surface and the amount of precipitation for a period of about one month between March and April 1999.

Figure 9 shows the water contents and the corresponding hydraulic heads at 9 March and 12 March after a period of rainfall (5 March – 7 March: 18.2 mm). Since the reference level for the gravitational potential was fixed at the soil surface

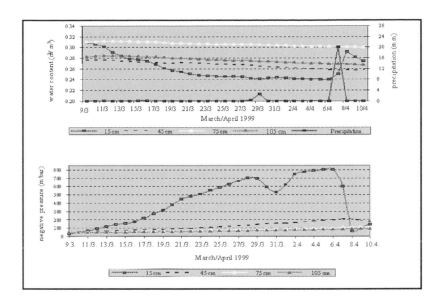

Figure 8. Measured soil water contents, soil water pressures and precipitation at the experimental field of the IHLW between 9 March and 10 April, 1999.

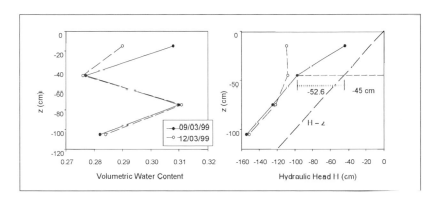

Figure 9. Distributions of water contents and hydraulic heads between 9 March and 12 March, 1999.

the gravitational potential below the soil surface are negative. In Figure 9 the summation of the gravitational head (potential) at -45 cm below soil surface and the tensiometer pressure head of -52.66 cm giving the hydraulic head (potential $= -97.6$ cm) is shown. With the exception of 12. March at the depth between -15 cm and -45 cm the gradient of the total head H is positive ($\partial H/\partial z > 0$). That means the direction of water flow is downward. Between -15 cm and -45 cm at 12 March the gradient is almost zero, there is no water movement. The gradient at the depth of -105 cm is responsible for the deep seepage. The Buckingham equation allows the calculation of the amount of deep seepage if the hydraulic conductivity is given. On the other hand it is possible to calculate the hydraulic conductivity if the amount of deep seepage is known. This value may be estimated from the difference of the water contents between 9 March and 12 March. Of course the presented interval is not ideal to perform such an estimation.

The distribution of water contents and of total potential heads between 17 March and 23 March (Figure 10) after a long period of no rain is much better suited for this case. Below the depth of −45 cm the gradient is positive and above the depth of −45 cm negative, that means the direction of water flow below −45 cm depth is downward and above −45 cm upward. With help of the corresponding distribution of the water contents (Figure 10) it is possible to estimate the amount of water removed. The difference of the water contents above −45 cm was removed upward by evapotranspiration, the difference of the water contents below −45 cm was removed by deep seepage. From Figure 10 it is evident that deep seepage is rather

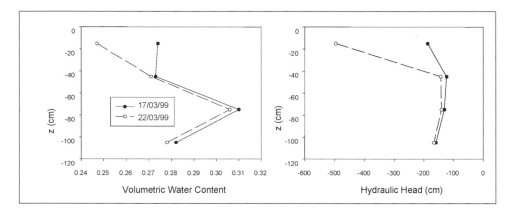

Figure 10. Distributions of water contents and hydraulic heads between 17 March and 22 March, 1999.

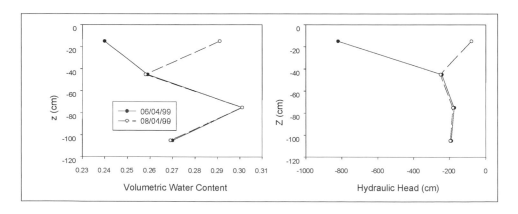

Figure 11. Distributions of water contents and hydraulic heads between 6 April and 8 April, 1999.

290

small, but nevertheless it gives a chance to calculate the hydraulic conductivity at
−105 cm depth. In our case we did not performed this calculation because the water content at −75 cm seems incorrect (Sensor not working properly?). But even if
the amount of deep seepage seems meaningless it may not be neglected in the
water balance of the soil body.

In Figure 11 distributions of water contents and total heads between 6 April and 8
April during a rain event are plotted. Distributions of the water contents give an im-
pression of the water stored immediately below soil surface, but it can be seen, that
water only infiltrate to about −45 cm, deeper layers are not effected by this rain
event. From differences of water contents it can be calculated, that the amount of
water infiltrated into the soil was about 18 mm. In our case this estimation is very
rough, because we do not know the water contents close to the soil surface. A
more detailed spatial resolution of measured water contents is a very important basic requirement for further virtual lysimeters. By comparison of the infiltrated
amount of water and the measured amount of rain (with a separate rain gauge) interception losses may be calculated. Below −45 cm there is still a gradient upward,
representing water flow direction to the soil surface. This gradient originates from
the period of dryness before the rain event.

Conclusions

The concept of virtual lysimeters consist of a special arrangements of sensors to
measure soil water contents and soil water pressures in the field, which allows to
estimate the soil water balance (or a balance of another substance) of a soil body,
without hydrological separation with a container and with only little disturbances of
the soil. In contrast to traditional lysimeters the boundary conditions correspond to
the real flow pattern in the considered soil body. New technologies to measure soil
water content and matric potential allow to realize the concept of the virtual lysimeters, but nevertheless there are still some problems using these new technologies.
TDR as well as FDR sensors should be calibrated in the field. Tensiometers are
restricted to measure matric potentials in a range from 0 to −0.7 bar, the accuracy
of blocks to measure matric potentials less than about −0.7 bar is rather poor. Accurate measurements of matric potentials less than −0.7 bar are still an unsolved
problem, blocks are only an imperfect alternative (Kastanek,1996).

According to our experience we propose the layout of a virtual lysimeter shown in
Figure 12.

The depth of the soil body covered by a virtual lysimeter should be as deep as
possible, because at greater depth the gradient of the total potential does not
change too much with time (gradient needed to estimate the deep percolation) and
matric potentials at this depth are usually within the measuring range of tensiometers. To reduce disturbances of the soil tensiometers as well as TDR or FDR sensors should be buried into the soil from the surface without cutting deep ditches.
Close to the soil surface measurements of water content and matric potential are
needed with little spatial distance. It is essential to measure and to record the components of the soil water balance continuously using electronic data loggers. Simul-

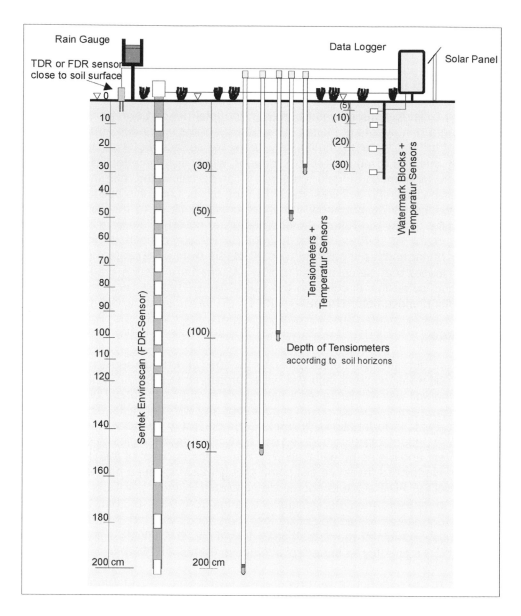

Figure 12. Recommended layout of a virtual lysimeter.

taneous measurement of the temperature distribution in the soil body is essential to increase accuracy, because some sensors (e.g. blocks) are temperature sensitive. From the examples shown we learned more spatial dense measurements of the soil water content is an important prerequisite to evaluate the soil water balance, the water content close to the soil surface has to be included. For this reason a TDR or FDR sensor with short rods should be inserted vertically at the soil surface. Accurate registration of deep seepage is still a problem. It is possible to estimate interception losses if the precipitation is known. Otherwise interception losses may be included in the soil water budget. Wall effects are simple not existing and the development of the root system is not restricted. Due to the non linearity of the Richards equation it is better to use actual measurements at distinct daytimes instead of mean values of days.

To evaluate the measured data a profound soil physical background is essential. It is even possible to estimate changes of soil physical parameters with time, like the hydraulic conductivity. In combination with a strong theoretical background virtual lysimeters are more than a field device to estimate the actual evapotranspiration, they allow to study and help to understand soil water behavior in the field with only little disturbances.

References

ISSS (1976). "Soil physics terminology" (Editor G.H. Bolt). *Bull. International Soil Science Society,* 49:26-36.

Cepuder, P., Zartl, A., and Hauer, G. (2001). *"Experiences with different soil water measuring systems on diverse locations in Lower Austria".* First Report, IHLW, University of Agricultural Sciences Vienna, Project number 302-D1-AUS-11184.

Jury, W.A., Gardner, W.R., and Gardner, W.H. (1991). Soil Physics, 5th Edition, John Wiley & Sons.

Kastanek, F. (1995). "Kritische Bemerkungen zur Verwendung von Lysimetern". 5. Gumpensteiner Lysimetertagung, Bundesanstalt fuer alpenlaendische Landwirtschaft, Gumpenstein, Styria, Austria, 93-102 (in German).

Kastanek, F. (1996) "Kritische Bemerkungen zur Bestimmung der Wasserspannung im Boden mit der Bohrlochmethode". Die Bodenkultur, 47, 223-233 (in German).

Loiskandl, W., Kastanek, F., and Cepuder P. (1998). "Simulation of the flow of water in a lysimeter". 16th World Congress of Soil Science, Montpellier, France, Paper Nr. 361.

(1987). Glossary of Soil Science Terms, Soil Science Society of America, Madison.

Section 5: Infiltration

Surface Water Hydrology, Singh, Al-Rashed & Sherif (eds)
© 2002 Swets & Zeitlinger, Lisse, ISBN 90 5809 363 8

Effect of clay dispersion on infiltration into bare soils: Multi-stage infiltration equations

João L.M.P. de Lima
IMAR – Institute of Marine Research, Coimbra Interdisciplinary Centre
Department of Civil Engineering, Faculty of Science and Technology –
Campus 2, University of Coimbra, 3030-290 Coimbra, Portugal
e-mail: plima@dec.uc.pt

M. Isabel P. de Lima
IMAR – Institute of Marine Research, Coimbra Interdisciplinary Centre
Department of Forestry, Agrarian Technical School of Coimbra,
Polytechnic Institute of Coimbra, Bencanta, 3040-316 Coimbra, Portugal
e-mail: iplima@mail.esac.pt

Abstract

A correct description of the infiltration process into soil is important in hydrologic and soil erosion studies. Often, rainfall simulators are used to obtain measurements of infiltration. The objective of this paper is to report results of infiltration measurements from laboratory experiments making use of a rainfall simulator and a soil flume and to present infiltration equations that can account for breaks in infiltration rate. The soil material used was from a saline and sodic regolith profile developed on marine marls in a badland area of S.E. Spain. Breaks in infiltration rate can occur due to, for example, solute release and transfer into overland flow, or changes in soil texture and structure. The simplified, empirical, multi-stage infiltration equations proposed in this study, based on the Horton equation, were fitted to the experimental data. These equations describe adequately infiltration under the referred laboratory conditions.

Introduction

Infiltration is often described as the process by which water crosses the air-soil interface into the soil (e.g., Eagleson, 1970; Hillel, 1971). Due to an increasing concern in estimating runoff and soil erosion, there has been renewed interest in studying infiltration. A considerable amount of research has been conducted on the infiltration of water into soils and overland flow (e.g., Morin and Benyamini, 1977; Gerits et al., 1990; Dam and Feddes, 2000), which have led to a better understanding of the basic principles of soil-water flow. Water-flow in the surface layer of the soil is predominantly vertical, and can be simulated as one-dimensional flow in many applications (e.g., Lima, 1992; Romano et al., 1998).

Horton (1933) presented the classical view of infiltration as the process that separates rainfall into two major hydrologic components: surface runoff and subsurface recharge (see Figure 1). Figure 1 utilizes constant rainfall intensity to illustrate that

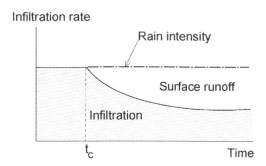

Figure 1. Classical view of infiltration as the process that separates rainfall into two major hydrologic components: surface runoff and subsurface recharge (Horton, 1933).

infiltration may be governed by rainfall intensity (prior to the critical time, t_c) and by soil conditions (after t_c). Infiltration capacity is perceived to approach a constant minimum, after a long time.

The Richards equation for variably saturated soil-water flow has a clear physical basis. Soil hydraulic data that have been collected (e.g., Bruand et al., 1996; Leij et al., 1996 and Wösten et al., 1994) enhance the applicability of the Richards equation. In the last two or three decades, various numerical routines to solve the Richards equation were developed. The Richards equation is difficult to solve because of its parabolic form in combination with the strong non-linearity of the soil hydraulic functions, which relate water content, soil-water pressure head and hydraulic conductivity (e.g., Dam and Feddes, 2000). Also the abrupt changes of moisture conditions near the soil surface, causing steep wetting fronts in dry soils, or causing steep drying fronts in wet soils, may pose problems. The result is that calculated soil-water fluxes may depend largely on the structure of the numerical scheme and the applied time and space steps (e.g., Milly, 1985; Celia et al., 1990; Zaidel and Russo, 1992; Miller et al., 1998).

Several problems arise when trying to apply the infiltration theory to field conditions. First, infiltration is greatly affected by the conditions at the soil surface. These surface conditions are affected by a number of factors including plant growth and other biological activities, rainfall, and irrigation. Often, soil surface conditions change during the precipitation process as well as from rainfall event to rainfall event and from season to season. This makes it particularly difficult to apply infiltration theory because parameters (e.g., permeability, porosity, moisture content, and bulk density) are changing.

The effect of swelling and dispersion of clay on the hydraulic conductivity and infiltration rate of soils has been the subject of several studies (e.g. McNeal and Coleman, 1966; Shainberg et al., 1981; Agassi et al., 1981). Most of these studies involved laboratory experiments under saturated conditions with an initial homogeneous chemical composition of the soil solution. Only a few attempts have been made to model the combined effect of solute movement and physico-chemical processes on the (un)saturated hydraulic conductivity of soils (e.g., Russo and Bresler, 1977; Russo, 1988).

The dynamic interaction between solute transport in runoff and physico-chemical processes at the soil surface can cause a reduction of the infiltration rate and increase the rate of overland flow during a rainfall event (e.g., Gerits and Lima, 1990), thereby increasing the erosion potential and decreasing the water available for plant growth. Moreover, because infiltration rates control not only the amount of surface runoff, but also the downward rate of removal of solute within the surface layer, infiltration rate should exert a significant effect on the chemical exchange process. Ahuja et al. (1981), Ahuja and Lehman (1983) and Ahuja (1986) verified this effect.

It is important to find a way to characterize infiltration before and after the chemical thresholds are crossed, i.e., by considering multi-stage infiltration. Thus, the main objective of this study was to find a way to describe infiltration under varying soil surface physico-chemical conditions. The procedure was tested on experimental data from laboratory experiments conducted on a soil flume using simulated rainfall. A Horton-type formula was applied by fitting an exponential function to measured infiltration data. Horton's formula is, probably, the most simple approach amongst the available in the literature. Also, it is one of the most commonly used infiltration formula.

Laboratory experiments

Sprinkling infiltrometers and rainfall simulation on soil flumes are means of obtaining measurements of infiltration, and are used in hydrologic and soil erosion studies. Usually, these infiltration measurements are implicitly based on Horton's classical view (Figure 1). It is an indirect method by which infiltration is assumed to be the difference between rainfall application rate and runoff rate (e.g., Amerman, 1983).

Experimental procedure

The soil material used in this study was a mixture of "Domed" crust and sub-crust material (Gerits and Lima, 1990), from a saline and sodic regolith profile developed on marine marls in a badland area of S.E. Spain (Finlayson et al., 1987). Badlands are attractive to geomorphologists and surface hydrologists because their predominantly barren surfaces facilitate observation of overland flow and a variety of erosion processes (e.g., Bryan and Yair, 1982; Howard, 1994; Cantón et al., 2001).

The electrical conductivity (EC_{25}) and the Sodium Adsorption Ratio (SAR_p) of saturated paste extracts from the crust and sub-crust layers were, respectively, 36.02 and 49.02 mS/cm and 76.4 and 69.6 $(mmol/l)^{1/2}$. The particle size distribution of the mixture consisted of 27.3% clay, 64.5% silt and 8.2% sand (weight percentage of fine earth). The $CaCO_3$ content of the fine earth fraction was 60%. The major part of the sand and silt fraction was composed of carbonates. The clay fraction contained only 20% carbonates. The clay mineralogy was estimated by X-ray diffraction to consist of 30 % kaolinite, 45% illite and 20% montmorillonite. More data on the chemical and physical characteristics of the regolith material are given in Gerits et al., (1987).

Air-dried soil aggregates (4-8 mm) were packed in a flume to a bulk density of 1250 kg/m^3. The 0.12 m thick soil layer was packed on the top of a sand layer to allow free drainage. The flume was 4.0 m long and 0.5 m wide with a slope of 25 degrees. Rainfall (demineralized water) was applied with a simulator equipped with capillary tubes. The kinetic energy of the simulated rainfall was 19 J/m^2, per mm of rain. Prior to the first experiment a test run was made to create a crust on the soil surface. In the flume experiments, rainfall was applied at uniform rates (17, 18, 19 and 30 mm/hr) during 60 minutes. After each experiment the soil was allowed to dry for several days before the next experimental run was performed.

During the experiments the rate of overland flow was recorded continuously and runoff was sampled periodically for chemical analyses. The electrical conductivity (EC$_{25}$) of the runoff samples was measured immediately after sampling and several hours later, after filtration. The differences between both measurements were always within 7%.

Experimental results

Typical test results for infiltration are shown in Figures 2A (dots), respectively for rainfall rate applications of 18, 19 and 17 mm/hr. Vertical ordinates on the diagrams represent the rate of application of rainfall (P) and the rates at which infiltration (I, below the solid line) and surface runoff (Q, above the solid line) occurred, expressed in mm/hr over the area of the soil flume. At the start of all runs, the infiltration capacity exceeded the rainfall intensity and, as expected, runoff did not commence immediately, due to wetting and surface storage effects. The time lag between the start of rainfall and the start of runoff varied according to the initial soil moisture conditions and the rainfall intensity.

As the runs proceeded, it was possible to observe changes in the rate of infiltration into the soil. Runoff data showed a distinct increase of the runoff rates during the second part of the experiments. Thus, two infiltration phases could be distinguished.

An explanation for the sudden decrease of the infiltration rates from the first to the second phase is the swelling and dispersion of clay during the runoff event. According to Gerits and Lima (1990) this is caused by changes in the chemical composition of the soil solution due to salt leaching by infiltrating water and the solute transport in runoff.

The evolution of the electrolyte concentration (Ce) and SAR$_p$ of the soil solution at the surface during each experiment is shown in Figures 2B together with the flocculation curve of the soil clays (Gerits and Lima, 1990). The chemical

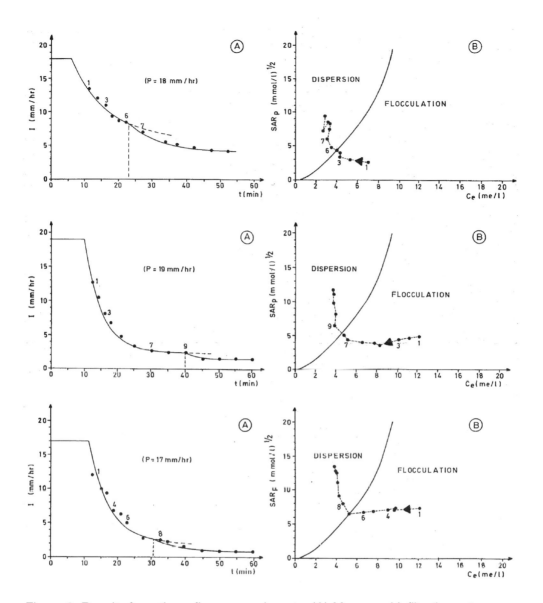

Figure 2. Results from three flume experiments: (A) Measured infiltration rates as a function of time (dots) and the fitted infiltration rates (solid line); and (B) Evolution of the electrolyte concentration (Ce) and the Sodium Adsorption Ratio (SAR$_p$) of the runoff, where the solid line is the flocculation curve of the soil clays (Gerits and Lima, 1990).

thresholds for flocculation of the soil clays (CaCO$_3$ removed) were determined according to the procedure described in Oster et al. (1980). The data in Figures 2A and 2B show that during the first infiltration phase the clay at the soil surface was in a flocculated state. During the second infiltration phase, once the chemical thresholds for the flocculation and dispersion of clay (Ce and SAR$_p$ values) were crossed, clay became dispersed and the pores of the surface soil became clogged with dispersed clay, resulting in a reduction of the infiltration rate (Gerits and Lima, 1990).

The chemical thresholds for flocculation and dispersion of clay established by laboratory batch experiments proved to be applicable in the analysis of results obtained from the runoff experiments under totally different conditions (Gerits and Lima, 1990). This enabled the prediction of the instant when a decrease in infiltration rate took place provided the solute transport in runoff can be accurately modelled. A more detailed observation of Figure 2 reveals a close agreement between the predicted onset of clay dispersion and the observed decrease of the infiltration rates. In fact, infiltration shows a somewhat delayed response compared to the predicted dispersion of clay throughout the various runs, as presented in Figure 3 for various experiments.

Figure 3 shows a plot of the time of occurrence of chemical thresholds established in batch experiments (i.e. clay dispersion estimated from runoff samples selected for chemical analyses) against the time of sudden decrease of infiltration rates of the soil flume experiments (estimated by linear regression on experimental data from phase 2). The agreement between the results obtained from the batch and flume experiments is remarkable because both experiments were performed under entire different energetic conditions. Generally, the raindrop impact and flowing water will enhance the dispersion of clay (Agassi et al., 1985).

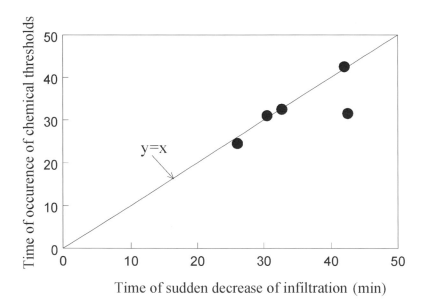

Figure 3. Consistency of time of occurrence of chemical thresholds established in batch experiments (i.e. clay dispersion estimated from runoff samples) and time of sudden decrease of infiltration rates of the soil flume experiments.

Infiltration equations and example of application

Under certain conditions, the infiltration rate into soil as a function of time from the beginning of rainfall can be subdivided into different phases. Figure 4 illustrates a case with three phases, showing the amount of rainfall that infiltrates into the soil and the amount of surface runoff. These phases, when existing, are clearly observed from the overland flow hydrographs (Q) and on the solute chemistry in time (see also Figures 2A). All phases end with a temporary final infiltration rate (I_f).

The horizontally hatched area in Figure 4 represents surface storage (the sum of detention and depression storage). Surface storage was supposed to be filled up in phase 1.

Infiltration rates (I) were calculated from the difference between runoff rates recorded during the experiments and the applied rainfall rates (P), with the exception of phase 1, where surface storage had to be taken into account. Because rainfall was uniformly distributed over the entire soil surface of the flume, the volume of detention storage was estimated using velocity measurements (dye measurements). In some of the experiments, surface detention and depression storage were not significant and therefore neglected.

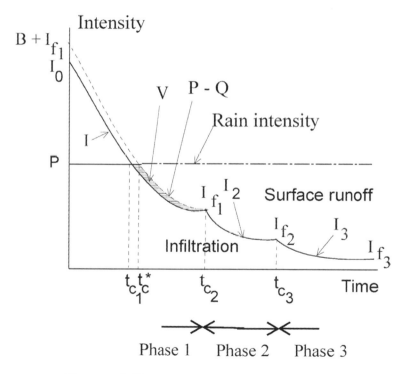

Figure 4. Infiltration curves with three phases.

Infiltration equations for phase 1

Horton (1933) postulated a three-parameter exponential decay equation as a model of the infiltration capacity of soils (see Figures 1 and 4). Two of the parameters in the equation are the initial and the minimum or long-term infiltration capacities. The third parameter is a constant multiplier in the negative exponential index and is dependent on the soil type and vegetation cover (see Equation 1).

For modelling the infiltration process in phase 1, this study used a Horton-type equation (e.g., Morin and Benyamini, 1977):

$$I(t) = I_{f_1} + (I_0 - I_{f_1}) \ e^{-a_1 t} \qquad \text{with } I_0 < P < I_{f_1} \tag{1}$$

where $I(t)$ is the infiltration rate as a function of time (m s^{-1}), I_0 is the initial infiltration rate into the soil (m s^{-1}), I_{f_1} is the final infiltration rate into the soil for phase 1 (m s^{-1}), P is the rainfall intensity (m s^{-1}) and a_1 is a parameter (s^{-1}).

Surface storage starts when $P = I(t_{c_1})$, where t_{c_1} is the critical time (or ponding time), which is the time between the initiation of rainfall and the initiation of overland flow on a surface without depression storage:

$$t_{c_1} = \frac{1}{a_1} \ln \left(\frac{I_0 - I_{f_1}}{P - I_{f_1}} \right) \tag{2}$$

By assuming that detention storage as a function of time is described by some exponential function, we may assume that the difference between the constant rainfall intensity and the measured runoff is described also by an exponential function (Figure 3):

$$P - Q = I_{f_1} + B \ e^{-a_1 t} \tag{3}$$

where Q is the discharge at the end of the soil flume (m s^{-1}) and B is a parameter (m s^{-1}).

For small surface detention storage values, we assume that $t_c^* = t_{c_1}$ and the volume of surface storage, will be given by (see Figure 1):

$$V = \int_{t_c^*}^{t_{c_2}} (P - Q) \ dt - \int_{t_c^*}^{t_{c_2}} I \ dt \tag{4}$$

where V is the volume of surface storage (depression and detention) (m), t_c^* is the time at which runoff starts (s), and t_{c_2} is the critical time for phase 2 (s) (see Figure 3).

Solving the integrals and rearranging, an expression for I_0 is obtained:

$$I_0 = (B + I_{f_1}) + \cfrac{a_1 V}{\left[e^{-a_1 t} - \cfrac{P - I_{f_1}}{I_0 - I_{f_1}} \right]}$$

(5)

where I_{f_1} and V are measured, and parameters B and a_1 are obtained by linear regression applied to a logarithmic transformation of equation 3.

If time t_c' cannot be considered approximately equal to the critical time t_{c_1}, then the volume of surface storage, V, will be given by (see Figure 1):

$$V = \int_{t_{c_1}}^{t_c'} P \ dt + \int_{t_c'}^{t_{c2}} (P-Q) \ dt + \int_{t_{c_1}}^{t_{c2}} I \ dt$$

(6)

Solving the integrals, and eliminating t_c', an expression for the initial infiltration rate of the soil, I_0, is obtained:

$$I_0 = \cfrac{a_1 V - (P - I_{f_1}) \left[\ln \left(\cfrac{B}{I_0 - I_{f_1}} \right) + 1 \right] + e^{-a_1 t_{c2}}}{e^{-a_1 t_{c2}} - \cfrac{P - I_{f_1}}{I_0 - I_{f_1}}} + I_{f_1}$$

(7)

where I_{f_1} and V are measured, and parameters B and a_1 are obtained by linear regression applied to a logarithmic transformation of equation 3. Equation 7 is not explicit with respect to I_0 but can be solved easily with the help of some iterative procedure.

To conclude, for phase 1 the infiltration rate is calculated from Equation 1; in this equation I_{f_1} and V are measured, parameters B and a_1 are obtained by linear regression and I_0 is calculated using Equations 5 and 7, respectively, for small and large values of surface storage.

Infiltration equations for phase 2, 3,..., j

For phases 2, 3,..., j, the following equations are used:

$$I_j(t) = I_{f_j} + \left(I_{f_{j-1}} - I_{f_j} \right) e^{-a_j (t - t_{c_j})}$$

(8)

305

where a_j is a parameter for phase j (s^{-1}), t_{c_j} is the critical time for phase j (where $I_{0_j} = I_{f_{j-1}}$) (s), and I_{f_j} is the final infiltration rate of the soil for phase j (m s^{-1}). Parameters t_{c_j} and a_j are estimated by linear regression applied to a logarithmic transformation of Equation 8.

Results

The equations derived in the last sections were fitted to the experimental results described in Section 2. The resulting infiltration curves are shown in Figures 2A (solid lines). The equations were found to give a reasonable representation of the observed infiltration curves, with the exception of a short period at the start of run-off for some runs. Johnston et al. (1980) also observed this deviation. An explanation for this behaviour is that the observed results are an average infiltration over the soil flume and do not take into account the areal variations in the start of overland flow, i.e. partial contribution to runoff at the lower end of the soil flume.

Figure 5 shows another result (see also Figure 2), here for a three-phase infiltration type of behaviour. For this case, the data are also presented in Table 1. The applied rainfall intensity was 30 mm/hr during 60 minutes. Also in this case the multistage infiltration equations proposed describe well the behaviour observed in the laboratory experiments.

The values of the time of initiation of runoff which were predicted by the equations fitted to the data (using all measured values) tended to be greater than the observed times. In these experiments this overestimation was dependent on the rain

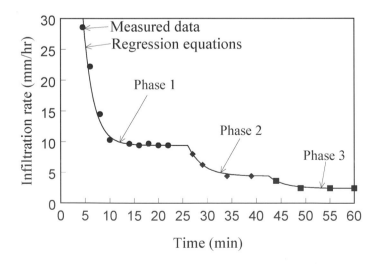

Figure 5. Example of an experimental run with three phase infiltration rates (see also Table 1): observed (dots) and theoretical (solid line).

Table 1. Summary of experimental results for three phase infiltration rates (see also Figure 5).

Phase 1

Regression equation	$I = 198.80\ e^{-0.492\ t}+9.42$
Initial infiltration rate	30.00 mm/hr
Final infiltration rate	9.42 mm/hr
Critical time	4.5 min

Phase 2

Regression equation	$I = 5.00\ e^{-0.333(t\ -26)}+4.42$
Initial infiltration rate	9.42 mm/hr
Final infiltration rate	4.42 mm/hr
Critical time	26.0 min

Phase 3

Regression equation	$I = 1.98\ e^{-0.334(t\ -2.5)}+2.44$
Initial infiltration rate	4.42 mm/hr
Final infiltration rate	2.44 mm/hr
Critical time	42.5 min

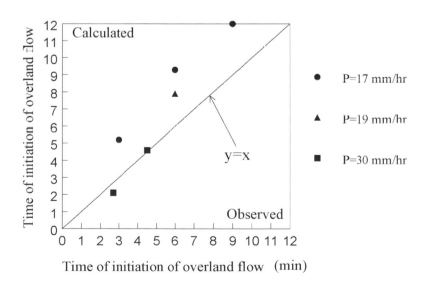

Figure 6. Time of initiation of runoff: calculated versus observed values for several experimental runs.

fall intensity as shown in Figure 6. For higher rainfall intensities (i.e., P = 30 mm/hr) deviations almost did not exist. The conclusion is that, in the determination of the infiltration equations for low rainfall rates, eventually the first one or two runoff measurements should not be considered.

To fit the infiltration equation to phase 1 often require the assumption of unrealistically high values of the initial infiltration capacity (I_0). This was also found by Beutner et al. (1940), Nassif and Wilson (1975) and Johnston et al., (1980).

Conclusions

Infiltration measurements into soil, conducted on a soil flume under simulated rainfall, showed the occurrence of breaks in infiltration rates. The soil material used in the experimental work was from a saline and sodic profile developed on marine marls.

In this study, simplified empirical multi-stage infiltration equations, based on the Horton equation, were proposed to take into consideration this type of behaviour. These equations account for discontinuities in the infiltration rate due to changes in the physico-chemical conditions of the soil surface during the runoff event. A possible explanation for the sudden decrease of the infiltration rates is the swelling and dispersion of clay during the runoff event. According to Gerits and Lima (1990) this is caused by changes in the chemical composition of the soil solution due to salt leaching by infiltrating water and the solute transport in runoff. The exponential equations proposed, which were fitted to the experimental data, represent well infiltration rate under the conditions described in this study and within the limits of applicability of the laboratory experiments.

Further research should be devoted to linking infiltration equations to a solute release model with chemical thresholds, which could account for the dynamic interaction between solute transport in runoff and physico-chemical processes that may take place in the soil.

Acknowledgements

This study was partly funded by the Foundation for Science and Technology (Research Project FCT - POCTI/35661/MGS/2000) of the Portuguese Ministry of Science and Technology.

References

Agassi, M., I. Shainberg and J. Morin (1981). "Effect of electrolyte concentration and soil sodicity on infiltration rate and crust formation". *Soil Sci. Soc. Amer. Journal*, 45, 848-851.

Agassi, M., J. Morin and I. Shainberg (1985). "Effect of raindrop impact energy and water salinity on infiltration rates of sodic soils". *Soil Sci. Soc. Amer. Journal*, 49, 186-190.

Ahuja, L.R. (1986). "Characterization and modeling of chemical transfer to runoff". In: J.B. Stewart (ed.), *Advances in Soil Science*, 4, 149-188.

Ahuja, L.R., A.N. Sharpley, M. Yamamoto and R.G. Menzel (1981). "The depth of rainfall-runoff-soil interaction as determined by P_2^{3}"*Water Resour. Res.*, 17(4), 969-974.

Ahuja, L.R. and O.R. Lehman (1983). "The extent and nature of rainfall-soil interaction in the release of soluble chemicals to runoff". *Journal of Environ. Qual.*, 12(1), 34-40.

Amerman, C.R. (1983). "Infiltration measurements". In: *Advances in Infiltration* (Proc. of the National Conference on Advances in Infiltration, December 12-13, 1983, Chicago, Illinois), Am. Soc. of Agric. Eng., 201-214.

Beutner, E.L., R.R. Gaebe and R.E. Horton (1940). "Sprinkled plane runoff and infiltration experiments on Arizona Desert soils". *Trans. Amer. Geophys. Union, Papers Hydrol.*, 21, 550-558.

Bruand, A., O. Duval, H. Wösten and A. Lilly (eds.) (1996). "The use of pedotransfer in soil hydrology research in Europe". Workshop Proc., October 10-12, 1996, Orleans, France, 211 p.

Bryan, R.B. and A. Yair (1982). "Perspectives on studies of badland geomorphology". In: Bryan, R.B. and A. Yair (Eds.), *Badland Geomorphology and Piping*. Geo Books, Norwich, 1-12.

Cantón, Y., F. Domingo, A. Solé-Benet and J. Puigdefábregas (2001). "Hydrological and erosion response of a badland system in semiarid SE Spain". *Journal of Hydrology*, 252, 65-84.

Celia, M.A., E.T. Bouloutas and R.L. Zarba (1990). "A general mass-conservative numerical solution for the unsaturated flow equation". *Water Resour. Res.*, 26, 1483-1496.

Dam, J.C. van and R.A. Feddes (2000). "Numerical simulation of infiltration, evaporation and shallow groundwater levels with the Richards equation". *Journal of Hydrology*, 233, 72-85.

Eagleson, P.S. (1970). "Dynamic hydrology". McGraw-Hill Book Co., New York, 462 p.

Finlayson, B.L., J. Gerits and B. van Wesemael (1987). "Crusted microtopography on badland slopes in southeast Spain". *Catena*, 14 (1/2), 131-144.

Gerits, J., A.C. Imeson, J.M. Verstraten and R.B. Bryan (1987). "Rill development and badland regolith properties". In: R.B. Bryan (ed.), Rill Erosion, *Catena Supplement*, 8, 141-160.

Gerits, J. and J.L.M.P. de Lima (1990). "Solute transport and wind action in relation to overland flow and water erosion". *Catena Supplement*, 17, 67-78.

Gerits, J., J.L.M.P. de Lima and T.M.W. van den Broek (1990). "Overland flow and erosion". In: M.G. Anderson and T.P. Burt (eds.), *Process Studies in Hillslope Hydrology*, John Wiley & Sons Ltd., 173-214.

Hillel, D. (1971). *"Soil and water physical principles and processes"*. Academic Press, New York, 288 p.

Horton, R.E. (1933). "The role of infiltration in the hydrologic cycle". *Trans. Amer. Geophys. Union*, 14, 446-460.

Howard, A.D. (1994). "Badlands". In: Abrahams, A.D. and Parsons, A.J. (Eds.), *Geomorphology of Desert Environments*. Chapman & Hall, London, 213-242.

Johnston, H.T., E.M. Elsawy and S.R. Cochrana (1980). "A study of the infiltration characteristics of undisturbed soil under simulated rainfall". *Earth Surface Processes*, 5, 159-174.

Leij, F.J., W.J. Alves, M.Th. van Genuchten and J.R. Williams (1996). *"The UN-SODA unsaturated soil hydraulic database"*. User's manual version 1.0 Soil Salinity Laboratory, Riverside, CA.

Lima, J.L.M.P. de (1992). "Model KININF for overland flow on pervious surfaces". In: *Hydraulics and Erosion Mechanics of Overland flow*, T. Parson and A. Abrahams (Eds.), UCL Press, 69-88.

McNeal, B.L. and N.T. Coleman (1966). "Effect of solution composition on hydraulic conductivity". *Soil Sci. Soc. Amer. Proc.*, 30, 308-312.

Miller, C.T., G.W. Williams, C.T. Kelly, and M.D. Tocci (1998). "Robust solution of Richards equation for nonuniform porous media". *Water Resour. Res.*, 34, 2599-2610.

Milly, P.C.D. (1985). "A mass conservative procedure for time-stepping in models of unsaturated flow". *Adv. Water Resour.*, 8, 32-36.

Morin, J. and Y. Benyamini (1977). "Rainfall infiltration into bare soils". *Water Resour. Res.*, 13(5), 813-817.

Nassif, S.H. and E.M. Wilson (1975). "Influence of slope and rain intensity on runoff and infiltration". *Hydrol. Sciences Bulletin dés Sciences Hydrol.*, 20(4), 539-553.

Oster, J.D., I. Shainberg and J.D. Wood (1980). "Flocculation value and structure of Na/Ca montmorillonite and illite suspensions". *Soil Sci. Soc. Amer. Journal*, 44, 955-959.

Romano, N., B. Brunone and A. Santini (1998). "Numerical analysis of one-dimensional unsaturated flow in layered soils". *Adv. Water Resour.*, 21, 315-324.

Russo, D. (1988). "Numerical analysis of the nonsteady transport of interacting solutes through unsaturated soil, 1. Homogeneous systems". *Water Resour. Res.*, 24(2), 271-284.

Russo, D. and E. Bresler (1977). "Analysis of the saturated-unsaturated hydraulic conductivity in a mixed sodium-calcium soil systems". *Soil Sci. Soc. Amer. Journal*, 41, 706-710.

Shainberg, I., J.D. Rhoades and R.J. Prather (1981). "Effect of low electrolyte concentration on clay dispersion and hydraulic conductivity of a sodic soil". *Soil Sci. Soc. Amer. Journal*, 45, 273-277.

Zaidel, J. and D. Russo (1992). "Estimation of finite difference interblock conductivities for simulation of infiltration into initially dry soils". *Water Resour. Res.*, 28, 2285-2295.

Wöslen, J.H.M., G.J. Veerman, and J. Stolte (1994). "*Water retention and conductivity characteristics of topen subsoils in the Netherlands*". The Staring Series. Tech. Docu. 18, DLO Winand Staring Centre, Wageningen, The Netherlands, in Dutch, 66 p.

Surface Water Hydrology, Singh, Al-Rashed & Sherif (eds)
© 2002 Swets & Zeitlinger, Lisse, ISBN 90 5809 363 8

An infiltration test to evaluate the efficiency of groundwater recharge dams in arid countries

Gerhard Haimerl, Franz Zunic and Theodor Strobl
Institute of Hydraulic and Water Resources Engineering
Technische Universität München
80290 München, Germany
e-mail: g.haimerl@bv.tum.de

Abstract

To evaluate the effectiveness of groundwater recharge, a field test has been carried out in the Sultanate of Oman. The test is a large scale infiltration test with the possibility to measure the suction in the unsaturated soil in different depths. The measurement data serves as a basis for a comprehensive numerical model to simulate the recharge process downstream of dams. Nevertheless, the evaluation of the data confirms some theoretical ideas of the infiltration process. Based upon them some general recommendations for the recharge operation can be given. However, some of the results are quite unexpected and arouse curiosity for further investigations.

Introduction

The Institute of Hydraulic and Water Resources Engineering of the Technische Universität München, Germany, is carrying out a research project to evaluate the effectiveness of groundwater recharge dams and to develop strategies for the optimisation of the operation of the recharge dams (Strobl, Haimerl, 1999).

The project is carried out in co-operation with the Ministry of Regional Municipalities, Environment and Water Resources of the Sultanate of Oman. It includes a series of infiltration tests in Oman in the year 2001. Based on the evaluation of hydrogeological and topographical data, the Wadi Ahin catchment area (1054 km²) at the coast in the north of Oman has been chosen for the observation representing a typical wadi in this region (Fig. 1). The wadi flows seawards in north-eastern direction from the mountains through the Batinah plain, a recharge dam is located in the main wadi channel in 11 km distance to the sea.

The concept of the field tests is to follow the way of the infiltrating water through the soil. To evaluate the amount of recharging water the infiltration rate into the soil as well as the way and the velocity of the water through the pores of the soil matrix has been observed. Therefore, the intention of these tests is to simulate the infiltration in a wadi channel compared to the infiltration during a recharge event as real as possible. For this purpose, contrary to the common infiltrometer test (e.g. double ring infiltrometer), an infiltration basin has been set up (Haimerl, 2001).

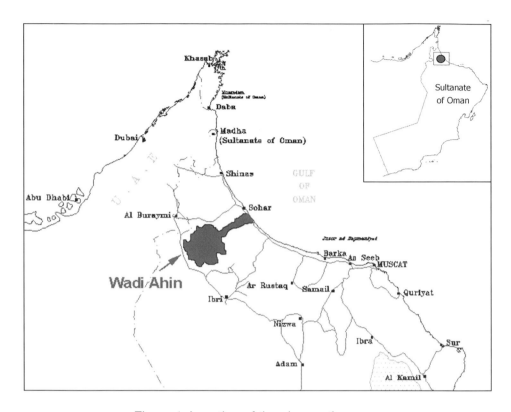

Figure 1. Location of the observation area.

Test Field and Measurement Instrumentation

The infiltration from the surface to the aquifer passes the unsaturated layer of soil where the pores are only partially filled with water. The water movement in the unsaturated soil is directly depending on the water tension or suction, which is the negative water pressure in the soil matrix. The water tension complies with the sum of the water retaining forces of the soil, the matrix potential. A gradient of the matrix potential is the driving force of water in the soil since water will move from areas of high potential to those of low potential (De Wiest, 1969).

The measurement of this matrix potential by measuring the water tension gives information about the water movement. The suction in coarse material like gravel decreases rapidly when the soil is wetted (Bouwer, 1978) so that the moment of an increasing water content can be determined. By measuring this process in different depths, the advance of the wetting front in the soil can be followed. To measure the suction in the soil, tensiometers are implemented in different depths.

The field tests have been carried out approximately 4 km downstream of wadi Ahin dam in the main recharge area. The location has been selected in a typical reach of the wadi and it has been paid special attention to preserve the natural condition of the surface and the soil.

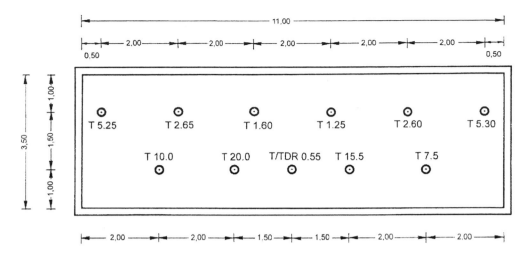

Figure 2. Layout of the test basin.

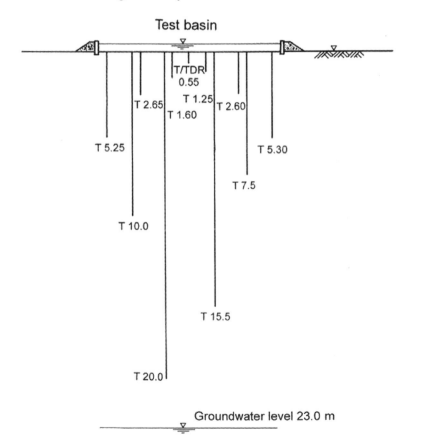

Figure 3. Vertical cross section of the test basin and subsoil.

In a rectangular infiltration basin of 11.0 m x 3.5 m, eleven holes have been drilled to install tensiometers (T) in a depth from 0.55 m to 20.0 m (Fig. 2 and 3). The boundary of the test basin has been built after installing the measurement instruments and backfilling of the boreholes as a concrete wall. To stabilise the wall, the outer side of the basin has been filled with wadi gravel (Fig. 4).

In the upper layers down to 2.5 m below the wadi bed, four tensiometers have been installed which are also equipped with a temperature probe. Between the depths of 5.0 m and 20.0 m seven tensiometers without temperature probes have been used. To also measure the soil moisture content below the surface, a Theta ML2 time domain reflectrometry (TDR) probe has been installed next to a tensiometer in 0.55 m depth (Fig. 2 and 3).

Close to the test basin three test pits have been excavated and a 30 m deep hole has been drilled with core recovery to take soil samples. The soil samples and the core show slightly silty sand, gravel and cobbles up to 6.5 m. Below, weak conglomerate and in places loose sand and gravel can be observed. Between 16.5 m and 18.0 m a layer of cemented gravel is located. The hydraulic permeability of the soil was determined to the range of 10^{-4} to 10^{-5} m/s. The cemented gravel layer has a permeability of approximately 10^{-7} m/s. The drilling hole has been prepared as a piezometer to measure the ground water level in a depth of about 23 m during the test period.

Figure 4. Test basin during impoundment of test 1.

Execution of the Tests

Wetting Phases

The water to fill the test basin has been taken from a well near the test site. As soon as the defined test area had been wetted, the proceeding of the infiltration front through the soil has been followed by the measurement of the suction in the-different depths. The impoundment of the basin (Fig. 4) has been done in three test phases of 4 days wetting each and 10 days drying in between to simulate different initial moisture contents and recharge conditions for each test. A fourth impoundment of the test basin has been done after a dry period of ten weeks with a short duration of 1.5 hours to simulate a small recharge event. The different boundary conditions of the tests are displayed in Table 1.

Table 1. Boundary conditions of infiltration tests 1 to 4.

	test 1	test 2	test 3	test 4
Duration of wetting [h]	99.5	99.5	98.5	1.5
Initial soil condition	dry	wet	wet	dry
Water level in the basin [cm]	50	50	20	15

To evaluate the water balance and to determine the infiltration rate from the basin, the inflow to the basin and the water level in the basin have been recorded continuously during the wetting phases.

The record of the tensiometers in different depths clearly shows the proceeding of the wetting front after starting the test (Fig. 5). The beginning of the drop down of the suction curve is the moment of the arrival of the infiltrating water at the tensiometer.

Drying Phases

After each infiltration test the drying phase has also been observed. The duration was ten days after test 1 and 2, 75 days after test 3 and 35 evaluated days after test 4, by now. The rising suction in all depths during the drying of the soil can be seen in Fig. 6. It displays the suction at depth 0.55 m, 2.65 m, 7.5 m, 10.0 m and 20.0 m. The time of the wetting phases 1-4 is marked at the bottom of the diagram.

Fig. 6 also shows the drop down of the suction during the wetting phases and its delay in greater depths, especially at 20.0 m. The high initial values before the beginning of test 1 do not necessarily represent the initial conditions of the soil, because they are remarkably influenced by the installation of the tensiometers in the boreholes refilled with dry sand.

317

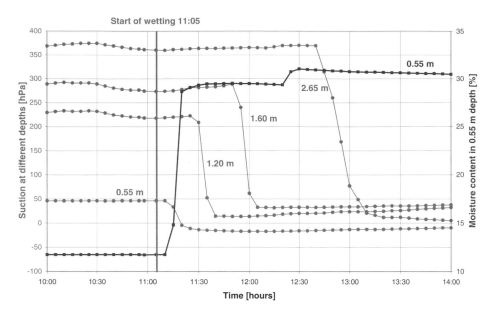

Figure 5. Measurements at the beginning of test 1.

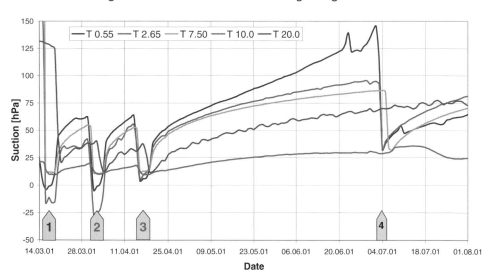

Figure 6. Measured suction during the test period.

Evaluation of the Measurement

Suction

The measured suctions have been interpreted qualitatively. The installation of the tensiometers in boreholes with the necessary refilling by sand around the instruments ensures the capillary contact to the soil matrix. On the other hand the different pore size of the refill material and the soil leads to a different absorption of water in the sand and to a delay of the water movement, especially during drying. Moreover, the natural inhomogenities of soil do not legitimate qualitatively exact conclusions concerning the value of the suction. Nevertheless, the order of magnitude and the gradient of suction along the soil profile are essential clues to describe the water movement.

At all measurement points, during steady state percolation the suction was bigger than the theoretical value assuming the hydrostatic pressure. This suggests that the soil is not fully saturated, even near the surface and after several days of infiltration. To scrutinise this presumption and to quantify the influence of the installation conditions of the tensiometers, additional tests are carried out in the Oskar-von-Miller Research Laboratory of the Institute of Hydraulic and Water Resources Engineering of the Technische Universität München.

A further important observation is that the value of the suction oscillates by a frequency of one day. This oscillation is induced by the evaporation potential at the surface. It depends largely on the capillary connection of the tensiometers to the atmosphere at the surface. Nevertheless, regarding the accuracy of the tensiometers, oscillations with an amplitude of 25 hPa to 50 hPa in up to 5 m depth have been measured. This suggests that the evaporation potential has an effect on the water movement up to a depth of several meters.

Water Movement

To balance and to compare the water movement under different boundary conditions the measurement data has been evaluated regarding the *infiltration rate* from the test basin, the *infiltration velocity* to the several depths and the *infiltration volume* that penetrated into ground. The results are displayed in Table 2 and Fig. 7. Based on them, the influence of the boundary conditions can be explained.

One of the boundary conditions is the different initial moisture content. Test 1 has been carried out after a dry period of about two years, test 4 after a dry period of 2.5 month. Test 2 and 3 have been conducted 10 days after the last wetting. Thus, the initial moisture condition can be classified as *dry* for test 1 and 4 and as *wet* for tests 2 and 3 (Table 1). The water level in the basin (Table 1) can be differentiated into a higher water level during test 1 and 2 and a lower level during test 3 and 4.

Comparing the results of the tests in dry and wet soil (Table 2) it can be observed that the infiltration rate at the surface is higher when the soil is dry. This proves the theory that a higher suction in dry conditions leads to a higher infiltration. For this reason, the total infiltration volume is also bigger in dry soil. However, in this case the infiltrating water has to fill the pores on its way to the aquifer. Thus, compared to the pre-wetted soil, it takes a longer time to reach the groundwater level. In contrast, in wet soil the water percolates faster and the rise of the groundwater level is higher even though the infiltrated volume is less. The test proves that the efficiency of groundwater recharge (groundwater rise rated to the used amount of water) is increasing when the initial moisture content of soil is higher.

319

Table 2. Summary of test 1 to 4.

	test 1	test 2	test 3	test 4
Infiltration time up to 20 m depth [h]	113.0	54.2	68.3	-*
Infiltration time to the groundwater table [h]	126.0	93.4	-*	-*
Groundwater rise [cm]	11	19	-*	-*
Average infiltration rate [cm/h]	14.9	12.5	7.3	10.8
Infiltration volume [m³]	568	484	276	6.8

*   not measured

Another observation aims at the influence of the water level in the basin which represents the flow depth during a recharge event in wadi channels. It can be seen that the infiltration rate and volume as well as the percolation velocity is lower when the water level at the surface is lower. This proves that the infiltration depends on the hydraulic pressure head. A higher pressure head causes a higher infiltration.

The last different boundary condition of the tests is the duration of the wetting. To simulate a small recharge event a very short test (test 4) has been carried out. Despite the low water level during this test, the infiltration rate was quite high due to a dry soil condition. The water passed quickly up to 2.5 m depth, but then the impoundment had already been finished and the infiltration slowed down immediately. This can clearly be seen in the graphical depiction of the infiltration times in Fig. 7. In the further record of the test, it could be measured that the water reached

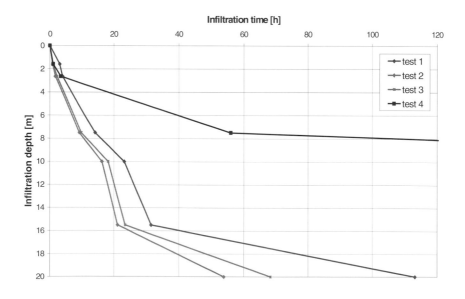

Figure 7. Infiltration time during test 1 to 4.

15.5 m depth after 350 hours. A further proceeding could not be observed, the water did not reach the aquifer. Thus very short infiltration times do not enable recharge.

Moreover, Fig. 6 shows the influence of the different soil layers on the infiltration. The layer of cemented gravel between 16.5 m and 18.0 m retards the proceeding of the wetting front. The percolation from 15.5 m to 20.0 m took a longer time than from the surface up to 15.5 m.

Conclusions and Perspective

The evaluation of the infiltration test results can be summarised as follows. The efficiency of groundwater recharge is increasing with

– the moisture content of soil,
– the water level of the surface flow,
– time of infiltration.

From this, a general recommendation for the use of stored flood water for groundwater recharge can be given. Within the limits of the maximum storage time and other scheme specific restrictions, the water should be released to those recharge channels which provide a small wetted area and a water level in the order of magnitude of some decimetres. A large wetted area with a flow depth of only a few centimetres should be avoided. Moreover, a limitation of the wetted area provides a long time of infiltration. Nevertheless, attention has to be paid of course to choose areas with a high hydraulic permeability and suitable geological conditions for recharge.

Despite the mentioned uncertainties of the exact measurement readings of the tensiometers the results show two specialities that demand further investigations. First, even after a few days of steady state infiltration the pores are not saturated with water. The measured water pressure is below the theoretical pressure gradient in all depths. Second, remarkable daily oscillations of the suction can be measured at least up to a depth of 5 m. Thus it is obvious that the evaporation potential at the surface has a considerable influence on the water and vapour movement.

To specify these current general results from the field tests, the next step of work is to imply a numerical model to simulate the recharge event. Besides the described direct evaluation of the results, the test has been conceived to obtain the opportunity to calibrate such a model. After a successful preparation of the model, a greater variety of parameters can be studied. Besides the observed parameters of the field tests, different soil properties, longer wetting and drying phases, and different depths of the aquifer will be investigated. Moreover, the measured suctions and temperatures will be used to take the vapour transport in soil into consideration. Thus it will be possible to quantify the influence of the evaporation more detailed.

References

Bouwer, H. (1978). "Groundwater Hydrology". McGraw-Hill Kogakusha Ltd., Tokyo, Japan.

De Wiest, R.J.M. (1969). "Flow Through Porous Media". Academic Press, New York, USA.

Haimerl, G. (2001). "Talsperren zur Grundwasseranreicherung in ariden Gebieten – Bewirtschaftungsstrategien und Optimierungsmöglichkeiten". Institute of Hydraulic and Water Resources Engineering, Rep. Nor. 91 p. 227–235, Technische Universität München, Germany (in German).

Oaksford, E.T. (1985). "Artificial Recharge: Methods, Hydraulics, and Monitoring". In: Artificial Recharge of Groundwater, editor Takashi Asano, Part I, Chapter 4, Butterworth Publishers, Boston, USA.

Strobl, Th. and Haimerl, G. (1999). "Optimization of the Recharge Process Downstream of Groundwater Recharge Dams". Civil and Environmental Engineering Conference, 8-12 November 1999. Conference Proceedings, Volume 5, Part II, p. V~87–V~96, Bangkok, Thailand.

Section 6: Watershed modeling

Surface Water Hydrology, Singh, Al-Rashed & Sherif (eds)
© *2002 Swets & Zeitlinger, Lisse, ISBN 90 5809 363 8*

Hydrologic modeling in arid areas: Concepts, theory and data

Leonard J. Lane and Mary R. Kidwell
United States Department of Agriculture
Agricultural Research Service
2000 East Allen Road, Tucson, Arizona, 85719 USA
e-mail: ljlane@tucson.ars.ag.gov

Abstract

Water resources are strategic national resources and development of knowledge to enable their sustainable management and use must be a high national priority. It is impossible to monitor all of our watersheds, yet we need the ability to predict the consequences of land use and management decisions on watershed resources before use and management is undertaken. Simulation models provide the mechanism for this prediction ability. Water balance modeling is used to illustrate the application of simulation modeling to management of arid watersheds. Arid environments are characterized by extreme hydrologic variability. Simulation models can mimic this variability, but have their conceptual and practical limitations. Major challenges to overcome these limitations and improve our ability to predict water balances in arid environments are identified and discussed.

Introduction

Water resources are of strategic national importance. Development of secure, adequate, and sustainable water resources can only be accomplished in two ways. First is development of additional water resources of such abundance that their wise use is sustainable. Second is conservation of existing water resources coupled with development of new knowledge and technology to utilize existing resources more efficiently. In the very simplest terms, the first way is called the water "supply" approach and the second is called the water "demand" approach. Both approaches require our best science and technology development and transfer. However it should be noted that the first approach often cannot be accomplished because of physical and political constraints limiting sources of new water resources. Even where new water resources are available, they can never be sustainable without coupling water "supply" approaches with water "demand" approaches. For example, Jackson et al., (2001) conclude that in the next 3 decades the world's accessible runoff is unlikely to increase by more than about 10% while the world's population may increase by about a third. Additional information on water resources at the global scale is available from a number of sources (e.g. see Dickinson, 1991; Gleick, 1998, 2000; and Jackson et al., 2001).

Increased competition for water and watershed resources is resulting in wider recognition of regional, national, and global water shortages and the need to understand land use impacts on water supply and quality. The wider recognition of water shortages and deterioration of water quality will increase the global impacts of watershed research and the value of new concepts, theory, and data. This wider recognition of global impacts of watershed research, in turn, increases the need for hydrologic modeling in arid areas.

All land is composed of watersheds (also called catchments or drainage basins) and thus understanding and predicting the current status and future trends of our watershed resources are of great practical and broad societal importance. Therefore, watershed management decisions must be based on the best possible science. There is a critical need for simulation model development as a means of integration, or synthesis, of science and decision-making and parallel (concurrent) technology transfer (e.g., NRC, 1999).

We know it is impossible to gage or monitor all of our watersheds, yet we need the ability to predict consequences of land use and management decisions upon watersheds and their resources before the proposed use and management are undertaken. Simulation models offer the potential to provide this prediction capability. Development of improved simulation models with valid prediction capabilities is thus essential to our understanding and management of water resources (e.g. for some recent compendia and general modeling sources and discussion see Singh, 1995; Mays, 1996; Hoggan, 1997; NRC, 1999; Bates and Lane, 2000; and Anderson and Bates, 2001).

Water resources are particularly important in arid and semi-arid environments because of their scarcity and because the structure and function of watersheds and their ecosystems in these environments are fragile and subject to dramatic changes (often deleterious to water supply and quality) from unwise utilization and management. Knowledge of the water balance, or budget, is essential to management of water resources. Therefore, modeling the water balance on our watersheds is central to bringing our best science to bear on decision making for joint water supply and water demand approaches.

Scope and Limitations

In this paper we:

1) Briefly describe arid climates,
2) Use time series to illustrate hydrologic extremes and their importance in the water balance,
3) Discuss and illustrate selected examples of water balance modeling, and
4) Identify gaps in knowledge, modeling, and data that limit our ability to develop sustainable water resources in arid areas.

This paper is neither a compendium of hydrologic models, nor is it used to develop new models. Rather, it is a call for improved simulation models and their coupling with decision making to provide new concepts, theory, and data for sustainable management of water resources in arid areas of the subtropics.

Arid Climates

The most complete definitions of climate include interactions of weather with topography, soil, vegetation, and land use to produce a physically based method, incorporating aspects of land use, to describe the long-term expectations of precipitation, temperature, etc. for a region. However, for the limited purposes herein, a climate definition scheme based on precipitation, temperature, and their seasonal distributions will suffice. The following description of arid climates is derived in part from Lane and Nichols (1999) and Trewartha and Horn (1980).

Arid (desert) areas generally receive too little precipitation to support dryland agricultural or domestic livestock grazing. In contrast, in semi-arid (steppe) areas adequate moisture is usually available at some time during the year to produce forage for livestock, and there are even some years when dryland crop production is successful. It should be noted that both climates are characterized by extreme variability with commonly occurring droughts and infrequent periods of above average rainfall resulting in flooding (see the section below entitled Hydrologic Time Series). Most discussion hereafter will emphasize arid climates, but some of the concepts and examples thought to have direct application in arid areas will be derived from observations and modeling in semi-arid areas.

The majority of the world's arid areas occur along two wide belts at approximately 30 degrees latitude north and south of the equator. In these subtropical belts the air is usually descending, and dry much of the time. Semi-arid areas associated with the arid deserts are mostly north or south of the deserts (Africa, Asia, and Australia) or inland at higher elevations (North America, South America, Middle East, Africa, and Asia). On a more localized scale, a combination of mountains and prevailing wind direction can cause "rain shadow" effects, resulting in arid and semi-arid areas downwind of major mountain features.

About a third of the world's land surface is either arid, normally with less than 250 mm of annual precipitation, or semi-arid with between 250 mm and 500 mm of annual precipitation. As described below, somewhat more precise definitions of desert and semi-arid areas are given by climatic classifications based on precipitation, temperature, and their seasonal distributions. For example, Lane and Nichols (1999) following Trewartha and Horn (1980) and the classifications of Koppen (1931), presented upper and lower mean annual precipitation limits defining arid climates.

The semi-arid climates, where annual precipitation is not strongly seasonal, were defined by equations linking mean annual values of precipitation, P (mm), and temperature, T (degrees C). The upper limit for semi-arid climates, in terms of mean annual precipitation given a specific value of mean annual temperature, is defined by

$$P = 20T + 140 \tag{1}$$

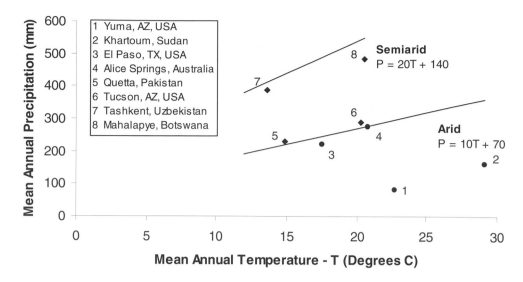

Figure 1. Mean annual precipitation and temperature for arid and semi-arid sites around the world. Also shown are the lines (Eqs. 1 & 2) defining semi-arid and arid climates.

The corresponding lower limit that separates arid and semi-arid climates (or alternatively desert and steppe climates) was defined as 1/2 the value of the upper limit from Equation 1, or

$$P = 10T + 70 \tag{2}$$

Thus the arid, or desert, areas of the world are those areas where the combination of mean annual precipitation, P, and mean annual temperature, T, fall below the line defined by Eq. 2. For example, long-term mean annual precipitation and temperature at Khartoum, Sudan (15.6^0 N, 32.6^0 E) are 162 mm and 29.1^0 C respectively. For a value of T of 29.1, Eq. 2 requires that P be less than 10 x 29.1 + 70 = 361 mm. Since the mean annual precipitation at Khartoum at 162 mm is less than half of the computed value of 361 mm, Khartoum is classified as an arid climate. The long-term annual precipitation and temperature values at Yuma, AZ, USA (32.7^0 N, 114.6^0 W) are P = 81 mm and T = 22.7^0. While not as hot as Khartoum, the P and T values plot even farther below the line defined by Eq. 2, and therefore Yuma is even more arid than Khartoum. Mean annual precipitation and temperature data for a number of locations around the world, as well as the lines and regions defined by Eqs. 1 and 2, are shown in Fig. 1.

Evapotranspiration is defined as the sum of water evaporation from soil, litter, etc. and transpiration from living plants. Annual potential evapotranspiration significantly exceeds precipitation in arid and semi-arid areas and can be accurately predicted with a number of techniques. In contrast, actual evapotranspiration is nearly equal to precipitation and is difficult to calculate under field conditions. Although actual evapotranspiration differs little in magnitude from precipitation on an annual basis, these small differences are extremely important because they by and large determine soil moisture status, runoff, and groundwater recharge. In brief periods during and then following large precipitation events, precipitation can exceed evapotranspiration and it is in these periods when surface runoff, soil moisture recharge, and even groundwater recharge can occur.

The vegetation growing in arid areas is adapted to lack of moisture, extreme variations in precipitation and temperature, soil characteristics, competition, and herbivory. Seasonal distributions of precipitation and temperature also play a dominant role, but it is the absolute amount of seasonal rainfall, particularly the seasonal distribution of rainfall that is most important for the vegetation of arid regions. The flora of these regions is extremely variable and can be rather distinctive as a result of factors such as continental position, rain shadow effect, proximity to cool ocean currents, high pressure air systems, and general air movement over the earth's surface (Brown, 1974). However, a notable feature common to all arid regions is the low density of vegetation that they support.

The soils in arid areas are notable for their variations with topographic features. Desert soils are characterized by their non-homogeneity in space and their generally close relationship with the parent geologic material due to their thinness, the lack of moisture, and the slowness of soil forming processes. The better, or more developed, soils are often formed on alluvial deposits or deposits of loess. Vertical differentiation of soil profiles may also be indistinct or lacking due to weak chemical activity resulting from the dryness. A typical exception to this generalization occurs where there has been deposition and leaching of calcium, in the form of calcium carbonate, and other soluble salts which form hard, impermeable subsoil (Lane and Nichols, 1999). Variation in soil properties from undifferentiated profiles to impermeable formations have a significant influence on the water balance in desert areas and significantly affect hydrologic processes, erosion and sedimentation, biological productivity, and thus land use and management.

Hydrologic Time Series

Examination of hydrologic time series can add insight to historical trends, illustrate current status of the variables of interest, and suggest possible future trajectories in the trends. The upper portion of Fig. 2 shows some global trends adapted from data presented by Gleick (1998) and Jackson, et al. (2001). During the last century, world population and land area used for irrigated agriculture increased almost exponentially. However, notice that the "demand" time series of world population is out pacing the "supply" time series of irrigated area for agriculture. As stated in the Introduction, a wider recognition of water shortages and deterioration of water quality will increase the global impacts of hydrologic research and the value of new concepts, theory, and data.

As insightful as world or global analyses can be, they do not easily lead to water resources solutions because water resources decision-making is conducted at na-

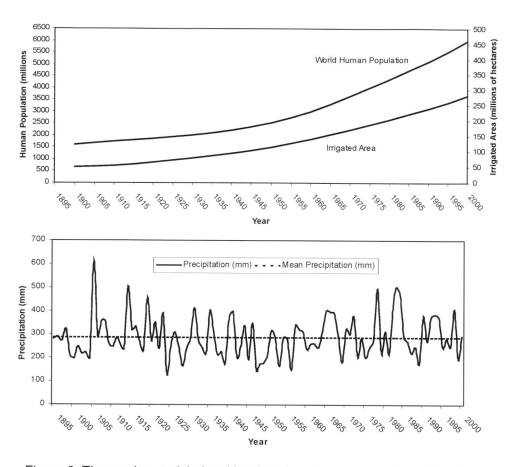

Figure 2. Time series at global and local scales. The upper portion of the figure uses time series to illustrate the "supply" and "demand" features of world water resources and the lower portion illustrates the extreme variability in annual precipitation in a semi-arid area.

tional, regional, and local scales. Lane, et al (1994) summarized analyses of hydrologic series from global to regional to point scales. Such analyses are instructive in illustrating the features and variability of hydrologic time series.

The lower portion of Fig. 2 shows a time series of precipitation at Tucson, AZ, USA (University of Arizona gage, 32^0 14′ N, 110^0 57′ W, 734 m MSL, data from 1895 to 2000). Mean annual precipitation at Tucson is 288 mm with a standard deviation of 87 mm resulting in a coefficient of variation (standard deviation divided by the mean) of 0.30 or 30%. The lowest annual precipitation was 129 mm in 1924 and the highest was 614 mm in 1905. This variation of annual precipitation from 129 to 614 mm represents a range of 485 mm, which is 168% of the mean and 557% of the standard deviation of annual precipitation. Finally, the ratio of the maximum daily precipitation to the minimum annual precipitation is 106 mm/129 mm = 0.82 which means that in 106 years of record the maximum daily precipitation was 82%

330

of the measured annual precipitation in the driest year. The significant variation in precipitation in arid and semi-arid areas such as Tucson is obvious.

Time series analyses of the data from Tucson indicated that there were no significant linear trends, serial correlation, or cycles or periods in annual precipitation at Tucson for the period 1895 to 2000. Subsets of the data (i.e. 1905 to 1947 and 1947 to 1984) might show statistically significant decreases or increases, but no trends or patterns are apparent in the entire 106-year record. There is an important point to make in interpreting time series analyses of short periods of record from arid areas – analyses of short periods of record are risky and statistical inferences from them are subject to high levels of uncertainty. In the next section of this paper we present examples of water balance calculations in arid and semi-arid areas. The reader should keep in mind the high variability of climate in such areas and the certainty that this high climatic variability induces comparable variability in all components of the water balance.

Examples of Water Balance Modeling

The term hydrologic cycle is the most general of those describing the cycling or movement of water through the lands, oceans and atmosphere. The term water balance as used herein has a similar meaning to the term hydrologic cycle but it connotes a budgeting or balancing of components in the hydrologic cycle. As such, water balance usually connotes a specific spatial scale such as a watershed, field, plot, or point and a specific time period such as annual, seasonal, or daily. Some analyses of the hydrologic cycle/water balance on a global and continental scale suggest that in general we are using too much groundwater relative to surface water resources (see Slutsky and Yen, 1997). In many arid areas groundwater withdrawals far exceed groundwater recharge resulting in the "mining" of historical groundwater resources – a demonstrably unsustainable practice.

To discuss water balances for arid areas at the point to watershed scale, it is necessary to compare and contrast water balances in humid and sub-humid areas with those in semi-arid and arid areas. For generalized discussions of water in arid areas see Roberts (1993), Savenije (2000), and DePauw et al. (2000). Some comparisons and contrasts between humid and arid water balances are summarized in Table 1. The comments are for dominant factors or generalized relationships. There are exceptions, most of which are time, space, or intensity scale dependent and of great scientific interest. However, for the discussions in this paper, we compare and contrast based on generalizations or the usual case.

Water balance models scan a range of complexity and applications dependent upon their intended use, needed precision in predictions, and treatment of temporal, spatial, and process intensity scales. Examples range from a simple, largely data-base model, as presented by Evans and Jakeman (1998), to simplified daily water balance models used in vegetation production modeling at a point or plot scale (Lane et al., 1984, 1995), to simulation of water balance across vegetation communities (Kremer and Running, 1996), and to complex, process based models (Refsgaard and Storm, 1995).

Table 1. Comparison of water balance components for humid and arid areas.

Component	Humid	Arid	Comments
Precipitation, P	Abundant	Scarce	See Fig. 1 herein
Potential Evapotranspiration, PET	PET < P	PET > P	
Actual Evapotranspiration, AET	AET < P	AET ~ = P	In arid areas, P does exceed AET during storms and during brief periods of high soil moisture storage
Runoff, Q	Perennial streams and rivers are abundant. Watersheds "store" precipitation and release it as runoff throughout the year	Most streams are ephemeral; perennial streams can rise in wetter regions and flow through arid areas	Runoff often generated only during prolonged wet periods or from intense storms in arid areas
Groundwater, GW	Groundwater-surface water connection direct and obvious, i.e. gaining perennial streams	Groundwater-surface water connection less obvious, i.e. losing ephemeral streams	Groundwater resources in arid areas often "mined" because of low recharge rates
Soil Moisture, SM	Well developed soils with adequate water holding capacity support rain-fed agriculture	Poorly developed soils and inadequate soil moisture result in desert landscapes. Cultivated agriculture possible with irrigation	Water erosion, floods, and poor drainage common in humid areas. Droughts, salinity, wind and water erosion common in arid areas.
Quantification of Water Balance	Measurements Possible for P, PET, AET, Q, SM, and GW over relatively short time periods, i.e. years to decades	Longer time periods required to measure P, Q, etc. because of higher natural variability, i.e. infrequent hydrologic events	In arid areas, if some terms of the water balance are obtained by subtracting measured terms that are on the same order of magnitude, then large
Modeling the Water Balance	Space, time, and process intensity scale problems plague modeling efforts	Same comment as for humid climates, but longer time series of measured data required to obtain same relative level of precision for model calibration and validation	errors result

Example Water Balance Calculations Using a Simple Model

Our objective in these example calculations was to compute water balances in three very different climatic regimes and to contrast and compare the results from the three locations. Because of data limitations at the sites, we selected a simple water balance model that could be operated based on limited available climatic, soils, vegetation, and land use data. The CREAMS Model (Knisel, 1980) met the selection criteria and had been previously applied at these and similar sites.

The one-dimensional water balance equation for a unit area, to plant rooting depth, ignoring runon and assuming subsurface lateral flow is zero, can be written as

$$dS/dt = P - Q - AET - L \qquad (3)$$

where dS/dt is the change in soil moisture (mm), P is precipitation (mm), Q is runoff (mm), AET is actual evapotranspiration (mm), L is percolation or leaching below the rooting depth (mm), and t is time (days, months, years, etc.). The CREAMS Model solves Eq. 3 for a daily time step and then sums the results for monthly and annual values.

Renard, et al (1993) applied the CREAMS Model on the Walnut Gulch Experimental Watershed in Arizona, USA, Lane and Osterkamp (1991) applied the Model in the Mojave Desert of Nevada, USA, and Osterkamp et al. (1995) applied the Model using data from the Al Ain Agromet Station in the UAE. The following summary of water balance calculations is adapted from these three studies.

Results of the CREAMS water balance modeling for three locations are shown in Tables 2 4. In each of these tables, Column 1 lists the month or the annual period, Column 2 lists precipitation in mm, Column 3 lists surface runoff in mm, Column 4 lists the actual evapotranspiration in mm, Column 5 lists percolation below the plant rooting depth in mm, and Column 6 list the average plant available soil moisture in mm. Notice that the annual values in Columns 2-5 are annual summations whereas the annual value for plant available soil water is an average annual value.

The CREAMS Model was applied to a small semi-arid watershed on the Walnut Gulch Experimental Watershed in Arizona, USA (see Renard et al., 1993 and Goodrich et al., 1997 for descriptions of Walnut Gulch). Rainfall and runoff data were available for 17 years (1965–1981), and were used to optimize the model parameters for runoff simulation. As P and Q were measured, the model was calibrated to match observed values of runoff, Q, and then AET and L were estimated using a form of Eq. 3. Values of Q, AET and L in Tables 2–4 do not exactly sum to P because dS/dt was not zero over the simulation period. However, dS/dt was relatively small, −1.4 mm for the data shown in Table 2.

The mean monthly precipitation distribution at Walnut Gulch is bi-modal (Table 2) with a strong summer peak from July through September and a small secondary

Table 2. Average annual water balance for Watershed 63.103 at Walnut Gulch, Arizona, USA as calculated with the CREAMS Model calibrated using 17 years of rainfall and runoff data, 1965–1981. All values are in mm.

Month (1)	Precipitation (2)	Runoff (3)	AET (4)	Percolation (5)	Plant Available Soil Water (6)
January	18.0	0.58	18.6	0.03	22.2
February	14.2	0.28	18.0	0.17	20.7
March	15.0	0.18	21.2	0.0	16.1
April	3.8	0.0	11.8	0.0	6.6
May	5.3	0.13	7.4	0.0	2.0
June	8.3	0.28	8.4	0.0	1.3
July	87.9	7.24	62.2	0.0	9.8
August	63.3	4.78	63.7	0.0	14.9
September	39.1	3.45	34.8	0.0	15.7
October	21.0	1.70	16.5	0.0	16.0
November	7.7	0.05	9.7	0.0	16.0
December	19.3	1.02	12.1	0.0	18.8
Annual	302.9	19.7	284.4	0.20	13.3

peak from December through March. Soil moisture storage (plant available soil water) follows this trend with recharge occurring July through October and again in December and January. Rapid soil moisture depletion occurs from February through June (Table 2, last column).

In contrast with Walnut Gulch, Rock Valley, Nevada (Table 3) is dominated by winter precipitation and soil moisture recharge (November through March) with rapid soil moisture depletion from April through August. Mean annual precipitation at Rock Valley is about half of what it is at Walnut Gulch while mean annual runoff is estimated to be about an order of magnitude less than at Walnut Gulch (Tables 2 and 3).

Mean monthly precipitation at Al Ain is strongly dominated by the winter months. Soil moisture recharge is estimated to occur from January through March with soil moisture depletion and low levels of soil moisture throughout the remainder of the year (Table 4).

Although mean annual precipitation at Al Ain is about 1/3 as much as at Walnut Gulch, mean annual runoff is about 60% as much as at Walnut Gulch (Tables 2 and 4). The relatively high values of runoff estimated by the CREAMS model at Al Ain are due to the combination of large rainfall events, soil properties, and the short

Table 3. Average annual water balance for Rock Valley, Nevada, USA as calculated with the CREAMS Model partially calibrated using mean monthly soil moisture data. Simulations for 1965–1976. All values are in mm.

Month (1)	Precipitation (2)	Runoff (3)	AET (4)	Percolation (5)	Plant Available Soil Water (6)
January	15.0	0.08	13.0	0.0	24.0
February	27.0	2.01	15.0	1.60	31.0
March	17.0	0.05	23.0	0.46	34.0
April	6.0	0.0	19.0	0.0	21.0
May	6.0	0.0	12.0	0.0	11.0
June	5.0	0.0	8.0	0.0	7.0
July	9.0	0.0	10.0	0.0	6.0
August	13.0	0.0	11.0	0.0	6.0
September	11.0	0.10	11.0	0.0	8.0
October	10.0	0.05	9.0	0.0	8.0
November	13.0	0.0	9.0	0.0	10.0
December	19.0	0.28	9.0	0.0	16.0
Annual	151.0	2.57	149.0	2.06	15.0

(20 years) period of simulation. The model estimated 70.7 mm of runoff in 1972 and 51.5 mm in 1990. Together these two years account for about half of the total runoff in 20 years, or half the long term mean. This illustrates the high level of variability of precipitation and thus runoff, AET, percolation, and soil moisture in the data summarized in Table 4. Finally, only precipitation data were available at Al Ain so the model results are predictions only and do not include any calibration or validation. Given this level of uncertainty, the mean annual runoff estimated for Al Ain may be considerably in error.

The modeling results summarized in Tables 2–4 illustrate several important features of water balance modeling and water resources management in arid areas. Even in arid cmates (not including the hyper-arid deserts devoid of vegetation), one may expect some times during the year when soil moisture may be recharged and thus water is available for plant growth. This is especially true in semi-arid areas such as Walnut Gulch (Table 2). Conversely, one may also expect that there will be long dry periods when soil moisture is near zero or exhausted. Although infrequent and small compared with humid regions, some groundwater recharge may occur in upland areas due to percolation of water below the plant rooting depth. However, percolation estimates are infrequent, making statistical inference risky. This is especially true if percolation is obtained by subtracting AET and Q from P. Small dif-

Table 4. Average annual water balance for Al Ain, UAE as calculated with the CREAMS Model without calibration. Simulations are for 1971–1990. All values are in mm.

Month (1)	Precipitation (2)	Runoff (3)	AET (4)	Percolation (5)	Plant Available Soil Water (6)
January	6.1	0.20	4.4	0.0	6.5
February	36.2	6.47	11.8	0.33	17.5
March	22.5	4.17	15.8	3.05	23.7
April	11.7	1.04	15.2	0.0	22.4
May	2.1	0.25	8.2	0.0	16.3
June	1.7	0.03	5.5	0.0	10.6
July	1.9	0.0	5.0	0.0	7.4
August	6.0	0.03	5.6	0.0	6.4
September	5.6	0.28	4.9	0.0	7.4
October	0.0	0.0	2.1	0.0	5.7
November	1.3	0.08	1.8	0.0	4.6
December	3.6	0.13	2.6	0.0	4.5
Annual	98.7	12.7	82.9	3.38	11.1

ferences (percolation estimates) obtained from subtracting relatively large, nearly equal, numbers are notoriously error prone. Thus, percolation estimates from short periods of record in arid areas should be viewed as qualitative numbers (suggesting percolation), not as quantitative estimates. Similar logic applies to surface runoff estimates in arid areas.

Nonetheless, it is possible to draw some generalizations from data such as presented in Tables 2–4. Plant available soil water is seen to vary from zero (monthly averages approach zero but remain positive because of averaging, monthly values in individual years are often zero) to something on the order of a few 10's of mm. This sparse and highly variable soil moisture limits short-term management options. Unless irrigation is used, management schemes should be tailored to long-term objectives involving well-adapted flora and fauna able to withstand climatic extremes. As a corollary of this, management practices may take long periods to positively impact the water balance and thus monitoring and evaluation may require correspondingly long time periods. Again, the converse is true for mismanagement or over utilization of resources. Plant available soil water is so erratic and limited that fragile ecosystems once damaged may take very long periods to recover; or, they may not recover at all.

Discussion

Current research and scientific discussion clearly illustrate that global water demand is increasing faster than water supply. Even where some new water resources are available, they can never be sustainable without coupling water "supply" approaches with water "demand" approaches.

It is also clear that increased competition for water and watershed resources is resulting in wider recognition of regional, national, and global water shortages and the need to understand land use impacts on water supply and quality. The wider recognition of water shortages and deterioration of water quality will increase the global impacts of watershed research and the value of new concepts, theory, and data. This wider recognition of global impacts of watershed research in turn increases the need for hydrologic modeling in arid areas.

Analyses of trends in global water supply and demand lend insight into generalized water resources problems. However, as insightful as world or global analyses can be, they do not easily lead to water resources solutions because water resources decision-making is done at national, regional, and local scales. Therefore, it is appropriate for scientists and engineers conducting hydrologic modeling to fully recognize, and to report, the high levels of climatic variability present in arid and semi-arid areas at all spatial scales, especially point to regional scales. Further, these high temporal and spatial variabilities in arid climates induce comparable levels of variability into hydrologic modeling results. High variability in hydrologic modeling results translate directly into high levels of uncertainty and these too should be recognized and reported in hydrologic modeling activities. Hydrologic models are essential decision-making and management tools, but they can only realize their full potential as such by fully tracking and reporting the levels of uncertainty present in the models and their predictions.

Water balance models are critical in addressing water supply problems and providing basic input to water demand analyses. Water balance modeling in arid areas presents additional challenges. First, the inherent climatic variability is high in arid areas and this requires long periods of hydrologic monitoring to produce baseline data for modeling and analyses. Second, given the infrequent nature of runoff events in arid areas, it may take very long time periods to analyze the impacts of land use and management practices on water supply and water quality. Both the baseline data and the impacts data are essential for developing, calibrating, and validating hydrologic models. Substantial and long-term investments of scientific and monetary resources are required to obtain these data. Third, components of the water balance are not well quantified in arid areas and thus additional knowledge and understanding required for this quantification can only come from additional hydrologic research. This, too, requires substantial long-term investments. Finally, sound social policy must be science-based and this requires integration of biophysical and social science research within our modeling and analyses to bring the best science to bear on decision-making.

Some Gaps in Knowledge, Modeling and Data

The Discussion section presented some needs and gaps in hydrologic modeling especially pertinent to arid areas. However, some specific gaps are listed here based upon the above analyses, current discussions in the literature, and our personal experiences.

- Hydrologic modeling is an essential step in bringing science to bear on decision-making for water resource development and management in arid areas. However, problems of temporal, spatial, and process-intensity scales and the associated high variability limit application of current hydrologic models.

- High variability in climatic inputs (see Fig. 2) results in high variability of model results that in turn produce high levels of uncertainty in model predictions. An important challenge is to incorporate this uncertainty into hydrologic modeling, document and communicate that uncertainty, and quantify the resulting impacts on decision-making.

- Spatial variability of hydrologic model inputs and processes introduce high levels of uncertainty into hydrologic modeling. It is also important to incorporate this uncertainty into hydrologic modeling, document and communicate that uncertainty, and quantify the resulting impacts on decision-making.

- We feel that process-intensity scale effects and their impacts on thresholds and nonlinear responses are under-appreciated by the hydrologic modeling community. For example, there are thresholds and nonlinearities in rainfall-runoff relationships, in flow regime (subcritical vs. supercritical flow), and in suspended solids concentrations (entrainment vs. deposition). Inadequate representation of these process intensity-scale dependent thresholds and nonlinearities limits the ability of hydrologic models to accurately predict responses to a broad range on input values.

- Our hydrologic modeling abilities have outpaced our abilities to conduct field experiments. Simulation models in three spatial dimensions and time can be constructed but we still lack the ability to measure water flow in space and time. Often, hydrologic time series at a point are used to develop, calibrate, and validate distributed models with the result that it is possible to get the "right answer" for all the wrong reasons.

- There is an imbalance in large hydrological experiments with an over emphasis on quantifying components of the water balance (e.g. rainfall-runoff relationships, land surface-atmosphere interactions, groundwater flow) at the expense of systematic research on the entire water balance and the rich interactions and feedback among its components. Component research should continue and be strengthened, but it should be guided and focused by overarching systematic experiments on the water balance.

Acknowledgments

We are pleased to acknowledge the helpful comments of reviewers, H. D. Fox and H.E. Canfield, as well as the anonymous reviewers. We gratefully acknowledge the USDA-Agricultural Research Service for supporting this research and much of the research infrastructure upon which it is based. Finally, we gratefully acknowledge the Conference organizers and sponsors for giving us the opportunity to prepare this manuscript and participate in the Conference.

References

Anderson, M.G., and Bates, P.D. (Eds.) (2001). Model Validation: Perspectives in Hydrological Science, John Wiley and Sons, Ltd., Chichester, UK, 500 pp.

Bates, P.D. and Lane, S.N. (Eds.) (2000). High Resolution Flow Modelling in Hydrology and Geomorphology, John Wiley and Sons, Chichester, UK, 374 pp.

Brown, G.W. (Ed.) (1974). Desert Biology, Volume II, Academic Press, New York, 601 pp.

DePauw, E., Gobel, W., and Adam, H. (2000). Agrometeorological aspects of agriculture and forestry in the arid zones. *Ag. And Forest Meteorology*, 103: 43-58.

Dikinson, R.E. (1991). Global change and terrestrial hydrology – a review. *Tellus*, 43 AB: 176-181.

Evans, J.P. and Jakeman, A.J. (1998). Development of a simple, catchment-scale, rainfall-evapotranspiration-runoff model. J. *Environmental Modeling & Software*, 13: 385-393.

Gleick, P.H. (1998). Water in crisis: Paths to sustainable water use. *Ecological Applications*, 8(3): 571-579.

Gleick, P.H. (2000). The World's Water 2000–2001. The Biennial Report on Freshwater Resources, Island Press, Washington, D.C., 315 pp.

Goodrich, D.C., Lane, L.J., Shillito, R.A., Miller, S.N., Syed, K.A., and Woolhiser, D.A. (1997). Linearity of Basin Response as a Function of Scale in a Semi-Arid Ephemeral *Watershed. Water Res. Res.*, 33(12): 2951-2965.

Hoggan, D.H. (1997). Computer-Assisted Floodplain Hydrology and Hydraulics, McGraw-Hill, 676 pp.

Jackson, R.B., Carpenter, S.R., Dahm, C.N., McNight, D.M., Naiman, R.J., Postel, S.L., and Running, S.W. (2001). Water in a changing world. Issues in Ecology, Technical Report. Ecological Applications, 11(4):1027-1045.

Knisel, W.G. (1980). *CREAMS*: A field-scale model for chemicals, runoff and erosion from agricultural management systems: USDA Conserv. Res. Report 26.

Koppen, W. (1931). Grundriss der Klimakunde. Berlin: De Gruyter.

Kremer, R.G. and Running, S.W. (1996). Simulating seasonal soil water balance in contrasting semi-arid vegetation communities. *Ecological Modelling.* 84:151-162.

Lane, L.J. and Osterkamp, W.R. (1991). Estimating Upland Recharge in the Yucca Mountain Area. Proc. ASCE Irrig. and Drain. Conf., Honolulu, HI, pp. 170-176.

Lane, L.J. and Nichols, M.H. (1999). Semi-Arid Climates and Terrain. In: Encyclopedia of Environmental Science (Ed. D.E. Alexander and R.W. Fairbridge), Encyclopedia of Earth Sciences Series, Kluwer Academic Publications, The Netherlands, pp. 556-558.

Lane, L.J., Romney, E.M. and Hakonson, T.E. (1984). Water balance calculations and net production of perennial vegetation in the northern Mojave Desert. *J. Range Mgt.* 37(1):12-18.

Lane, L.J., Nichols, M.H. and Osborn, H.B. (1994). Time Series Analyses of Global Change Data. *J. Environmental Pollution*, 83:63-68, Huntington, UK.

Lane, L.J., Hakonson, T.E., and Bostick, K.V. (1995). Applications and limitations of the water balance approach for estimating plant productivity in arid areas. *In:* Proc. Symposium on Wildland Shrub and Arid Land Restoration, Las Vegas, NV, USA, Oct 19-21m 1993 (Roundy, B.A., McArthur, E.D., Haley, J.S., and Mann, D.K., Eds.), pp. 335-338.

Mays, L.W. (Ed.) (1996). Water Resources Handbook. McGraw-Hill, New York, USA.

NRC (National Research Council). (1999). New Strategies for America's Watersheds. National Academy Press, Washington, D.C., 311 pp.

Osterkamp. W.R., Lane, L.J. and Menges, C.M. Techniques of Ground-water Recharge Estimates in Arid/semi-arid Areas, with Examples from Abu Dhabi. *J. Arid Environments*, 31:349-369. 1995.

Refsgaard, J.C. and Storm, B. (1995). MIKE SHE. In: Singh, V.P., (Ed.), Computer Models of Watershed Hydrology. Water Resources Pubs., CO, USA, pp. 806-846.

Renard, K.G., Lane, L.J., Simanton, J.R., Emmerich, W.E., Stone, J.J., Weltz, M.A., Goodrich, D.C., and Yakowitz, D.S. (1993). Agricultural impacts in an arid environment: Walnut Gulch studies. *Hydrol. Sci. & Tech*, 9: 145-190.

Roberts, B.R. (1993). Water Management in Desert Environments. A comparative Analysis. Springer-Verlag, Berlin, 337 pp.

Savenije, H.H.G. (2000). Water scarcity indicators; the deception of the numbers. *Phys. Chem. Earth (B)*, 25:199-204.

Singh, V.P. (Ed.). (1995). Computer Models of Watershed Hydrology. Water Resources Pubs., CO, USA, 1130 pp.

Slutsky, A.H. and Yen, B.C. (1997). A macro-scale natural hydrologic cycle water availability model. *Journal of Hydrology*, 201:329-347.

Trewartha, G.T. and Horn, L.H. (1980). An introduction to climate. 5th ed. New York: McGraw-Hill Book Company, 415 pp.

Surface Water Hydrology, Singh, Al-Rashed & Sherif (eds)
© *2002 Swets & Zeitlinger, Lisse, ISBN 90 5809 363 8*

Watershed model to study hydrology, sediment, and agricultural chemicals in rural watersheds

Deva K. Borah, Renjie Xia and Maitreyee Bera

Illinois State Water Survey, 2204 Griffith Drive, Champaign, IL 61820, USA

e-mail: borah@uiuc.edu

Abstract

A Dynamic Watershed Simulation Model (DWSM) is being developed at the Illinois State Water Survey, USA to simulate surface and subsurface runoff, soil erosion, and entrainment and transport of sediment and agricultural chemicals in rural watersheds. The model has routing schemes developed using approximate analytical solutions of the physically based equations preserving the dynamic behaviors of water, sediment, and the accompanying chemical movements within a watershed. An overview of the model and its theoretical basis and application to a 100-km^2 watershed in Illinois, USA are presented. The model reasonably predicted the water and sediment discharges at the watershed outlet, however, the agricultural chemical transport simulations of the model needs further improvements for watersheds of this size.

Introduction

Flooding, upland soil and streambank erosion, sedimentation, and contamination of water from agricultural chemicals are critical environmental, social, and economic problems in Illinois and other states of the U.S., and throughout the world. For example, Lake Decatur, a drinking water supply reservoir for the city of Decatur in Illinois, USA receives water containing high nitrate-nitrogen (nitrate-N) concentration from the Upper Sangamon River basin draining a 2,400 km^2 agricultural watershed. The lake often violates the state and national drinking water standard of maximum 10 milligrams per liter (mg/L) of nitrate-N (Demissie et al., 1996). The lake also has a high sedimentation rate that is gradually reducing its water supply capacity.

Understanding and evaluating the watershed processes and problems are continued challenges for scientists and engineers. Mathematical models simulating these processes are useful tools to analyze these complex processes, to understand the problems, and to find solutions through land-use changes and best management practices (BMP). The models help in evaluating and selecting from alternative land-use and BMP scenarios. Implementation of these practices can help reduce the damaging effects of storm water runoff on water bodies and the landscape. Developing reliable watershed simulation models and validating them on real world watersheds with measured and monitored data is also challenging.

A number of watershed simulation models exist today. Most of the models were developed in the 1970s and 1980s and since the early 1990s, most modeling research focussed on development of the graphical user interfaces (GUI) and integration with geographic information systems (GIS) and remote sensing data. While enormous progress has been made in developing and refining interfaces, greater efforts are now needed to focus on model formulations – conceptualization and description of hydrologic and water quality processes, efficient algorithms and computational techniques, including both new developments and enhancement of existing codes (Chen, 2001; Committee on Watershed Management, 1999).

Some of the well-known watershed-scale nonpoint source pollution models for rural basins are Soil and Water Assessment Tool or SWAT (Arnold et al., 1998), Hydrological Simulation Program – Fortran or HSPF (Bicknell et al., 1993), Agricultural NonPoint Source pollution or AGNPS model (Young et al., 1987), Areal Nonpoint Source Watershed Environment Response Simulation or ANSWERS (Beasley et al., 1980), Precipitation-Runoff Modeling System or PRMS (Leavesley et al., 1983), KINematic runoff and EROSion or KINEROS model (Woolhiser et al., 1990), Dynamic Watershed Simulation Model or DWSM (Borah et al., 1999; 2000; 2002), and a European Hydrological System or MIKE SHE model (Abbott et al., 1986). SWAT and HSPF are long-term continuous simulation models useful for analyzing long term effects of hydrological changes and watershed management practices, specially, agricultural practices. AGNPS, ANSWERS, KINEROS, and DWSM are single-event models useful for analyzing severe single-event storms and evaluating watershed management practices, specially, structural practices. PRMS and MIKE SHE have both long-term and single-event simulation capabilities.

Theoretical (mathematical) bases, the most important elements of mathematical models, and computational algorithms used in the above models are somewhat different. These models have some physically based concepts and formulations and some empirical relations with varying degrees. Physically based models are better suited to predict future conditions and evaluate land use changes and BMPs than empirical models. However, the models having the most physically based formulations with multi-dimensional governing equations are often impractical to apply to large watersheds because of their computational intensity and possible numerical instabilities. Based on physically based formulations and efficient algorithms, DWSM was found to be the most dynamic and promising watershed-scale single-event model for rural basins among the above-cited models having all the three nonpoint-source pollution model components – hydrology, sediment, and chemicals.

The Illinois State Water Survey (ISWS) has been developing the DWSM through improving and expanding a model developed earlier by Borah et al. (1980), Borah (1989a, b), and Ashraf and Borah (1992). The DWSM uses physically based governing equations to simulate surface and subsurface storm water runoff, propagation of flood waves, soil erosion, and entrainment and transport of sediment and agricultural chemicals in agricultural watersheds. The model has three major components: (1) DWSM-Hydrology (Hydro) simulating watershed hydrology, (2) DWSM -Sediment (Sed) simulating soil erosion and sediment transport, and (3) DWSM-Agricultural chemical (Agchem) simulating agricultural chemical (nutrients and pesticides) transport. Each component has routing schemes developed using approximate analytical solutions of the physically based equations preserving the dynamic behaviors of water, sediment, and the accompanying chemical movements within a watershed. Different components of the DWSM have been applied and tested on watersheds in Illinois (Borah et al., 1999; 2000; 2001a, b; 2002; Borah and Bera, 2000).

In this paper and presentation, an overview of the DWSM, its concepts and formulations, and application (calibration-validation) of its three components to the Big Ditch watershed, a 100 km^2 tributary subwatershed of the upper Sangamon River, using monitored data are briefly described and discussed. The hydrology and sediment components of the model performed reasonably well in predicting the water and sediment discharges at the watershed outlet, however, the agricultural chemical component of the model needs further improvements for watersheds of this size, the preliminary results are promising.

In addition to the modeling challenges, the primary purpose of developing the DWSM is to use it as a watershed management tool. The DWSM may be used to evaluate alternative land use and BMPs in reducing flooding, soil erosion, and sediment and agricultural chemical discharges into water bodies, such as the Lake Decatur resulting from severe storm events, responsible for moving large quantities of sediment and pollutants across a watershed. Having physically-based concepts and formulations, the DWSM would be useful to other parts of the USA and the world.

The DWSM Scheme and Components

The watershed is divided into subwatersheds, specifically, into one-dimensional overland elements, channel segments, and reservoir units. For example, Figure 1 shows the Big Ditch watershed in Illinois, USA draining a 100 km^2 (38 square-mile) basin and divided into 26 overland elements, 13 channel segments, and no reservoir unit. The overland elements are identified with the numbers 1-26 and the channel segments 27-39. These divisions take into account the nonuniformities in topographic, soil, and land-use characteristics, which are treated as being uniform with representative characteristics within each of the elements or segments.

Figure 2 shows schematic model representations of six overland elements (1-6) contributing to three channel segments (27-29) of Big Ditch watershed (Figure 1). An overland element is represented as a rectangular area with the same area as in the field, width equal to the adjacent (receiving) channel length, length equal to area divided by the width, and representative slope, soil, cover, and roughness based on physical observations of these characteristics in the element. A channel segment is represented with a straight channel having the same length as in the field and having a representative cross-sectional shape, slope, and roughness based on physical observations and measurements. A reservoir unit is represented with a stage-storage-discharge relation (table) developed based on topographic data and discharge calculations using outlet measurements and established relations.

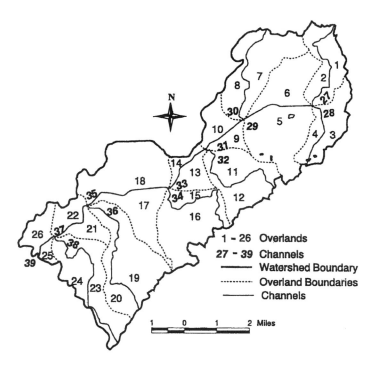

Figure 1. Big Ditch watershed in Illinois divided into DWSM segments
(1 mile = 1.609 kilometer).

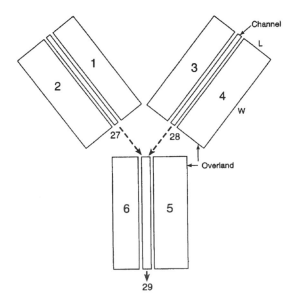

Figure 2. Schematic representations of DWSM overland and channel segments.

The same watershed subdivisions are used while simulating the watershed processes from each of the model components – hydrology, sediment, and agricultural chemical. Simulation starts at the uppermost overland element and continues through the adjacent overland and the channel, then two overlands and one channel, etc. throughout the watershed and finally ending in a channel or a reservoir at the watershed outlet using a predetermined computational sequence based on gravity flow logic. In each flow element, segment, or unit, the model component(s) is/are used to simulate the processes. If only watershed hydrology is simulated, only the DWSM-Hydro is used. If hydrology and sediment or only sediment is simulated, both the Hydro and Sed are used. If hydrology, sediment, and chemicals or only chemicals are simulated, all the three components – Hydro, Sed, and Agchem are used. An efficient sequencing scheme is used, in which the outflow discharges of water, sediment, and chemicals from a flow element, segment, or unit are stored until those are used as inflows while simulating through the following flow element, segment, or unit. Once discharges from an element, segment, or unit are used, those are erased to make the storage space available for discharges from another element, segment, or unit.

DWSM-Hydro: Hydrologic Simulations

Rainfall is the major input to the model. It initiates all the processes simulated in the model and, therefore, is the driving force of the model. Single or multiple raingage records may be used. The overland elements are the primary sources of runoff (flowing water) in which rainfall turns into surface runoff after losing first to interception at canopies and ground covers, then to infiltration through the ground surface and depression storage above it. Simulation starts with the computation of rainfall excess rates, the rainfall available for surface runoff, from given breakpoint rainfall record (rainfall recorded at different times during a storm) or synthetic rainfall depths or rates generated using analytical procedures.

Two alternative procedures are used to compute rainfall excess. The first one is a simple procedure using Soil Conservation Services or SCS (1972) runoff curve number method. The second one is a detailed procedure involving computations of interception losses using a procedure used by Simons et al. (1975) and infiltration rates using an algorithm developed by Smith and Parlange (1978) from a simplified solution of the equation for one-dimensional diffusion of water under gravity. The first method computes rainfall excess rates, which are subtracted from rainfall rates to compute infiltration rates assuming other losses negligible during a storm event. The second method computes interception and infiltration rates, which are subtracted from rainfall rates to compute rainfall excess rates.

The excess rainfall is routed over the overland elements beginning at their upstream edges (ridges), in which flows are zeros, up to their downstream edges, coinciding with the receiving channel banks using the analytical and approximate shock-fitting solutions of the kinematic wave equations, developed by Borah et al. (1980). Figure 3 shows schematic representations of overland and channel flows. As shown in this figure, routing of excess rainfall over only a unit width of the element resulting in "flow per unit width" is required because the physical and mete-

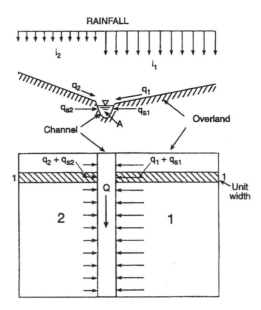

Figure 3. Schematic representations of DWSM overland and channel flows.

orological characteristics of an overland element are assumed uniform. The unit-width flow is uniform along the overland width and discharges uniformly along the channel length.

A portion of the infiltrated water flows laterally towards downstream as subsurface flow sometimes in accelerated mode in the presence of tile drains. Similar to surface runoff, subsurface water from infiltration is routed through a unit width of the soil matrix underneath the overland surface beginning at its upstream edge (ridge), in which flow is assumed zero, up to their downstream edge, coinciding with the receiving channel bank. The lateral subsurface flow is computed using a kinematic storage equation introduced by Sloan et al. (1983). Currently, the tile drain flows are lumped with the lateral subsurface flow using an effective lateral saturated hydraulic conductivity concept presented in Borah et al. (2000). The conservation of subsurface water or spatially uniform and temporarily varying continuity equation is solved to keep track of (updating) the subsurface water volume.

Two overland elements contribute surface and subsurface flows to one channel segment laterally from each side of the channel as shown in Figure 3. The receiving waters from the overland elements and upstream channels are routed through the channel segment using the analytical and shock-fitting solutions of the kinematic wave equations (Borah et al., 1980). The channel network carries the receiving waters from overland elements and upstream channel segments downstream of the watershed and ultimately to the watershed outlet. During its journey, the runoff water may be intercepted by lakes or reservoirs, which release it again to downstream channels at reduced rates after temporary storage. The standard storage-indication or Puls method (SCS, 1972) is used to route floodwater through the reservoirs.

348

The above procedures, and their sources, used in computations of infiltration and rainfall excess rates and routing these over and under the overland surfaces, and routing their contributions through the channels and reservoirs are described in Borah (1989a), Borah et al. (1999), and Borah et al. (2000, 2002).

DWSM-Sed: Soil Erosion and Sediment Transport Simulations

Similar to the hydrologic component, soil erosion and sediment transport are simulated along with water through the overland elements and stream segments. The eroded soil or sediment is divided into number of particle classes (size groups). Agricultural watersheds having extensive aggregates, the sediment is divided into five size groups: sand, silt, clay, small aggregate, and large aggregate after Foster et al. (1985). Each size group is dealt individually during the simulation of each of the processes, and total response, in the form of sediment concentration and discharge is obtained by integrating the responses from all the size groups.

The model computes soil erosion due to raindrop impact using concepts and expressions derived by Meyer and Wischmeier (1969) and Mutchler and Young (1975). The eroded (detached) soil is added to an existing detached (loose) soil depth from where entrainment to runoff takes place with sufficient velocity and shear (capacity). Erosion due to flow shear stress and deposition depends on sediment transport capacity of the flow and the sediment load (amount of sediment already carrying by the flow). Based on past research (Alonso et al., 1981), the bed load formula of Yalin (1963) is used to compute sediment transport capacities in overland elements under any flow condition and for all size groups. In computing capacities in the channel segments, the total load formula of Yang (1973) is used for sediment sizes ≥ 0.1 mm (fine to coarse sands) and the total load formula of Laursen (1958) is used for sediment sizes < 0.1 mm (very fine sands and silts). If the capacity is higher than the sediment load, erosion takes place and the flow picks up more materials from the bed. If the loose soil volume at the bed is sufficient, sediment entrainment takes place from the detached soil depth. Otherwise, the flow erodes additional soil from the parent bed material, which is calculated as described in Borah (1989b). If the sediment transport capacity is lower than the sediment load, the flow is in a deposition mode and the potential rate of deposition is equal to the difference of the two. The actual rate of deposition is computed by taking into account particle fall velocities. Deposited sediment is added to the loose soil volume. If the sediment transport capacity and the sediment load are equal, an equilibrium condition is assumed where there is neither erosion nor deposition.

All the above processes are interrelated and must satisfy locally the conservation principle of sediment mass expressed by the sediment continuity equation. The continuity equation is solved with an approximate analytical procedure to keep track of erosion, deposition and sediment discharges along the flow elements and segments. Detailed descriptions of these procedures are given in Borah (1989b), Ashraf and Borah (1992), and Borah et al. (1999, 2002).

At present, the model does not route sediment through reservoir units and assumes deposition of all the sediment carried by the flow. Therefore, the model is applicable to large detention ponds, lakes, and reservoirs where most of the sediment is trapped and sediment bypassed is negligible.

DWSM-Agchem: Agricultural Chemical Transport Simulations

The agricultural chemical transport component of DWSM involve simulations of mixing and exchange of nutrients and pesticides between soil and soil solutions on the ground and in the soil matrix with rain and runoff waters on the ground surface as well as water infiltrated through it. These processes are simulated on the overland elements. The simulated chemicals are transported with surface runoff in dissolved forms and with sediment in adsorbed forms through the overland elements and channel segments along with water and sediment simulated in the hydrology and sediment components. At present, the model does not simulate the chemical processes and route those in reservoir units.

The model assumes equilibrium between dissolved and adsorbed phases of the chemicals, governed by a linear adsorption isotherm. The soil profile is divided into small homogeneous soil increments. The model routes infiltrating rainwater and solutes through the soil increments using the concept of complete mixing (Ingram and Woolhiser, 1980) and computes water contents and chemical concentrations in each of the increments as described by Ashraf and Borah (1992). When runoff begins, exchange of chemicals from a mixing soil layer of the soil profile, containing the chemicals in dissolved form, with surface runoff and vice versa are simulated using the concept of non-uniform mixing of runoff with the mixing soil layer (Ahuja and Lehman, 1983; Heathman et al., 1985). Exchange of chemicals in adsorbed forms with the eroded and deposited sediments are computed based on preference factors of the individual size groups as described in Ashraf and Borah (1992). The entrained chemicals are routed along slope lengths in dissolved form with surface runoff and in adsorbed form with the transported sediment using approximate analytical solutions of the continuity (mass conservation) equations. Descriptions of these procedures are given in detail in Ashraf and Borah (1992) and Borah et al. (1999; 2002).

Application of the DWSM on the Big Ditch Watershed in Illinois, USA

All the components of the DWSM are being tested (calibrated and validated) on the Big Ditch watershed in east central Illinois, USA. The Big Ditch watershed, shown in Figure 4, drains 100 km^2 (38 square-mile) agricultural land with almost 100 percent row crop production equally divided between corn and soybean. Soils are mostly silt loam and silty clay loam, poorly drained, and are fertile with high organic content and high resistance to drought. The topography is generally flat, streams having gradients 0.0003-0.0046.

The Big Ditch watershed is a subwatershed of the 2,400 km^2 (925 square-mile) Upper Sangamon River basin draining into the Lake Decatur. Lake Decatur is a public water supply reservoir for the city of Decatur in Illinois, USA. Lake Decatur has a history of high nitrate-N concentration, periodically exceeding 10 mg/L, and

350

Figure 4. Big Ditch watershed in Champaign County, East Central Illinois, USA
(1 mile = 1.609 kilometer).

violating state and national drinking water standards. The lake also has a high sedimentation rate that is gradually reducing its water supply capacity. Past study by Demissie et al. (1996) indicated that watershed based solutions using BMPs would reduce nitrate-N concentration in the lake. In that study, the watershed was extensively monitored and stream flow and nitrate-N concentration at five tributary and three mainstem gaging stations were continuously measured. The Big Ditch was one of those stations. Borah et al. (1999) monitored the Big Ditch streamgage (Figure 4) for flow, suspended sediment, and agricultural chemicals (nitrate-N, phosphate-phosphorous or phosphate-P or PO4-P, atrazine and metolachlor) during the 1998 and 1999 spring rainfall events. Breakpoint rainfall was recorded in raingage 6 (Figure 4) during both years and in 5 only in 1999. Rainfall, flow, suspended sediment, and chemical concentration data from this monitoring study are being used to test (calibrate and validate) the DWSM.

Applications of DWSM-Hydro & Sed

While applying the DWSM-Hydro & Sed to the Big Ditch watershed, the watershed was divided into 26 overland and 13 channel segments as shown earlier in Figure 1. Rainfall, flow, and sediment data from storms of May 1998 were used to calibrate the model and June 1998 to validate it. Figure 5 shows the calibration and Figure 6 shows the validation results. As described earlier, the DWSM uses two alternative procedures to compute rainfall excess: runoff curve number and an interception-infiltration routine. In this application, the interception-infiltration routine was

351

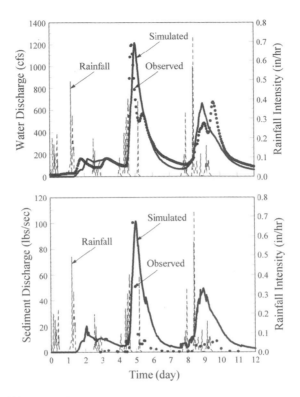

Figure 5. Water and sediment discharges from Big Ditch watershed
resulting from the May 1998 storms: Model calibration
(1 cfs = 0.028 m³/s, 1 in = 25.4 mm, 1 lb = 0.454 kg).

used. The most sensitive model parameters are the vertical saturated hydraulic conductivity, Manning's roughness coefficient, the effective lateral saturated hydraulic conductivity, and the flow detachment coefficient. These parameters were calibrated while running the May 1998 storm events (Figure 5), and kept these and other parameters and input data the same while running the June 1998 storms (Figure 6).

As shown in Figures 5 and 6, the overall model performance in simulating water and sediment discharges is good, specially, during the intense storms. The model over predicted water discharges during some less intense storms. Such discrepancies may be attributed to the single event nature of the model, not accounting for losses of runoff water and soil moisture through evapotranspiration between storms, especially after less intense storms. The over prediction of sediment discharges during less intense storms is mostly due to over prediction of runoff. There may be other factors involved, which needs further investigations.

Intense storms are the most critical storms when sediment and agricultural chemicals move in large amounts across a watershed. Therefore, as a single event

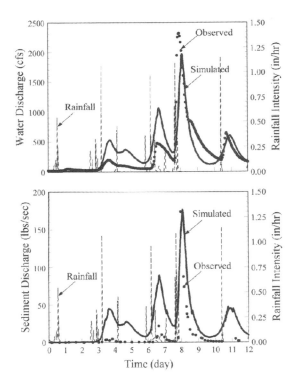

Figure 6. Water and sediment discharges from Big Ditch watershed
resulting from the June 1998 storms: Model validation
(1 cfs = 0.028 m³/s, 1 in = 25.4 mm, 1 lb = 0.454 kg).

model, the DWSM provides a useful tool in predicting water and sediment discharges from agricultural and rural watersheds during intense rainfall events.

Application of DWSM-Agchem

Application of the DWSM-Agchem to the 100 km^2 Big Ditch watershed is currently in progress. This is the first attempt to apply this model component to a bigger watershed of this size. Preliminary results of phosphate-P (PO4-P) simulations are presented here. The model was successfully applied before to experimental laboratory boxes (Ashraf and Borah, 1992), experimental field plots (Borah and Ashraf, 1993a), and a 1.4 hectare field-sized watershed (Borah and Ashraf, 1993b).

Similar to DWSM-Hydro & Sed applications, the May 1998 storms were used to calibrate the model and June 1998 storms to validate it on the Big Ditch watershed. During calibration, only three parameters – partition coefficient, interaction constant, and mixing soil layer depth were adjusted and during validation, these parameters were kept the same. Figure 7 shows the calibration and validation results of phosphate-P simulations where simulated phosphate-P concentrations are compared with observed phosphate-P concentrations. As shown in this figure

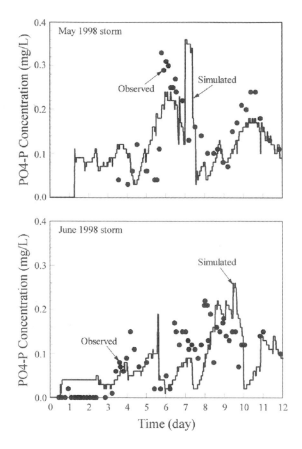

Figure 7. Simulated and observed phosphate-phosphorous (PO4-P) concentrations from Big Ditch watershed resulting from the May (calibration) and June (validation) 1998 storms.

although there are discrepancies between individual simulated and observed values, the simulated concentrations of phosphate-P follow the observed concentration patterns. The model needs further improvements and refinements to adequately simulate chemical concentrations in a large watershed like the 100 km^2 Big Ditch watershed.

Summary and Conclusions

An overview of the DWSM – a Dynamic Watershed Simulation Model, its concepts and formulations, and applications (calibration-validation) of its three components to the 100 km^2 Big Ditch watershed in Illinois, USA, using monitored data are presented. The hydrology and sediment components of the model performed reasonably well in predicting the water and sediment discharges at the watershed outlet, however, the agricultural chemical component of the model needs further improvements for watersheds of this size, the preliminary results are promising. The DWSM may be used as a watershed management tool to evaluate alternative

354

land use and best management practices in reducing flooding, soil erosion, and sediment and agricultural chemical discharges into water bodies resulting from severe storm events. Such storms are responsible for moving large quantities of sediment and pollutants across a watershed. Having physically-based concepts and formulations, the DWSM is useful to other parts of the USA and the world.

Acknowledgements

The study is currently funded by the Illinois State Water Survey (ISWS) and Illinois Council on Food and Agricultural Research (C-FAR) under the Water Quality Strategic Research Initiative (WQ-SRI) program. Illinois Groundwater Consortium (IGC) supported the initial development of the DWSM and monitoring of the Big Ditch watershed. Susan Shaw and Laura Keefer from the ISWS contributed in the monitoring.

References

Abbott, M.B., J.C. Bathurst, J.A. Cunge, P.E. O'Connell, and J. Rasmussen (1986). "An introduction to the European Hydrological System – Systeme Hydrologique Europeen "SHE" 2: Structure of a physically based distributed modelling system". *J. of Hydrology,* **87**: 61-77.

Ahuja, L.R. and O.R. Lehman (1983). "The extent and nature of rainfall-soil-interaction in the release of soluble chemicals to runoff". *Journal of Environmental Quality,* **12**(1): 34-40.

Alonso, C.V., W.H. Neibling and G.R. Foster (1981). "Estimating sediment transport capacity in watershed modeling". *Transactions of the ASAE,* **24**(5): 1211-1220, 1226.

Arnold, J.G., R. Srinivasan, R.S. Muttiah and J.R. Williams (1998). "Large Area Hydrologic Modeling and Assessment Part I: Model Development". *Journal of the American Water Resources Association,* 34(1): 73-89.

Ashraf, M.S. and D.K. Borah (1992). "Modeling pollutant transport in runoff and sediment". *Transactions of the ASAE,* 35(6): 1789-1797.

Beasley, D.B., L.F. Huggins and E.J. Monke (1980). "ANSWERS: A model for watershed planning". *Transactions of the ASAE,* **23**(4): 938-944.

Bicknell, B.R., J.C. Imhoff, J.L. Kittle, Jr., A.S. Donigian, Jr. and R.C. Johanson (1993). *Hydrologic Simulation Program - FORTRAN (HSPF) User's Manual for Release 10.* Report No. EPA/600/R-93/174.

Borah, D.K. (1989a). "Runoff simulation model for small watersheds". *Trans. of the ASAE,* 32(3): 881-886.

Borah, D.K. (1989b). "Sediment discharge model for small watersheds". *Trans. of the ASAE,* 32(3): 874-880.

Borah, D.K. and M.S. Ashraf (1993a). "RUNOFF3: A hydrologic and nonpoint source pollution model". *Advances in Hydro-Science and Engineering*, Vol. I, Sam S.Y. Wang (ed.), Center for Computational Hydroscience and Engineering, Univ. of Miss., University, MS: 367-372.

Borah, D.K. and M.S. Ashraf (1993b). "Nonpoint source pollution model for agricultural and rural watersheds". *Proceedings of the Engineering Hydrology Symposium*, July 25-30, 1993, San Francisco, CA, ASCE: 395-400.

Borah, D.K. and M. Bera (2000). *"Hydrologic Modeling of the Court Creek Watershed".* Contract Report 2000-04, Illinois State Water Survey, Champaign, IL.

Borah, D.K., M. Bera, S. Shaw and L. Keefer (1999). "Dynamic Modeling and Monitoring of Water, Sediment, Nutrients, and Pesticides in Agricultural Watersheds during Storm Events". Contract Report 655, Illinois State Water Survey, Champaign, IL.

Borah, D.K., S.N. Prasad and C.V. Alonso (1980). "Kinematic wave routing incorporating shock fitting". *Water Resources Research*, 16(3): 529-541.

Borah, D.K., R. Xia and M. Bera (2000). "Hydrologic and water quality model for tile drained watersheds in Illinois". ASAE, Paper No. 002093, St. Joseph, MI.

Borah, D.K, R. Xia and M. Bera (2001a). "Hydrologic and Sediment Transport Modeling of Agricultural Watersheds". *Proceedings of the World Water & Environmental Resources Congress*, Ed. D. Phelps and G. Sehlke, ASCE-EWRI, Reston, VA: CD-ROM.

Borah, D., R. Xia and M. Bera (2001b). "DWSM – A Dynamic Watershed simulation Model for Studying Agricultural Nonpoint Source Pollution". ASAE, Paper No. 012028, St. Joseph, MI.

Borah, D.K., R. Xia and M. Bera (2002). *"DWSM – A Dynamic Watershed Simulation Model".* Chapter in: Mathematical Models of Small Watershed Hydrology, Volume 2, Edited by V.P. Singh, D. Frevert, and S. Meyer, Water Resources Publications, LLC. In press.

Chen, Y.D. (2001). "Watershed Modeling for Nonpoint Source Water Quality Simulation: History, Recent Development and New Trends". *Proc. of 5th Int'l. Conf. on Diffuse/Nonpoint Pollution and Watershed Management*, IWA and Marquette Univ., Milwaukee, WI: CD-ROM.

Committee on Watershed Management (1999). *"New Strategies for America's Watersheds".* National Research Council, National Academy Press, Washington, D.C.

Demissie, M., L. Keefer, D. Borah, V. Knapp, S. Shaw, K. Nichols and D. Mayer (1996). *"Watershed Monitoring and Land Use Evaluation for the Lake Decatur Watershed"*. Miscellaneous Publication 169, Illinois State Water Survey, Champaign, IL.

Foster, G.R., R.A. Young and W.H. Neibling (1985). "Sediment composition for nonpoint source pollution analyses". *Transactions of the ASAE,* **28**(1): 133-139, 146.
Heathman, G.C., L.R. Ahuja and O.R. Lehman (1985). "The transfer of soil applied chemicals to runoff". *Transactions of the ASAE,* **28**(6): 1909-1915, 1920.

Ingram, J.J. and D.A. Woolhiser (1980). "Chemical transfer into overland flow". In *Proceedings Symposium Water Management,* 21-23 July 1980, Boise, ID, ASCE: 40-53.

Laursen, E. (1958). "The total sediment load of stream". *Journal of the Hydraulics Division,* ΛSCE, **54**(HY 1): 1-36.

Leavesley, G.H., Lichty, R.W., Troutman, B.M. and Saindon, L.G. (1983). *Precipitation-runoff modeling system – User's Manual,* USGS Water Resources Investigative Report 83-4238.

Meyer, L.D. and W.H. Wischmeier (1969). "Mathematical simulation of the process of soil erosion by water". *Transactions of the ASAE,* 12(6): 754-758, 762.

Mutchler, C.K. and R.A. Young (1975). "Soil detachment by raindrops". *Proceedings of the Sediment Yield Workshop,* Oxford, MS, 1972. USDA-ARS-S-40: 113-117.

Simons, D.B., R.M. Li and M.A. Stevens (1975). "Development of models for predicting water and sediment routing and yield from storms on small watersheds". Colorado State University, Fort Collins, CO.

Sloan, P.G., I.D. Moore, G.B. Coltharp and J.D. Eigel (1983). *Modeling Surface and Subsurface Stormflow on Steeply-Sloping Forested Watersheds.* Water Resources Institute Report 142, University of Kentucky, Lexington, KY.

Smith, R.E. and J.Y. Parlange (1978). "A parameter-efficient hydrologic infiltration model". *Water Resources Research,* **14**(3): 533-538.

Soil Conservation Service or SCS (1972). Hydrology. Section 4. In *National Engineering Handbook.* SCS, Washington, D.C.

Woolhiser, D.A., R.E. Smith and D.C. Goodrich (1990). *KINEROS, A kinematic runoff and erosion model: Documentation and User Manual.* ARS-77, U.S. Department of Agriculture, Agricultural Research Service, Fort Collins, CO.

Yalin, M.S. (1963). "An expression for bed-load transportation". *Journal of the Hydraulics Division,* **ASCE, 89**(HY 3): 221-250.

Yang, C.T. (1973). "Incipient motion and sediment transport". *Journal of the Hydraulics Division*, **ASCE, 99**(HY 10): 1679-1704.

Young, R.A., C.A. Onstad, D.D. Bosch and W.P. Anderson (1987). *"AGNPS, Agricultural nonpoint source pollution model: A watershed analytical tool".* U.S. Department of Agriculture Conservation Research Report 35.

Surface Water Hydrology, Singh, Al-Rashed & Sherif (eds)
© 2002 Swets & Zeitlinger, Lisse, ISBN 90 5809 363 8

Integrated solution to flash floods problems in wadi Watier, South Sinai, A.R.E

A. Hassan Fahmi and M. Samir Farid
National Water Research Center (NWRC), Egypt

Mohsen Sherif
Civil Engineering Department, Faculty of Engineering
United Arab Emirates University, Al Ain, UAE
e-mail: msherif@uaeu.ac.ae

Abstract

Proper hydrology evaluation is essential for flood protection to guarantee the public safety of the surrounding environment. Infrastructure of wadi Watier such as the international road Nuweibaa/Ras El-Naqab with length of 70 km, the crossing with the international road Dahab/Taba with length of 200 m, Nuweibaa City, and several tourist villages are seasonally subjected to distortion at the main stream and at the delta, Figure 1. The onrush of the surface runoff can be minimized at the intersections, with the international roads, by means of detention and storage works and at Nuweibaa City by means of dikes, dissipation works, and Irish crossings. The main problem is the ruin of these infrastructures with flooded water several times per year and the lost of tenths millions of cubic meters from water flowing to the Gulf of Aqaba. These ruins resulted in death of passengers and damage several trucks and autos. Consequently, maintenance of this road costs several millions Egyptian pounds every year. The main purposes of this research are to minimize the harmful distortion effects from flash floods on the international roads sharply and to maximize the use of runoff water to achieve protectorate surrounding environmental of the area. Other benefits can be achieved such as store some volumes of million cubic meters of direct water for seasonal agriculture sufficient to irrigate about 4500 feddans/year. Moreover, reuse of recharging water for groundwater aquifers, as indirect contribution, can cover the requirements of about 3000 feddans/year. Another positive impact is the creation of a good agriculture-land upstream the proposed structures amount to about 500 feddans. Results showed that, with a proper hydrological analysis, a number of positive impacts could be achieved. Furthermore, costs of structures can be compensated within few years after executions of the proposed control works.

Introduction

The area of wadi Watier, about 3600 km^2, represents a big horizon of southeastern Sinai. It represents a significant importance of the political, economical, and social for South Sinai. The main stream of the Wadi has the international road with length of 70 km, which considered vital artery to the area in view of the connection of the

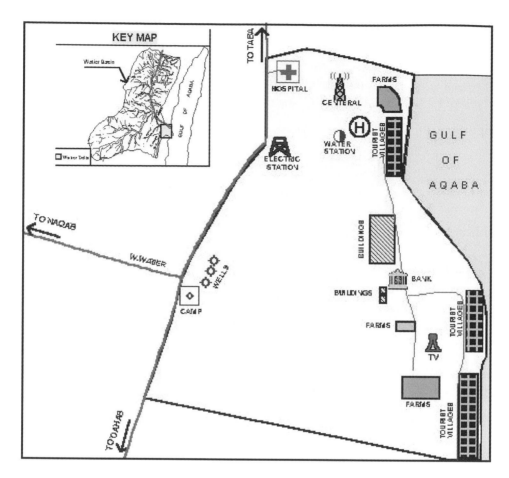

Figure 1. Location map of wadi Watier delta.

Arab countries with Egypt through Nuweibaa harbor. Due to the receiving of large amount of rainfall within the wadi catchment, it is distinguished with flooded water, appropriate fruitful soils for cultivating, inhabitance with Bedouins, natural springs, constructed wells, and tourism Canyon area. Moreover, at the delta of the Wadi, there are Nuweibaa City, many tourism villages, and roads network.

In 1986, Watier road was constructed, it was planned first and the drainage system was subsequently design without detailed hydrological studies. The potential debris flows was not recognized or considered in the design of this highway or its drainage system. Rainfall data were collected and analyzed for different return periods. The geomorphological and surface geological parameters were estimated. For economical reasons, the probable floods of 25 years were selected to design the different kinds of control works. Two criteria were established in the hydraulic design of dams. The protection criterion, which was based on the maximum daily rainfall, and the development criterion, which was based on the average seasonally rainfall. The study includes the flood hydrographs of all the subcatchment areas, which are given herein. HEC-1 model was used to obtain flood hydrographs. Channel routing and reservoir routing procedures which were carried out in order to get the final outflow from the different locations of dams at any channel section. Such information is essential and vital for the development and protection purposes. Water resources development includes storage of water for irrigation and groundwater recharge.

Topography

The study area has, at least, 34 subcatchments, Figure 2, which affect the morphological characteristics. These subbasins were classified to two groups; Eastern group which include 14 subbasins and Western group which include 20 subbasins. Eastern subbasin's areas range from 7 km^2 to 450 km^2 while western subbasin's areas range from 6 km^2 to 1400 km^2. For each subbasin, drainage area, length of the longest watercourse, and average slope were measured as list in Table 1.

Surface Geology

This part of the study is concerned with the determination of the losses parameters for each subbasin to get the excess rainfall or runoff depths. The catchment area is covered by basement rocks, mainly granites which are highly fractured and intruded by basic dikes trending in NE-SW direction. The basement rocks are nonconformably over lain by Cretaceous rocks, mainly sandstones followed by shales and limestones. Figure 3 shows the general classifications of the land cover in wadi Watier. Alluvial deposits derived from local rocks fill the drainage streams of the Wadi. The most well developed fault trend is the NNE or Aqaba trend. It is most apparent in the basement areas to the southeast and extends for more than ten kilometers. There are also NW-SE and E-W trends in the sedimentary areas to the North and West.

Rainfall losses could be, then, classified as either surface retention loss or infiltration. Losses are often separated into accumulated losses prior to initiation of surface runoff (initial abstraction) and losses during surface runoff. However, initial abstraction is not identical to surface retention loss because it includes some amount of infiltration losses prior to the start of surface runoff. So, the average distributions of the losses for wadi Watler were estimated as follows:

Start loss   = from 5.00 to 20.6 mm (seepage + infiltration, it depends on soil textures).

Retention loss = from 1.30 to 12.7 mm.

Constant loss   = from 0.33 to 25.4 mm/h (start from the end of the initial losses).

These analyses are used to estimate the rainfall excess and then amount of floods at any point of interest.

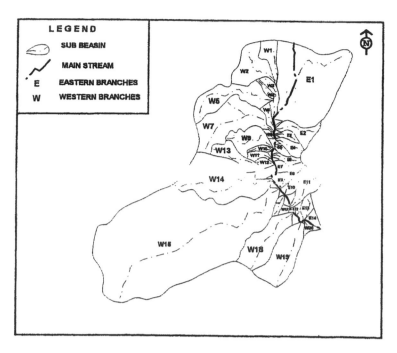

Figure 2. Subbasins of wadi Watier.

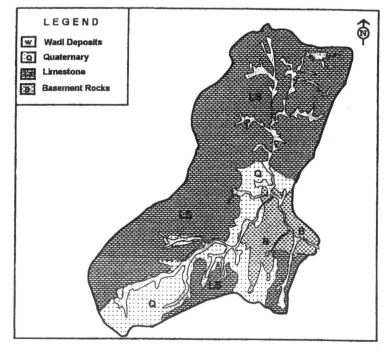

Figure 3. General geologic map.

Table 1. General topographical characteristics.

Code	Area (km²)	Length (km)	Slope (m/km)	Code	Area (km²)	Length (km)	Slope (m/km)
E1	450	44	6	W1	56	20	12
E2	55	21	18	W2	142	30	7
E3	14	8.5	20	W3	14	8	19
E4	32	14.6	24	W4	5.2	6	26
E5	7.1	5	60	W5	164	45	10
E6	23	13.5	30	W6	11	4	30
E7	12.2	6	42	W7	200	40	12
E8	17.2	4.5	28	W8	6.5	5	29
E9	20.5	9	28	W9	85	24	17
E10	9.3	7.3	37	W10	6.2	5.5	33
E11	33	13	28	W11	14.5	9	27
E12	5.7	4.6	108	W12	9	5	25
E13	12.5	7	10	W13	83	25	22
E14	10	3.5	122	W14	290	45	16
E : means Eastern Wadis				W15	1400	80	9
				W16	16	10	48
				W17	13	8	67
W: means Western Wadis				W18	175	32	18
				W19	135	23	35
				W20	6.5	2.2	170

Rainfall

The analysis of rainfall data is an important element in the hydrologic study. It is used as an input to determine the amount of infiltration and runoff volumes for each subbasin in the study area. Evaporation during rainfall storms is generally small and, therefore, was neglected, as it does not affect the water balance of the system. The hydraulic design of the different type of structures could be determined.

Since any hydraulic structure should operate under the future needs, it is important to predict the maximum daily rainfall by means of valuable historical data. The maximum daily rainfall was considered for protection calculation and assessment, while the average seasonally rainfall was considered for development plans. Selection of the appropriate level of risk for design, that will be accepted, is based on economic or policy aspects. So, rainfall data from seven rainfall gage stations were collected. Graphical and analytical methods were used to check the accuracy of the predicted and extrapolated values.

The results of the analytical solution were nearly the same as those of the graphical solution. Rainfall depths distribution for the average seasons and for the 25 year return period were drawn, as shown in Figure 4 and Figure 5 respectively.

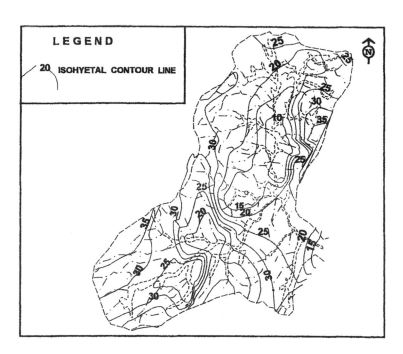

Figure 4. Rainfall distribution of the average seasonally depths.

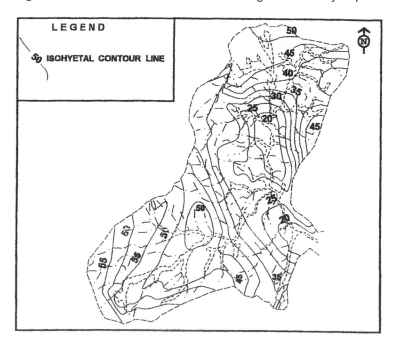

Figure 5. Rainfall distribution of the 25 year return period.

Each subbasin in the study area has been located on the foregoing distributions were used as inputs to determine the average annual floods and the maximum floods for the 25 year return period.

Flood Evaluation and Control Structures

The HEC-1 model was fed by the analyzed results of meteorological, morphological, and geological studies. The hydrographs at the outlets of the subbasins, according to rainfall depths at different return periods were estimated. For the purposes of retention and detention designs, rainfall depth at 25 year return period with duration of 2 hour was considered. The maximum daily rainfall was considered for protection calculation and assessment, while the average yearly rainfall was considered for the development plans.

Dams locations were selected upon field investigations according to the most practical and suitable sites. This includes construction process and transportation of the equipments and materials. The hydrological aspects of reservoir planning deal with:

1. Water availability in the area on which the dam is proposed to be constructed;
2. Determination of storage capacity to serve the target pattern of demand; and
3. Operation of reservoir with the given target pattern of demand.

Two types of dams were proposed in this respect, the detention and the storage dams. According to the amount of runoff volume from each subbasin, seventeen detention dams and five storage dams were proposed to be constructed on ten subcatchments as shown in Figure 6. Rating curve for each dam site was drawn using field-surveying data to determine reservoirs capacities and spillways levels. Storage reservoirs were classified according to the economic purpose, water man-

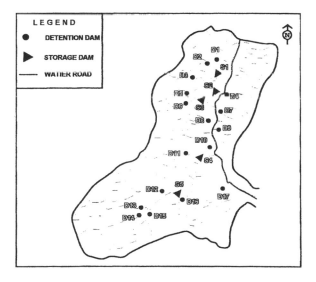

Figure 6. Proposed dam sites.

agement, and flood control. The estimated peak discharges, runoff volumes, and proposed reservoirs capacities are given in Table 2.

Flood Routing

One of the methods of avoiding flood damages and of recharging groundwater aquifer is by reducing surface flood discharges by storage. The main function of the proposed storage dams is to raise the water level upstream of the reservoir to store the seasonal flood and to spill maximum flood of the 25 year return period. In practice, it is a common problem that a flood hydrograph is known at one site in the stream of a channel or reservoir, while it is required to determine the hydrograph for the same flood further downstream of that site. The resulted hydrograph contains lower peak, which occurs at later time, with broader base. Lower peak indicates the peak attenuation as a result of passage through a reservoir or a channel reach. According to the subcatchments that have more than one dam, the confluence points and channel routing were determined, using Maskingum-Cunge method, and the combination of the different hydrographs was then estimated. The stream network model schematic was developed for each group of dams, as shown in Figure 7.

Table 2. Runoff estimations and possible reservoirs capacity.

Dam's Type		Average Flood		Max. Flood		Reservoir Capacity $(1000m^3)$
Detention	Storage	Q_{peak} (m^3/sec)	Volume $(1000m^3)$	Q_{peak} (m^3/sec)	Volume $(1000m^3)$	
D1		37	500	89	1.296	147
D2		57	1.118	144	2.994	199
	S1	19	268	49	694	900
D3		47	638	104	1.511	147
	S2	150	2.873	398	7.675	7.810
D4		36	354	69	756	230
D5		46	779	105	1.882	87
	S3	20	358	48	864	1.125
D6		14	195	34	495	145
D7		25	245	26	276	200
D8		57	611	87	1.002	204
D9		15	132	29	289	180
D10		32	474	66	1.040	312
D11		140	2.057	157	2.450	200
	S4	69	1.058	188	2.932	3.400
D12		156	2.704	253	4.633	367
D13		114	1.912	222	3.971	55
D14		41	509	165	2.231	130
D15		112	2.032	279	5.370	213
	S5	86	1.954	125	2.839	3.800
D16		112	2.032	279	5.370	173
D17		0	0	42	487	4.5

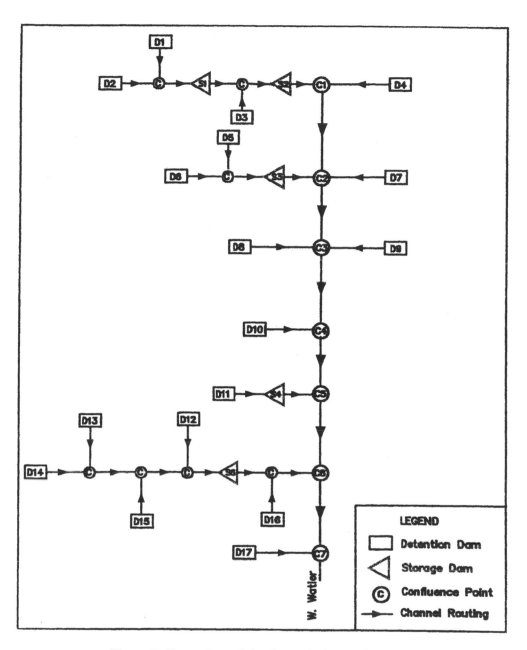

Figure 7. Network model schematic for wadi Watier.

Table 3. Maximum depths and velocities along wadi Watier.

Wadi's Code	Confluence Point	Before Dams Construction		After Dams Construction	
		Water Depth (m)	Velocity (m/sec)	Water Depth (m)	Velocity (m/sec)
W1 to W6 & E1	C_1	1.05	1.63	0.20	0.50
W7 & E2	C_2	1.18	1.76	0.28	0.71
W8 & E3	C_3	1.10	1.68	0.32	0.76
W9	C_4	1.15	1.73	0.36	0.82
W14	C_5	3.70	3.65	1.50	2.03
W15	C_6	4.86	4.34	1.60	2.13
W18	C_7	2.50	3.00	1.10	1.65

Considerable amounts of water are stored behind the dams, while the excess will flow out and intersect the main road of wadi Watier. The stream channel routing for each subbasin was analyzed and peak discharges were determined using HEC-1 model. Another stream channel routing for the main road of the Wadi was analyzed to determine water depths and velocities at different along the road. Table 3 shows the results of the simulation process that determine water depths and velocities at the main confluence points along the main Wadi before and after the control works, due to flood hazard.

Evaporation, Sedimentation and Seepage from Reservoirs Sites

The loss of water from the open reservoirs by evaporation must be considered. Unless evaporation losses from storage reservoirs are evaluated, management can not be accurately defined. Evaporation from the study area was estimated at each location of the storage sites. Mean daily and mean monthly potential evaporations were multiplied by the surface area of each reservoir at different water depths to evaluate the potential evaporation volume in wet and dry seasons. It is recognized that the floodwater is subject to huge losses due to evaporation, which, in turn, reduces the amount of the available water. As such, water must be utilized in a short period of time to minimize evaporation losses.

Sedimentation takes place in the upper layer of the catchment. Due to sediment accumulation, reservoirs capacities are reduced, water evaporation is increased, and water spillage is increased as well. Many factors affect the processes of sedimentation. These factors encompass hydrology, topography, soil erodibility, land cover, incorporated residue, residual land use, subsurface effects, tillage, roughness, and tillage marks. The modified universal soil loss equation was applied to predict the volume of sediment at each location of the proposed seventeen sites for detention dams and for storage dams. Estimated sediment yield at storage dams includes sediment routing from detention dam sites to the storage areas.

Deposition is the counterpart of erosion. The products of erosion may deposit immediately below their sources, or may transport to considerable distances and then deposit in channels, on flood plains, in lakes, reservoirs, and Aqaba gulf. In wadi Watier, a system of seventeen retarding reservoir and five-storage reservoir has been adapted to minimize movement of the sediments to the inter-state highway and Nuweibaa area. The estimated sediment yield could be calculated and listed in Table 4.

With time, as infiltration and percolation continue through a uniform soil, five infiltration experiments were carried out in the storage sites in wadi Watier area. The seepage to the groundwater from the different sites of storage reservoirs was estimated. The seepage was evaluated as a percentage from the storage on daily and monthly bases. Table 5 presents the percentages of seepage at different sites of storage reservoirs.

Table 4. Effect of sediments on reservoirs.

Dam Site	Reservoir Capacity (1000 m^3)	Trapped Sediment (1000 m^3)	No. of Events for Complete Silting
D_1	147.3	31.1	5
D_2	199.0	61.4	4
D_3	147.8	55.0	3
D_4	480.8	21.7	23
D_5	86.5	50.7	2
D_6	145.0	9.7	15
D_7	200.0	9.9	21
D_8	204.5	45.4	5
D_9	180.0	12.7	15
D_{10}	312.0	45.7	7
D_{11}	200.0	73.4	3
D_{12}	366.7	79.8	5
D_{13}	55.0	111.8	1
D_{14}	130.9	88.1	2
D_{15}	213.1	95.5	3
D_{16}	174.1	68.4	3
D_{17}	4.5	29.4	1
S_1	900.0	33.0	28
S_2	7810.0	83.0	95
S_3	1125.0	123.0	10
S_4	3400.0	48.0	71
S_5	3800.0	131.0	30

Table 5. Percentage of seepage at different sites.

Dam Site	S_1	S_2	S_3	S_4	S_5
Daily Seepage	0.31 %	0.48 %	0.52 %	1.36 %	0.41 %
Monthly Seepage	9.23 %	14.5 %	15.7 %	40.9 %	12.4 %

Analysis and Conclusions

Wadi Watier (3600 km^2) receives large amounts of rainfall which can support some activities enhance the sustainable development in the area. Protection and utilization structures are essential to prevent possible destruction due to floods and to ensure the maximum use of the available runoff water.

Seventeen detention dams and five storage dams are proposed to develop the area and to prevent possible disasters. Locations of dams were selected according to peak discharges as well as total available volumes. Sites with high peaks but small water volumes or with large water volumes but small peaks were not selected to reduce the overall cost of the protection and storage system. Selected sites have both high peaks and large volumes of water. Two criteria were established in the hydraulic design of dams. The protection criterion, which was based on the maximum daily rainfall, and the development criterion which was based on the average seasonally rainfall.

Based on the maximum daily rainfall, the 25 year return period with rainfall duration of 2 hour was considered. At dams' locations, the runoff water is estimated is about 50 million m^3/season. Based on average seasonal rainfall depths, the average seasonal flood volume is about 22 million m^3/season.

The total capacity of the dams is about 20 million m^3. The efficiency of the system for development is about 90%, defined as the storage capacity of the system divided by the average volume available per season at dams' sites.

Evaporation losses from storage reservoirs are considerable and will reduce the period of water availability after rain storms. Therefore, water must be utilized efficiency as soon as possible to limit such evaporation losses.

Due to sediment accumulation, capacities of different reservoirs will be reduced. Eight of the retarding reservoirs will be completely filled with sediment after three rainfall events or less. Four need fifteen events or more to be completely filled with sediments. One storage reservoir accommodates ten events before being filled with sediments. All other storage reservoirs accommodate about thirty to ninety five events. The maintenance of the retarding system against sediment accumulation is thus vital to ensure the efficiency of the system.

Seepage in reservoirs sites is evident, however, this process takes place only in the first few years of dams construction. The recharge (per month) from different storage reservoirs sites is varied between 10% to over 40% of the total reservoir storage.

Recommendations

1. Detailed hydrological studies must be performed before plans of the infrastructure.
2. Decontaminate the trapped sediment from dams' reservoirs periodically.

3. Maintenance of the proposed structures must be done continuously before the rainy season and directly after the flood occurrence; and
4. Warning systems should be established to notify the international road of wadi Watier when hazardous conditions exist that would justify temporary highway closure.

References

Barron, T. (1907). "The Topography and Geology of the Peninsula of Sinai (Western Portion)." Egypt Survey Dept., Cairo, 241 p.

Bernard, M.M. (1932). "Formulas for Rainfall Intensities of Long Durations." ASCE, Vol. 96, 592-624.

Cunnane, C. (1978). "Unbiased Plotting Positions-a review", J. Hydrol., 37, pp. 205-222.

Chen, C.L. (1983). "Rainfall Intensity-Duration-Frequency Formulas." J. Hydr. Eng., ASCE, 109(12), 1603-1621.

Clark, C.O. (1945). "Storage and the Unit Hydrograph." Transactions of the ASCE 110, pp. 1419-1446.

Elshinnawy, I.A. (1993). "Evaluation of Transmission Losses in Ephemeral Streams with Compound Channels". Ph. D. Thesis, Civil Engineering Department, University of Arizona, Tucson, AZ, USA.

Gumbel, E.J. (1958). "Statistic of Extremes", Colombia University Press, New York.

Mein, R.G. and Larson, C.L. (1973). "Modeling Infiltration During a Steady Rain." Water Res., V17(4), 1005-1013.

Ministry of New Communities in Egypt (1993). "Protection of Nuweiba/Ras El-Naqab International Road from Flood Hazards and Use of Flood Water in the Development", Egypt.

Salas, J.D., Smith, R.A. and Tabios, G.Q. (1990). "Statistical Computer Techniques in Hydrology and Water Resources", Department of Civil Engineering, Colorado State University, U.S.A.

U.S. Army Corps of Engineers (1987). "HEC-1 Flood Hydrograph Package". Users Manual, Hydrologic Engineering Center, Davis, California, 32 p. plus appendices.

U.S. Weather Bureau of Reclamation (1987). "Design of Small Dams." Third Edition, Denver, Colorado.

William J.R. and Ber, H.D. (1977). "Sediment Yield Prediction Based on Watershed Hydrology". Transaction ASCE, Vol. 20, no. 6, pp. 1100-1104.

Surface Water Hydrology, Singh, Al-Rashed & Sherif (eds)
© 2002 Swets & Zeitlinger, Lisse, ISBN 90 5809 363 8

Assessment of runoff from arid terrain

David Stephenson
Water Systems Research Group, University of the Witwatersrand
PO Box 277, WITS, 2050, South Africa
e-mail: steph@civil.wits.ac.za

Abstract

It is difficult to calculate runoff from arid areas for various reasons. Rivers will generally only flow after storms and rapidly infiltrate into the ground. Runoff can occur along the sandy beds of the rivers undetected from the surface. The estimation and modelling of flood flow is even more difficult because the infiltration capacity of the ground is an important factor and a completely different set of criteria applies in determining runoff and flood. The spectrum of extreme floods is not as directly related to the rainfall spectrum as in the case of temperate climates. Flow records are invariably scarce and unreliable. This is because for most of the time there is no flow in the river and most of the flow would occur during floods. Gauges may be overtopped or wash away or unobserved when the rare flow does occur. The patching and extension of records in order to obtain a reliable time-series is therefore a difficult task and there is even less data for calibrating models. Experience in Botswana on synthesis of monthly flow records and prediction of extreme floods is described. The applicability of hydrograph methods and even the rational method is queried and instead hydraulic-type modular modelling is proposed as it has been used successfully on a number of catchments in the Kalahari Desert region of Southern Africa.

Introduction

The characteristics of arid catchments pose unique problems in estimating runoff. The hydrology of these catchments is highly variable, without runoff or yield for long periods, followed by intense storms.

There are many more variables, or degrees of freedom, in arid catchments, which make calculation of runoff suspect, i.e.

Antecedent moisture content is low so infiltration is initially high;
Catchment cover is highly absorbent;
Catchment surface roughness is difficult to estimate and varies depending on vegetation or denudation;
There are few existing gauging stations to calibrate models against;
Infrequent rains make preparation of intensity – duration – frequency relationships – difficult;
River channel beds are suspended during flood;
Low flow is below the sand surface in channels.

Construction of roads and dams in Southern Africa, particularly the arid countries to the west (Botswana, Namibia), requires the calculation of floods. Rainfall records are generally more abundant than streamflow records, and rarely is there a suitable streamgauge at the correct location, which picks up high flows as well as low flows. So the most common method is to use rainfall-runoff models. These vary from simple rational methods or unit hydrographs, through to models including time area, kinematic and hydraulic.

Data Problems

A problem peculiar to arid regions is the difficulty of obtaining accurate surface runoff data. There is often a lack of gauges due to expense of installation, difficulty of collecting data from remote installations, damage due to flash floods, etc. Even if conventional gauges are installed, such as weirs, level recorders and gauge plates, they may not respond correctly. Blocking of connecting pipes leads to recording failure. Sand deposits behind weirs renders the rating curve incorrect.

In the simplest of gauging stations, gauge plates may be installed on the banks of a river channel. Moving river beds require frequent recalibration. Flow through the sandy bed may be undetected. Beds may be suspended during floods.

Daily reading of gauge plates is of little use. Flows may practically all occur immediately after storms, and then rapidly after the water seeps away into the sand. In fact, flows downstream of the storm cell can be less than upstream at the cell, which is contrary to normal hydrological appreciation.

It is unlikely the gauge observer will be on site at the peak of the flood, due to set times of observing and due to inconvenience of travel during the storm. So the reading will inevitably miss the peak and may miss the complete hydrograph.

Alternative devices are needed to record, viz. a recorder which switches on when water is detected, although the clock must run continuously.

The probability of undermeasuring flow by gauge plate reading is a function of flood duration or catchment area, and frequency of rain and frequency of recording. Thus, a 3 000 km^2 catchment has a critical storm duration of about 6 hours and rain may occur 50 times a year. If recording is once in 24 hours, there is a 75% chance of missing any one storm completely, and even if the reading is during the storm runoff, the average estimate will be half the peak. The average runoff and peak flow are thus drastically underestimated which affects dam design.

Flood Calculation

Linear methods of flood calculation can lead to large errors. They assume a direct relationship between rainfall volume and runoff and that the system is in a state of equilibrium. In fact there is often no such thing as a concentration time or time to equilibrium for large catchment (over a few km^2). This is because infiltration capacity can exceed precipitation for long, low intensity storms. It is generally the short sharp storm which may not cover the entire catchment which results in the biggest flood. Hence storm distribution must be considered in flood calculations.

The unit hydrograph procedure is used extensively in South Africa for flood determination (Hydrological Research Unit, 1972). The method assumes a linear excess rainfall runoff relationship and excess rain is taken as a proportion of rainfall. The basic time-area concept behind the rational method is extended in the unit hydrograph method. Both rational and unit hydrograph methods do not account for hydraulic or physiographic characteristics and therefore may be unsafe to extrapolate for extreme events (Stephenson, 1983).

Simplistic correlations for flood estimates have been proposed by various authorities, e.g. Institute of Hydrology for Eastern Botswana (1986):

$$Q_p = 7.2\ A^{0.45}T^{0.36} \tag{1}$$

where Q_p is the flood peak in m^3/s with a return period T years off a catchment of area A km^2.

Op ten Noort and Stephenson (1982) allowed for rainfall rate and produced equations of the form

$$Q_p = (0.0012M - 0.21)\ A^{0.69}T^{0.49} \tag{2}$$

where M is the mean annual precipitation in mm. Their results were published in nomograph form (1982). Similar relationships were proposed for Zimbabwe (Mitchell, 1974) and Zambia (Balek, 1977).

Correlation of Flood Magnitude with Catchment Area

The major factor relating to flood magnitude is reputed to be catchment area (Kovacs, 1971). However, unless basic characteristics are accounted for there is a wide band in the plot of peak observed flood and catchment area. For example, the maximum observed flood of 1 929 m^3/s on the Kafue catchment (150 000 km^2) is only 12% of that on the Save of 13 800 m^3/s (catchment area 102 000 km^2), despite the fact that the mean annual precipitation is 1 500 mm in the Kafue and 1 100 mm on the Save catchment.

Table 1 presents data showing the effect of catchment characteristics on flood peaks. Normalised flood refers to the value read off the respective line in Fig. 1, i.e. expected maximum observed flood, for average catchments.

The fact that mean annual rainfall does not have as marked an effect on flood peaks as catchment characteristics is demonstrated by comparing, for example, floods in the Fish river (MAP 200 mm, area 46 400 km^2, max flood 8 300 m^3/s with the Vaal river (MAP 800 mm, area 19 300 km^2, max flood 7 900 m^3/s).

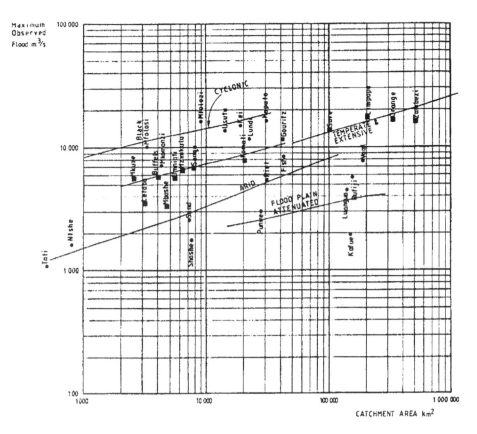

Figure 1. Some maximum observed flood peaks in Southern Africa over the period 1900-1990 and fitted lines for various conditions.

Table 1. Table of selected catchments and the effect on flood peaks (Stephenson, 1993).

River	Effect	Catchment area km^2	Normal-ised max flood m^3/s	Max ob-served	Modelled 100 yr flood m^3/s
Orange	Long channel	342 000	18 000	16 000	15 000
Kafue	Wide marshes	153 000	14 000	2 000	3 000
Pungwe	Wide flood plane	29 000	9 000	3 000	3 000
Motlhoutse	Flat arid	18 000	4 000	–	1 200
Shashe	Flat arid	7 800	3 000	1 800	–
Mbashe	Steep lower reaches	5 000	6 000	3 000	5 000

Peak Flow Charts

Stephenson and Meadows (1986) presented a set of nomographs for calculating flood peaks off plains, using the kinematic method. If it can be assumed that the rainfall intensity-duration relationship for a specified frequency of exceedance can be approximated by a formula of the form.

$$i = \frac{a}{(c+t_d)^p} \tag{3}$$

where i is the rainfall rate in mm/h and t_d is the storm duration in hours, then a simple equation for peak flow can be derived. Thus for inland catchments in South Africa, it was found that c = 0.24h and p = 0.89, a is a function of rainfall region, mean annual precipitation, MPA, in mm and recurrence interval T in years, e.g.

$$a = (b \ \imath \ e.MAP)T^{0.3} \tag{4}$$

b and e are regional constants. Losses are subdivided into an initial loss u in mm and a uniform infiltration loss rate f in mm/h (see Fig. 2).

The rate of excess rainfall is:

$$i_e = i - f \tag{5}$$

For small catchments the maximum peak runoff rate occurs when the duration of excess rain equals the concentration time, t_c. For plain rectangular catchments, the concentration time is a function of excess rainfall rate

$$t_c = (L/\alpha i_e^{m-1})^{1/m} \tag{6}$$

where

$$t_c = t_e \text{ (both in hours here)} \tag{7}$$

and

$$\alpha = \sqrt{S}/n \text{ (Manning equation)} \tag{8}$$

i_e is excess rainfall intensity, L is catchment length, S is the downstream slope, m is an exponent, 5/3 in Manning's equation, n is the Manning roughness.

The following expression may then be derived for i_e/a from equations (3) to (6):

$$i_e = \frac{1}{\left[c + \frac{u/a}{i_e/a + f/a} + \left(\frac{L}{\alpha a^{m-1}} \right)^{1/m} \left(\frac{a}{i_e} \right)^{1-1/m} \right]^p} - \frac{f}{a} \tag{9}$$

377

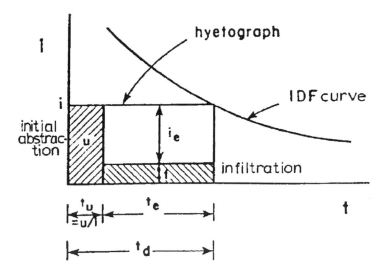

Figure 2. Excess flow hyetograph derived from IDF curve.

The maximum runoff rate per unit width of catchment is

$$q = i_e L \qquad (10)$$

Large Catchments

For very long or absorbent catchments, the theoretical concentration time t_c is high. In such cases, the corresponding excess rainfall rate for a duration equal to t_c is low and in fact could conceivably be less than the infiltration rate f. It is thus apparent that in such cases the maximum runoff rate may coincide with a storm of shorter duration than the concentration time of the catchment. The entire catchment will thus not be contributing at the time of the peak in the hydrograph. If a local, intense storm turns out to be the design storm, the areal reduction factor applied to point rainfall intensity relationships may also be less significant.

Before equilibrium is reached the runoff per unit width at the mouth of the catchment at any time t_e after the commencement of excess rain or runoff is

$$q = \alpha(i_e t_e)^m \qquad (11)$$

where

$$t_e = t_d - u/i \qquad (12)$$

$$< t_c = (L/\alpha i_e^{m-1})^{1/m} \qquad (13)$$

and

$$i_e = \frac{a}{(c+t_d)^p} - f \tag{14}$$

hence

$$q = \alpha \left\{ \left(t_d - \frac{u}{i} \right) \left(\frac{a}{(c+t_d)p} - f \right) \right\}^m \tag{15}$$

or

$$\frac{q}{\alpha a^m} = \left[\left\{ t_d - \frac{u}{a}(c+t_d)^p \right\} \left\{ \frac{1}{(c+t_d)^p} - \frac{f}{a} \right\} \right]^m \tag{16}$$

$q/\alpha a^m$ is plotted against t_d in Fig. 3 (the full lines) (for different values of the dimensionless parameters $U = u/a$ and $F = f/a$). For all cases of $F > 0$, the lines exhibit a peak runoff and the corresponding storm duration t_d for an infinitely long catchment. For most catchments, it is necessary to establish whether t_e is less than t_c, i.e. whether the peak occurs before the catchment has reached equilibrium.

In fact, $t_e = t_d - t_u$ (17)

Therefore for $t_c = t_e$

$$t_d = t_u + t_c = u/i + (L/\alpha)^{1/m}/l_e^{1-1/m} \tag{18}$$

or

$$\frac{L}{\alpha a^{m-1}} = \left\{ t_d - \frac{u}{a}(c+t_d)^p \right\}^m \left\{ \frac{1}{(c+t_d)^p} - \frac{f}{a} \right\}^{m-1} \tag{19}$$

$L/\alpha a^{m-1}$ may therefore be plotted against t_d for selected values for u/a and f/a as on the right hand side of Fig. 3.

Now the peak runoff will be the maximum of either

(a)   that corresponding to $t_e = t_c$ for short catchments, or
(b) $t_e < t_c$ for long catchments.

In order to identify which condition applies, enter the chart for the correct $U = u/a$ with $L/\alpha a^{m-1}$ with right hand side and using the dotted line corresponding to the correct F, read off the corresponding $t_d = t_c + t_u$ on the abscissa. It may occur that the equilibrium t_c is off the chart to the right, in which case it is probably of no interest since the following case applies. Select the full line with the $F = f/a$ and decide whether its maximum lies at or to the left of the value t_d previously established. If

379

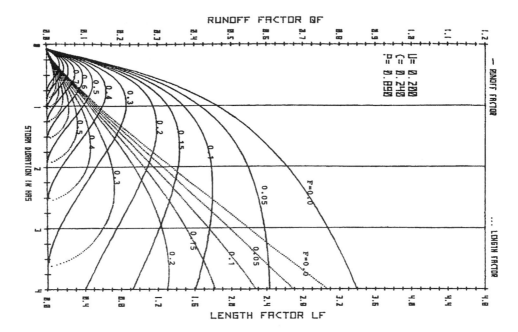

Figure 3. Peak runoff factors for U = 0.20.

the peak lies to the left, read the revised design storm duration t_d corresponding to the peak, and the corresponding peak flow parameter $q/\alpha a^m$ on the left hand ordinate.

Modification for Practical Units

The preceding equations assume dimensional homogeneity. Unfortunately both the Manning resistance equations and the I-D-F relationships are empirical and the coefficients depend on the units employed. In the Manning form of Equation (8), q is in m^2/s if $i_e t_e$ is in metres. α is $\sqrt{S}/n$ in S.I. units where S is the dimensionless slope and n is the Manning roughness.

It is most convenient to work with t_d in hours and i_e and a in mm/h. The numbers are then more realistic. In Equation (6), if q is in $m^3/s/m$, α in m-s units, a in mm/h and t_d in hours, then the left hand side should be replaced by

$$\frac{q 1000^m}{\alpha a^m} = \frac{10^5 q}{\alpha a^{5/3}} = \frac{10^5 Q}{B \alpha a^{5/3}} \tag{20}$$

where Q is total runoff rate off a catchment of width B metres. Note that the right hand side of (16) is in h^m if t_d is in h, so no correction is made to the above factor to convert a to secs, only to convert mm to m. This is what the left hand axis of Fig. 3 represents if a is in mm/h. It is referred to as the runoff-factor, QF. Similarly, the left

380

hand side of Equation (19) is $L/\alpha a^{m-1}$ in homogeneous units, or if a is in mm/h, L in metres and α in m-s units, then it should be replaced by:

$$\frac{L}{\alpha}\frac{a}{3600000}\frac{1000^m}{a^m} = \frac{L}{36\alpha a^{2/3}} \quad \text{(the length factor LF)} \tag{21}$$

Example

Consider a plane rectangular catchment with the following characteristics:

Overland flow length L = 8000 m, width B = 9000 m, slope S = 0.01, Manning roughness n = 0.1, inland region, MAP = 620 mm/annum, 20 year recurrence interval storm, rainfall factor a = $(7.5 + 0.034 \times 620)20^{0.3}$ = 70 mm/h, c = 0.24 h, p = 0.89, infiltration rate f = 7 mm/h, initial abstraction u = 14 mm, so $\alpha = \sqrt{S}/n$ = 1.0, F = f/s = 0.1, U = u/a = 0.2

$$LF = \frac{L}{36\alpha a^{2/3}} = \frac{8000}{36 \times 1 \times 70^{2/3}} = 13$$

From Fig. 3, concentration time t_c = >> 4h. The time to equilibrium is longer than the storm resulting in the peak runoff. In this case, it appears that the peak runoff corresponds to a 2.5 hour storm (shorter than the time to equilibrium) and the corresponding runoff factor is

$$QF = 0.45 = 10^5 Q/B\alpha a^{5/3}$$

therefore Q = 0.45 x 9000 x 1 x $70^{5/3}/10^5$ = 48 m^3/s.

The corresponding precipitation rate is:

$$i = a/(c + t_d)^p = 70/(0.24 + 2.5)^{0.89} = 28 \text{ mm/h}.$$

The equivalent rational coefficient C is 48/(9000 x 8000 x 28/3600000) = 0.1.

It may also be confirmed that the storm duration corresponding to time to equilibrium of the catchment is over 20 hours:

Try t_d = 10 h.

if $i_e = 70/(0.24 + 10.0)^{0.89} - 7 = 1.9$ mm/h = 0.5×10^{-6} m/s

$t_c = (L/\alpha i_e^{m-1})^{1-m} = \{8000/1 \times 0.5 \times 10^{-6})^{2/3}\}^{3/5}$ = 20.0 h

$t_u = u/i = 14/28$ = 0.5 h

Storm duration t_d = 20.5 h

Conclusions

Conventional methods of storm analysis and flood determination often fall short in the case of extreme floods off arid catchments. It is not sufficient to extrapolate short duration records unless they include samples of extreme events.

Standard methods of estimating floods such as the rational method and unit hydrographs can not be used to generate hydrographs for extreme storms because of the preconceived hydrograph shape, and the rainfall-runoff relationship is not linear.

The critical storm duration is not necessarily related to the basin lag time as the runoff from short sharp storms is often more severe than for long duration storms in the case of flat arid land.

References

Balek, J. (1977). *Hydrology and Water Resources in Tropical Africa*. Elsevier.

Hydrological Research Unit (1972). *Design Flood Determinations in South Africa*. University of the Witwatersrand.

Institute of Hydrology (1986). *A review of flood estimates for Kolobeng and Metsemotlhaba dams*. For Alexander Gibb and Partners, Botswana. Wallingford.

Kovacs, G. (1971). Relationship between characteristic flood discharges and catchment area. *Symp. Role of hydrology in developing Africa II*, WMO No. 301, pp. 18-23.

Kovacs, Z.P. (1988). *Regional maximum flood peaks in Southern Africa*. TR137, Dept. Water Affairs, Pretoria.

Mitchell, T.B. (1974). A study of Rhodesian floods and proposed flood formulae. *The Rhodesian Engineer*, November.

Op ten Noort, T. and Stephenson, D. (1982). *Flood peak calculation in South Africa*. Water Systems Research Group, University of the Witwatersrand, Report No. 2/1982.

Stephenson, D. (1983). Hydrological myths. *The Civil Engineer in SA*, July, pp. 337-383.

Stephenson, D. and Meadows, M.E. (1986). *Kinematic Hydrology and Modelling*, Elsevier, NY.

Stephenson, D. (1993). Estimation of extreme floods with particular reference to Southern Africa. *Proc. Intl. Conf. Hydrology and Water Ress.*, New Delhi.

Surface Water Hydrology, Singh, Al-Rashed & Sherif (eds)
© 2002 Swets & Zeitlinger, Lisse, ISBN 90 5809 363 8

Stability and accuracy of the finite element solution to the kinematic wave overland flow

Fouad H. Jaber and Rabi H. Mohtar
Department of Agricultural and Biological Engineering
Purdue University, West Lafayette, IN 47907, USA
e-mail: mohtar@purdue.edu

Abstract

The numerical solution of the kinematic wave equations for overland flow is known for its numerical oscillations. The conventional Galerkin finite element consistent formulation results in violation of physical reality and numerical oscillations. Another cause of oscillations is the first derivative terms in the kinematic wave equations that result in non-symmetric operators that are very oscillatory in nature. The lumped formulation is tested in this paper as an alternative to the consistent formulation. It was found that the lumped formulation significantly improves stability of the solution with no reduction in accuracy. The upwind finite element method is also tested to remedy the oscillations caused by the first derivative. The latter scheme is applied to both, lumped and consistent formulations, using upwind factors of 0.1 and 1.0 and is evaluated in terms of stability and accuracy of the solution. The upwind method did not provide any improvement to the stability of both the lumped and the consistent formulation. It required smaller or equal time steps compared to a non-upwinded formulation for the same level of accuracy. The lumped formulation is recommended to solve overland flow problems as it was found to be the most efficient formulation to solve the 1-D kinematic wave equation for overland flow. The dynamic time step for the lumped formulation developed in this study can be easily integrated in overland flow routing models to guide the choice of the optimal time step with minimum user input.

Introduction

The kinematic wave theory has been applied in many areas and is now well established for modeling a variety of hydrologic processes. Growing environmental and ecological concerns have increased the role of the kinematic wave theory in describing and modeling environmental and hydrologic processes (Singh 1996, p. xv). Higher computational power and the development of spatial data analysis tools such as Geographic Information Systems (GIS) have made the use of numerical methods to solve these problems much easier.

Several research articles have been reported addressing numerical oscillations in the conventional Galerkin finite element solution of flow problems. Christie et al. (1976) stated that central difference approximations of first derivatives are to be avoided since they give rise to numerical oscillations for reasonable grid sizes. Zhang and Cundy (1989) used a MacCormak finite difference scheme to solve overland flow problems. They reported that numerical oscillations were evident in most hydrographs and in some cases the solution was aborted. Chan and Williamson (1992) report that partial differential equations, which contain a first derivative term, exhibit numerical oscillation in the Galerkin finite element solution if the size of the mesh exceeds a critical value. This critical value is based on the Peclet number being smaller than two.

Numerical methods require discretization in both space and time. The size of the element and the time step are crucial to the stability and the accuracy of the numerical formulation used. Previous studies (Courant et al., 1956, Vieux and Segerlind, 1989) have shown that the actual time step used in the time integration scheme must not be longer than the time during which a gravity wave front may propagate through the system nor longer than the time step variation in the forcing function. The prior condition is known as the Courant condition (Courant et al., 1956). This condition, however, is a stability criterion and not an accuracy criterion. Lyn and Goodwin (1987) showed that although a given difference scheme may be stable; its convergence may be poor. In other words, if a time step has low convergence, then it is too large for accuracy (Mohtar and Segerlind, 1995). Limited research has been conducted on criteria for a time step that would ensure stability as well as accuracy of the kinematic wave solution. The task of choosing the proper time step has often been considered a matter of experience.

In this study, three Galerkin finite element formulations will be tested for oscillations and accuracy: The consistent formulation for coupling time derivative terms that assumes that the variation in time derivative of h varies linearly between nodal values (Segerlind, 1984, pp. 178). The lumped formulation of the Galerkin finite element solution of transient problems that assumes that the variation in the time derivative of the depth between the midpoints of adjacent elements is constant (Segerlind, 1984, pp. 180); and the upwinding formulation in finite elements that consists of using weighting functions of non-symmetric forms different from those used as shape functions.

The objectives of this paper are:

1) To improve the numerical solution of the one-dimensional kinematic wave equation for overland flow by evaluating the stability and accuracy of the consistent, lumped and upwind formulations of the Galerkin finite element method.
2) To develop and evaluate an accuracy-based dynamic time step estimate for the most stable and accurate solution using numerical experimentation.

Numerical Formulation for Solving the Kinematic Wave Equation

Governing equations

The one-dimensional kinematic wave equation is governed by the continuity equation:

$$\frac{\partial h}{\partial t} + \frac{\partial q}{\partial x} = r_e(x,t) \qquad (1)$$

384

and the conservation of momentum equation that is reduced to:

$$S_o = S_f \tag{2}$$

q and h in equation (1) will then be related by the following equation

$$q = \alpha_x h^\beta \tag{3}$$

where, using Manning's equation, $\alpha_x = \dfrac{S_{ox}^{1/2}}{n}$; $\beta = \dfrac{5}{3}$ in which n is the Manning roughness coefficient and SI units are used; q is the unit width flow; h is the vertical flow depth; $r_e(x,t)$ is the excess rainfall rate; S_o is the element bottom slope; S_f is the friction slope; and x and t represent the space and time domains, respectively.

The consistent formulation

In the consistent Galerkin formulation ∂h/∂t is assumed varying linearly between nodal values and yields the system of ODE's of the form:

$$[C]\{\dot{h}\} + [B]\{q\} - \{f\} = 0 \tag{4}$$

where [C] and [B] and {f} are global matrices constructed using the direct stiffness method from the following element matrices, respectively:

$$\left[C^{(e)}\right] = \frac{L}{6}\begin{bmatrix} 2 & 1 \\ 1 & 2 \end{bmatrix} \; ; \; \left[B^{(e)}\right] = \frac{1}{2}\begin{bmatrix} -1 & 1 \\ -1 & 1 \end{bmatrix} \text{ and } \left\{f^{(e)}\right\} = \frac{L}{2}\begin{Bmatrix} r(x,t) \\ r(x,t) \end{Bmatrix} \tag{5}$$

where L is the length of the element.

The lumped formulation

The lumped formulation is defined as ∂h/∂t constant between the midpoints of adjacent elements. The element matrix for the lumped formulation is:

$$\left[C^{(e)}\right] = \frac{L}{2}\begin{bmatrix} 1 & 0 \\ 0 & 1 \end{bmatrix} \tag{6}$$

The upwind finite element formulation

An "upwind" finite element formulation to solve one-dimensional kinematic wave equations is presented in this paper in function of X_i and X_j, which are respectively the coordinates of the nodes of an element ij based on Heinrich et al. (1977) who developed an upwind finite element method for convective transport equation. Element matrices for this formulation are developed in this paper. The one dimensional upwind formulation of Heinrich et al. (1977) requires that the weighting function on the advective term of the equation $\partial q / \partial x$ for node i be

$$W_i = W_i(x, \gamma) = N_i + \gamma \ F(x) \tag{7}$$

and for node j it is

$$W_j = W_j(x, \gamma) = N_j - \gamma \ F(x) \tag{8}$$

where γ is the upwinding coefficient (0 is no upwinding and 1 is fully upwinded). $F(x)$ is some positive function such that it is equal to 0 at the nodes and

$$\int_{X_i}^{X_j} F(x) dx = \frac{1}{2} L$$

where L is the length of the element. A function that satisfies these conditions and was used is the following:

$$F(x) = -3 \frac{(x - X_j)(x - X_i)}{(X_j - X_i)^2} \tag{9}$$

The residuals of the advective term in the kinematic wave equation $\dfrac{\partial q}{\partial x}$ for one element are:

$$R_i^{(e)} = \int_{X_i}^{X_j} W_i \frac{\partial N}{\partial x} q dx \ \text{and} \ R_j^{(e)} = \int_{X_i}^{X_j} W_j \frac{\partial N}{\partial x} q dx \tag{10}$$

Solving these integrals for $\gamma = 1.0$ results in the following:

$$R_i^{(e)} = 0, \quad R_j^{(e)} = -q_i + q_j \tag{11}$$

or in matrix form

$$\begin{Bmatrix} R_i \\ R_j \end{Bmatrix} = \begin{bmatrix} 0 & 0 \\ -1 & 1 \end{bmatrix} \begin{Bmatrix} q_i \\ q_j \end{Bmatrix} \tag{12}$$

For an upwind factor $\gamma = 0.1$ the results becomes

$$\begin{Bmatrix} R_i \\ R_j \end{Bmatrix} = \begin{bmatrix} -0.45 & 0.55 \\ -0.55 & 0.55 \end{bmatrix} \begin{Bmatrix} q_i \\ q_j \end{Bmatrix} \tag{13}$$

For $\gamma = 0$ the matrix reduces to the [B] matrix of Equation 9.

Methodology for Developing the Dynamic Time Step Criteria

The one dimensional kinematic wave overland flow equation is solved using constant and varying rainfall and Manning's α for a certain mesh using different time steps. For each time step the error between the analytical solution and the numerical solution was computed. A time step versus error graph was generated for this series of solutions. As the value of the time step increased, the error was expected to grow, provided that the spatial discretization error is small. There exists a time step value that will integrate the problem within a specified error. Using a smaller time step might result in an unnecessarily longer computational time, while a larger time step might result in large errors and might even lead to instability in the solution. The specified problem was solved using several mesh sizes generating a relationship between mesh size and optimal time step. The results were summarized in a regression equation that was used to define a time step estimate. The time to equilibrium, which depends on the storm, the slope and the roughness factor of the watershed, was introduced in the equation as a problem specific factor so the criteria can be used for other problems. The time step estimate equations were tested using a different problem to ensure the validity of the results.

This procedure, originally developed by Mohtar and Segerlind (1998, 1999a, 1999b), is used for the most stable formulation that is identified in this study. Dynamic time steps are developed for the identified formulation and the results are compared to the dynamic time steps of the conventional Galerkin method.

The development problems used are the same as the one used by Jaber and Mohtar (2001). In order to illustrate the solution accuracy requirement in terms of time steps and element sizes, nodal values of water depths were computed for 5, 10, 20, 25, 50, and 75 elements for each case. Various time steps were used to integrate the system of ordinary differential equations. Each scenario was solved using the following conditions: contributing area of unit width L is equal to 152.4 m (500 ft), the average rainfall excess intensity is 2.74 cm/hr (2.5E-5 ft/sec), the average bottom slope is 0.05, and the average Manning's coefficient is 0.035. For cases where r and α varied in space the following relations were used:

$$r_e = -0.036x + 5.49$$
$$\alpha = 6.015e^{0.00984x} \tag{14}$$

The time of concentration of each simulation was calculated according to the following equation (Singh, 1996, p. 798):

$$t_c = \frac{1}{\beta} \int_0^L \left[\frac{1}{\alpha(\eta)} \right]^{1/\beta} \left[\int_0^\eta r_e(\xi)\,d\xi \right]^{(1-\beta)/\beta} d\eta \tag{15}$$

where $\beta = 5/3$; L is the slope length; and η and ξ are dummy variables.

387

The storm duration was set to exceed the time to concentration so that pre-equilibrium, equilibrium and post-equilibrium conditions could be evaluated. Jaber and Mohtar (2001) showed that the developed dynamic time steps were still applicable for cases where no equilibrium is reached such as unsteady rainfall. The storm length during which rainfall was continuously falling was set to be 0.4 hours. The total simulation time was 1 hour.

The evaluation problems used were the following: slope of L = 144 m (472.5 ft), S_o = 0.02, roughness n = 0.009. The rainfall event simulated was r_e = 19 cm/hr (1.731E-4 ft/s) applied for 0.133 hours. The above values of r_e and α were used in the cases were r_e and α were constant. For the spatially varying r and α the following equations were used:

$$r_e = -0.2736x+39.51$$
$$\alpha = 8.0903e^{0.0079x}$$

(16)

This α accounts for a variation in Manning's roughness coefficient n from a value of 0.02 to 0.04. It also allows for a variation in slope from 0.04 to 0.13. t_c was found to be 0.054 hours for case 1 and 0.07 hours for cases 2, 3 and 4. The time step was calculated using the generated equations for each scenario as appropriate. The equations were also tested for time varying rainfall and high Manning's roughness coefficient problems with n value of 0.1.

Results and Discussion

Upwinded consistent and lumped formulations

Upwinding formulations using factors $\gamma = 0.1$ and $\gamma = 1.0$ were used to evaluate the effect of this method on the lumped and consistent finite element solutions. The same problem used to evaluate the dynamic time steps was used to assess the upwind finite element method for the overland flow problem. The fully upwinded formulation ($\gamma = 1.0$) resulted in a higher error than the consistent and lumped formulation (Figure 1). The solution was stable for time steps equal to the dynamic time step but it became unstable when the time step used was 1.25 times the dynamic time steps for large mesh sizes. The solution was unstable for all scenarios for time steps equal to 1.5 times the dynamic time step were used. Contrary to what it was expected upwinding did not improve on the stability of this problem. Further investigations show that the instability in the solution is mainly caused by the transition between the different phase of the hydrograph especially the transition between the rising phase and the equilibrium phase. The $\gamma = 0.1$ upwinding formulation, had accurate results but did not result in any improvement on the stability of the solution (Figure 2).

Lumped formulation dynamic time step equation

Preliminary runs of the evaluation problems using the lumped formulation instead of the consistent formulation showed significant improvement in the results. The same level of accuracy than the consistent formulation was reached using larger time steps. This showed the necessity to develop new dynamic time steps for the

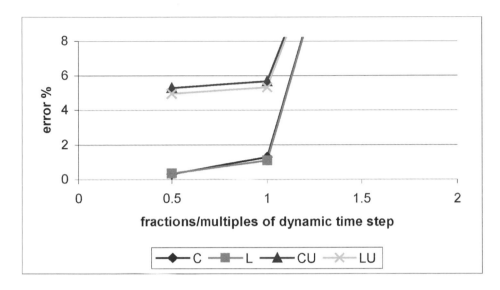

Figure 1. Error variation with multiples/fractions of the dynamic time step for the consistent – C, the lumped – L, the fully upwind consistent – UC, and the fully upwind lumped formulations.

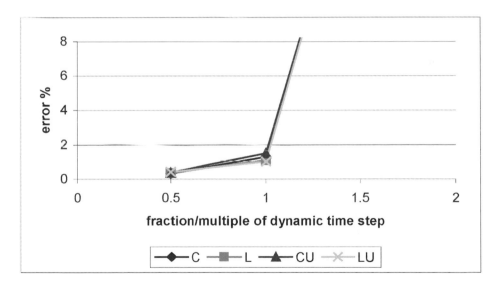

Figure 2. Error variation with multiples/ fractions of the dynamic time step for the consistent – C, the lumped – L, the 0.1 upwind consistent – UC, and the 0.1 upwind lumped formulations.

lumped formulation. The dynamic time step equations were developed using the methodology described above for four cases. The resulting regression equations are:

For case 1: r_e and α constant:

$$\Delta t = \frac{0.89 t_c}{N^{0.96}} \qquad (17)$$

For case 2: time and space varying r_e and constant α:

$$\Delta t = \frac{0.79 t_c}{N^{0.86}} \qquad (18)$$

For case 3: constant r_e and space varying α:

$$\Delta t = \frac{1.27 t_c}{N^{1.16}} \qquad (19)$$

For case 4: space and time varying r_e and space varying α:

$$\Delta t = \frac{0.79 t_c}{N} \qquad (20)$$

The above developed dynamic time steps were evaluated using a different problem and showed to be adequate for solving kinematic wave problems for the lumped formulation. The use of half the dynamic time step did not result in a significant reduction in error, while the use of 1.5 times the dynamic time step resulted in non-converging solutions for the 4 cases. The results for case 4 are shown in Figure 3.

Figures 4 shows the change of the dynamic time steps of case 4 with the number of elements for both the consistent and lumped formulations. Similar trends were observed for the other cases. By definition, the dynamic time step is the largest time step that ensures the accuracy and thus the stability of the solution. Figure 4 shows that the lumped formulation dynamic time steps are double the size of the consistent formulation on average. The lumped formulation solution remains stable for time steps double the size of the ones used in the consistent formulation solution with the error remaining under 2%. Gresho and Lee (1979), while analyzing the transient heat equation, argued that the lumped formulation is inaccurate in regions where it is more stable than the consistent formulation. Our results indicate that this statement does not hold for the kinematic wave equation for overland flow. The lumped formulation improved on the stability of the solution with no effect on the accuracy. This result coincides with Yue (1989) in his analysis of the three dimensional Navier-Stokes equation for open channel flow.

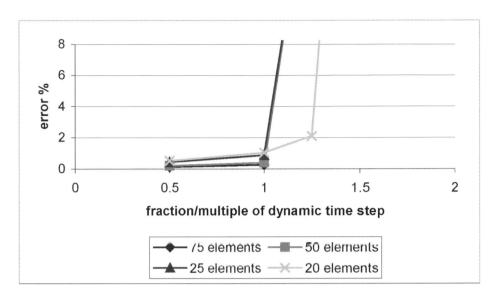

Figure 3. Evaluation of dynamic time step for case 4 lumped formulation.

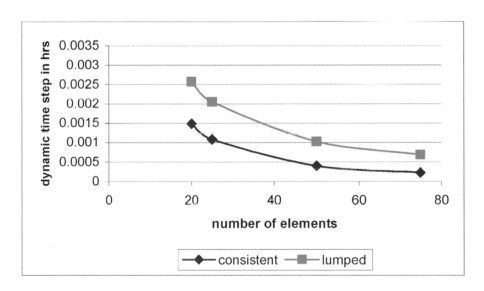

Figure 4. Comparison between dynamic time steps of the lumped and consistent formulation case 4.

Conclusions

Stability and accuracy of three finite element formulations to solve the overland flow problem were analyzed namely: the consistent, the lumped and the upwind formulations. Results show that the upwind formulation did not reduce oscillations and had a lower accuracy than the non-upwinded formulations. The lumped formulation was found to significantly improve the stability of the solution with no reduction in accuracy.

The study also provided a new dynamic time step criteria for the solution of the one-dimensional overland flow problem for the lumped formulation. These time steps are functions of the grid size and the time to equilibrium, the latter being a function of rainfall, slope and Manning's roughness. All these parameters are readily available and needed to run the hydrologic processes. The dynamic time steps were found to be very good criteria to use to determine the size of time step that will generate an accurate and stable solution. This is in contrast with the effect of the lumped formulation solution for transient heat problems, in which the improvement in stability cost deterioration in accuracy. The lumped formulation of the Galerkin finite element method for transient problems should be used when solving the one-dimensional overland flow equation.

The dynamic time step equations developed could be integrated within any hydrologic computer programs utilizing the lumped formulation of the one-dimensional overland flow kinematic wave theory. The time step can be calculated automatically depending on the mesh size that the user chooses.

Acknowledgment

The authors would like to thank Purdue University Agricultural Research Program Office, Environmental Sciences and Engineering Institute and Purdue Research Foundation for their support for this research.

References

Chan, E.K.C. and Williamson, S. (1992). "Factors influencing the need for upwinding in two dimensional field calculation". *IEEE transactions on magnetics*. 28, 1611-1614.

Christie, I., Griffiths, D.F., Mitchell, A.R. and Zienkiewicz, O.C. (1976). "Finite element methods for second order differential equations with significant first derivatives". *International journal of numerical methods in engineering,* 10, 1389-1396.

Courant, R., Friedrichs, K. and Lewy, H., (1956). *On the Partial Difference Equations of Mathematical Physics.* New York University Institute of Mathematics.

Gresho, P.M. Lee, R.L. (1981). "Don't suppress the wiggles-they're telling you something!" *Computers and fluids.* 9(2): 223-255.

Heinrich, J.C., Huyakom, P.S., Zienkiewicz, O.C. and Mitchell, A.R. (1977). "An upwind finite element scheme for two-dimensional convective transport equation". International journal of numerical methods in engineering, 11,131-143.

Jaber, F.H. and Mohtar, R.H. (2001). "Dynamic time step for the 1-D Overland flow kinematic wave solution". Journal of Hydrologic engineering, ASCE. In press.

Lyn, D.A. and Goodwin, P.G. (1987). "Stability of a general Preissmann scheme." *Journal of Hydraulic Engineering.* ASCE 113(1), 16-28.

Mohtar, R.H. and Segerlind, L.J. (1995). "Accuracy based time step criteria for solving parabolic equations." In *Modeling, Mesh Generation, and Adaptive Numerical Methods for Partial Differential Equations*, IMA Volume 75, (J. Flaherty, I. Babuska, W. Henshaw, J. Oliger, and T. Tezduyar, eds.). Springer-Verlag, NJ, USA.

Mohtar, R.H. and Segerlind, L.J. (1998). "Dynamic time step estimates for two dimensional transient field problems using square elements". *International Journal of Numerical methods in engineering*, 40, 1-14.

Mohtar, R.H. and Segerlind, L.J. (1999a). "Dynamic time step estimates for one-dimensional linear transient field problems". *Transactions of the ASAE*, 42(5), 1477-1484.

Mohtar, R.H. and Segerlind, L.J. (1999b). "Dynamic time step and stability criteria comparison". *International Journal of Thermal Sciences.* 38, 475-480.

Segerlind, L.J. (1984). *Applied finite element analysis*, Second edition. John Wiley and Sons New York.

Singh, V.P. (1996). *Kinematic Wave Modeling in Water Resources: Surface-Water Hydrology.* John Wiley & Sons, Inc.

Vieux, B.E. and Segerlind, L.J. (1989). "Finite Element Solution Accuracy of an Infiltrating Channel." In *Proceedings of the 7th International Conference on Finite Element Methods in Flow Problems*, University of Alabama, Huntsville, April 3-7.

Yue, J. (1989). "Selective lumping effects on depth-integrated finite element model of channel flow". *Advances in water resources.* 12(2), 74-78.

Zhang, W. and Cundy, T.W. (1989). "Modeling of two-dimensional overland flow". *Water resources research,* 25(9), 2019-2035.

Surface Water Hydrology, Singh, Al-Rashed & Sherif (eds)
© 2002 Swets & Zeitlinger, Lisse, ISBN 90 5809 363 8

Dew keeps deserts alive

Witold G. Strupczewski
Water Resources Department
Institute of Geophysics Polish Academy of Sciences
Ksiecia Janusza 64, 01-452 Warsaw, Poland
e-mail: wgs@igf.edu.pl

Vijay P. Singh
Department of Civil and Environmental Engineering
Louisiana State University, Baton Rouge, LA 70803-6405, USA
e-mail: cesing@lsu.edu

Stanislaw Węglarczyk
Institute of Water Engineering and Water Management
Cracow University of Technology, Warszawska 24, 31-155 Cracow, Poland
e-mail: sweglar@lajkonik.wis.pk.edu.pl

Abstract

Condensation of the atmospheric vapor into water occurs in nature near ground surface in two different forms: fog and dew. While for the formation of fog a high relative humidity (in practice close to 100%) is necessary, the formation of dew requires a cold well-insulated surface and air-saturation is not necessary. Precipitation in deserts tends to be episodic and localized, with aridity erratically interrupted by torrential rains. Although rainfall exceeds in quantity other forms of precipitation, it is dew which is the main source of water available for plants and animals. The present investigations on dew are oriented mainly toward agricultural practices and to some extent toward water recovery from dew. An investigation of arid natural ecosystems requires the information concerning water and the thermal regime of the soil and the atmospheric boundary layer. To fulfil this requirement, an improvement of *in-situ* measuring techniques relevant for harsh climatic and soil moisture conditions of deserts and development of monitoring systems, including remote sensing techniques, are necessary. There are inherent characteristics of the desert environment to which life in the desert has adapted. Any change in the dew and fog regime caused by local or global human impact can easily destroy the fragile natural ecosystems created. Monitoring of all forms of precipitation, especially dew and fog, as well as both air and soil humidity can provide information necessary for environmental control of deserts.

Introduction

In basic climatic terms, a desert can be defined as an area which receives little or no rainfall and experiences no season of the year in which rain regularly occurs (Nicholson, 1995). What distinguishes deserts from non-deserts is the amount of moisture available to the ecosystem, or the difference between water availability through precipitation and the potential evapotranspiration. According to estimates based on this concept, true deserts make up roughly 20% of the Earth's land surface and another 15% of the area is semiarid. Most desert environments share several common climatic characteristics. These include episodic and localized rainfall, low atmospheric relative humidity, large fluctuations of temperature (and humidity) between day and night, low cloudiness and high insolation.

The common perception of a desert is dry region with inhospitable conditions for life. In fact, desert ecosystems are characterized by a remarkable diversity in vegetation and most teem with life well adapted to the harsh climatic and soil moisture conditions. The availability of water in space and time controls the biota of deserts. The moisture availability is the principal determinant of the type and amount of vegetation because water is central to the growth process of plants. Thus the environmental characteristics to which biosphere adapts include rainfall and other forms of precipitation, atmospheric moisture, solar radiation, surface temperature, and ground chemistry including salinity and soil moisture conditions. To understand the interactions between climate, biosphere and biogeochemical processes, the monitoring of the various forms of water supply; water storage, including soil moisture processes; and thermal regime is necessary. They are of primary interest for hydro-ecological studies of the desert biosystems.

Problems of Monitoring

The problems of monitoring hydrometeorological processes in arid areas have been a subject of several publications (e.g., Hughes, 1999; Görgens and Hughes, 1982; McMahon, 1979). One of the major problems is that the relative harshness of the environment, low population and consequently poor communication systems make it difficult to establish instrumentation networks. The high range of variability of the hydrometeorological variables means that the instruments used have to be able to measure over wider ranges than in humid regions. Episodic and localized rainfall requires longer periods of records and the denser raingauge networks. All of these contribute to the need for specialized monitoring equipment, which can be expensive and difficult to maintain. It is no accident that of the many experimental research catchments that do exist or have existed, relatively few are located within arid zones. The lack of understanding of some of the processes involved and a lack of sufficient data hamper the application of hydrological estimation methods to arid zones.

Rainfall and Runoff

Perhaps the greatest amount of research into moisture sources in arid regions has been directed at problems relating to the liability, frequency, duration and intensity of precipitation, and consequently to surface runoff. Generalizations about these matters are difficult because of the different kinds of atmospheric conditions that prevail in different arid regions (Durrenberger, 1999). The distribution of rainfall in a desert is erratic in both space and time, especially in regions of convective rainfall. The amount of rainfall varies tremendously from year to year; many years may pass without a drop. Rainless stretches of 10-14 years have been recorded at locations such as Swakopmund, Namibia, and Lima, Peru, in the coastal deserts of Southern Africa and South America. Usually rain occurs only a few days within the year and most of the rain, which falls, occurs within a few hours. Nicholson (1995, 1998) reports several extreme rainfall events in deserts. At Lima, where the mean annual rainfall is 46 mm, 1524 mm fell during one storm in 1925. In the Sahara at Nouakchott, Mauritania, over twice the annual mean, or 249 mm, fell on one day. In Helwan, Egypt, where the mean annual rainfall is about 20 mm, seven storms produced a quarter of the rain, which fell during the entire 20-year period. In Sidi bou, Tunisia, nearly 800 mm fell during September and October while the mean rainfall is of the order of 10-20 mm. Since the desert rainfall is localized in space, it is a matter of good luck to record the maximum point depth by means of routine rain-gauge networks. Studies by Sharon (1981) in the Namib desert have shown that for individual storms, rainfall at stations just 2-3 km apart can differ by a factor of 10-20.

The runoff from intense desert storms is an active agent in sculpting the desert and forming some pools in the dry river bed, which provide temporary water storage. Dry wadis may be lined by trees or shrubs utilizing water remaining deep under the stream bed. In dryland regions, the dominant form of runoff is the overland flow of Hortonian character. It is because soils and mantles are too thin to maintain sub-surface flow and the groundwater table is too deep to supply water to streams. The low infiltration and the high proportion of overland flow promote the quickest response in stream discharge. The low water storage capacity of the soil makes that the surface layer becomes dry soon after runoff ceased. The groundwater table is usually too deep to be accessible for tree roots. Both the coverage and biomass fluctuate in response to varying moisture during the year or over longer periods. To keep the continuity of desert life, alternative sources of water must have existed and have been accessible in a more or less systematic manner to the biota of deserts. It was discovered in the Peru-Atacama and Namib deserts that some plant and animal species have special adaptations which allow them to utilize fog water (Nicholson, 1999). Certain plants can absorb water directly through the leaf surface.

Near-Surface Humidity

Despite the low bio-mass productivity, the interest in desert ecosystems seems to be rising. It may be caused by the growing interest in investigation of the forms of life existing in extreme terrestrial conditions, which is driven by search for outer space life and ecological concern. Some news arising from reconnaissance surveys into the mystery of desert life are amazing and captivating. Some hydroscopic elements found in the desert food-chain (Broza, 1979; Kidron, 1999) are likely to be able to absorb water from humid air promoting the primary production. Investigation in the Negev Desert, Israel (Lange, et al., 1970; Kappen et al., 1979) has shown that dew, fog and high air humidity are a major water source for microorganisms, snails (Shachak et al., 1987) and insects (Broza, 1979) largely controlling their distribution (Kappen et al., 1980; Shachak et al., 1995). Since snails feeding on endolithic lichens have been found to ingest limestone (Shachak et al., 1987,

1995) and to redistribute nitrogen (Jones and Shachak, 1990), air moisture distribution patterns may also assist in the study of soil forming processes. In conclusion, monitoring of air- and soil-moisture processes is indispensable for investigation of desert's ecosystems. Despite its importance, global measurements and analyses of soil moisture still remain deficient.

Nocturnal Surface Cooling

The subtropical location of many deserts, clear skies and the scant cloud cover produce a regime of strong solar insolation and high surface temperatures during day time. Temperature is enhanced by the dryness of the soil and the bare ground. The heat is concentrated at the surface because the dry ground transports little heat to deeper layers of soils. The ground temperature in excess of 70 °C has been recorded at several locations. At Port Sudan on the Red Sea, a sand temperature of 83.5 °C was recorded (Nicholson, 1999). The temperature decreases rapidly with depth into the ground and with height above the surface. Nicholson (1999) reported that in Kara Kum desert in Asia the daytime temperature typically drops over 15 °C within the first few centimeters and 20-30 °C within the first 10 cm below the surface. The air temperature can drop by more than 10 °C within the first 10 cm above the surface. The temperature ranges of 15-22 °C are common and objects in the shade are much cooler than those in the sun. Therefore the air temperature measured at meteorological stations is not relevant for the surface temperature.

The concentration of heat at the surface means that there is no subsurface thermal reservoir. As a consequence of this and the sparse vegetation cover, the surface cools extremely rapidly and efficiently at night. Most of the heat stored during the day escapes to the upper atmosphere during the night (Garratt, 1992). The result is a large daily range of both ground and air temperatures. Nicholson (1999) gives some examples of daily extremes. At a station in the Sahara, the air temperature fell from a daytime maximum exceeding 37 °C to −1 °C at night; at Tucson, Arizona, a daily range of 56 °C was recorded. Since one can assume no change in the vapor content of the air, the fall in the air temperature gives rise to the relative humidity, and surface layer of air will be close to a full saturation state.

Estimation of Near-Surface Humidity

There is a general lack of dependable, daily humidity data from the current global network of weather stations (Kimball et al., 1997). In the US, humidity is generally measured only at primary National Weather Service stations. Running et al. (1987) estimated this humidity measurement network to be less than 1 station per 100,000 km^2 throughout the western U. S. A. Reliable, direct measurements of humidity require sensitive dew point hygrometer, chemical or electrical instruments. These sensors are prone to error and the measurement accuracy may degrade over time without frequent sensor calibration (Marks et al., 1992; Glassy and Running, 1994). Minimum daily air temperature (T_n) is often used as a surrogate for mean daily dew point ($T_{d,day}$) to estimate near surface humidity. Assumptions of T_n and $T_{d,day}$ equivalency are generally valid when condensation occurs during the night. This method is unreliable under arid conditions where nightly minimum temperature may remain well above the dew point. Kimball et al. (1997) developed the empirical model to improve the accuracy of T_n-based humidity estimates using daily air temperature, annual precipitation and estimated daily potential evaporation. Taking the above into account, a dewpoint hygrometer or a psychrometer with appropriate tables should be used in the desert regions to measure dewpoint and relative humidity daily run.

One should note that it is the daily minimum surface temperature, which is of main interest for biospheric investigations (e.g. Luo & Goudriaan, 2000), and it may differ considerable from the minimum daily temperature measured at a height of 2m above the surface. Moreover, the soil heat flux into the atmosphere depends also on the thermal conductivity of the surface layer, i.e., it can vary in space due to inhomogeneous soil and consequently the surface temperature can show high spatial variability.

The hope for improvement of the present situation is credited to remote sensing measurements of the surface temperature. Cloudless sky, often encountered in desert regions, offers good conditions for measurement in the thermal infrared spectrum. Recent studies have demonstrated that remote sensing techniques can be also applied to measure soil moisture at the ground surface. The microwave sensors offer the potential of truly quantitative measurements from airborne or spaceborne instruments (e.g., De Troch et al., 1997).

Dew and Fog

Rainfall is the dominant form of precipitation in most deserts. In desert areas where virtually no rain occurs, dew and fog drip represent the only forms of precipitation available for biota. Fog is a cloud of already condensed water droplets. A high relative humidity (in practice saturated air) is necessary for formation of fog. Many of the coastal deserts are relatively moist environments with a high frequency of fog and relatively high humidity. In the Namib, the water condensed from the frequent fog is several times greater than rainfall depth. In the coastal deserts of Peru and Chile, the moisture obtained from the fog that moves inland across the coastal ranges sustain a moderately dense shrub forest. In summer the atmosphere above Sahara is very humid. Kidron's (1999) investigation of the Negev Desert, Israel shows a considerable increase of the average dew and fog amount with elevation despite the greater distance from the sea of the more elevated sites and consequently their lower moisture supply. An increase in the percent of days with dew and fog with altitude was also noted.

The formation of dew requires a cold surface, but 100% humidity is not necessary. Thus, dew is common even in dryer zones. Because of the great amount of radiational cooling that occurs in deserts, the formation of dew is a frequent occurrence. The onset of dewfall occurs when the surface cools to the dew-point temperature and will coincide with a reversal in the sign of the humidity gradient. Dewfall is an important contributor to the water budgets of many desert regions. Observations of dew amounts tend to give values less than 0.5 mm per night (Garratt and Segal, 1988).

Physics of Condensation

A coherent and organized introduction to the theory of condensation can be found in Monteith (1957), Garratt (1992) and Nikolayev et al. (1996). It is interesting to learn the external factors upon which dewfall, in particular, depends. To answer this important question, use can be made of the Penman combination type equation of evaporation (Gerratt, 1992). According to the Penman model of evaporation from open water surfaces (Shuttleworth, 1993), the energy used for evaporation ($\lambda \cdot E$), where E is the evaporation depth and λ – latent heat of vaporization, is expressed as the weighted total of the radiation term (R) and the aerodynamic term (A):

$$\lambda \cdot E = \frac{\Delta}{\Delta + \gamma} R + \frac{\gamma}{\Delta + \gamma} A \tag{1}$$

where γ is the psychrometric constant and Δ is the gradient of the saturated vapor pressure curve at the air temperature;

$$R = R_{N0} - G_0 \tag{2}$$

is the difference between the net radiation at the surface level and the soil heat flux at the interface, and

$$A = \rho \lambda \delta q / r_{aV} \tag{3}$$

where ρ is the air density, δq is the saturation deficit, and r_{aV} is the aerodynamic resistance dependent upon the wind speed and thermal stability.

Dealing with the process opposite to evaporation, i.e., with condensation, both $\lambda \cdot E$ and R_{N0} are negative at night and G_0 is positive as heat flux towards surface, so the maximum negative values of E will occur when the air is saturated and $\delta q = 0$. In Equation (1) these maximum values are given by the first term on the RHS. Garratt (1992) estimated peak values of this term of about -60 W m^{-2} occur at $T = 293$ K which corresponds to the dew formation of 0.08 mm hr^{-1}.

The values of the soil heat flux at the interface depend on many factors, including the soil type (hence, physical properties), soil moisture content and solar radiation (hence, the time of the day). In case of condensation, the flux may be a significant fraction of the net radiation and must be taken into account. Spatially, soil anisotropy results in spatial variability of G_0 and hence of dewfall depth as well. Therefore, while working with *in-situ* measurement of dew, one should decide whether to make it for real conditions at the point, i.e., to focus on real value of dewfall depth at the measurement's location, or to apply artificial standard surface accepting measurement bias. It is just what distinguishes the dewfall from rainfall measurements, where the local structure of soil seems to have negligible effect on the measured value. Obviously, for the desert biosphere's investigation, the interest is in water supplied to the surface, while for the atmospheric boundary layer studies may be convenient to limit the influence of local soil conditions on measured values. In fact, the underestimation of the significance of the soil heat flux in the radiation balance was one of the main causes of failure of the condensers in Southern France built for water recovery from dew (Chaptal, 1932; Jumikis, 1965). Instead of following the designs of ancient dew ponds, the concrete shells with massive foundations were used (Nikolayev et al. 1996; Herschy, 1998).

Dewfall, as an atmospheric source of water is only one of two important contributions to dew formation; the other has been termed *distillation* by Monteith (1957) and it is observed on the surface covered by canopy. In distillation, soil is the additional source of dew formation on canopies.

Measurement of Condensation

Dew and fog precipitation is measured all together. In order to classify precipitation as dew and/or fog, the visibility in the morning can be used. For example, mornings characterized by <1000 m visibility, lasting for at least half an hour were classified by Kidron (1996) as foggy mornings.

The weighing methods serve for dew and fog measurement (Kidron, 1998; Luo & Goudriaan, 2000). Blotting paper or velvet-like cloths are used to collect the condensed water and then weighed. Since data regarding dew precipitation are very scarce, it seems reasonable to describe here a simple weighing method applied by Kidron (1998), which is called the cloth plate method (*CPM*). The device consists of 6 cm x 6 cm x 0.15 cm velvet-like cloth (Universal Comp., Germany) attached to the center of 10 cm x 10 cm x 0.2 cm glass plates. The glass plates are glued on the top of 10 cm x 10 cm x 0.5 cm plywood plates, providing an identical 0.7 cm thick depositional surface. The cloth serves to absorb dew and fog precipitation. A new cloth is attached to the plate each in the afternoon and carefully collected during the following morning, approximately 0.5 h after sunrise into separate pre-weighted flasks which are immediately sealed. At that time, the dew condensation is maximal (Kidron, 1988). The cloth is then weighed for dew and fog quantities. A value of 0.03 mm is used as a threshold for both dew and fog precipitation.

Conclusions

The desert hydrology still concentrates on the surface hydrology, i.e., on the traditional rainfall-runoff approach. Studies of other forms of precipitation, i.e., dew and fog, are oriented mainly toward agricultural practices and to some extent toward water recovery from dew.

Dew, as a main contributor to leaf wetness, is considered to be an important factor in plant disease epidemics and also in the deposition of acidic air pollutants on plant surface (Luo and Goudriaan, 2000). It provides the free water that is essential to the development of many foliar bacterial and fungal plant pathogens (Wallin, 1963) and may enhance the deposition of the pollutants (Hughes and Brimble-combe, 1994). For instance, the leaf wetness is used to forecast rice blast epidemics (one of the most severe diseases of rice crops).

The influence of dew on agricultural production in arid climates is the subject of a few publications (see for references: Kidron, 1999). The beneficial effect of dew for citrus trees as well as a variety of plants has been stated. Exposed to dewfall, they showed a substantially high recovery following dewfall and a statistically higher growth rate and yield than unbedewed plants. Dew and fog may increase seeding survival, plant growth and crop yield in arid zones. Despite the important role of dew and fog in arid ecosystems, and increasing agricultural practices, the quantitative information on both condensation forms is scarce.

The recovery of clean water from dew has remained a longstanding challenge in many places all around the world (Nikolayev et al. 1996; Herschy, 1998). During the nineteenth century, the widespread extension of piped water supplies caused dew ponds to fall into disuse. In the late twentieth century a sociological trend towards simplicity of life systems led to the revival of the old facilities.

Resulting from the complexity of the subject, investigation of arid natural ecosystems requires the integration of a wide range of specialist skills. Water science experts, meteorologists, hydro-meteorologists and hydrogeologists are supposed to supply biologists with the information concerning water and the thermal regime of the soil and of the atmospheric boundary layer. To fulfill this requirement, an improvement of *in-situ* measuring techniques relevant for harsh climatic and soil moisture conditions of deserts and development of monitoring systems, including remote sensing techniques, are necessary.

There are inherent characteristics of the desert environment to which life in deserts has adapted. Any change in dew and fog regime caused by local or global human impact can easily destroy the fragile natural ecosystems created. Monitoring of all forms of precipitation, especially dew and fog, as well as of both air and soil humidity can provide information necessary for environmental control of deserts.

References

Broza, M. (1979). "Dew, fog and hygroscopic food as a source of water for desert arthropods". *J. Arid. Envinom.,* 2, 43-49.

Chaptal, L. (1932). "La captation de la vapeur d'eau atmosphérique", *La nature,* 60(2893): 449-454.

De Troch, F.P., Troch, P.A., Su, Z. (1997) "Use of remote sensing data from airborne and spaceborne active microwave sensors towards hydrological modeling". In: *Integrated approach to environmental data management systems,* ed. Harmancioglu et al., NATO ASI Series, 2. Environment – V.31, Kluwer Ac. Pub.: 171-188.

Durrenberger, R.W. (1999). "Arid climates". In: *Encyclopedia of Hydrology and Water Resouces.* R.W. Herschy & R.W. Fairbridge, eds. Kluwer Ac. Pub., 70-78.

Garrat, J.R. (1992). *The atmospheric boundary layer.* Cambridge Univ. Press.

Garrat, J.R., Segal, M. (1988). "On the contribution of atmospheric moisture to dew formation". *Bounnd. Layer Meteor.,* 45, 209-236.

Glassy, J.M., and Running, S.W. (1994). "Validating diurnal climatology logic of the MT-CLIM model across a climatic gradient in Oregon". *Ecol. Appl.,* 4(2): 248-257.

Görgens, A.H.M. and Hughes, D.A. (1982). "Synthesis of streamflow information relating to the semi-arid Karoo Biome of South Africa. *South African J. of Science,* 78(2), 58-68.

Herschy, R.W. (1998). "Dew ponds". In: *Encyclopedia of Hydrology and Water Resouces.* R.W. Herschy & R.W. Fairbridge, eds. Kluwer Ac. Pub., 176-183, 186-196.

Hughes, D.A. (1999). "Arid zone hydrology". In: *Encyclopedia of Hydrology and Water Resouces.* R.W. Herschy & R.W. Fairbridge, eds. Kluwer Ac. Pub., 87-89.

Hughes, R.N. and Brimblecombe, P. (1994). "Dew and guttation: formation and environmental significance". *Agric. For. Meteorol.,* 67, 173-190.

Jumikis, A.R. (1965). "Aerial wells: secondary sources of water." *Soil Sci.* 100: 83-95.

Kappen, L., Lange, O.L., Schultze E.D., Evenari, M., Bushborm, V. (1979). "Ecophysiological investigations on lichens of the Negev Desert, IV: Annual course of the photosynthetic production of Ramalina maciformis (Del.). Bory." *Flora,* 168, 85-105.

Kappen, L., Lange, O.L., Schultze E.D., Evenari, M., Bushborm, V. (1980). "Ecophysiological investigations on lichens of the Negev Desert, VII: The influence of the habitat exposure on dew imbibition and photosynthetic productivity." *Flora,* 169, 216-229.

Kidron, G.J. (1998). "A simple weighing method for dew and fog measurements". *Weather,* 53, 428-433.

Kidron, G.J. (1999). "Altitude dependent dew and fog in the Negev Desert, Israel". *Agric. and Forest Meteorol.,* 96, 1-8.

Kimball, J.S., Running, S.W., Nemani, R. (1997). "An improved method for estimating surface humidity from daily minimum temperature". *Agric. For. Meteorol.,* 85, 87-98.

Lange, O.I., Schultze, E.D., Koch, W. (1970). "Ecophysiological investigations on lichens of the Negev Deser III: CO_2 gas exchange and water metabolism of crus-

tose and foliose lichens in their natural habitat during the summer dry period."
(*Technical translations 1656 of the National Resaearch Council of Canada), Flora*,
159, 525-538.

Luo, W., Goudriaan, J. (2000). "Dew formation on rice under varying durations of
nocturnal radiative loss", *Agric. For. Meteorol.*, 104, 303-313.

Marks, D., Dozier, J., Davis, R.E. (1992). "Climate and energy exchange at the
snow surface in the alpine region of the Sierra Nevada, 1. Meteorological measurements and monitoring". *Water Resour. Res.*, 28(11): 3029-3040.

McMahon, T.A. (1979) "Hydrological characteristics of arid zones*", IAHS Publ.* No.
128, pp. 105-123.

Monteith, J.L. (1957). "Dew". *Quart. J. Roy. Met. Soc.* 83, 322-341.

Nicholson, S.E. (1995). *Dryland Climatology.* Oxford University Press, Oxford.

Nicholson, S.E. (1998). "Desert Hydrology" and "Deserts". In: *Encyclopedia of Hydrology and Water Resouces.* R.W. Herschy & R.W. Fairbridge, eds. Kluwer Ac.
Pub., 176-183, 186-196.

Nikolaev, V.S., Beysens, D., Gioda, A., Milimouk, I., Katiushin, E., Morel J.P.
(1996). "Water recvovery from dew". *J. of Hydrol.*, 182, 19-35.

Running, S.W., Nemani, R.R. and Hungerford, R.D. (1987). "Extrapolation of synoptic meteorological data in mountainous terrain, and its use for simulating forest
evaporation". *Can. J. For. Res.*, 17: 472-483.

Shachak, M., Jones, C.G., Granot, Y. (1987). "Herbivory in rocks and the weathering of a desert." *Science,* 236, 1098-1099.

Shachak, M., Jones, C.G., Brand, S., (1995). "The role of animals in an arid ecosystem: snails and isopods as controllers of soil formation, erosion and desaliniza-
tion." *Advances GeoEcol.,* 28, 37-50.

Sharon, D. (1981). "The distribution of rainfall in space and time in the Namib desert". *J. Climatol.*, 1, 69-75.

Shuttleworth, W.J. (1992). "Ch.4. Evaporation". In: *Handbook of hydrology, ed.
Maidment D.R.,* McGraw-Hill. Inc.

Wallin, J.R. (1963). "Dew, its significance and measurement in phytopathology".
Phytopathology, 53, 1210-1216.

Surface Water Hydrology, Singh, Al-Rashed & Sherif (eds)
© 2002 Swets & Zeitlinger, Lisse, ISBN 90 5809 363 8

Laboratory experiments on the influence of storm direction on soil loss from sloping areas

João L.M.P. de Lima

IMAR – Institute of Marine Research, Coimbra Interdisciplinary Centre
Department of Civil Engineering, Faculty of Science and Technology –
Campus 2, University of Coimbra, 3030-290 Coimbra, Portugal
e-mail: plima@dec.uc.pt

Vijay P. Singh

Department of Civil and Environmental Engineering
Louisiana State University, Baton Rouge, LA 70803, USA
e-mail: cesing@lsu.edu

Isabel M. Barreira

IMAR – Institute of Marine Research, Coimbra Interdisciplinary Centre
Department of Civil Engineering, Faculty of Science and Technology –
Campus 2, University of Coimbra, 3030-290 Coimbra, Portugal
e-mail: imbar@dec.uc.pt

M. Isabel P. de Lima

IMAR – Institute of Marine Research, Coimbra Interdisciplinary Centre
Department of Forestry, Agrarian Technical School of Coimbra,
Polytechnic Institute of Coimbra, Bencanta, 3040-316 Coimbra, Portugal
e-mail: iplima@mail.esac.pt

Abstract

The soil material transported by overland flow is important in such activities as agricultural soil management, evaluation of soil nutrient losses, and sediment transport to water reservoirs. This paper reports the results of laboratory experiments,
which were undertaken to study the effect of storm direction on the water erosion
process. The experiments were carried out using a soil flume adjustable to different
slopes and a rainfall simulator. To simulate moving rainstorms, the rainfall simulator was moved upstream and downstream over the soil surface. The results show
that the storm direction affects the soil loss. The soil loss due to downstream moving rainstorms is higher than that due to identical upstream moving rainfall storms.

Introduction

Soil erosion is a natural phenomenon influencing soil genesis and landscape dynamics. An understanding of the water erosion dynamics and the determination of
soil material transported by overland flow are needed in engineering studies and

such activities as agricultural soil management, evaluation of soil nutrient losses, and sediment transported to rivers and water reservoirs.

Soil erosion represents the combined effect of the processes of soil detachment and transport by raindrop impact and surface flow (e.g., Römkens et al., 1997). Thus, rainfall drives both the overland flow and soil erosion processes. Any factor influencing surface flow characteristics also influences soil erosion.

Soil erosion is highly affected by rainfall intensity and surface runoff (e.g., Meyer, 1981; Foster, 1982). Although infiltration, runoff and soil erosion have been extensively studied in the field and laboratory, most studies using simulated rainfall have applied rainfall at a constant rate. This contrasts with natural rainfall, which is highly variable, in both time and space (e.g., Huff, 1967; Eagleson, 1978; Sharon, 1980; Lima, 1998; Willems, 2001).

The spatial and temporal distributions of rainfall are amongst the main factors affecting watershed and hillslope runoff. Nevertheless, most methods used in hydrologic studies assume that the storm arrives instantaneously over the drainage area and then remains stationary. Therefore, these hydrologic studies do not take into account the effect of the runoff response caused by the movement of the storm across the drainage area. Ignoring storm movement can result in (considerable) over- and under-estimation of runoff peaks (e.g., Maksimov, 1964; Yen and Chow, 1968; Wilson, 1979; Stephenson, 1984; Singh, 1998; Lima and Singh, 1999). Because of the inter-relation between rainfall and runoff, the movement of storms is expected to affect runoff and the associated soil loss; for moving storms, the distribution of rainfall intensity in space and time is continuously changing.

The main objective of this laboratory study is to quantify the influence of the storm direction on soil loss from sloping areas. Experiments were carried out using a soil flume adjustable to different slopes and a rainfall simulator. The simulated rainfall moved upstream and downstream over the soil surface. Several bed slopes were used for carrying out the experiments.

Experimental set-up

The laboratory experiments described in this study were conducted in the Department of Civil Engineering of the University of Coimbra (Portugal). The experiments were carried out using a soil flume and a rainfall simulator.

Characteristics of the soil flume

The soil flume was constructed with metal sheets and had the following dimensions: 2.0 m length x 0.1 m width x 0.12 m height. Surface runoff and drainage water were collected at the end of the flume. Because slope is also one of the critical factors controlling soil erosion by overland flow (Bryan and Poesen, 1989), the structure had two slope adjusting screws allowing the control of the flume slope (Figure 1).

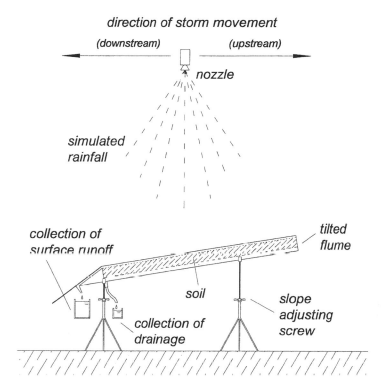

Figure 1. Schematic representation (side view) of the soil flume and hydrological variables involved in the laboratory experiments.

Characteristics of the soil

The soil used in the laboratory experiments was collected from the right margin of River Mondego in the city of Coimbra, Portugal. The selection was made because this soil was readily available in large quantities and, under natural rainfall, it exhibited extensive soil erosion (e.g., sheet, gullies). The soil material consisted of 11% clay, 10% silt and 79% sand.

After being collected from the original place the soil was submitted to a standard procedure involving pre-sieving through a 4.75 mm aperture square-hole sieve to remove coarse rock and organic debris. The soil was packed into the flume by uniformly spreading the soil material. To obtain a flat surface, a sharp, straight edged blade that could ride on the top edge of the side walls of the flume was used to remove excess soil. The blade was adjusted such that the soil level in the flume equalled the retaining bar at the bottom end of the flume. Afterwards, the soil was tapped with a wooden block. The soil presented a uniform thickness of 0.1 m. Before starting the experimental runs, the soil was wetted up to field capacity. These procedures were repeated for the different slopes analysed. The same soil material was maintained throughout the various replicates of the simulations carried out for each slope.

Characteristics of the rainfall simulator

Laboratory-and field-based rainfall simulation studies have been widely used to investigate soil erosion (e.g., Bryan and De Ploey, 1983; Bowyer-Bower and Burt, 1989; Morgan, 1995). The basic components of the sprinkling-type rainfall simulator used in this study were a nozzle, a support structure in which the nozzle was installed, and the connections with the water supply and the pump. The laboratory experiments were conducted using one single downward-oriented full-cone nozzle spray (3/4 HH – 4 FullJet Nozzle Brass-Spraying Systems Co.). The nozzle height was 1.5 m, measured above the geometric centre of the soil surface.

A flexible rubberised hose distributed water from de pump to the nozzle. A pressure gauge monitored the pressure at the nozzle. The working pressure on the nozzle was maintained constant at 0.5 bar. The maintenance of a stable pressure avoided variations in rain intensity during the simulated rainfall events (Figure 2).

The storm movement was obtained by moving on wheels, back and forth, the support structure of the nozzle, as shown in Figures 1 and 2. This was achieved using two electric motors. The moving-storm velocity remained constant for all the experiments, presenting the value of 0.33 m/s.

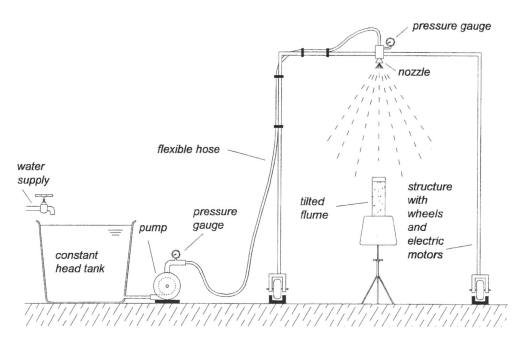

Figure 2. Schematic representation (front view) of the rainfall simulator
including the movable support structure and the connections with
the water supply and the pump.

Methodology

Simulated rainstorms moving upstream and downstream over the soil surface were applied to the following bed slopes: 5, 10, 15, 20 and 25%. The storm movement velocity and storm duration remained constant for all the experimental runs. The initial soil water conditions were approximately identical for all runs.

Rainfall measurements

The laboratory experiments were limited to one rainstorm velocity and one simulated rainfall pattern. During storm simulations the total amount of rain was kept constant because the rainfall intensity pattern and storm velocity did not vary during each event. Thus, the storm duration was also constant.

The rainfall intensity and distribution are dependent on the nozzle size and type, water pressure at the nozzle and the height above the plot surface. The rainfall distribution was measured on a horizontal plane with equally sized gauges for a time period of 30 seconds, maintaining the nozzle static. The gauges consisted of 0.1 m diameter cylindrical containers. The measurements were repeated five times and mean values were calculated for the five replicates.

Since the width of the soil surface was small (0.1 m), the pattern of the simulated rainstorm can be simplified, as shown in Figure 3. The average rainfall intensity was 8.3 mm/min and the water application length was 2.3 m. As in natural spatial rainfall fields, a high intensity rainfall area is embedded within the areas of lower intensity, as described by Bras and Rodrigues-Iturbe (1976), Sivapalan and Wood (1987), Willems (2001), among others.

The rain-drop sizes produced by the nozzle are shown in Figure 4 (Lima, 1997). The average drop-size (equivalent drop diameter) was approximately 1.5 mm. The measurements were done using the stain method (Hall, 1970).

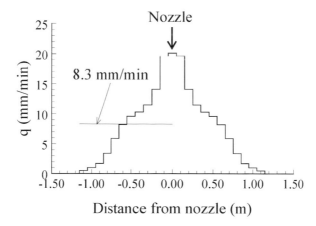

Figure 3. Distribution of rainfall intensities supplied by the nozzle on a horizontal surface, for an operating pressure of 0.5 bar, at a height of 1.5 m.

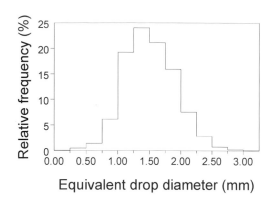

Figure 4. Measured drop size distribution under the nozzle (Lima, 1997).

Soil loss measurements

The fixed simulated rainfall pattern (see Figure 3) moved in an alternating sequence, upstream and downstream, over the soil surface. The consecutive rainfall
events were generated at periodic time intervals, such that all the overland flow
from the previous event had ceased, maintaining approximately the same moisture
conditions in the superficial layer of the bed soil. The overland flow caused by each
rainfall event was collected in a metal container placed in the bottom end of the soil
flume for determination of the runoff volume and soil loss. The sediment weight
was estimated by low-temperature oven-drying the samples.

Results and discussion

For each slope (5, 10, 15, 20 and 25%) tested, thirteen runs were conducted, each
with run representing a cycle of two rainfall events: one corresponding to the
movement of the storm upstream and the other downstream.

The Hortonian overland flow occurred clearly on the flume when the rain intensity
exceeded the infiltration rate. Because the erosive rainfall was moving, the Hortonian overland flow occurred only on parts of the soil surface. The topsoil saturation
overland flow did not take place due to the short period of time for which the moving rainfall was effectively falling on the soil surface. Seepage (interflow) also did
not take place.

The transport of fine erodible soil material was mainly due to overland flow. The
sediment transport by rain splash had a relatively minor contribution. The greatest
effect was caused by rain falling on a thin overland flow layer, when present, leading to both strong sediment detachment and transport.

The results of these experiments showed significant differences in the soil loss between identical simulated rainfall moving downstream and upstream. The soil loss
for downstream moving rainfall was higher than the soil loss for identical rainfall

Table 1. Summary of soil loss data for the five experimental soil surface slopes.

| Slope (%) | Runs | Soil loss (g/m^2) | | | |
| | | Downstream | | Upstream | |
		Mean	STDV	Mean	STDV
5	13*	2.15	0.64	1.70	0.62
10	13*	5.06	2.40	3.33	1.37
15	13*	8.10	3.30	4.10	1.70
20	13*	12.07	4.50	4.82	2.32
25	13*	15.90	5.40	5.10	1.60

* 13 runs for downstream moving storms and 13 runs for upstream moving storms;
STDV – Standard deviation.

moving upstream. A summary of the results obtained for all the slopes tested is shown in Table 1.

Plots of soil loss obtained for an alternating sequence of downstream and upstream moving storms are shown in Figures 5 to 9.

Figure 5 illustrates the soil loss observed during an alternating sequence of rainstorms moving downstream and upstream for a 10% soil bed slope. A similar behaviour was observed for other slopes (5%, 15%, 20% and 25%). The results show that the downstream moving storms yielded higher soil loss than did upstream moving storms. As the number of simulated rainstorm events increased, the difference of soil loss between downstream and upstream moving storms decreased due to changes of the characteristics of the surface layer of the soil, namely, the reduction of fine sediment materials transported by overland flow in previous runs.

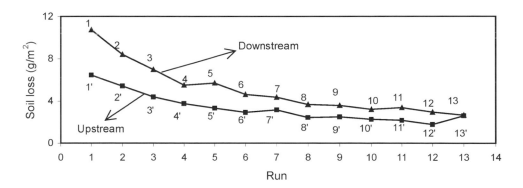

Figure 5. Soil loss for an alternating sequence of downstream and upstream moving rainstorms, for the 10% slope.

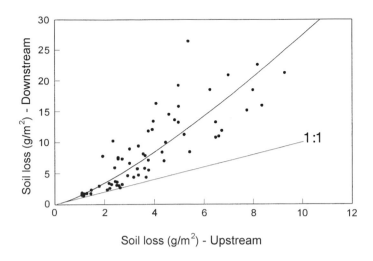

Figure 6. Soil loss (all 130 runs) for downstream moving storms (65 runs) against upstream moving storms (65 runs).

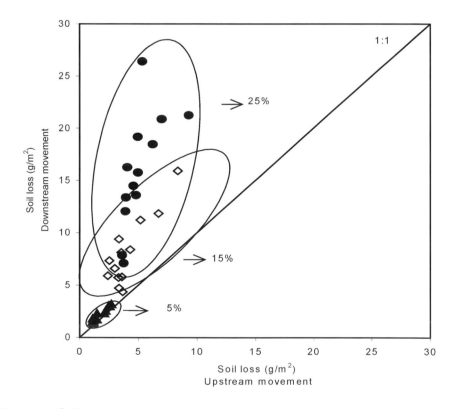

Figure 7. Soil loss produced by downstream moving storms against upstream moving storms for 5%, 15% and 25% slopes.

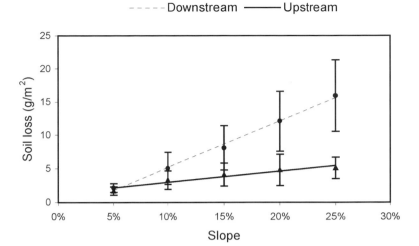

Figure 8. The mean soil loss and standard deviation (STDV) as a function of soil surface slopes for storms moving in the upstream and downstream directions.

The behaviour observed in Figure 6 shows some variability, which increases with slope. This variability can be explained not only by the simulated rain variability, but also by the experimental procedure related to the preparation of the soil material and to the filling of the flume with soil.

The curve fitted to the data in Figure 6 was used only to show the trend of the behaviour observed in the comparison of soil loss for storms moving upstream and downstream (power curve fit by the least square regression, with $r^2 = 0.89$):

$$Ed = 1.40 \, Eu^{1.29} \tag{1}$$

Where Ed is the soil loss for downstream moving storms (g/m^2) and Eu is the soil loss for upstream moving storms (g/m^2).

Soil loss is small for slopes up to 5%, as shown in Figures 7 and 8. Also the difference between the soil loss for downstream and upstream moving storms is small. For higher slopes, the soil erosion increases strongly (Figures 7 and 8).

The absolute and relative differences between soil loss for storms moving in the downstream and upstream directions increased with slope, as shown in Figure 9.

These soil loss results, presented in Table 1 and Figures 6 to 9, are clearly linked with the characteristics of overland flow hydrographs resulting from rainstorms moving in the upstream and downstream directions. Singh (1998) and Lima and Singh (1999) identified distinct hydrologic responses for storms moving upstream and downstream. When compared to storms moving downstream, storms moving upstream are characterised by hydrographs with: (1) earlier rise, (2) lower peak

413

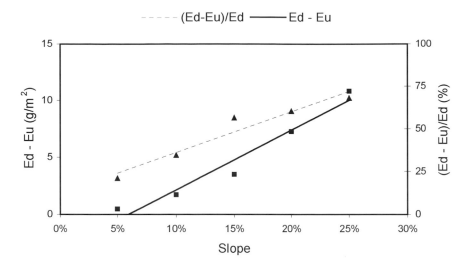

Figure 9. Absolute and relative differences between soil losses, for storms moving in the downstream and upstream directions as a function of soil surface slopes, where Ed is the soil loss for downstream moving storms and Eu is the soil loss for the upstream moving storms.

discharge, (3) less steep rising limb, and (4) longer base time. These authors also reported that for storms moving in both the downstream and upstream directions the relative peak discharge was greatest for storm velocity equal to the mean overland flow velocity. These experiments were obtained theoretically (Singh, 1998) and experimentally (Lima and Singh, 1999) for overland flow on an impermeable plane.

Conclusions

The laboratory experiments described in this work show that the spatial and temporal distributions of rainfall have a marked influence on the soil loss. The results showed that downstream moving storms yielded higher soil loss than identical upstream moving storms. It should be noted that besides the storm direction all other parameters were kept constant (e.g., rainfall intensity, bed slope, and soil).

The results also show that an increase in bed slope causes: (1) an increase in soil loss for both upstream and downstream moving storms and (2) an increase in the relative differences between soil loss for identical storms moving, respectively, in the downstream and upstream directions.

This study contributes to improving the understanding of water erosion factors and processes. Future laboratory experiments will consider a wider range of conditions, including mainly the storm velocity, rainfall patterns, and storm lengths. The evolution of sediment transport during the runoff event should also be characterized.

Acknowledgments

This study was funded by the Foundation for Science and Technology (Research Project FCT – POCTI/35661/MGS/2000) of the Portuguese Ministry of Science and Technology, Lisbon, Portugal. Isabel Matias Barreira was funded under the referred research project. Some of the experiments reported in this study were conducted with the help of Mariano A. Lopes da Silva, student of the Civil Engineering Course of the Department of Civil Engineering, Faculty of Science and Technology, University of Coimbra, Portugal.

References

Bowyer-Bower T.A.S. and Burt, T.P. (1989). "Rainfall simulators for investigating soil response to rainfall". *Soil Technology*, 2, 1-16.

Bras, R. and Rodrigues-Iturbe, I. (1976). "Rainfall generation: A nonstationary time-varying multidimensional model". *Water Resources Research*, 12(3), 450-454.

Bryan, R.B. and De Ploey, J. (1983). "Comparability of soil erosion measurements with different laboratory rainfall simulators". In: De Ploey, J. (Ed.), *Rainfall Simulation, Runoff and Soil Erosion, Catena Supplement*, 4, 33-56.

Bryan, R.B. and Poesen, J. (1989). "Laboratory experiments on the influence of slope length on runoff, percolation and rill development". *Earth Surface Processes and Landforms*, 14, 211-231.

Eagleson, P.S. (1978). "Climate, soil and vegetation: The distribution of annual precipotation derived from observed storm sequences". *Water Resources Research*, 14(5), 713-721.

Hall, M.J. (1970). "Use of the stain method in determining the drop-size distributions of coarse liquid sprays". *Transactions of the American Society of Agricultural Engineer*, 41, 33-37.

Huff, F.A. (1967). "Time distribution of rainfall in heavy storms". *Water Resources Research*, 3(4), 1007-1019.

Lima, J.L.M.P. de (1997). "Modelação de simuladores de chuva com nebulizadores de cone preenchido". *Actas do 3º Simpósio de Hidráulica e Recursos Hídricos dos Países de Língua Oficial Portuguesa* (3º SILUSBA), APRH/AMCT/ABRH, 15 a 17 de Abril de 1997, Maputo, Moçambique, Vol. II, 10 pp. (in Portuguese)

Lima, M.I.P. de (1998). "Multifractals and the temporal structure of rainfall". Doctoral dissertation, Wageningen Agricultural University, Wageningen, The Netherlands.

Lima, J.L.M.P. de and V.P. Singh (1999). "The influence of storm movement on overland flow – Laboratory experiments under simulated rainfall". In: Hydrologic

Modeling (Eds. V.P. Singh, Il Won Seo and J.H. Sonu), *Water Resources Publications*, 101-111.

Maksimov, V.A. (1964). "Computing runoff produced by a heavy rainstorm with a moving center". *Sov. Hydrol.*, 5, 510-513.

Meyer, L.D. (1981). "How rainfall intensity affects interrill erosion". *Transactions of the American Society of Agricultural Engineer*, 24(6), 1472-1475.

Morgan, R.P.C. (1995). "*Soil Erosion and Conservation*". Longman (second edition), London.

Römkens, M.J.M, Prasad, S.N. and Gerits, J.J.P. (1997). "Soil erosion modes of sealing soils: a phenomenological study". *Soil Technology*, 11, 31-41.

Sharon, D. (1980). "The distribution of hydrologically effective rainfall incident on sloping ground". *Journal of Hydrology*, 46, 165-188.

Singh, V.P. (1998). "Effect of the direction of storm movement on planar flow". *Hydrological Processes*, 12, 147-170.

Sivapalan, M. and Wood, E.F. (1986). "A multidimensional model of nonstationary space-time rainfall at the catchment scale". *Water Resources Research*, 22(7), 1289-1299.

Yen, B.C. and Chow, V.T. (1968). *A study of surface runoff due to moving rainstorms*. Hydraulic Engineering Series No. 17, Department of Civil Engineering, University of Illinois, Urbana, USA.

Willems, P. (2001). "A spatial rainfall generator for small spatial scales". *Journal of Hydrology*, 252, 126-144.

Wilson, C.B., Valdes, J.B. and Rodrigues-Iturbe, I. (1979) "On the influence of the spatial distribution of rainfall on storm runoff". *Water Resources Research*, 15(2), 321-328.

Surface Water Hydrology, Singh, Al-Rashed & Sherif (eds)
© 2002 Swets & Zeitlinger, Lisse, ISBN 90 5809 363 8

Integration of SCS-CN and USL equations for determination of sediment yield

Surendra K. Mishra
National Institute of Hydrology, Roorkee-247 667, Uttaranchal, India

Vijay P. Singh and John J. Sansalone
Department of Civil and Environmental Engineering
Louisiana State University
Baton Rouge, LA 70803-6405, USA.
e-mail: cesing@lsu.edu

Abstract

For determination of the sediment yield, the Soil Conservation Service Curve Number (SCS-CN) method (SCS, 1956) is coupled with the universal soil loss equation (USLE). The coupling leads to an analytical method that use a new $C = S_r = DR$ concept, where C is the runoff coefficient, S_r is the degree of saturation, and DR is the sediment delivery ratio, and requires data on using rainfall and watershed characteristics. The method permits to interpret the potential maximum retention from USLE. An application of the method to rainfall-runoff-sediment data of a micro paved-watershed (area = 300 m^2) located in Cincinnati, Ohio, USA, yields satisfactory results.

Introduction

The Soil Conservation Service Curve Number (SCS-CN) method (SCS, 1956) and the universal soil loss equation (USLE) (Wischmeier and Smith, 1965) are widely used in hydrology and environmental engineering for computing, respectively, the amount of direct runoff from a given amount of rainfall and the potential soil erosion from small watersheds. A great deal of published material on these methods along with their applications is available in hydrologic literature. The texts of Ponce (1989), Singh (1992), Novotny and Olem (1994) are but a few examples. Since USLE was developed for estimation of the annual soil loss from small plots of an average length of 22 meters, its application to individual storm events and large areas is subject to errors. Its accuracy increases if it is coupled with a hydrological rainfall-excess model (Novotny and Olem, 1994). The current practice is to derive hydrologic information from a rainfall-runoff model and utilize it in computation of the potential erosion using USLE for determining the sediment yield which is important in watershed management. The work of Clark et al. (1985) is an authoritative treatise on the impact of erosion and sedimentation on environment in general, and water quality in particular.

SCS-CN Method

The SCS-CN method couples the water balance equation with two hypotheses which, respectively, are:

$$P = I_a + F + Q \tag{1}$$

$$\frac{Q}{P - I_a} = \frac{F}{S} \tag{2}$$

$$I_a = \lambda S \tag{3}$$

where P is the total rainfall, I_a is the initial abstraction, F is the cumulative infiltration, Q is the direct runoff, S is the potential maximum retention, and λ (= 0.2, a standard value) is the initial abstraction coefficient. Eq. 2 is a proportionality concept (Fig. 1). Combination of Eq. 2 and Eq. 1 leads to the SCS-CN method:

$$Q = \frac{(P - I_a)^2}{P - I_a + S} \tag{4}$$

which is valid for $P \geq I_a$, Q = 0, otherwise. Coupling of Eq. 4 with Eq. 3 for $\lambda = 0.2$ enables a determination of S from P-Q data. In practice, S is derived from a mapping equation expressed in terms of the curve number (CN):

$$S = \frac{1000}{CN} - 10 \tag{5}$$

where S is in inches. CN is derived from the tables given in the National Engineering Handbook (Section-4) (NEH-4) (SCS, 1956) for catchment characteristics such

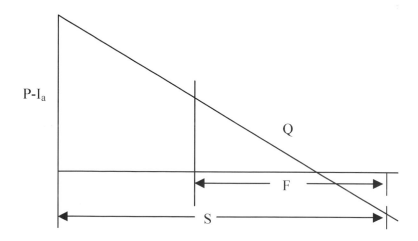

Figure 1. The proportionality concept.

as soil type, land use, hydrologic condition, and antecedent soil moisture condition. Thus, CN is a function of the catchment characteristics. The higher the CN value, the greater the runoff factor, C, or runoff potential of the watershed and vice versa.

Universal Soil Loss Equation

The universal soil loss equation (USLE) estimates the potential soil loss from upland (sheet and rill) erosion. Mathematically, it is expressed as (Wischmeier and Smith, 1965):

$$A = (R)(K)(L_s)(V_c)(M_p) \tag{6}$$

where A is the potential soil loss for a given storm or period (tonnes/ha), R is the rainfall factor, K is the soil erodibility factor, L_s is the slope-length factor, V_c is the vegetative cover (or cropping management) factor, M_p is the erosion control practice factor. Eq. 6 expresses the soil loss per unit area due to erosion by rainfall-runoff, excludes wind erosion, and does not yield direct sediment yield estimates. The rainfall factor, R, is computed as (Foster et al., 1977):

$$R = 0.5R_r + 7.5Qq^{1/3} \tag{7}$$

where R_r is the rainfall energy factor, Q is the runoff volume (cm), and q is the peak runoff rate (cm/hr). Since q is more related to detachment than Q (Williams and Berndt, 1977), a reduction in peak discharge by the vegetation cover will also reduce the sediment transport. The rainfall energy factor, R_r, is computed as

$$R_r = \Sigma[(2.29 + 1.15 \ln X_i) D_i] I \tag{8}$$

where i is the time interval of the rainfall hyetograph, D_i is the rainfall volume during i^{th} time interval (cm), I is the maximum 30 min rainfall intensity of the storm (cm/hr), and X_i is the rainfall intensity (cm/hr).

The soil erodibility factor, K, is a measure of potential erodibility of soil. It is a function of soil texture. The slope length factor, L_s, accounts for overland runoff length and slope. For slopes > 4%, it can be determined as:

$$L_s = L^{1/2} (0.0138 + 0.00974 Y + 0.001138 Y^2) \tag{9}$$

where Y is the percent slope over the runoff length and L is the length in meters from the point of origin of the overland flow to the point where the slope decreases to the extent that sedimentation begins. The vegetation cover factor or cropping management factor, V_c, estimates the effect of ground cover conditions, soil conditions, and general management practices on erosion rates. The erosion control practice factor, M_p, takes into account the effectiveness of erosion control practices such as land treatment by contouring, compacting, establishing sedimentation basins, and other control structures. Generally, V_c reflects protection of the soil surface against the impact of rain droplets and subsequent loss of soil particles, whereas M_p includes treatments that retain eroded particles and prevent them from

419

further transport. The experimentally derived values of the above factors for various soil-vegetation-land use complexes are available elsewhere (Singh, 1992; Ponce, 1989; Novotny and Olem, 1994).

Computation of Sediment Yield

The above computed potential erosion is used for determination of the sediment yield using the sediment delivery ratio, DR. Erosion is distinguished from the sediment yield in that the former represents the potential erosion that is taken equal to the sum of sheet (upland) and channel erosion and the latter refers to the sediment measured in the receiving body of water in a given time period. Vegetation dissipates rainfall energy, binds the soil, increases porosity by its root system, and reduces soil moisture by evapotranspiration to affect the sediment yield.

DR is a dimensionless ratio of the sediment yield (Y) to the total potential erosion (A) in the contributing watershed. Expressed mathematically,

$$DR = \frac{Y}{A} \tag{10}$$

Eq. 10 is used to compute the sediment yield. Thus, the delivery ratio acts as a scaling parameter, varying from 0–1. It generally decreases with the size of the basin (Roehl, 1962).

Novotny and Olem (1994) equated the nonpoint pollution with soil loss and calibrated for DR by minimizing the difference between the USLE-computed upland erosion and measured sediment yield estimates. According to Wolman (1977), the DR concept conceals a number of processes that contribute to temporal or permanent deposition of sediments in an eroding watershed, for these are highly variable, intermittent, and describable only statistically. The correlation of DR with the runoff coefficient (Novotny and Olem, 1994) indicates a significant effect of infiltration and other hydrologic losses on the magnitude of DR which is affected by rainfall impact, overland flow energy, vegetation, infiltration, depression and ponding storage, change of slope of overland flow, drainage, and so on (Novotny et al., 1979; Novotny, 1980; Novotny et al., 1986; Novotny and Chesters, 1989). Since these factors vary with time throughout the year, the sediment yield also varies with time. Walling and Kane (1983) showed that certain minerals of eroded sediment, such as Si, Al, Fe, Ti and K, do not vary in a certain range of hydrologic conditions, while others, particularly those associated with organic fractions, may vary significantly.

The soil texture determines both permeability and erodibility of soils. Permeability describes infiltration, which, in turn, determines hydrologic activeness of the soil surface in terms of both runoff generation and soil erosion. Erosion is primarily driven by surface runoff (Langbein and Schumm, 1958; Gottschalk, 1964; Leopold et al., 1964; Walling and Webb, 1983), if wind effects are ignored, and according to the SCS-CN method, the runoff generation is closely linked with infiltration. Thus, the processes of runoff generation and soil erosion are closely interrelated. However, the equations of SCS-CN method and USLE have not yet been investigated for their interrelationship. Thus, the objective of this paper is to integrate these equations for (a) better interpretation of the parameters of both the methods and (b) determination of sediment yield of a micro-experimental watershed (area = 300 m^2) located in Cincinnati, Ohio, USA.

Analytical Derivation

To integrate the SCS-CN method with USLE analytically it is necessary to first describe the SCS-CN method as a $C = S_r$ concept using a volumetric principle and then to signify the SCS-CN parameter S using USLE.

(a) C = S$_r$ Concept

For $I_a = 0$ (i.e. immediate ponding situation), the SCS-CN proportionality hypothesis (Eq. 2) equates the runoff factor C (= Q/P) to the degree of saturation, S_r, that is defined using the terms shown in Fig. 2 as:

$$S_r = \frac{F}{S} = \frac{V_w}{V_v} \tag{11}$$

which is valid for the completely dry antecedent moisture condition (AMC) of the soil, for which $V_a = V_v$. In Eq. 11, V_v is the void space and V_w is the space occupied by the infiltrated moisture. In Fig. 2, V_a is the air space that represents S and V_s is the volume of solids. Thus, the total volume V is the sum of V_w and V_s. These quantities can also be expressed in terms of depth for a unit surface area.

(b) Physical Significance of S

Eq. 6 computes the potential soil loss from a watershed in tonnes per hectare of the watershed. Since A is the potential mass of soil per unit watershed area which

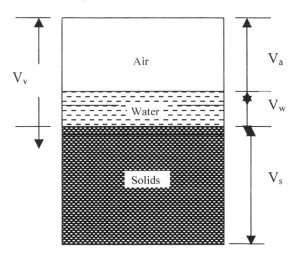

Figure 2. Schematic diagram showing soil-water-air.

421

can be observed at the outlet of the watershed, it is equal to the product of the density of solids, ρ_s, and the volume of solids per unit surface area, V_s. Expressed mathematically,

$$A = V_s\, \rho_s \tag{12}$$

Eq. 12 can also be expressed in terms of V and V_v as:

$$A = (V - V_v)\, \rho_s \tag{13}$$

Its division by S yields

$$\frac{A}{S} = \frac{(V - V_v)\rho_s}{V_a} \tag{14}$$

As above, for a completely dry soil, $V_a = V_v$. Thus, the substitution of V_v for V_a into Eq. 14 yields

$$\frac{A}{S} = \frac{(V - V_v)}{V_v}\;\rho_s = \frac{(1-n)}{n}\rho_s \tag{15}$$

where n is the soil porosity (dimensionless). It follows that

$$S = \frac{n}{(1-n)\rho_s}A \tag{16}$$

Coupling of Eq. 16 with Eq. 12 leads to a definition of S as

$$S = \frac{n}{(1-n)\rho_s}(R)(K)(L_s)(V_c)(M_p) \tag{17}$$

Eqs. 16 and 17 define S to be directly proportional to A and n and inversely proportional to ρ_s. It implies that S depends on all the factors affecting the potential erosion of the watershed; the higher the potential erosion from a watershed, the higher will be the value of S and vice versa. It is of common knowledge that highly porous soils (of high n), such as sands, yield high S-values whereas nonporous soils (of low n), such as loams and clays, yield low S-values. According to Eq. 16, heavy sediments (of high mass density) will yield low S-values and vice versa.

Eq. 17 can also be written in terms of the potentially erodible soil depth, V, as:

$$V = (1/\rho_s)(R)(K)(L_s)(V_c)(M_p) + S \tag{18}$$

where S holds for a completely dry soil or it can be taken equal to S of AMC I. Thus, USLE aids in computation of the potential erodible depth of the watershed. Since Eq. 16 also describes A in terms of the watershed characteristic explained by S and the soil type, 'A' can be better interpreted in terms of the potential maximum erodible depth of the watershed. Such an interpretation can also be supported as (Novotny and Olem, 1994): "For unconsolidated geological materials (soils, river deposits, sand dunes etc.), erodibility depends on particle size and texture of the material, water content, composition of the material, and the presence or absence of protective surface cover such as vegetation. Furthermore, loose soils with low chemical and clay content have highest erodibility."

(c) Coupling of SCS-CN Method with USLE

Similar to the SCS-CN proportional equality (or $C = S_r$) concept, it is possible to extend it for sediment yield as

$$C = S_r = DR \tag{19}$$

in which all variables range (0, 1). Using the usual definitions and $I_a = 0$, Eq. 19 can be expanded as:

$$\frac{Q}{P} = \frac{F}{S} = \frac{P}{P+S} = \frac{Y}{A} \tag{20}$$

Thus, Eq. 20 defines Y as

$$Y = C\,A \tag{21}$$

Eq. 21 implies that the sediment yield is directly proportional to potential maximum erosion 'A' and the runoff factor C is the proportionality constant. Alternatively,

$$Y = \frac{AP}{P+S} \tag{22}$$

For given watershed characteristics or A and S, the actual sediment yield Y increases with the rainfall amount, which is in conformity with the general notion that the higher the rainfall, the higher will be the sediment erosion and its transport and hence the higher the sediment yield, and vice versa. As $S \to 0$ (or $CN \to 100$), $Y \to A$ since $Q \to P$. Similarly, as $S \to \infty$ (or $CN \to 0$), $Y \to 0$ since $Q \to 0$. It is consistent with the general notion that the direct surface runoff (or surface water) primarily drives sediment erosion. Eq. 22 integrates the SCS-CN method with USLE and can be modified for the following elements of the rainfall-runoff-erosion process.

(i) Incorporation of I_a

The initial abstraction I_a can be incorporated in Eq. 22 as:

$$Y = \frac{(P-I_a)A}{P-I_a+S} \tag{23}$$

Taking $I_a = 0.2S$ which is a standard practice, Eq. 23 can be recast as

423

$$Y = \frac{[P - 0.2S]A}{P + 0.8S} \tag{24}$$

Eq. 24 infers that the sediment yield reduces with the increasing initial abstraction and vice versa.

(ii) Incorporation of Antecedent Moisture

For the incorporation of the antecedent moisture M which represents the amount of moisture in the soil profile before the start of a storm, the SCS-CN proportionality (Eq. 2) can be modified according to $C = S_r$ concept as:

$$\frac{Q}{P - I_a} = \frac{F + M}{S + M} \tag{25}$$

Combination of Eq. 25 with the water balance equation (Eq. 1) leads to

$$\frac{Q}{P - I_a} = \frac{P - I_a + M}{P - I_a + M + S} \tag{26}$$

Thus, similar to the derivation of Eq. 23, Y can be derived using Eq. 26 as:

$$Y = \frac{(P - I_a + M)A}{P - I_a + S + M} \tag{27}$$

It is noted that, according to the above volumetric concept, $S + M = V_a + V_w = V_v$. Thus, Eq. 17 is revised for M as

$$S = \frac{n}{(1 - n)\rho_s}(R)(K)(L_s)(V_c)(M_p) - M \tag{28}$$

Similar to Eq. 24, Eq. 28 can also be expressed as:

$$Y = \frac{(P - 0.2S + M)A}{P + 0.8S + M} \tag{29}$$

Eq. 29 infers that the sediment yield will increase with the increase in the antecedent moisture amount, M, and vice versa.

(iii) Incorporation of initial flush

Similar to the concept of initial abstraction, the concept of initial flush, I_f, is quite popular in environmental engineering (Foster and Charlesworth, 1996; Sansalone and Buchberger, 1997). The initial abstraction is a loss of water primarily due to evaporation and it does not contribute to runoff. On the other hand, the initial flush is not a loss of sediments, rather it appears at the outlet of the watershed. For the most part, it is contributed by the initial runoff generated at the start of rainfall, after meeting the initial abstraction requirements. To incorporate I_f in Eq. 29 it is necessary to redefine the sediment delivery ratio, DR, as

$$DR = \frac{Y - I_f}{A - I_f} \tag{30}$$

For simplicity, following the second SCS-CN hypothesis (Eq. 3), I_f can also be related to the potential maximum erosion, A, as

$$I_f = \lambda_1 A \tag{31}$$

Combination of Eq. 31 with Eq. 30 yields

$$DR = \frac{Y - \lambda_1 A}{(1 - \lambda_1)A} \tag{32}$$

Thus, as above, Eq. 29 can be further modified for the initial flush as

$$Y = \left[\frac{(1 - \lambda_1)[P - 0.2S + M]}{P + 0.8S + M} + \lambda_1 \right] A \tag{33}$$

Here, it is possible to show that, for M = 0, as $S \to 0$, $Y \to A$ and as $S \to \infty$, $Y \to 0$, for $Y \geq 0$.

As shown by Novotny and Olem (1994) using field data, there exists a power relation between C and DR:

$$DR = \alpha C^\beta \tag{34}$$

where α and β are respectively the coefficient and exponent of the power relation. The relationship depicted by Novotny and Olem (1994) for a particular data set yields $\alpha = 0.97$ and $\beta = 1.03$. These values suggest that the above derived relations assuming C = DR should generally hold, for both the parameters are approximately equal to 1. However, for completeness, a relationship for determining the sediment yield, Y, using the power relation (Eq. 34) is given as:

$$Y = \left\{ (1 - \lambda_1)\alpha \left[\frac{P - 0.2S + M}{P + 0.8S + M} \right]^\beta + \lambda_1 \right\} A \tag{35}$$

This is the general form of the model for computation of the sediment yield from rainfall amount and catchment characteristics.

Application

Study Area

Rainfall-runoff-water quality data (Table 1) were collected during 1995–1997 by Sansalone and Buchberger (1997) on a 15 x 20 m asphalt pavement at milestone 2.6 of I-75 that is a major north-south interstate in Cincinnati, Ohio, U.S.A. The details of the site are available elsewhere (Sansalone and Buchberger, 1997). The runoff from the selected stretch is contributed by the four southbound lanes, an exit lane, and a paved shoulder, all draining to a grassy v-section median at a transverse pavement cross-slope of 0.020 m/m. The runoff from the highway site (longitudinal slope = 0.004) finally drains to Mill Creek. The flow of the highway is primarily characterized by sheet flow, and the land use as urban (industrial, commercial, and residential) (Sansalone and Buchberger, 1997). The storm water runoff diverted through the epoxy-coated converging slab, a 2.54 cm diameter.

Parshall flume, and a 2 m long 25.4 cm diameter PVC pipe to a 2000 liter storage tank, was measured at a regular 1 min interval using an automated 24 bottle sampler with polypropylene bottles. Rainfall was recorded in increments of 0.254 mm using a tipping bucket gauge. Cincinnati receives an annual average 1020 mm of rainfall and 420 mm of snow. High rainfall occurs in March (average = 106 mm) and July (average = 104 mm) and highest snowfall occurs in January (average = 150 mm) (Soil, 1982). The average winter and summer temperatures are 0 °C and 23.4 °C, respectively. The water quality data at every 2 min interval were collected during rainfall-runoff events at the experimental site and samples were analyzed for dissolved and particulate bound metals for several rainfall-runoff events (Sansalone and Buchberger, 1997; Sansalone et al., 1998; and Li et al., 1999). It is worth emphasizing that for each discrete sample, dissolved and particulate heavy metal concentrations were obtained after sample preparation and digestion through Inductively Coupled Plasma-Atomic Emission Spectroscopy (ICP-AES) analysis.

Table 1. Rainfall-runoff-sediment yield data of Cincinnati watershed.

Date	Rainfall duration (min)	Rainfall depth (mm)	Runoff depth (mm)	IPRT (min)	TDS mass (g)	TSS mass (g)	TS mass (g)
Apr. 8, 1995	295	25.0	18.13	8	235.5	688.8	924.3
Apr. 30, 1995	109	1.0	0.70	10	43.9	17.9	61.8
Jul. 15, 1995	16	0.4	0.14	10	10.8	10.6	21.4
Sept. 8, 1995	202	4.0	1.34	13	132.8	66.8	199.6
Oct. 3, 1995	393	8.5	4.20	11	113.3	172.3	285.6
May 21, 1996	35	0.9	0.32	4	34.8	17.5	52.3
Jun. 18, 1996	63	11.3	9.26	5	104.8	257.7	362.5
Jul. 7, 1996	50	40.4	32.15	4	210.5	275.6	454.9
Aug. 8, 1996	51	14.1	12.92	7	166.8	637.5	804.3
Oct. 17, 1996	616	29.1	12.31	5	454.1	400.5	854.6
Nov. 25, 1996	150	3.1	0.72	8	47.5	39.7	87.2
Dec. 16, 1996	340	3.4	0.89	14	59.6	37.9	97.4
Jun. 12, 1997	20	2.0	1.55	3	60.2	32.9	93.1

Notations: IPRT = initial pavement residence time, TDS = total dissolved solids, TSS = total suspended solids, and TS = total solids = TDS + TSS = sediment yield.

Each dissolved or particulate sample concentration at time i, c_i, was multiplied by the runoff volume during 1 min. interval, v_i, to determine solid mass (m_i) as: $m_i = c_i v_i$. In Table 1, total solids (TS) is the sum of dissolved (DS) and suspended (TSS) solids and is taken to represent the sediment yield in the forgoing analysis.

Results and Discussion

Based on the above analytical development for determination of the sediment yield, nine models can be formulated as shown in Table 2. Model 1 excludes the initial abstraction I_a, antecedent moisture M, and initial flush I_f components. Model 2 accounts only for initial abstraction with $\lambda = 0.2$ (a standard value) and in Model 3, λ is allowed to vary. Model 4 accounts for both initial abstraction and antecedent moisture but allows the variation of λ. Model 5 is distinguished from Model 4 for $\lambda = 0.2$. Models 6 and 7 include all I_a, M, and I_f but the former assumes $\lambda = 0.2$. Models 8 and 9 correspond to Eq. 34 with the difference that the former assumes a constant value of λ (= 0.2) and the latter allows it to vary.

These models were applied to the data of the Cincinnati watershed (Table 1) and their performance is evaluated using the Nash and Sutcliffe (1970) efficiency computed in percent as:

$$\text{Efficiency} = (1 - D_1/D_u) \times 100 \tag{36}$$

where D_1 is the sum of the squares of deviations between computed and observed data:

$$D_1 = \Sigma \ (Y_o - \hat{Y})^2 \tag{37}$$

Table 2. Simulation results of various model formulations.

Model	Model formulation for sediment yield	Parameters							Eff.[a] (%)	Eff.[b] (%)	A/S ratio (g/mm)
		S (mm)	λ	λ₁	A (g)	M (mm)	α	β			
1	$Y = \dfrac{AP}{P+S}$	15.10	-	-	1090	-	-	-	75.79	88.93	72.19
2	$Y = \dfrac{A(P-0.2S)}{P+0.8S}$	8.81	-	-	961	-	-	-	77.98	88.26	109.08
3	$Y = \dfrac{A(P-\lambda S)}{P+(1-\lambda)S}$	8.15	0.23	-	942	-	-	-	70.00	88.14	115.59
4	$Y = \dfrac{A[P-\lambda S+M]}{P+(1-\lambda)S+M}$	8.23	0.25	-	944	-	0.21	-	78.02	88.12	111.85
5	$Y = \dfrac{A[P-0.2S+M]}{P+0.8S+M}$	8.72	-	-	961	-	0.0	-	77.98	88.27	110.21
6	$Y = \left[\dfrac{(1-\lambda_1)[P-0.2S+M]}{P+0.8S+M}+\lambda_1\right]A$	8.64	-	0.01	947	-	0.0	-	77.80	88.27	109.61
7	$Y = \left[\dfrac{(1-\lambda_1)[P-\lambda S+M]}{P+(1-\lambda)S+M}+\lambda_1\right]A$	7.45	0.37	0.05	909	-	0.51	-	77.92	87.70	114.20
8	$Y = \left\{(1-\lambda_1)\alpha\left[\dfrac{P-0.2S+M]}{P+0.8S+M}\right]^\beta+\lambda_1\right\}A$	4.05	-	0.03	929	0.0	1.02	2.0	78.56	82.85	229.38
9	$Y = \left\{(1-\lambda_1)\alpha\left[\dfrac{P-\lambda S+M]}{P+(1-\lambda)S+M}\right]^\beta+\lambda_1\right\}A$	2.43	0.68	0.07	913	0.19	1.01	3.0	79.51	81.18	348.47

Note: Superscript 'a' and 'b' stand for efficiency in computation of sediment yield and direct runoff, respectively.

and D_o is the model variance, which is the sum of the squares of deviations of the observed data about the observed mean:

$$D_o = \Sigma \ (Y_o - \overline{Y} \ )^2 \tag{38}$$

where Y_o is the observed data, Y with '$\wedge$' and '–' (bar) stand for computed data and mean of the observed data, respectively. The efficiency varies on the scale of 0 to 100. It can also assume a negative value if $D_1 > D_o$ implying that the variance in the observed and computed data values is greater than the model variance. In such a case, the mean of the observed data fits better than the model. The efficiency of 100 implies that the computed values are in perfect agreement with the observed data.

The parameters of the aforementioned models were estimated using the non-linear Marquardt algorithm of the least squares procedure of the Statistical Analysis System (SAS, 1988). The parameters were initially set as zero in all cases and the lower and upper limits were decided by trial and error. If the computed value of a parameter in a run did not fall in the prescribed range, the limit was extended accordingly for the next run. If the subsequent runs produced the estimate within the prescribed range, the parameter estimate was assumed to be optimal globally. Approximately, three to five runs or a few more in some cases were required to compute the final estimates of model parameters.

The estimated parameters and the efficiencies resulting from computation of the sediment yield are shown in the third last column of Table 2. Here, it is noted that Models 1–9 determine the sediment yield using rainfall P and ignore a direct involvement of the observed runoff. In application of these models, variables A and M were also taken as parameters due to lack of their observations. It is seen from the table that parameter S varies from 2.43 to 15.10 mm or CN (Eq. 5) varies from 94 to 99, λ from 0.23 to 0.68, λ_1 from 0.01 to 0.07, A from 909 to 1090 grams, M from 0.0 to 0.51 mm, α from 1.01 to 1.02, and β from 2 to 3. Clearly, the parameters vary in a narrow range and, therefore, support their credibility. The estimated high values of CN exhibit a high runoff potential of the watershed, which is consistent with the paved nature of the Cincinnati watershed. The resulting efficiency varies from 75.79 to 79.51%, which is quite high to indicate a satisfactory performance of the models in computing the sediment yield from the Cincinnati watershed.

Model 1 is the simplest of all the above 9 models and the resulting efficiency of 75.79% is indicative of (a) the dependence of the sediment yield on the SCS-CN-generated runoff and (b) the applicability of $C = S_r = DR$ concept (Eq. 19). It is further supported by the efficiency (= 77.98%) of Model 2 that is directly based on the existing SCS-CN method (Eq. 4 with $I_a = 0.2S$). If λ is allowed to vary as in Model 3, the efficiency improves to 78.03%. It is noted that λ in Model 3 is estimated as 0.23 which is well within the range of (0.1, 0.3) (Chen, 1982; SCD, 1972). The incorporation of the antecedent moisture M, as in Models 4 and 5, marginally reduces the resulting efficiency. It is because of its variation from one rainfall event to another. The assumption of constant M is, therefore, not appropriate. The inclusion of the concept of initial flush (Models 6 and 7) and allowing the variation of λ in optimization (Model 7) slightly reduces the efficiency to 77.92%. This reduction actually undermines the validity of Eq. 31 and, therefore, requires further improvement in future. However, a high magnitude of efficiency supports the general applicability of the concept. The efficiency is improved to 78.56% (Model 8) and 79.51% (Model 9) if DR is allowed to vary nonlinearly with the runoff factor C (Eq. 34). The sediment yield computed using Models 1–9 is compared with the observed value in Fig. 3. The closeness of data points to the line of perfect fit (LPF) indicates a satisfactory model performance.

The above evaluation, however, excludes the observed runoff, Q, (Table 1) in estimation of the sediment yield. In practice, the sediment yield from upland areas is generally correlated with the observed runoff (Singh and Chen, 1983), rather than the rainfall. Therefore, to further support the validity of the coupling of the SCS-CN method with USLE, it is in order to check the credibility of the estimated values of the parameters by computing the direct surface runoff using these values and comparing the computed runoff with the observed runoff.

For computation of the direct runoff, SCS-CN-based models, designated as Models R1–R9 corresponding to Models 1–9, respectively, were used. Models R1–R3 are equivalent to Eq. 4 with $I_a = 0$, $I_a = 0.2S$, and $I_a = \lambda S$, respectively. Models R4 and R5 are equivalent to Eq. 26 with $I_a = \lambda S$ and $I_a = 0.2S$, respectively. Models R4, R7, and R7 are the same in formulation, and so are Models R5, R6, and R8.

Using the parameters of Table 2, the direct runoff was computed and the resulting model efficiencies are shown in the second last column of Table 2. It is seen that Model R1 that is the simplest in form exhibits the highest efficiency (= 88.93%) and the most complicated Model R9 shows the lowest efficiency (= 81.18%). Fig. 4 compares graphically the computed runoff with the observed runoff. Although these efficiencies are reasonably high to adopt any model formulation for determination of the sediment yield and direct runoff, a more appropriate model based on the computed efficiencies can be chosen. As a trade-off, Models 2–7 for determination of the sediment yield and the corresponding runoff Models R2–R7 appear to be more suitable for adoption, for their efficiencies in computation of sediment yield vary from 77.80 to 78.03%, and these vary from 88.12 to 88.27% in runoff computation. Investigations in future are, however, necessary for checking the validity of Eq. 31.

It is also appropriate to investigate the estimated parameter values for A/S ratio which, according to Eq. 15, should be constant for a watershed. The computed values of this ratio are given in the last column of Table 2. Apparently, all the recommended Models 2–7 yield A/S ratio in the range of (109.08, 115.58) g/mm, which is quite narrow to assume a constant value of this ratio as 111.76 g/mm, an average value, for the Cincinnati watershed. Determination of A/S ratio for various watersheds may prove to be significant in future for estimating the sediment yield using NEH-4 tables.

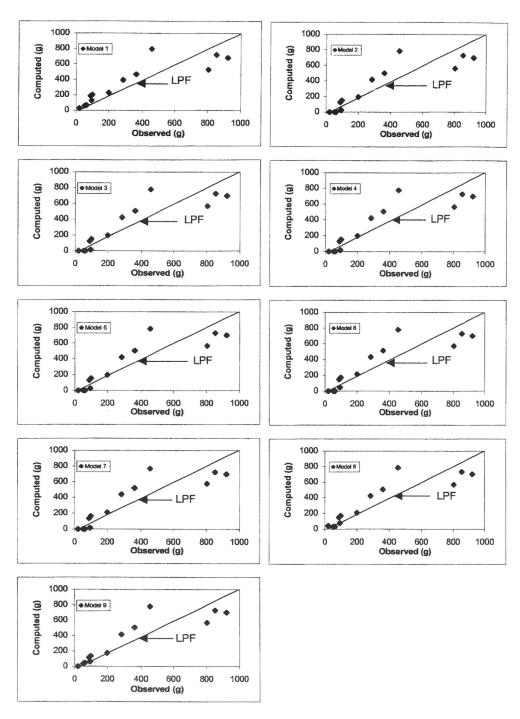

Figure 3. Observed and computed sediment yield (g) using models 1–9 (Table 2).
LPF stands for the line of perfect fit.

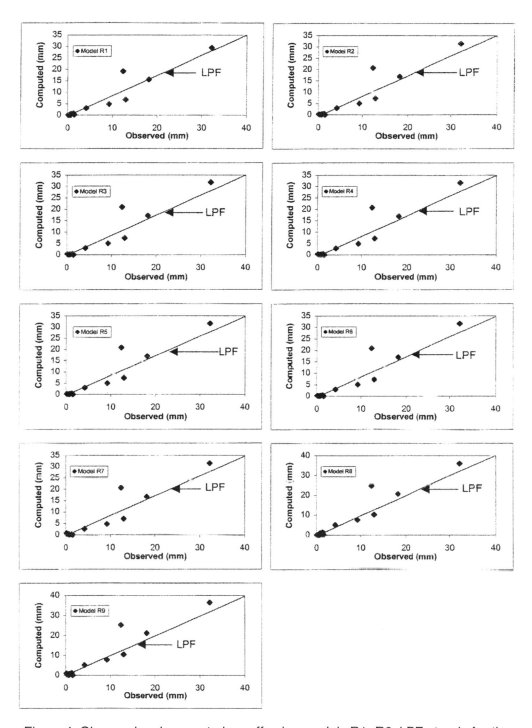

Figure 4. Observed and computed runoff using models R1–R9. LPF stands for the line of perfect fit.

Conclusions

The following conclusions are drawn from the study:

1. The introduced $C = S_r = DR$ concept enables a determination of the direct runoff and sediment yield using rainfall data and watershed characteristics.
2. The SCS-CN parameter potential maximum retention, S, can be described in terms of the components of USLE and the potential erodible depth can be physically determined from USLE and S.
3. In application to the rainfall-runoff-sediment yield data of the Cincinnati watershed, the recommended Models 2–7 compute the sediment yield with efficiencies ranging from 77.80 to 78.03% and Models R2-R7 determine the direct runoff with efficiencies varying from 87.70 to 88.27%. These values suggest a satisfactory performance of the models.
4. The average value of A/S ratio for the Cincinnati watershed is equal to 111.76 g/mm. Determination of this ratio for watersheds of varying complexity may be of significance in future for estimating the sediment yield using NEH-4 tables.

References

Chen Cheng-lung (1982). "An evaluation of the mathematics and physical significance of the soil conservation service curve number procedure for estimating runoff volume". In: Rainfall-Runoff Relationship, edited by V.P. Singh, Water Resour. Pub., Littleton, Colo. 80161.

Clark, E.H., J.A. Haverkamp and W. Chapman (1985). "Eroding soils-The Off-Farm Impacts". The Conservation Foundation, Washington D.C.

Foster, I.D.L. and S.M. Charlesworth (1996). "Heavy metals in the hydrological cycle: Trends and explanation". Hydrological Processes, 10, 227-261.

Foster, G.R., L.D. Meyer and C.A. Onstad (1977). "An erosion equation devised from basic erosion principles". Trans. ASAE, 20, 678-682.

Gottschalk, L.C. (1964). "Sedimentation, Part I: Reservoir sedimentation". In: Handbook of Applied Hydrology, V.T. Cow (ed), McGraw Hill, New York.

Langbein, W.B. and S.A. Schumm (1958). "Yield of sediment in relation to mean annual precipitation".Trans. Am. Geophys. Union, 39, 1076-84.

Leopold, L.B., M.G. Wolman and J.P. Miller (1964). *Fluvial Processes in Geomorphology*. W.H. Freeman, San Francisco.

Li, Y., S.G. Buchberger and J.J. Sansalone (1999). "Variably saturated flow in storm-water partial exfiltration trench". J. Environmental Engineering, ASCE, 125, 556-565.

Nash J.E. and J.V. Sutcliffe (1970). "River flow forecasting through conceptual models, Part I-A discussion of principles". J. Hydrology, 10, 282-290.

Novotny, V. (1980). "Delivery of suspended sediment and pollutatnts from non-point sources during overland flow". Water resources Bull., 16 (6), 1057-65.

Novotny, V. and H. Olem (1994). *Water Quality: Prevention, Identification, and Management of Diffuse Pollution*. John Wiley & Sons, New York, N.Y.

Novotny, V. et al. (1979). "Simulation of Pollutant Loadings and Runoff Quality". EPA 905/4-79-029, U.S. Environmental Protection Agency, Chicago, IL.

Novotny, V. and G. Chesters (1989). *Handbook of Nonpoint Pollution: Sources and Management*. Van Nostrand Reinhold, New York.

Novotny, V., G.V. Simsima and G. Chesters (1986). "Delivery of pollutants from nonpoint sources". In: Proc., Symp. on Drainage Basin Sediment Delivery, Int. Assoc. Hydrol. Sci., Publ. No. 159, Wallingford, England.

Ponce, V.M. (1989). *Engineering Hydrology: Principles and Practices*. Prentice Hall, Englewood Cliffs, New Jersey.

Roehl, J.W. (1962). "Sediment Source Areas, Delivery Ratios, and Influencing Morphological Factors". Publ. No. 59, Int. Assoc. Hydrol. Sci., pp. 202-213.

SCD (1972). *Handbook of Hydrology*. Soil Conservation Department, Ministry of Agriculture, New Delhi.

SCS (1956). *National Engineering Handbook, Hydrology*. Section 4, Soil Conservation Service, U.S. Dept. of Agriculture, Washington D.C.

Sansalone, J.J. and S.G. Buchberger (1997). "Partitioning and first flush of metals in urban roadway storm water". J. Env. Engrg., ASCE, 123(2), 134-143.

Sansalone, J.J., J.M. Koran, J.A. Smithson and S.G. Buchberger (1998). "Physical characteristics of urban roadway solids transported during rain events". J. Environmental Engineering, ASCE, 125(6), 556-565.

SAS (1988). *SAS/STAT User's Guide*. Release 6.03 Edition, Statistical Analysis System Institute Inc., SAS Circle, Box 8000, Cary, NC 27512-8000.

Singh, V.P. (1992). *Elementary Hydrology*. Prentice Hall, Englewood Cliffs, New Jersey.

Singh, V.P. and V.J. Chen (1983). "The relationship between storm runoff and sediment yield". Tech. Rep. 3, Water Resources Program, Dept. of Civil and Environmental Engineering, Louisiana State University, Baton Rouge, LA 70803.

Soil survey of Hamilton County (1982). U.S. Dept. of Agri., Washington, D.C., pp. 1-220.

Walling, D.E. and P. Kane (1983). "Temporal variation of suspended sediment properties". Proc. of Exter Symp. on Recent Developments in the Explanation and Prediction of Erosion and Sediment Yield, IAHS Pub. No. 137, Washington D.C., pp. 409-419.

Walling, D.E. and B.W. Webb (1983). "Patterns of sediment yield". In: Background to Paleohydrology, K.J. Gregory (ed.), Wiley, New York.

Williams, J.R. and H.D. Berndt (1977). "Sediment yield prediction based on watershed hydrology". Trans. ASAE 20, 1100-04.

Wischmeier, W.H. and D.D. Smith (1965). "Predicting Rainfall-Erosion Losses from Cropland East of Rocky Mountains". USDA Agricultural Handbook No. 282, Washington, D.C.

Wolman, M.G. (1977). "Changing needs and opportunities in the sediment field". Water Resources Res., 13, 50-54.

Surface Water Hydrology, Singh, Al-Rashed & Sherif (eds)
© *2002 Swets & Zeitlinger, Lisse, ISBN 90 5809 363 8*

Estimating the cross-sectional mean velocity in natural channels by the entropy approach

T. Moramarco
Research Institute for Hydrogeological Protection in Central Italy, CNR
Via M. Alta 126, 06128 Perugia, Italy
e-mail: T.Moramarco@irpi.pg.cnr.it

C. Saltalippi
Department of Civil and Environmental Engineering, University of Perugia
Via G. Duranti 93, 06125 Perugia, Italy
e-mail: saltalip@unipg.it

V.P. Singh
Department of Civil and Environmental Engineering
Louisiana State University
Baton Rouge, LA 70803-6405, USA
e-mail: cesing@lsu.edu

Abstract

The applicability of the linear relationship between the mean flow velocity and the maximum flow velocity, based on the entropy concept (Chiu, 1991), is investigated using the data collected during 20 years in three gauged river sections in the upper Tiber River basin in Central Italy. The error in estimating the cross-sectional mean velocity from the observed maximum velocity is analyzed and is found to be normally distributed. The mean value of the error is very close to zero and it does not exceed 0.013 ms^{-1}, while the maximum value of the standard deviation is about 0.07 ms^{-1}. Furthermore, since it is difficult to measure velocity at too many points in the flow cross-sectional area during high floods, a simple method is proposed for constructing the velocity profiles at a river section. The method is based on a velocity distribution law derived by Chiu (1988) using the principle of maximum entropy for wide open-channels. It is assumed that the mean and maximum velocities within different portions of the flow section have a parabolic shape which is easily derived by three simple constraints. Comparing the reconstructed velocity profiles with velocity points measured along verticals, the method was found capable of estimating with a reasonable accuracy the shape of the observed velocity profiles of high flood events simply by knowing the flow area and the maximum velocity of the cross-section along with the dimensionless entropy parameter.

Introduction

One of the basic properties of an open channel flow is the velocity distribution in a river cross section. This aspect has been also studied through a probabilistic approach based on the entropy concept (Chiu, 1991; Barbé et al., 1991). Considering the probabilistic formulation the mean velocity, $\bar{u}$, can be expressed as a linear function of the maximum velocity, u_{max}, through a dimensionless entropy parameter M (Chiu and Said, 1995). The M value is a fundamental information measure about the characteristics of the channel section such as changes in bed form, slope and geometric shape (Chiu and Murray, 1992) and it might be derived from the pairs of u_{max} and $\bar{u}$ measured at a channel section. Xia (1997) investigated the relation between mean and maximum velocities by using the velocity data collected in some river sections of the Mississippi river, and he found that the relation was perfectly linear both along straight reaches and river bends (Xia, 1997). The M value was constant and equal to 2.45 and 5 on the straight reaches and on river bends, respectively.

Although Xia's results were preliminary because of the limited amount of data, the relation between mean and maximum velocity, as introduced by Chiu (1991), if tested on others natural channels, might be useful for investigating the flood process in rivers (Moramarco and Singh, 2001). This relationship was introduced by Chiu (Chiu and Chiou, 1986) who derived the velocity spatial distribution equations for open channel flow. For this reason, Chiu (1987) also developed a geometrical technique for modeling the spatial distribution of velocity. However, this technique, based on the isovel curvilinear coordinates, uses the Cartesian coordinate transformation requiring the estimation, not always simple, of about six parameters. The objective of this paper is to analyze the accuracy of the relationship between the mean velocity and the maximum velocity by using the data collected during 20 years in three gauged river sections in the upper Tiber basin, in Central Italy. The probability distribution of the errors in estimating the mean velocity from the measurement of the maximum velocity is investigated. Besides, a simple method is proposed for modeling the spatial distribution of velocity at a river section. The method can also be conveniently adopted during high floods when it is difficult to carry out velocity measurements at too many points in the flow cross-sectional area.

Relation Between Mean and Maximum Velocity

To test the relationship between mean and maximum velocity in a channel cross section, the velocity data collected during 20 years of observations were analyzed for three gauged sections located at 137.4 km (P. Nuovo) along the Tiber River, at 82 km (Rosciano) on the Chiascio River, tributary of the Tiber River, and at 61 km (Bettona) on the Topino River, tributary of the Chiascio River. The number of velocity measurements and the flow characteristics are summarized in Table 1. The selected sections are equipped with remote ultrasonic water level gauge, while the measurements of velocity were made by current meter from cableways. By comparing the point velocities sampled along different verticals and applying the well-known velocity-area method, the maximum and mean velocities were estimated (Chiu and Said, 1995; Xia, 1997). Topographical surveys of the instrumented sections were also available in different years of sampling.

Table 1. Flow characteristics, discharge, Q, and water depth, D, of the available velocity measurements, N, for the three gauged sections in the upper Tiber River basin.

Location	N	Q $(m^3 s^{-1})$	D (m)
P. Nuovo	54	2÷537	1.1÷6.7
Rosciano	38	3÷160	1.3÷3.3
Bettona	48	3÷250	0.6÷4.7

The relations between the mean velocity, $\bar{u}$, and the maximum velocity, u_{max}, in the river cross sections at P. Nuovo, Rosciano and Bettona are shown in Fig. 1. The relations are perfectly linear and can be expressed in Chiu's form as:

$$\bar{u} = \Phi(M)\, u_{max} \tag{1}$$

in which

$$\Phi(M) = \left(\frac{e^M}{e^M - 1} - \frac{1}{M} \right) \tag{2}$$

and M is the entropy parameter. Eq. (1) shows that $\bar{u}$ and u_{max} together can determine $\Phi(M)$ and then the entropy parameter M. Through the best-fit line the values of $\Phi(M)$ and M were estimated for the three gauged sections as also shown in Fig. 1, wherein is also reported the correlation coefficient, R^2, relative to Eq. (1).

Error Analysis

The accuracy of the linear relationship between the mean and maximum velocities for the three gauged sections is investigated by evaluating the errors ε and $\varepsilon\%$ defined as follows:

$$\bar{u} = \bar{u}_{bf} + \varepsilon \tag{3}$$

$$\varepsilon\% = \frac{\bar{u}_{bf} - \bar{u}}{\bar{u}} \cdot 100 \tag{4}$$

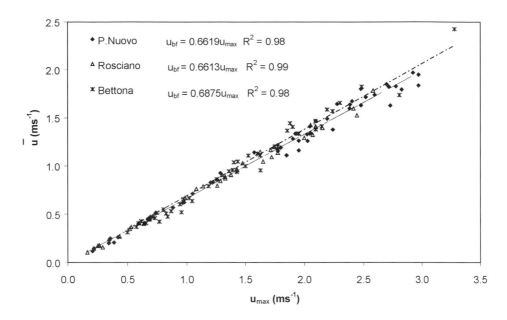

Figure 1. Relation between mean and maximum velocities in three gauged river sections.

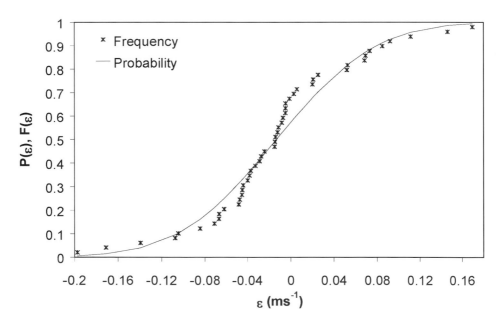

Figure 2. Cumulative frequency (F) and normal probability (P) of the error defined by Eq. (3) at Bettona.

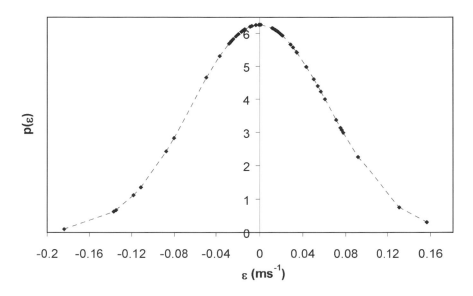

Figure 3. Normal probability density function of the error at the P. Nuovo section estimated by Eq. (3).

in which $\bar{u}_{bf}$ is the mean velocity calculated by the linear relationship shown in Fig. 1.

Both the errors ε and $\varepsilon\%$ at the three river sections are distributed according to a normal distribution (see Figs. 2 and 3).

For each sample, constituted by the pairs of $\bar{u}$ and u_{max} for the three gauged sections, the parameters for the distribution can be estimated from the mean and the standard deviation and additional probabilistic information might be carried out.

The expected value of the error ε corresponding to 0.5, 0.8 or 0.9 probability for each section is given in Table 2, wherein are also reported the mean, $\mu(\varepsilon)$, and standard deviation, $\sigma(\varepsilon)$.

A similar analysis was done for the error defined by Eq. (4) and it was found that the mean did not exceed 4%, while the maximum value of the standard deviation was about 9%.

Velocity Distribution Based on the Entropy Concept

The velocity distribution developed by Chiu (1988) uses the orthogonal curvilinear coordinate system (ξ, η) and is based on the principle of maximum entropy:

439

Table 2. Mean error $\mu(\varepsilon)$, standard deviation $\sigma(\varepsilon)$ of the error and error corresponding to 0.5, 0.8 and 0.9 cumulative probability for the three gauged river sections.

Location	$\mu(\varepsilon)$ (ms^{-1})	$\sigma(\varepsilon)$ (ms^{-1})	ε (ms^{-1})		
			P = 0.5	P = 0.8	P = 0.9
P. Nuovo	$8.89 \cdot 10^{-5}$	0.064	0.043	0.081	0.105
Rosciano	$6.5 \cdot 10^{-4}$	0.037	0.025	0.047	0.061
Bettona	-0.013	0.072	0.050	0.094	0.121

$$u = \frac{u_{max}}{M} \ln\left(1 + \left(e^M - 1\right) \cdot \frac{\xi - \xi_0}{\xi_{max} - \xi_0}\right) \qquad (5)$$

where u is the horizontal velocity and ξ is the curvilinear coordinate which corresponds to the isovels (Chiu, 1989). The Eq. (5) involves six parameters and we refer the reader to Chiu's work for more complete treatment (Chiu, 1988; 1989; 1995). Eq. (5) was applied for reconstructing the velocity profiles sampled in the cross-sectional flow area during a flood event that occurred on June 1997 at the P. Nuovo section and whose measured discharge was 500 m^{3}s^{-1}. The parameters were estimated by using a nonlinear regression technique (IMSL STAT/LIBRARY; Fortran Subroutine for Statistical Analysis; Version 1.1, 1989) and the velocity points sampled along different verticals were considered as benchmarks. Figure 4 shows the velocity profiles estimated by Eq. (5) with the velocity points sampled along three different verticals. The vertical location where the maximum velocity sampled occurs is indicated with x = 0, whereas x = 20m and x = −16m represent the horizontal distance in meters of the other verticals from that at x = 0.

As can be seen, Eq. (5) performs better in the middle portion of the flow area, whereas in the regions close to the side walls the velocity is poorly estimated. As Eq. (5) involves six or seven parameters (Chiu, 1989), its solution is not always simple and besides the results could be less than accurate in some portions of the flow area, as before shown, a practical method was developed for estimating the velocity profiles starting from the measurement of the maximum velocity. It was assumed that Eq. (5) written for a wide-open channel can be applied locally for reconstructing the velocity profiles along each vertical (Chiu and Said, 1995; Greco, 1999):

$$u_i = \frac{u_{max_i}}{M} \ln\left(1 + \left(e^M - 1\right)\frac{y}{D_i - h}\exp\left(1 - \frac{y}{D_i - h}\right)\right) \qquad i = 1, 2, \ldots, N_v \qquad (6)$$

with u_i and D_i are the horizontal velocity and water depth along the i-th vertical, respectively; y is the vertical distance measured from the channel bed and h is the depth below the water surface where the maximum velocity sampled, u_{max}, occurs; N_v is the number of verticals sampled in the cross-sectional flow area; u_{max_i} is referred to as the maximum of the velocity points value along each vertical.

For investigating the sensitivity of Eq. (6) to the entropy parameter M, three tests were carried out. The first assumes that the M value is constant along each vertical and its value is given from the best-fit line shown in Eq. (1). In the second test the M value is variable along each vertical as:

$$\Phi(M_i) = \frac{\bar{u}_i}{u_{max}} \qquad (7)$$

where $\bar{u}$ is the mean velocity along the i-th vertical.

In the last test, considering Eq. (2), M can be defined through Eq. (7) but u_{max} is referred to as u_{max_i}. Two velocity data sets collected during the flood events that occurred on June 1997 and on November 1996 at the P. Nuovo gauged section were used for the investigation. For the latter the measured discharge event was 550 m^3s^{-1}. Figure 5 shows, for the tests investigated, the comparison between the velocity profiles simulated through Eq. (6) and the velocity points sampled along

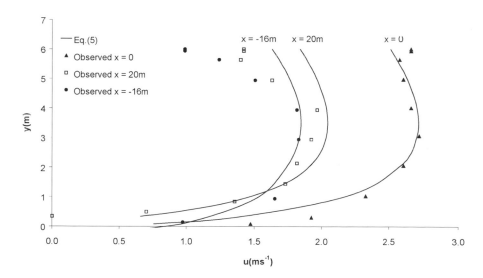

Figure 4. Velocity profiles estimated by Eq. (5) plotted against velocity points sampled along verticals at the P. Nuovo gauged section during the flood event that occurred on June 1997; x is the horizontal distance from the vertical where u_{max} occurs (x = 0).

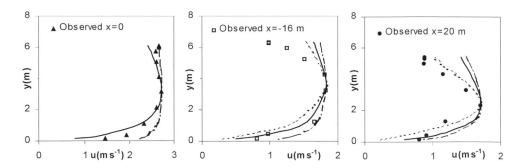

Figure 5. Velocity profiles estimated by Eq. (6) for tests n. 1 (—), n. 2 (- - - - -) and n. 3 (–·–·) plotted against velocity points sampled along verticals at P.Nuovo gauged section during the flood event that occurred on June 1997; x is the horizontal distance from the vertical where u_{max} occurs (x = 0).

the same verticals as shown in Figure 4, for the data collected during the flood events of June 1997. Therefore, comparing Figs. 4 and 5 it is evident that Eq. (6) improves the accuracy of the velocity profile estimation also in the portions close to the side walls. In particular, the test n. 2 gave the best results mainly near the bank. This aspect suggests that the entropy parameter M is also a signal of the boundary effects on the velocity distribution (Chiu and Said, 1995). In fact, in the middle flow area its value can be given from Eq. (1), whereas close to the banks, where the effects of the side walls are consistent, the M value should be estimated considering the ratio between the maximum value of the velocity points sampled in the flow area and the mean velocity along verticals, as it was shown through test n. 2. The M values used for the three verticals investigated and for each test are shown in Table 3, wherein is also reported the sampled mean and maximum velocity in verticals.

Table 3. Flood event that occurred on June 1997. Depth, D_i, mean and maximum velocity sampled along verticals, $\bar{u}_i$ and u_{maxi}, respectively, and M values for the three tests investigated.

Vertical	D_i (m)	$\bar{u}_i$ (ms^{-1})	u_{maxi} (ms^{-1})	M Test n. 1	M Test n. 2	M Test n. 3
x = 0	6.07	2.47	2.72	2.01	10.8	10.8
x = 20 m	5.66	1.66	1.97	2.01	-055	3.1
x = -16m	6.16	1.50	1.83	2.01	0.58	5.3

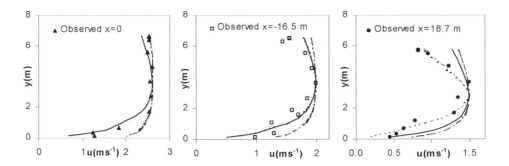

Figure 6. As in Figure 5, but for the flood event that occurred on November 1996.

Analogous results were obtained for the velocity data collected during the flood event of November 1996, as shown in Fig. 6.

Therefore if the topographical survey of the gauged section is available and the location of the maximum velocity sampled is known, Eq. (6) can be easily used for reconstructing the velocity profile along any vertical of the flow area. The M parameters can be derived through Eq. (7) considering as u_{max} the maximum value of the velocity points sampled, as derived from test n. 2. For this reason Eq. (6) might be applied also for new sites where the velocity data are not enough for estimating an accurate best fit line. Besides, in order to use Eq. (6) during high flood when it is not possible to carry out direct measurements of velocity mainly in the lower portion of the flow area, a simple method based on the maximum velocity sampled is developed for estimating the mean velocity along verticals and then the corresponding M values.

Alternative Estimate of the Cross-Sectional Mean Velocity

Test n. 2 gave the best results in estimating the velocity profile. This means that if the maximum velocity is sampled and the mean velocity along different verticals is known, the velocity profile can be reconstructed adequately through Eq. (6) in any portion of the flow area. Generally the true maximum velocity is unknown, but the maximum value in the data set of the velocity points can be assumed for it (Chiu, 1988). During high floods the main difficulty is to sample point velocities in the lower portion of the flow cross-sectional area, whereas the maximum velocity that usually occurs near the water surface can be sampled more easily, as for instance by a current meter from cableways. Once the maximum velocity is estimated, the mean velocity is derived by Eq. (1) and then the discharge if the flow area is known. The estimated discharge has high probability to be accurate, however, as previously shown, in some cases substantial errors might occur. Besides for ungauged river sections there are a few velocity data for determining the M value. For this case, Chiu proposed a simple technique to estimate the discharge by a single velocity profile passing through the point of maximum velocity and whose distribution is given as (Chiu and Said, 1995):

$$u = \frac{u_{max}}{M} \ln\left[1+\left(e^M - 1\right) \cdot \frac{y}{D-h} \exp\left(1-\frac{y}{D-h}\right)\right] \qquad (8)$$

where u is the velocity along the profile, D is the water depth, y is the vertical distance measured from the channel bed, and h is the depth below the water surface where the maximum velocity, u_{max}, occurs. If u_{max} and h are determined by the velocity profile sampled and they are substituted in (8), the M value can be determined through a least-squares estimate (Chiu and Said, 1995). Once M is estimated on the same profile, the location at which the velocity is equal to the cross-sectional mean, $\bar{u}$, is given by (Chiu and Said, 1995):

$$\bar{\xi} = \frac{\bar{y}}{D-h} \exp\left(1-\frac{\bar{y}}{D-h}\right) \qquad (9)$$

where

$$\bar{\xi} = \frac{\exp\left[Me^M \left(e^M - 1\right)^{-1} - 1\right] - 1}{e^M - 1} \qquad (10)$$

and $\bar{y}$ is the distance from the channel bed, on the profile investigated, where $u = \bar{u}$.

However, as will be shown afterwards, this approach allows us to overcome the lack of velocity data for new sites but it does not improve substantially the accuracy of the linear relationship (1).

Therefore, in order to have a more accurate estimate of the cross-sectional mean velocity also when velocity points data are not enough, an alternative and practical approach is proposed. It is assumed that the behavior of the mean velocity along the verticals sampled in the cross-sectional flow area can be expressed by a parabolic curve:

$$\bar{u}_v(x) = a \cdot x^2 + b \cdot x + c \qquad (11)$$

where a, b and c are parameters; $\bar{u}_v(x)$ is the mean velocity along the vertical located at the x coordinate in the transverse direction. x = 0 is selected for the vertical, where the maximum velocity, u_{max}, occurs (see Fig. 7).

The parameters of the parabolic curve can be estimated by two simple conditions. The first supposes that the maximum value of the mean velocity along the vertical occurs at x = 0, yielding:

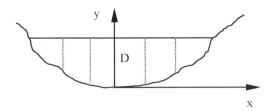

Figure 7. Sketch of the cross-sectional flow area with the verticals where velocity points are sampled (¦); D is the flow depth at the vertical located at x = 0 where the maximum velocity occurs.

$$\frac{d\bar{u}_v(x)}{dx}\bigg|_{x=0} = 0 \quad \text{and} \quad \bar{u}_v(0) = c \tag{12}$$

The first of Eq. (12) gives b = 0. $\bar{u}_v(0)$ can be estimated through the velocity points sampled along the vertical or by using the above-mentioned least-squares method applied to Eq. (8) when, as for instance during high floods, it was not possible to carry out velocity measurements in the lower portion of the vertical. The second condition allows to estimate "a" setting that at the banks (left side, $x = x_{LS}$, and right side, $x = x_{RS}$) the flow velocity is zero:

$$\bar{u}_v(x)\big|_{x=x_{LS};x=x_{RS}} = a \cdot x^2 + \bar{u}_v(0) = 0 \tag{13}$$

Obviously, if the flow area section is asymmetric with respect to x = 0, the "a" value for the right flow portion is different from that on the left. The same condition might be also used if velocity points are sampled along a vertical close to the bank; in this case the x coordinate is referred to the location where the vertical is sampled, thus permitting a more accurate estimate of the parabolic shape. The method has been applied to the velocity data collected during three flood events at the P. Nuovo gauged section. For two of them, the percentage error, $\varepsilon\%$, in estimating the flow cross-sectional mean velocity by Eq. (1) was 15% greater in magnitude as shown in Table 4 where the main characteristics of the three selected data set are also reported.

Table 4. Error, $\varepsilon\%$, in estimating the mean flow velocity observed $\bar{u}$, for the linear relationship, Eq. (1), at the P. Nuovo gauged section by using the maximum velocity, u_{max}, sampled along the vertical at x = 0 and whose flow depth is D.

Data Set n.	D (m)	$\bar{u}$ obs. (ms⁻¹)	u_{max} (ms⁻¹)	$\bar{u}$ eq.(1) (ms⁻¹)	$\varepsilon\%$
1	2.9	1.08	2.03	1.32	22.3
2	3.6	1.14	2.06	1.34	17.6
3	6.1	1.82	2.72	1.77	-4.1

The mean velocity along verticals sampled in the flow cross-sectional flow area, was considered as a benchmark. x = 0 is selected at the vertical where u_{max} occurs. Defined through the velocity points sampled, the depth below the water surface where u_{max} occurs, h, the M value is determined by the least-squares estimate applied to Eq. (8) (Chiu and Said, 1995). Once M is determined the velocity profile can be reconstructed and then the vertical mean velocity at x = 0 can be estimated. Fig. 8 shows the velocity profile at x = 0 determined by Eq. (8) and the velocity points sampled for the data set n. 3.

The estimated M value is equal to 4, obviously different from that previously obtained by Eq. (1). Known the mean velocity along the vertical at x = 0, it is possible to estimate parameters a and c (b = 0) by Eq. (12) and Eq. (13), then the values of the mean velocity along each vertical in the flow area. Figs. 9 and 10 compare the observed mean velocity along each vertical and that estimated by Eq. (11) for data set n. 1 and n. 3.

As can be seen, the parabolic shape represents adequately the distribution of the mean velocities in different portions of the flow area. Analogous results are also obtained for data set n. 2. This means that the parabolic shape implicitly takes into account the effects of the side walls on the velocity distribution in the cross-sectional flow area.

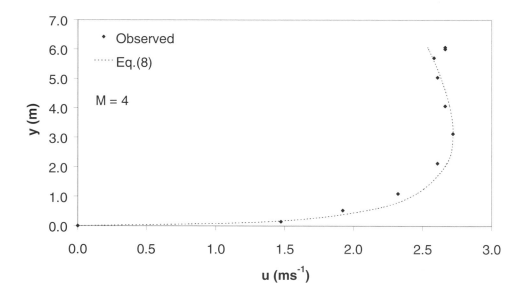

Figure 8. Entropy velocity distribution, Eq. (8), along the vertical where u_{max} occurs, plotted against the observed velocity profile, at the P. Nuovo gauged section for the data set n. 3. The entropy parameter value, M, estimated by least-squares, is also shown.

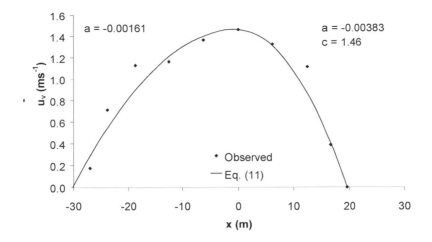

Figure 9. Parabolic distribution of the mean velocity in verticals, $\bar{u}_v$, plotted against observed mean velocity in sampled verticals, at the P. Nuovo gauged section, for the data set n. 1. Parameters of the distribution a and c are also shown (b = 0); x = 0 is located along the vertical where u_{max} occurs.

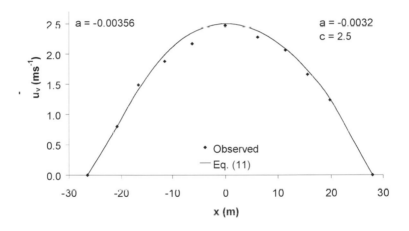

Figure 10. As in Figure 9, but for the data set n. 3.

Once the mean velocity along verticals was estimated by the parabolic shape, the cross-sectional mean velocity can be derived by applying the well-known velocity-area method (Herschy, 1985). This value with that obtained through Chiu's technique is shown in Table 5 wherein is also reported the error in determining the mean velocity for both approaches with the M value estimated by the least-squares. By comparing the errors in Table 4 and Table 5 it can be seen as the proposed method significantly improves the estimate of the cross-sectional mean velocity. However, as the mean width of the Tiber River reach investigated is about 40 meters, the method needs to be further tested on other natural wider channels.

Table 5. Error in estimating the mean flow velocity, by using Chiu's method, ε_{Chiu}, and the parabolic curve, ε_{parab}, for the three selected data set at the P. Nuovo gauged section. The M value computed by the least-squares method and the cross-sectional mean velocity estimated through Chiu's method, $\bar{u}_{Chiu}$, and the parabolic curve, $\bar{u}_{parab}$, are also shown.

Data Set n.	M	$\bar{u}_{Chiu}$ (ms^{-1})	$\bar{u}_{parab}$ (ms^{-1})	ε_{Chiu} (%)	ε_{parab} (%)
1	1	1.15	1.03	6.8	-5.0
2	1.5	1.28	1.12	11.3	-2.5
3	4	2.09	1.87	14.8	2.6

Conclusions

The linear relationship between the mean flow velocity and the maximum velocity was found accurate in three gauged river sections in the upper Tiber River basin in Central Italy. The error in estimating the mean velocity by the linear relationship was found to be normally distributed and only for a few data its magnitude exceeded 0.10 ms^{-1}. The practical method developed for reconstructing the velocity profiles at a river section which is based on the velocity distribution law derived by Chiu (Chiu and Said, 1995), was found capable of estimating with a reasonable accuracy the shape of the observed velocity profiles of high flood events. Finally, assuming that the behavior of the mean velocity within different portions of the flow area follows a parabolic shape, the cross sectional mean velocity might be easily estimated with an accuracy greater than that obtained using both the linear relationship and the Chiu method (Chiu, 1989).

Acknowledgments

The authors wish to thank B. Bani, C. Fastelli and R. Rosi for their technical assistance. This work was partly supported by the National Research Council of Italy, in part under the Special Project GNDCI (Publication n. 2447).

References

Barbé, D.E., Cruise, J.F. and Singh, V.P. (1991). "Solution of three-constraint entropy-based velocity distribution". J. Hydr. Engrg., ASCE, 117(10), 1389-1396.

Chiu, C.L. and Chiou, J.-D. (1986). "Structure of 3-D flow in rectangular open channels". J. Hydr. Engrg., ASCE, 112(11), 1050-1068.

Chiu, C.L. (1987). "Entropy and probability concepts in hydraulics". *J. Hydr. Engrg.*, ASCE, 113(5), 583-600.

Chiu, C.L. (1988). "Entropy and 2-D velocity distribution in open channels". *J. Hydr. Engrg.*, ASCE, 114(7), 738-756.

Chiu, C.L. (1989). "Velocity distribution in open channel flow". *J. Hydr. Engrg.*, ASCE, 115(5), 576-594.

Chiu, C.L. (1991). "Application of entropy concept in open-channel flow study". *J. Hydr. Engrg.*, ASCE, 117(5), 615-628.

Chiu, C.L. and Murray, D.W. (1992). "Variation of velocity distribution along non uniform open-channel flow". *J. Hydr. Engrg.*, ASCE, 118(7), 989-1001.

Chiu, C.L. and Said, C.A.A. (1995). "Maximum and mean velocities and entropy in open-channel flow". *J. Hydr. Engrg.*, ASCE, 121(1), 26-35.

Greco, M. (1999). "Entropy velocity distribution in a river". *Proc. of the IAHR. Symposium on River, Coastal and Estuarine Morphodynamics*, Genova, Vol. II, 121-130.

Herschy, R.W. (1985). *Streamflow Measurement*, Elsevier Applied Science Publishers LTD, London, England

Xia, R. (1997). "Relation between mean and maximum velocities in a natural river". *J. Hydr. Engrg.*, ASCE, 123(8), 720-723.

Moramarco, T. and Singh, V.P. (2001). "Simple method for relating local stage and remote discharge". *J. Hydrol. Engrg.*, ASCE, 6(1), 78-81.

Surface Water Hydrology, Singh, Al-Rashed & Sherif (eds)
© 2002 Swets & Zeitlinger, Lisse, ISBN 90 5809 363 8

The dynamics of main landscape forming factors of river mouth regions under different anthropogenic load

Natalia N. Mitina
Water Problems Institute Russian Academy of Sciences
3 Gubkin St., Moscow 119991, Russia
e-mail: natalia_mitina@mail.ru

Abstract

Landscape forming factors of the mouth areas of the sea coastal zone with the help of factor analyses based upon the method of the principal components is determined. It has been fixed that during the anthropogenic influence the landscape forming factors and their significance are changed. Proposed methodical approach permits to substantiate the forecasting evaluations of the variability of the components of the mouth areas bottom natural complexes under the different level of human impact and determine the threshold values, under which reforming of their landscape forming factors are happened.

Introduction

Sea shoal landscapes are zones of strong anthropogenic impact even near sparsely populated coasts as a result of river pollution influence. Similar situation is observed at the mouths of the rivers flowing into the Sea of Japan in the Far East coastal region. The 60-km-long section at depths of 0 to 40 m was selected in order to compare the main factors of formation and dynamics of undisturbed and anthropogenically polluted bottom natural complexes (BNC) in the mouth regions of the rivers flowing into the sea.

Materials and Technique

The coastal line of the shoal under study runs along the mountain range oriented mostly northeast, and smoothly winding, forms several bays with river mouths. Eleven streams with a low-flow discharge of 0.012 to 6.11 m^3/s flow into the shoal studied. The anthropogenic influence is maximum in the mouth areas[1] of the Rudnaya and Kamenka rivers with low-flow discharges of 5.76 and 6.11 m^3/s, respectively; whose catchment areas include mining works for enrichment of polimetallic ores. The waters of the above mouth regions have increased concentrations of *Pb, Zn, Mn, Cu, Fe*, and other metals.

Synchronous data of echo sounding, sampling of benthos and soils by grabbing and diving, hydrological surveying, as well as visual diver's observations, was ap-

---

[1] According to Mikhailov (1997), the sea boundary of a river mouth was distinguished by the maximum extension of the external (off-shore) zone of mixing of river and sea surface waters.

plied to the study of the BNC structure. The water temperature and salinity was measured at a depth interval of 1 m. The velocity and direction of the resultant long-shore constituent of drift-gradient currents were calculated for each point. As a result, each reference point characterized the uniform bottom section and was described by a set of quantitative and qualitative indicators including: geomorphological type of coastal and bottom topography, low-flow discharge of streams flowing into an adjacent BNC, point depth, soil mechanical composition, benthos communities, and a number of hydrological and hydrodynamical indexes of shoals water masses.

The detailed submarine landscape map of the coastal zone studied and, specifically, of the mouth zones of inflowing rivers was compiled on the basis of the data collected and the technique for BNC classification developed (Mitina, 1993). At the same time, the main problem of complex zoning and mapping, including the isolation of uniform territories or water areas with a clearly defined anthropogenic effect was solved.

To find out the total environmental quality of BNC by the Yu. Odum technique (1975), we have carried out the calculations using the indications of domination and species variety of algae communities of the characterized by the similar abiotic conditions: confinement to the same element of a topography mesoform, occurrence within certain depth interval and in identical hydrodynamical zone with equal indicators of hydrodynamical activity but with different hydrological and ecological indicators of a river flow. The comparative analysis of conditions of algal communities indicated that a coefficient of species variety (d) decreases and a domination coefficient (c) increases under conditions of contamination of freshened water bodies. Mouth areas with freshened non-contaminated (control) water bodies, characterized by maximal quantities of variety coefficients and by minimal values of domination coefficients, are in the most favorable conditions. As a result, non-contaminated and contaminated mouth areas was mapping (Mitina, 1998).

We have also constructed two matrices of the variables, characterizing the whole mouth areas region under study and non-contaminated mouth regions, in order to quantitatively substantiate structural changes in BNC components under the anthropogenic load. The dimensions of the initial matrix have been 8 x 92 for the non-contaminated areas and 8 x 159 for the whole area. The matrixes include eight variables (columns), quantitatively characterizing the properties of different BNC components, and 92 and 159 lines (observation points). The parameters measured, characterizing the properties of BNC components in the mouth regions, include: (x1) The deflection of the coastal line (in degrees) characterizing the configuration and geomorphological structure of the coast. (x2) The arithmetic mean size of the predominant fraction of sediments coming from a coast (in mm). (x3) The water discharge of streams flowing into the shoal area under study during the low-water period (in m^3/s). (x4) The depth of an observation point (in m). (x5) The projective covering of the bottom with coquina sediments (in %). (x6) The bottom velocity of the total component of drift-gradient currents (in m/s). (x7) The bottom water temperature (in °C). (x8) The salinity of the surface water (in ‰).[2]

The constructed matrixes have been analyzed by correlation and factor analysis. The factor analysis is based upon the method of principal components (Braverman and Muchnik, 1983). The application of this method makes it possible to distinguish the small set of latent factors, characterizing the basic mechanisms of formation and features of functioning of the coastal zone.

Result and Discussion

We selected main factors of formation and dynamics of mouth areas of rivers (summary variancy of included parameters is 72,7%). Table 1 is displayed the load of eight parameters on three principal factors for the non-contaminated and for the whole section of the shoal under study, obtained as a result of three rotations made in order to determine in more detail the loads of the individual variables of the principal components.

The analysis of a factor matrix constructed for the whole section under study

Factor A is most significant: it accounts for 33.8% of the total dispersion of variables and thus determines the highest degree of BNC dynamics of mouth areas. The heaviest load on the factor A have variables X1, X6, and X8 (*R* > 0.7) (Table 1). This factor characterizes the hydrodynamic and hydrochemical features of a water body at various geological – geomorphological types of coasts. Indeed, river runoff and sea currents are responsible for hydrochemical features of a water body in the mouth regions of a sea shoal. At the same time, the degree of coast opening affects a water salinity non-contaminatedregime. Isolated bays with re-tarded drift currents retain inflowing river waters. Water salinity is considerably lower in these bays. The interaction between sea hydrodynamics and geological – geomorphological structure of a coast and its lithology is responsible for the form of mouth regions at a coastal zone, specifically, for coast configuration, cliff height, and submarine slope dip.

Factor B is of a secondary significance in a factor matrix (22.6% of a total dispersion). After rotation the maximum loads on this factor have variables X4 and X7 (R > 0.9). Taking into account the sense of these parameters, we can suppose that factor A characterizes the temperature stratification of a water body. This environmental is very important for the biotic component of a submarine BNC. Thus, the presence of a temperature jump, established by us in the freshened water bodies

---

[2]Freshened and non-freshened water bodies are sharply distinguished by a type of temperature stratification of a water body. The non-freshened water bodies have a temperature gradually lowering down to bottom within 10°, and they occur in the coastal areas without large streams of a water discharge below 0.07 m³s⁻¹. Salinity is usually no less than 32.5‰. In the shallow-water zones adjacent to the issues of larger springs with water discharge exceeding 0.4 m³ s⁻¹, we observe a condition close to a homother-mie in the upper water layer from 10 to 14 m thick. A salinity of this freshened marine water is below 32.5 ‰.

(Mitina, 1996), points to the existence of the layer of mixing of marine and continental waters. The zones of increased biological variety and productivity of various marine organisms at different trophic levels are also confined to the layer with a temperature jump.

Factor C is accounts for 14,8% of a total dispersion. The high loads of variables X3 and X5 on factor C indicate that this factor characterizes hydrological conditions of a coast, governing the existence of the mass species of zoobenthos. This factor indicates that rivers supply biogenic elements from the entire catchment basin into the sea and govern the regime of salinity in the mouth regions. The conditions of pasturing of young and mature fish in the sea are formed under the influence of a biogenic discharge and salinity regime. Moreover, suspended organic matter is intensely consumed by filtering organisms, specifically, by mollusks, which are the feeding base for many fish species.

The analysis of a factor matrix obtained for the non-contaminated (control) river mouths areas

Factor A is most significant: it accounts for 33.8% of the total dispersion of variables (Table 1). After rotation, parameters X4 and X6 ($R > 0.9$) have high final loads on this factor. This factor A characterizes the temperature stratification of a water body, i.e. the hydrological regime of water mass of the sea shoal and is the second factor for the whole section.

Factor B is of a secondary significance in a factor matrix (24.7% of a total dispersion). Variables X1, X2, X5, and X7 has a maximum load on factor B. This factor

Table 1. Factoral matrix for data analysis (lines α – the whole section of the coastal sea zone, β – background sections).

Parameters		Factor A (1)	Factor B (2)	Factor C (3)
X1	α*	.79873	.20807	.04634
	β*	.16717	.81671	-.13670
X2	α	.63410	.06606	.52974
	β	.02273	.64193	.55330
X3	α	-.38803	-.08372	-.79657
	β	-.00066	-.08875	-.89780
X4	α	-.10202	.95481	.10750
	β	.96430	-.07153	.08752
X5	α	-.19212	.12873	.89324
	β	.19957	-.14339	.79776
X6	α	.74162	-.16974	-.19777
	β	-.37402	.75657	-.10886
X7	α	-.08830	-.95299	-.08288
	β	-.93610	-.14459	-.09732
X8	α	.65211	.41480	.28188
	β	.35355	.65169	.36553

characterizes the hydrodynamic and hydrochemical features of a water body at various geological – geomorphological types of coasts and is the first factor for the whole section.

Factor C is of a tertiary significance in a factor matrix (18,2%). The high loads of variables X3 and X5 on factor C indicate that this factor characterizes hydrological conditions of a coast, governing the existence of the mass species of zoobenthos. This factor is also the third for the whole volume of factors for the whole section, but in the case of background waters its input into total dispersion increases from 14.8% up to 18.22%.

Summary

The factor analysis of the consequences of the anthropogenic impact on BNC of the mouth areas of the Sea of Japan shoal has made it possible to determine the dominating processes of interaction between different submarine landscapes at river mouths and the environmental factors of its structural organization under the conditions of natural and polluted environment. The method has made it possible to distinguish a certain unique set of environmental factors, characterizing the functioning of mouth regions under different environmental conditions.

The three main factors were revealed, the summary variation of parameters included is 72.7%. Thus we have three significant BNC features of mouth regions of the rivers of coastal sea zone that determine the main features of their functioning and dynamics. The environmental factors distinguished are interrelated, as the links of an integral chain, but have different significance. The first factor is the most significant in a factor matrix. The second factor is intermediate between the first and the third ones.

In case of considerations of factors that determine the main features of functioning and dynamics of background sections in comparison to the whole section, the names of factors coincide, but factors A and B changed their places according to their meaning. Meanwhile the summary variation of parameters included increased and is 76.7% in comparison to 72.7%. Probably the exchange of priority of factors A and B in the BNC structure under conditions of non-contaminated water masses the state of their hydrochemical and hydrodynamic features is not significant as for evaluation of the whole coastal zone. In this zone the sections of local stresses can be mentioned. This supports the idea that the whole ecological state of aquatorium is stable and there are only insignificant local sections of anthropogenic pollution.

In sum the functioning of the aquatic ecosystems is changed under the anthropogenic influence. The loads in a factor matrix and, consequently, the loads on the submarine landscape components are changed. This is due to a selfregulating capacity of the aquatic ecosystem tending to sustain its stability. In combination with the additional studies, the methodical approach proposed makes it possible to substantiate the forecast estimates of variation in BNC components at mouths of rivers flowing into seas under different human impact and to establish the threshold values of reformation of the environmental factors of functioning and dynamics of the aquatic ecosystems at river mouth areas.

References

Braverman, E.M. and Muchnik. I.B. (1983). *Structural methods for empirical data processing*, Moscow: Nauka (in Russ.).

Khristoforova, N.K. (1989). Bioindication and Monitoring of the Sea Waters Polluted by Heavy Metals, Leningrad: Nauka, 192 p. (in Russ.).

Mikhailov, V.N. (1997). *River mouth of Russia and adjacent countries,* Moscow: Geos (in Russ.).

Mitina, N.N. (1993). "Structure and Classification of Bottom Natural Complexes of the Sea of Japan Coastal Zone", *Water Resources, Interperiodica Publishing, No. 3*, MAIK, 332-337.

Mitina, N.N. (1996). "The significance of the mouths of small rivers for the preservation of aquatic ecosystems", *Hydrology in a Changing Environment. BHS Occasional Paper, UK, No. 9*, 54-59.

Odum, Yu. (1975). *Principles of Ecology,* Moscow, USSR (in Russ.).

Surface Water Hydrology, Singh, Al-Rashed & Sherif (eds)
© 2002 Swets & Zeitlinger, Lisse, ISBN 90 5809 363 8

The modeling of dynamics of the Caspian Sea characteristics under oscillations of water inflow

Aleksey V. Babkin
State Hydrological Institute
23 Second Line, St. Petersburg, 199053 Russia
e-mail: abav@mail.ru

Abstract

The solution of differential equation of a lake water balance permitted to elaborate mathematical model, where dynamics of mutually related the Caspian Sea level, water surface area, evaporation and outflow to the Kara-Bogaz-Gol Gulf are the responses to inflow oscillation. The time lag of the sea characteristics to inflow and relation of their amplitudes are computed for different values of inflow and for large number of time periods. The observation data of water inflow and the sea level are approximated by periodical functions. The time between extreme values of the curves, which approximate the level and inflow, and relation of their amplitudes are compared, respectively, with the time delay of level to inflow and relation of amplitudes of their oscillation, computed by the model.

Introduction

The Caspian Sea is the largest endorheic lake of the Earth with surface area more than $380,000$ km^2. The catchment of the sea is a vast area of internal flow of $3,000,000$ km^2.

The variations of the sea level are caused by the dynamics of the Earth climatic system. In the geological past the sea level changed also due to transformation of the Eurasian flow systems. In the late Pleistocene, during the Wurmian glaciation the expansive water system of large interconnected basins from the Mansian Sea to the Mediterranean Sea was formed. This water system included the Aral, Caspian, Black Seas and the Sea of Marmara (Grosswald, 1989).

There was a time of the Early Khvalynian transgression of the Caspian Sea (Tshepalyga, 2000). The sea area reached $900,000–1,000,000$ km^2. The level exceeded its modern height by 76 m approaching to 50 m above m.s.l. and the water flowed from the Caspian to the Black Sea basin through the Manych Strait.

During the Late Khvalynian transgression of the Early Holocene the sea area and level were respectively $500,000$ km^2 and 0 m above m.s.l. The sea level was lower than a watershed between the Caspian and Black Sea basins, and the Caspian Sea existed as the endorheic lake. These transgressions were separated by the Enotaevka regression with the minimal sea level −64 m and water surface area $205,000$ thousand km^2.

The variations of the sea level in the historical past were also significant. In the time of the regression at the beginning of the second millenium A.D. the sea level was lower than −32 m, while in the 17 century its maximal height exceeded the mark of −23 m (Karpytchev, 2001).

The instrumental observations of the Caspian Sea level started in 30-th of 19 century. The changes of the sea level and its water balance components for observational period were analyzed by Zaikov (1946), Kritskii et al. (1975), Shiklomanov (1976), Georgievskii (1982), Ratkovich (1993), Malinin (1994) and other scientists.

The present study deals with the development of the mathematical model for research of general regularities in the variations of the Caspian Sea water surface area, level and water balance components, and exploration of the responses of each of these characteristics to the variation of water inflow. Such analyses may be actual for study of dynamics of the sea level and water balance in the present and past epochs and for estimation of their possible changes in the future.

The first step of modeling is evaluation of dependencies between the morphometric characteristics and the values of the water balance components of the Caspian Sea.

The Morphometric and Water Balance Characteristics of the Caspian Sea

The area of the sea *(S)*, evaporation from its surface *(E)*, and volume of outflow to Kara-Bogaz-Gol Gulf *(R)* grow with the increase of Sea level *(H)*. Let us assume the linear dependencies of sea level from water surface area, evaporation from area and outflow to the gulf from level

$$H = kS + k',$$ (1)

$$E = eS + e',$$ (2)

$$R = rH + r'.$$ (3)

The coefficients k and e are the changes of level and evaporation under unitary change of the sea area, r is the change of outflow to the Kara-Bogaz-Gol Gulf under unitary change of the sea level. k', e' and r' are the additional items, permitting, in some diapason of values of water surface area, evaporation and outflow to the gulf, to describe approximately by linear functions the dependencies between morphometrical and water balance characteristics of the sea.

The output of water Q is equal to the sum of evaporation and the volume of outflow to the Kara-Bogaz-Gol Gulf

$$Q = E + R.$$ (4)

Equations (1)–(4) permit to write the relation between the water output and the sea level as follows:

$$\frac{H}{1000} = \frac{Q}{\frac{e}{k}+r} - \frac{r'+e'+rk'}{\frac{e}{k}+r} + k'.$$

(5)

If the output of water is balanced by the inflow P (the sum of rivers discharge to the sea, groundwater inflow and precipitation to the sea surface), the sea is in some equilibrium state. The replacement of the water output by inflow in the Equation (5) permits to calculate the equilibrium levels of the Caspian Sea for different values of inflow. The relating to them magnitudes of water surface area, evaporation and out-flow to the gulf are calculated by Equations (1)–(3).

The values of coefficients k, e, r and items k', e' and r' may be estimated as coefficients and additional items of linear trends of corresponding dependencies.

Fig. 1 shows the dependencies between the sea evaporation and surface area (a), volume of outflow to Kara-Bogaz-Gol Gulf and the sea level (b), the sea level and area (c). These dependencies are shown with the lines of their trends.

The values of the Sea surface area under the different heights of its level are taken by Zaikov (1946) and Kritskii et al. (1975).

Coefficients k, e, r and items k', e' and r' are equal: $k = 6{,}87 \cdot 10^{-8}$ km^{-1}, $e = 1 \cdot 10^{-3}$ km year^{-1}, $r = 3{,}81 \cdot 10^3$ km^2 year^{-1}; $k' = -53{,}7 \cdot 10^{-3}$ km, $e' = -14{,}9$ km^3 year^{-1}, $r' = 117$ km^3 year^{-1}. The level in the formula (5) is divided on 1000 to express its value in meters.

The Caspian Sea water balance components under different levels are shown in Fig. 1 (d). According to Equation (5) the curve 1 describes the output of water Q, and respectively, the dependence between the equilibrium level height and inflow, curve 2 – the evaporation from the water surface area. The ordinate difference of magnitudes of curves 1 and 2 is equal to the volume of outflow to the Kara-Bogaz-Gol Gulf.

Parameterization of Dynamics of the Caspian Sea Characteristics

Let us propose that the variation of inflow to the Sea from its P_1 to

$$P_2 = P_1 + \Delta P$$

(6)

is around some mean value P_0, determining the average sea level

$$P_0 = 0{,}5(P_1 + P_2)$$

(7)

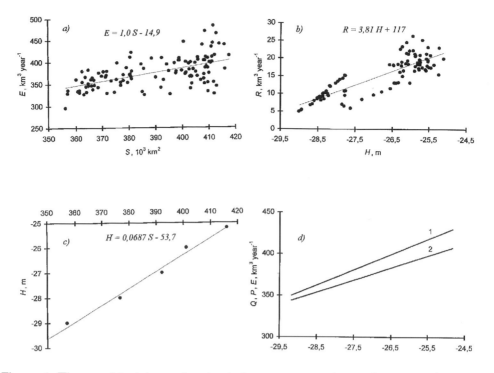

Figure 1. The empirical dependencies between evaporation and water surface area of the Caspian Sea (*a*), volume of outflow to the Kara-Bogaz-Gol Gulf and Sea level (*b*), sea level and area (*c*); components of the sea water balance under different height of its level (*d*).

and is described by the following equation

$$P = P_0 + 0,5\Delta P Sin\ \omega t\,,\tag{8}$$

where ΔP and ω are accordingly the amplitude and the frequency of inflow variation, t is the time.

$$\omega = \frac{2\pi}{T}\,,\tag{9}$$

where T is the period of oscillation.

According to Equations (1)–(3) and (5) the minimal and maximal inflow P_1 and P_2 and its magnitude P_0 presuppose the equilibrium states of the sea with values of its characteristics:

$$E = E_1,\ R = R_1,\ H = H_1,\ S = S_1,\ P_1 = E_1 + R_1,\tag{10}$$

$$E = E_2, \ R = R_2, \ H = H_2, \ S = S_2, \ P_2 = E_2 + R_2 \tag{11}$$

$$E = E_0, \ R = R_0, \ H = H_0, \ S = S_0, \ P_0 = E_0 + R_0. \tag{12}$$

The water balance equation in its differential form for the sea is

$$\frac{dH}{dt} = \frac{P_0 + 0,5 \cdot \Delta P \cdot Sin \ \omega t - (E + R)}{S}. \tag{13}$$

Expressing the dependencies of the sea level from water surface area (1), evaporation from the area (2), outflow of water to the Kara-Bogaz-Gol Gulf from level (3) as following

$$H - H_0 = k(S - S_0), \tag{14}$$

$$E - E_0 = c(S \ \ S_0), \tag{15}$$

$$R - R_0 = r(H - H_0), \tag{16}$$

taking into the consideration the relation (12), the water balance equation of the sea may be written as

$$\frac{dS}{dt} = \frac{cSin\omega t + a}{S} - b, \tag{17}$$

where

$$a = S_0(\frac{e}{k} + r), \tag{18}$$

$$b = \frac{e}{k} + r, \tag{19}$$

$$c = \frac{\Delta P}{2k}. \tag{20}$$

Equation (17) is related to the Abel type of differential equations (Kamke, 1990). It is expediently to solve this non-linear equation by its expansion to the follow power range:

$$S_{n+1} = S_n + Ih + \frac{IIh^2}{2} + \frac{IIIh^3}{6} + ... + \frac{Nh^N}{N!} \tag{21}$$

This expansion does not formally restrict the research of dynamics of the sea characteristics by some fixed time interval. The same solution may be used for modeling of response of the sea characteristics to the variation of water inflow in the broad range time periods.

Each consequent derivative of range (21) may be easy expressed through the previous derivatives. The first five derivatives may be written as follows:

$$I = \frac{1}{S}(a + c \, Sin \, \omega t) - b, \qquad (22)$$

$$II = \frac{1}{S}(c\omega \, Cos \, \omega t - (b + I)I), \qquad (23)$$

$$III = -\frac{1}{S}(c\omega^2 \, Sin \, \omega t + (3I + b)II), \qquad (24)$$

$$IV = -\frac{1}{S}(c\omega^3 \, Cos \, \omega t + (4I + b)III + 3II^2), \qquad (25)$$

$$V = -\frac{1}{S}(-c\omega^4 Sin \, \omega t + (5I + b)IV + 10II \cdot III). \qquad (26)$$

The expression of any consequent derivative is not more complicated than previous. If it is necessary to make solution of Equation (17) more precise, several additional derivatives may be easy formed and substituted to the range (21).

The initial condition for this equation was specified as follows:

$$S(t = 0) = S_0. \qquad (27)$$

The analysis of the response of the Caspian Sea characteristics to the inflow variation be will executed by the comparison of impact curve, reflecting the inflow dynamics

$$\lambda = \frac{P - P_0}{P_2 - P_1} = 0,5 \cdot Sin \, \omega t \qquad (28)$$

with the curves of dynamics of evaporation from the water surface area

$$\gamma = \frac{E - E_0}{P_2 - P_1}, \qquad (29)$$

outflow to the Kara-Bogaz-Gol Gulf

$$\chi = \frac{R - R_0}{P_2 - P_1} \qquad (30)$$

and with the curve of response, reflecting the variation of water output and other linearly related sea characteristics

$$\beta = \frac{Q - Q_0}{P_2 - P_1} = \frac{E - E_0}{E_2 - E_1} = \frac{R - R_0}{R_2 - R_1} = \frac{H - H_0}{H_2 - H_1} = \frac{S - S_0}{S_2 - S_1}. \tag{31}$$

The values of parameters λ, γ, χ and β may vary from $-0,5$ to $0,5$.

For the 1880–1996 the mean level of the Caspian Sea is approximately equal to -27 m. According to Equations (1)–(5) this level is related with the following values of Sea characteristics: $S_0 = 389,000$ km^2, $E_0 = 376$ km^3 year^{-1}, $R_0 = 14$ km^3 year^{-1} and $Q_0 = 390$ km^3 year^{-1}. Let us analyze the variation of the Caspian Sea characteristics when $P_0 = 390$ km^3 year^{-1}, $\Delta P = 64$ km^3 year^{-1} and $T = 140$ years.

Fig. 2a shows the dynamics of the Caspian Sea characteristics under variation of the water inflow. Curve 1 reflects the oscillation of water inflow, curve 2 describes by parameter β the variation of sea characteristics. Curves 3 and 4 show the changes of evaporation and outflow to Kara-Bogaz-Gol Gulf. The curves of dynamics of the sea characteristics under periodic inflow variation are also periodic. The periods of variation of level, water surface area, evaporation and outflow to the gulf are equal to the period of inflow oscillation.

The curve of parameter β is the sum of parameters γ and χ. The evaporation from the sea area, outflow to the gulf and output of water simultaneously reach their maximal and minimal values. The maxima and minima of parameter β are points of intersection of the curve of the sea characteristics variation and the curve of inflow variation. Comparing the Equations (28) and (31) it is possible to evaluate that at extreme points of β variation curve the inflow should be equal to water output. The increase of area, water level, evaporation from the sea surface and flow of water to the gulf is appeared if $P > Q$, when $P < Q$, these characteristics tend to decrease.

As it follows from Fig. 2a, the extremes of parameter β have a time lag τ if compared with parameter λ extremes. This means, that the area, level, evaporation from the sea and the volume of outflow to the gulf reach maximal and minimal values some times after the respective extreme of water inflow. The lag time of extremes of the sea characteristics to the inflow is within 0–$0,5\pi$.

To describe and analyze a response of the Caspian Sea characteristics to water inflow variation let us consider the amplitude of the response curve:

$$\delta\beta = \beta_{max} - \beta_{min} = \frac{H_{max} - H_{min}}{H_2 - H_1} = \frac{S_{max} - S_{min}}{S_2 - S_1} = \frac{E_{max} - E_{min}}{E_2 - E_1} = \frac{R_{max} - R_{min}}{R_2 - R_1}, \tag{32}$$

where β_{max} and β_{min} are the adjacent maximum and minimum of the response curve.

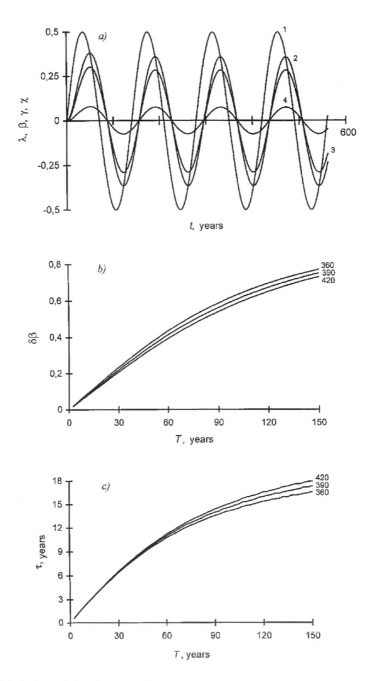

Figure 2. Variation of the Caspian Sea characteristics under oscillation of water inflow. The dynamics of parameters λ, β, γ and χ (curves 1, 2, 3 and 4) (*a*), dependencies of parameter β amplitude (*b*) and the time lag of the sea characteristics to inflow (*c*) from the period of variation (numbers near the curves show the values of inflow P_0).

δβ may vary from 0 to 1. The response curve amplitude shows the portion of the variation amplitudes of the sea level, water surface area, evaporation, volume of outflow to the gulf from a difference of levels, areas, evaporation and volumes of outflow of two equilibrium states, determined by extreme values of inflow P_2 and P_1 respectively.

By Equations (5)–(32) the program of calculation was elaborated. This program produces the curve of response upon the impact, calculates the amplitude of response curve δβ and the time lag τ consecutively for vast volume of values of inflow, period and amplitude of its variation. Computation shows that dependence of response curve amplitude δβ and time lag τ upon the amplitude of inflow variation ΔP is practically missing for typical values of inflow into the Caspian Sea, all possible periods and amplitudes of its variation.

Fig. 2b,c shows the dependencies of response curve amplitude and the time delay of maximal and minimal values of the Caspian Sea characteristics to corresponding inflow extremes from the period of variation. These dependencies were computed for the following inflow values: $P_0 = 360$ km^3 year^{-1}, $P_0 = 390$ km^3 year^{-1}, $P_0 = 420$ km^3 year^{-1}.

According to Equation (5) the inflow of 360 km^3 year^{-1} causes the sea level of −28,6 m. Such low height of the Caspian Sea level was observed at the 70-th of the 20 century. The inflow of 420 km^3 year^{-1} presupposes the sea level about −25,4 m. Such increased values of Sea level was at the 80-th of 19 century.

Fig. 2b illustrates that the parameter β amplitude grows with increase of variation period. For the same period the response amplitude δβ is than larger than water inflow to the Sea is lower.

The time lag of the sea characteristics to inflow also grows with increase of variation period. For the same period the time lag τ is larger under high inflow to the sea than under low inflow.

The Modeling Results and Observation Data

Fig. 3 represents the Caspian Sea level (a) and water inflow (b). Curve 1 reflects the observation ranges of sea level and inflow, which are approximated by periodical functions:

$$H = H_0 + \frac{dH}{2} Sin(\frac{2p}{T_H}t + \varphi_H),$$
(33)

$$P = P_0 + \frac{\Delta P}{2} Sin(\frac{2\pi}{T_P}t + \varphi_P).$$
(34)

465

The parameters of the level curve H_0, δH, T_H, φ_H and the inflow curve P_0, ΔP, T_P, φ_P are estimated by the method of the least squares:

$$S_H = \sum_{1}^{n}[H_i - H_0 - \frac{\delta H}{2}Sin(\frac{2p}{T_H}t_i + j_H)]^2 , \tag{35}$$

$$S_P = \sum_{1}^{n}[P_i - P_0 - \frac{\Delta P}{2}Sin(\frac{2\chi}{T_P}t_i + \varphi_P)]^2 . \tag{36}$$

There H_i and P_i – are the observation values of the sea level and water inflow in a year t_i, i – year number in the observed time series data, n – the length of time-series. S_H and S_P are the sums of squares of the differences of the sea level and water inflow from observation data and calculated by the Equations (33) and (34)

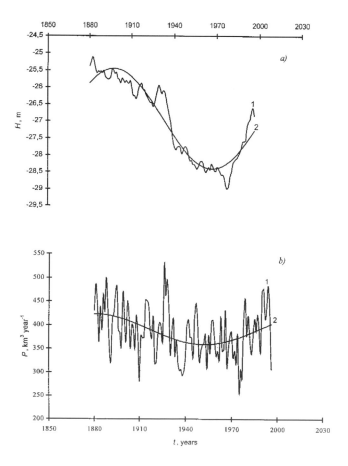

Figure 3. Variation of the Caspian Sea level (*a*) and inflow (*b*)
1 – observation data, 2 – approximation curves.

respectively. The Equations (33) and (34) are the true approximation curves when the sums S_H and S_P are the least (Linnik, 1962).

The sum of difference squares of observed and approximated by the Equation (33) sea levels was calculated successively for the large diapasons of values of level H_0, amplitude δH, period T_H and phase φ_H of its variation. The sum of difference quadrates S_H is the smallest when $H_0 = -26,93$ m; $\delta H = 2,96$ m; $T_H = 139$ years and $\varphi_H = 1,203\pi$. The approximation function (33) with this values of level variation parameters is represented in Fig. 3a by curve 2.

The sum of the difference squares of real and approximated by Equation (34) magnitudes of Sea inflow was calculated successively for the large diapasons of values of inflow, amplitude, period T_P and phase φ_P of its variation. The sum of difference quadrates S_P is the smallest when $P_0 = 391$ km^3 year^{-1}; $\Delta P = 64$ km^3 year^{-1}; $T_P = 143$ years and $\varphi_P = 0,197\pi$. The approximation function (34) with this values of inflow variation parameters is represented in Fig. 3b by curve 2.

The approximation curves of the Caspian Sea water inflow and level reveal the same properties as the impact and response parameters of the model. The period of sea level variation is approximately equal to the period of inflow variation. The extremes of approximation curve of level have a time lag to extremes of inflow approximation curve.

The minimal value of inflow approximation curve of 1952 is equal to 359 km^3 year^{-1}. According to Formula (5) this extreme inflow presuppose the equilibrium level H_1 −28,7 m.

The Sea level approximation curve gets the minimal value −28,41 m in 1967. The time lag of level approximation curve to inflow approximation curve is equal to 15 years. According to the model results when $P_0 = 390$ km^3 year^{-1} and $T = 140$ years the delay of response to impact is equal to 16–17 years (Fig. 2a, c).

The maximal value of inflow approximation curve is equal to 423 km^3 year^{-1}. According to Formula (5) this extreme inflow presuppose the equilibrium level H_2 −25,2 m. The difference of two equilibrium sea levels H_0 and H_1 is equal to 3,5 m.

When the difference of two equilibrated sea levels caused by maximal and minimal water inflow and amplitude of the sea level variation are determined, the Equation (32) permits to estimate the amplitude parameter β:

$$\delta\beta = \frac{\delta H}{H_2 - H_1} = 0,71, \tag{37}$$

The values of $\delta\beta$ estimated by comparison of the sea level and water inflow approximation curves and by the model computation are close. According to model

467

computation when inflow P_0 and variation period T are equal to 390 km^3 year^{-1} and 140 years respectively the parameter β amplitude is equal to 0,72 (Fig. 2a, b).

Summary

The variation of water inflow causes the variation of the Caspian Sea level, water surface area, evaporation and outflow to the Kara-Bogaz-Gol Gulf. According to the model the periods of sea characteristics and inflow variation are equal. The level, area, evaporation and outflow to the gulf have a time lag to inflow. The lag time of extremes of the sea characteristics to inflow extremes is within from 0 to $T/4$.

The dynamics of the sea characteristics was described through the relation of variation amplitudes of the sea level, water surface area, evaporation and outflow to the Kara-Bogaz-Gol Gulf to the difference of values of these components of two equilibrium states caused by maximal and minimal inflow. This relation and the time lag of the sea characteristics to inflow for periods from 2 to 150 years and for different values of inflow were computed on the special nomographs.

The relation of amplitude of the sea characteristics to the difference of their values of maximal and minimal inflow equilibrium states grows from 0 to 1 with increase of variation period. This relation is than larger than lower mean inflow to the sea.

The time lag of sea characteristics to inflow grows with increase of variation period. For the same value of period the time lag is larger under high water inflow than under low inflow.

The observed time-series data of the Caspian Sea level and water inflow were approximated by periodical functions with oscillation period up to 140 years. The time lag between extremes of approximation curves of level and inflow is equal to 15 years. According to the model results the lag of Sea level to inflow under its variation with period 140 years is equal to 16–17 years.

The relation of amplitude of level approximation curve to the difference of levels of two equilibrium states, caused by maximal and minimal values of inflow approximation curve is equal to 0,71. According to the model results the relation of level variation amplitude to the difference of equilibrium levels of maximal and minimal inflow values is equal to 0,72.

The modeling results, in general, concord with the observation data.

This model may be interesting for research of variation of the Caspian Sea water balance components in present and past. The indirect data about changes of the sea level in the historical past make it possible to reconstruct the curve of water inflow variation, which with some reservations, permits to describe the dynamics of moisture of the large endorheic Eurasian area.

References

Georgievskii, V.Iu. (1982). "Water balance of the Caspian Sea under the observation data". Leningrad. Proceedings of LGMI. vol. 79, 94–112 (in Russian).

Grosswald, M.G. (1989). *The last great glaciation of the USSR territory.* Znanie, Moscow, 49 pp. (in Russian).

Kamke, E. (1990). *The manual for ordinary differential equations.* Nauka, Moscow, 576 pp. (in Russian).

Karpytchev, Iu.A. (2001). "The variation of the Caspian Sea level in the historical period." *Water resources,* No. 1, 5–17 (in Russian).

Kritskii, S.N., Korenistov D.V., Ratkovich, D.Ya. (1975) *Variations of the Caspian Sea level.* Nauka, Moscow, 159 pp. (in Russian).

Linnik, Iu.V. (1962). *Method of the least squares.* Nauka, Moscow, 350 pp. (in Russian).

Malinin, V.N. (1994). *The problem of forecasting the level of the Caspian sea.* Izd-vo RGGMI, St. Petersburg, 160 pp. (in Russian).

Ratcovich, D.Ya. (1993). *Hydrological principles of water supply.* Izd-vo IVP, Moscow, 429 pp. (in Russian).

Shiklomanov, I.A. (1976). *Hydrological aspects of the Caspian Sea problems.* Gidrometeoizdat, Leningrad, 79 pp. (in Russian).

Tshepalyga, A. (2000). "Rapid inundations of the Ponto-Caspian shelves, their origin and impact on early man settlement." Proc. of Conference "Hydrological consequence of global climate changes, geologic and historic analogs of future conditions." Moscow, Institute of Geography of Russian Academy of Sciences, 30–31.

Zaikov, B.D. (1946). Water balance of the Caspian Sea and the causes of its level reduction. Proceedings of GUGMS, vol. 38, ser. 4, 5–50 (in Russian).

Section 7: Flood and drought frequency analysis

Surface Water Hydrology, Singh, Al-Rashed & Sherif (eds)
© 2002 Swets & Zeitlinger, Lisse, ISBN 90 5809 363 8

Nonstationarities in river flow time series

A. Ramachandra Rao
School of Civil Engineering, Purdue University
West Lafayette, IN 47907, USA

Khaled H. Hamed
Irrigation and Hydraulics Dept., Faculty of Engineering
Cairo University, Orman, Giza, Egypt
e-mail: hamedkhaled@hotmail.com

Abstract

Hydrologic time series often exhibit signs of nonstationarity. These non-stationarities may be due to natural or anthropogenic causes and may manifest as quasi-periodicities, trends, changes in mean and variance and in spectral characteristics. The nonstationarities may be analyzed by using, among other methods, (a) spectral analysis, (b) time-frequency analysis, (c) tests based on evolutionary spectra for nonstationarity in time series, and (d) time scale analysis based on wavelets. In this paper the monthly and annual river flow data from the Midwestern United States are analyzed by using the above-mentioned tests. The results are presented and compared. Perhaps the most important conclusion from this study is that there is a general agreement in the results about nonstationarity of the series. All the river flow sequences analyzed in the study exhibit nonstationarity characteristics. There is also evidence about changes in periodicities with time in these series. These changes indicate that the series may be nonlinear as well.

Introduction

Hydrologic time series often have changes in mean, variance and other statistical characteristics. These changes may be due to natural or man-made causes. In fact, climatic changes often appear as nonstationary behavior in the mean. Methods and tests to investigate the nonstationary behavior of time series are still evolving. Results from a single test may not be definitive enough to arrive at strong conclusions about the behavior of the series. Consequently, it is preferable to analyze the series by using a variety of tests and compare the results. If different tests, which are based on different assumptions, give similar results, then the conclusions based upon them would be reliable. This is the approach taken in the present study.

Nonstationarities in time series are analyzed in this paper by using methods based on (a) spectral analysis, (b) time frequency analysis, (c) tests based on evolutionary spectra for nonstationarities in time series and (d) time scale analysis based on wavelets. River flow data from the midwestern United States are analyzed and the results presented in the present study.

The paper is organized as follows. The data used in the study is discussed in the next section. The analysis by the four methods mentioned above and the corresponding results are presented next. The conclusions from the study are presented in the last section. The discussion of tests and computational details are kept brief because of space limitations and also because they are easily available in the literature. The details of these tests and related aspects are, however, found in the references.

Data Used in the Study

Flow data used are in this study. These data are obtained from the United States Geological Survey (USGS) database (USGS, 1992). Flow time series from 26 stations in the Midwestern region in the U.S. were obtained from the USGS database. These consist of 6 stations in Iowa, 7 stations in Illinois, 6 stations in Indiana, and 7 stations in Ohio. Table 1 gives the station information and length of the time series for each station as well as starting and ending years. Figure 1 shows the flow time series from these 26 stations.

Maximum Entropy and Multi-Taper Method of Spectral Analysis

The Maximum Entropy Method (MEM) (Burg, 1975) is a parametric method for spectral analysis. In the MEM, autoregressive parameters are estimated by using Burg's algorithm (Burg, 1968), which results in more accurate spectral estimates than those obtained by using ordinary autoregressive spectral estimation methods. The MEM offers high-resolution power spectra and provides great precision for estimating the frequency of individual harmonics.

The Multi-Taper Method (MTM) (Thomson, 1982) makes use of an extended version of the spectral representation in Eq. 1. The process x_t in this case may include a number of periodic components in addition to the underlying stationary process (Percival and Walden, 1993),

$$x_t = \sum_j C_j \cos\left(2\pi f_j t + \phi_j\right) + \zeta_t \tag{1}$$

Details of the multi taper method are found in (Thomson, 1990a; 1990b). In the MTM method, the multi-window approach provides a simple, yet very useful, likelihood ratio test for the significance of periodic components. One of the important features of the MTM is that their properties do not depend on the assumption of normality or stationarity of the time series (Thomson, 1990b).

Results From MEM Analysis

The MEM method offers no measure for the significance of different periodic components. However, the closeness of the roots of the characteristic equation to unity can be used as a guide for the significance of identified components. Bearing in mind that noise roots may be pushed towards the unit circle as a result of increase-

Table 1. USGS stations for flow time teries.

No	State	USGS ID Number	Station Name	Start Year	End Year	N
1	IA	05418500	Maquoketa R. n. Maquoketa, IA	1913	1992	79
2		05420500	Mississippi R. at Clinton, IA	1873	1992	119
3		05454500	Iowa R. at Iowa City, IA	1903	1992	89
4		05464500	Cedar R. at Cedar Rapids, IA	1902	1992	90
5		05484500	Racoon R. at Van Meter, IA	1915	1992	77
6		05490500	DesMoines R. at Keosa. , IA	1911	1992	81
7	IL	03345500	Embarasse R. at Ste. Marle, IL	1912	1992	80
8		05435500	Pecatonica R. at Freeport, IL	1915	1992	77
9		05527500	Kankakee R. n. Wilmington, IL	1916	1992	76
10		05570000	Spoon R. at Seville, IL	1915	1992	77
11		05572000	Sangamon R. at Monticello, IL	1911	1992	81
12		05592500	Kaskaskia R. at Vandalia, IL	1911	1992	81
13		05597000	Big Muddy R. at Plumfield, IL	1914	1992	78
14	IN	03276500	Whitewater R. at Brookville, IN	1921	1992	71
15		03324500	Salmonie R. at Dora, IN	1924	1992	68
16		03326500	Mississinewa R. at Marion, IN	1930	1992	62
17		03335500	Wabash R. at Lafayette, IN	1923	1992	69
18		03373500	E. Fork White R. at Shoals, IN	1915	1992	77
19		05518000	Kankakee R. at Shelby, IN	1923	1992	69
20	OH	03109500	L. Beaver Cr. E. Liverpool, OH	1916	1992	76
21		03202000	Racoon Cr. at Adamsville, OH	1925	1992	67
22		03261500	G. Miami R. at Sidney, OH	1915	1992	77
23		03262000	Loramie Cr. at Lockington, OH	1920	1992	72
24		03269500	Mad R. near Springfield, OH	1915	1992	77
25		03270500	G. Miami R. at Dayton, OH	1914	1992	78
26		04191500	Auglaize R. near Defiance, OH	1917	1992	75

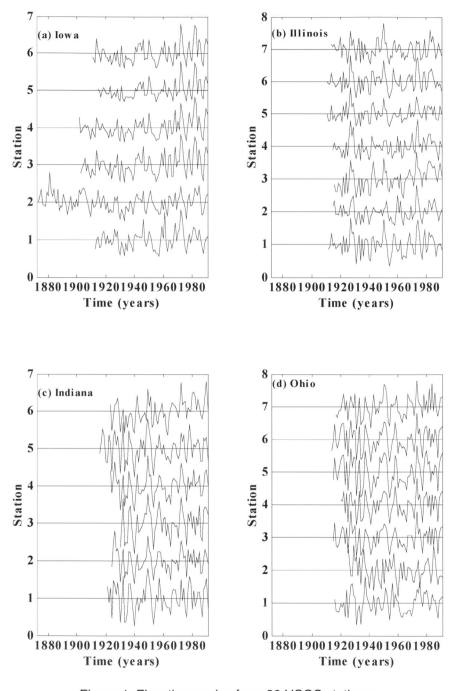

Figure 1. Flow time series from 26 USGS stations.

ing the AR order (Kay, 1988), only periodicities corresponding to roots larger than 0.99 in the case of AR order of N/2 (where N is the number of observations) are considered as potentially significant components. Whether a given component is spurious or actually significant is confirmed through the comparison of the different AR order results as well as the comparison with the results from the MTM analysis. Also, the existence of a component at a given period in a large number of records indicates that such periodicity is not likely to be spurious. In Fig. 2 the components are indicated at their corresponding frequencies with circles. The size of each circle is proportional to the closeness of its respective root to the unit circle, scaled between 0.98 and 1.0. The results from the MEM analysis of the USGS flow data using AR models of order of $N/2$ (N is the number of observations) are shown in Fig. 2. In Fig. 2 the longest periodicity is 106.7 years which is present in only one station, followed by 73.7 and 67.6 years in another two stations. Groups of periodicities found in the USGS flow data are summarized in Table 2. In Table 2 the first column is the approximate range of the group, the second and third columns are for AR order of $N/2$ (N is the number of observations) and give the number of stations n, which have significant components in that range and the mean period in that group. The fourth and fifth columns give the same information for AR order of $N/3$. According to Table 2, the most common periods for the USGS flow data are 2.6, 3.2, 4.4, 5.7, 7.6, and 10.9 years.

Results From MTM Analysis

One advantage of using MTM is that in addition to identifying periodic components, a significance level can also be estimated for each component. When a large number of regional data are available, additional inference can be made about such periodicities. Generally speaking, a periodicity which is significant in a large number of stations in a given region suggests that such a component can be concluded to be a regional periodicity. In statistical terms, if the number of stations with such a periodicity is significantly larger than the expected number under the null hypothesis of no periodicity, then this periodicity is statistically significant; otherwise it is spurious. The expected number of stations under the null hypothesis is simply equal to the significance level times total number of stations. A 95% confidence interval can then be constructed for the average number by considering the binomial distribution with a rate of failure of. The use of a significance level of 1/N recommended by Thomson (1990a) seems to be too strict for such an analysis, since the goal is not to assess significance for a single station but rather to compare the actual and expected number of occurrences of significant components in the studied region. Therefore a statistical significance level of of 0.1 (probability of obtained F-value is 0.90 or more) was considered for the significance of line components in addition to the 1/N level (probability equal to or exceeding 1-1/N), where N is the number of data points.

The results from the MTM analysis of the USGS flow data are shown in Fig. 3 for a bandwidth value of 2.5/N. Periodicities found in the USGS flow data by the MTM are summarized in Table 3. The first column in Table 3 gives the approximate range of each group of periodicities. Columns 2 to 5 pertain to the case of $W = 2.5/N$ and give the number of stations with significant components at the 0.1

477

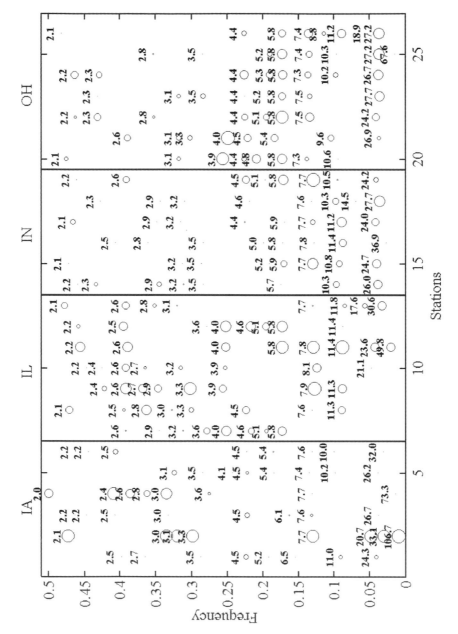

Figure 2. Periodicities in the USGS flow data using MEM analysis with N/2 AR order.

Table 2. Periodicities in the USGS flow data, MEM analysis.

Periodicity Group (Years)	p = N/2		p = N/3	
	No. of Stations	Mean Period	No. of Stations	Mean Period
30-40	4	33.1±2.7	2	31.6±1.9
20-30	17	25.2±2.2	21	26.3±2.2
10-12	18	10.9±0.6	23	10.9±0.4
7-8	20	7.6±0.2	17	7.6±0.2
5.3-6	17	5.7±0.2	23	5.7±0.1
4.9-5.3	10	5.2±.08	0	—
4-5	19	4.4±0.2	24	4.5±0.2
3.5-4	14	3.7±0.2	14	3.7±0.2
3-3.5	20	3.2±0.1	16	3.3±0.1
2.8-3	10	2.9±0.05	11	2.9±0.06
2.4-2.8	21	2.6±0.2	20	2.6±0.1
2.25-2.4	7	2.3±0.02	5	2.3±0.02
2.15-2.25	11	2.2±0.04	7	2.2±0.05
2-2.15	7	2.1±0.02	17	2.1±0.02

level, the number of stations with significant components at the $1/N$ level, the mean of the period in the respective group, and the average percentage of the total variance explained by components in that group. Columns 6 to 9 give the same information but for the case of $W = 4/N$. For this group of 26 stations, considering a 95% confidence interval, no more than five stations should have significant components at a given frequency at the 0.1 level under the null hypothesis of no periodicity. Furthermore, no more than one station should have significant components at a given frequency at the $1/N$ level under the null hypothesis of no periodicity. Based on the numbers in Table 3, there is not sufficient evidence of periodicities in the range from 89 to 121 years, although a component with such periodicity explains 8 to 15% of the total variance in about seven stations. However, it is seen from Figure 3 that most of these stations are found in Iowa, which raises the question of a possible spatial variation of periodicities. The group of 24 to 30 years with an average period of 25.7 years is found in more stations than would be only due to chance. Periodicities of 10–12, 7–8, and 4–5 also seem to be evident. Periodicities of 2–3 years seem to be less evident at the $1/N$ level, with the groups centered at

479

Table 3. Periodicities in the USGS flow data, MTM analysis.

Period Group (years)	W = 2.5/N				W = 4/N			
	n[1] at 0.1	n[1] at 1/N	Mean Period	% Var	n[1] at 0.1	n[1] at 1/N	Mean Period	% Var
89-121	2	0	93.3±6.0	8	7	0	109.2±8.5	15
24-30	17	7	25.7±1.0	10	13	8	27.3±1.2	9
10-12	18	5	11.1±0.3	12	18	1	10.9±0.4	12
7-8	19	6	7.6±0.2	10	4	2	7.8±0.1	13
5-6	21	3	5.7±0.2	13	14	4	5.7±0.2	14
4-5	24	6	4.5±0.2	9	14	7	4.5±0.2	10
3.5-4	13	1	3.7±0.2	4	12	1	3.7±0.2	5
3-3.5	14	2	3.2±0.1	5	16	4	3.3±0.1	5
2.8-3	9	2	2.9±0.03	5	10	0	2.9±0.05	6
2.4-2.8	22	0	2.6±0.1	4	22	3	2.6±0.09	4
2.25-2.4	6	0	2.3±0.04	4	6	0	2.3±0.03	4
2.15-2.25	8	0	2.2±0.03	4	11	1	2.2±0.03	4
2-2.15	16	1	2.1±0.04	4	20	4	2.1±0.03	4

[1] n is the number of stations with components that are significant in a given periodicity group at the specified significance level.

2.1 and 2.6 being the most common at the 0.1 level. The percentage of total variance explained by the identified periodicities range from 4 to 14%. The largest contribution comes from the 5.7 year component at 13–14%, followed by the 11.1 years component at about 12% of the total variance.

Evolutionary Spectral Analysis

Classical spectral estimation methods are based on the assumption that time series are either stationary or can be rendered stationary by using simple transformations (Priestley, 1984; 1988). Therefore, classical analysis when applied to data from nonstationary processes may give misleading results, depending on the degree of nonstationarity of the process. Priestley (1965) developed an approach to study the spectra of nonstationary processes, which has the same physical interpretation as the spectra of stationary processes. The main difference between classical and evolutionary spectra is that the evolutionary spectrum of a time series

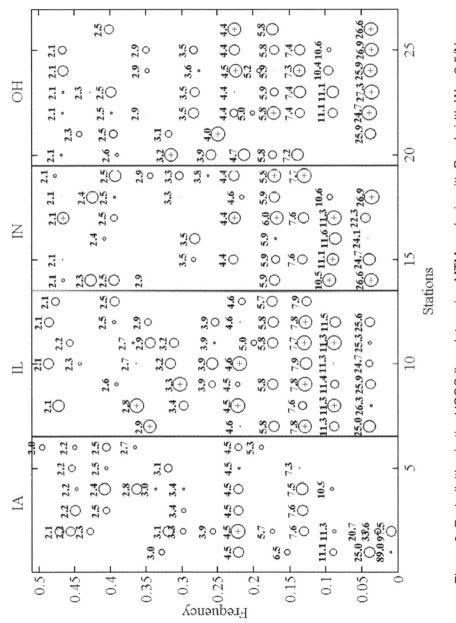

Figure 3. Periodicities in the USGS flow data using MTM analysis with Bandwidth W = 2.5/N.

is time dependent and describes the local power-frequency distribution at each instant in time (Priestley, 1984; 1988). In addition to having a physical interpretation similar to the spectra of stationary processes, evolutionary spectra also provide information on the stochastic structure of the process, which is useful in modeling, and can be estimated by using simple techniques, which do not require prior information about the structure of the process. The numerical computation of evolutionary spectra is roughly equivalent to splitting up a realization in several segments and then estimating individual spectra for each of the segments. Priestley (1965) suggests a method for calculating the evolutionary spectra $S_t(f)$ of a nonstationary time series based on a double windowing technique.

The technique suggested by Priestley (1965) is similar to that applied in the Multi Taper Method (MTM) for spectral analysis, with the difference being that in MTM the variance is reduced by averaging the spectra from the same data segment by using multiple data tapers, thus preserving, to some extent, time resolvability. For the analysis in this section the MTM is used for evaluating evolutionary spectra of the river flow data. In applying the MTM to study evolutionary spectra, the signal is divided into a number of (possibly overlapping) segments each of length T and MTM spectra are calculated for each segment.

Priestley and Subba Rao (1969) suggest a test for nonstationarity in time series based on evolutionary spectra. The test has the form of the analysis of variance for a two-factor design, based on the logarithms of the evolutionary spectral estimates, Y_{ij}, given in Eq. 2.

$$Y_{ij} = \log \hat{S}_{t_i}(f_j) \tag{2}$$

The subscripts i and j in Eq. 2 refer to time segments and frequencies, respectively. The test is based on the sums of uncorrelated variates. Therefore, time segments t_i used in the test should have little or, preferably, no overlapping. The same condition applies for frequencies f_j also, which should have a spacing of at least $2W$, the band width of the data tapers. Statistics used in the test are given in Table 4 below, where dot subscripts indicate averaging over the respective subscript. The distribution of these statistics is given in Eqs. 3 to 5.

$$S_{I+R}/\sigma^2 = \chi^2_{(I-1)(J-1)} \tag{3}$$

$$S_T/\sigma^2 = \chi^2_{(I-1)} \tag{4}$$

$$S_F/\sigma^2 = \chi^2_{(J-1)} \tag{5}$$

Based on the statistics calculated in Table 4, the test procedure is as follows:

1. If the test based on S_{I+R} is significant, then the time series X_t is nonstationary and non-uniformly modulated.

Table 4. Statistics for the Priestley and Subba Rao nonstationarity test.

Item	Deg. of Freedom	Sum of Squares
Between Times	$I.\ 1$	$S_T = J \sum_{i=1}^{I} (Y_{i\bullet} - Y_{\bullet\bullet})^2$
Between Frequencies	$J.\ 1$	$S_F = I \sum_{j=1}^{J} (Y_{\bullet j} - Y_{\bullet\bullet})^2$
Interaction + Residual	$(I.\ 1)(J.\ 1)$	$S_{I+R} = \sum_{i=1}^{I}\sum_{j=1}^{J} (Y_{ij} - Y_{i\bullet} - Y_{\bullet j} + Y_{\bullet\bullet})^2$

2. If the test based on S_{I+R} gives the result as insignificant, proceed to test S_T. A significant S_T indicates that the process is nonstationary, while an insignificant S_T indicates the process is stationary.

3. Similarly, S_F can be used to test for complete randomness. Therefore, a significant S_F indicates a correlated process (colored noise), while an insignificant S_F indicates a random process (white noise).

4. The test can be extended to test stationarity at different frequencies f_j based on the statistic $S_{TF}(f_j)$ in Eq. 6, which is distributed as $\chi^2(I-1)$.

$$S_{TF}(f_j) = \sum_{j=1}^{J} (Y_{ij} - Y_{ij})^2 / \sigma^2 \tag{6}$$

5. Similarly, the test can be extended to test for randomness at different times t_i based on the statistic $S_{TF}(t_i)$ in Eq. 7, which is distributed as $\chi^2(J-1)$.

$$S_{TF}(t_i) = \sum_{j=1}^{I} (Y_{ij} - Y_{i\bullet})^2 / \sigma^2 \tag{7}$$

Results From Evolutionary Spectral Analysis

Figure 4 shows the evolution of the frequencies as well as the significance of line components in the USGS flow data from station 2. This time series is the longest of the available flow data in this region with a length of 119 observations (1874 to 1992). A sliding window of length 60 years is applied at 2-year increments to produce the evolutionary spectra. Figure 5 shows the evolution of the amplitudes of line components for the same time series. In Fig. 5, two main features can be identified. The first feature is the drift in the frequency of a number of components such as the 4.51-year and the 11.3-year components. The second feature is the appar-

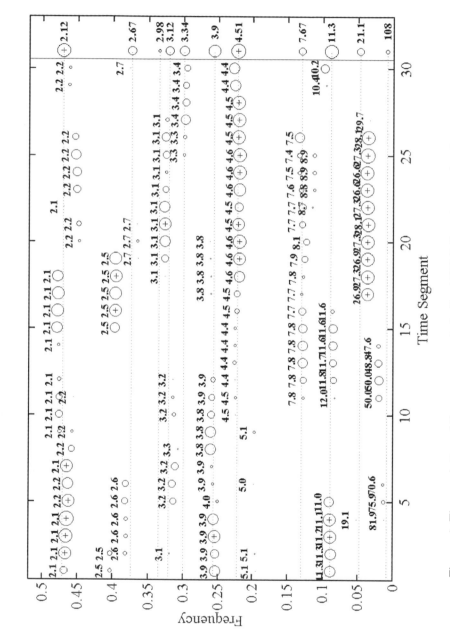

Figure 4. The evolution of line components in the USGS flow time series from station 2.

484

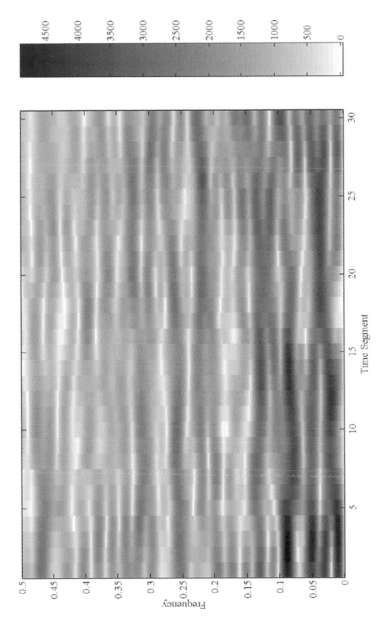

Figure 5. The evolution of the amplitude of line components in the USGS flow time series from station 2.

ent aggregation and splitting of components. This is very clear in Fig. 5 for frequencies near 0.01 cpy (100 years period) and 0.04 cpy (25 years period), which aggregate at segment 5 and proceed to aggregate with the 0.046 cpy frequency (22 years period) at segment 15. These frequencies split back to three distinct frequencies at approximately their original positions at segments 20 and 27. A similar behavior can be observed at segment 20 for the component at 0.13 cpy (7.6 years period). The behavior observed in Figs. 5 and 6 is not typical of stationary linear systems for which line components should appear more or less as straight lines parallel to the time axis at their respective frequencies.

Results of testing the USGS flow data series for stationarity are given in Table 5. For this data set, four time segments each of length $N/3$ at 30% overlapping are used. A bandwidth of $W = 2$ is used for the MTM to ensure three to five independent frequencies, depending on the length N of the time series. From the results in Table 5, we conclude that 13 of the time series are nonstationary at the 0.05 level, and 11 are nonstationary at the 0.01 level. For this group of 26 stations, the 95% confidence interval for the number of stations giving false indication of nonstationarity would be 0 to 3 at the 0.05 level, and 0 to 2 at the 0.01 level. Therefore, the

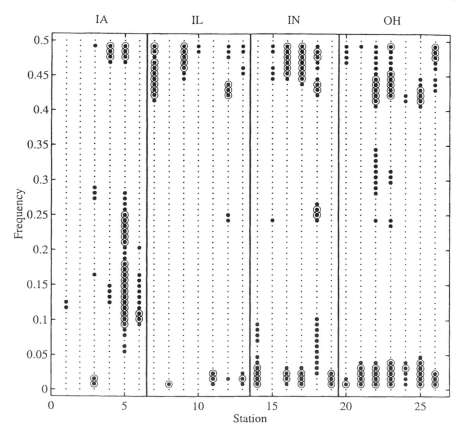

Figure 6. Results of the stationary test at different frequencies for the USGS flow data.

486

Table 5. Results of the Priestley and Subba Rao stationarity test for USGS flow data.

	Number of Significant Stations		
	Based on S_{I+R}	Based on S_T	Based on S_F
At the 0.05 Level	13	9	21
At the 0.01 Level	11	6	17

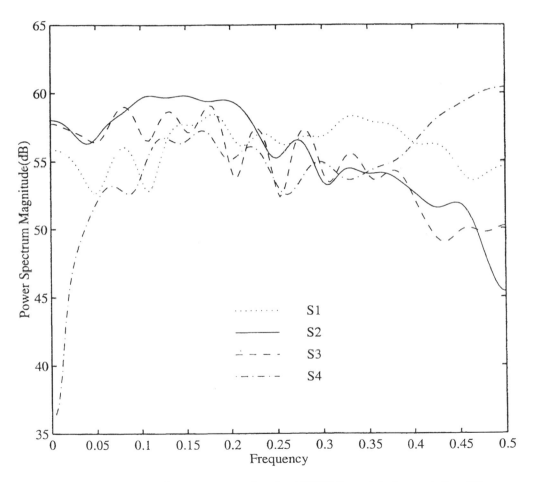

Figure 7. The evolutionary spectra for the USGS flow data from station 22.

number of nonstationary time series in this group is larger than that due to chance. Also, from the results of the test on S_F we conclude that 21 time series at the 0.05 level and 17 time series at the 0.01 level are autocorrelated and there spectra is significantly different than those of random processes.

The results of testing the USGS flow data for stationarity at different frequencies are shown in Figure 6. In Figure 6, points which are significant at the 0.05 level are shown as solid circles. Points that are significant at the 0.01 level are further indicated by a hollow circle surrounding the solid one. From Figure 6, it is observed that most of the stationary behavior appears at high (0.4 to 0.5 cycles per year, cpy) or low frequencies (0 to 0.05 cpy), but some stations (stations 5, 6, and 22) do exhibit nonstationarity at intermediate frequencies as well.

Figure 7 shows the evolutionary spectra of the USGS flow data from station 22 as an example of a nonstationary time series in this group. In Figure 7, there are discrepancies between different spectra especially at high and low frequencies. The stationarity test results for this particular station gave $F(S_{I+R})$ = 1.0000, $F(S_T)$ = 1.0000, and $F(S_F)$ = 1.0000. According to these values the time series from station 22 is highly nonstationary and its spectrum is significantly different from that of a random process. Further testing at individual frequencies supports nonstationarity at high and low frequencies seen in Fig. 5.

Wavelet Analysis

To overcome several well known difficulties associated with spectral analysis, a different approach known as Wavelet analysis has been introduced during the past 10 years (Daubechies, 1992; Combes et al., 1990; Mallat, 1996; Meyer, 1993; Strang and Nguyen, 1996; Chui, 1992). In Wavelet analysis, a different type of basis functions called Wavelets, are used. Wavelets, as opposed to sine and cosine functions, have a limited extent in time (a limited support), and are therefore localized in time as well as in frequency.

In the theory of Wavelet transforms, the notion of "frequency" is changed to an equivalent one of "scale", where large scales correspond to low frequencies and small scales correspond to high frequencies. Wavelet analysis provides a technique for analyzing the structure of a time series by using variable size windows. By using Wavelet analysis, longer time windows can be used to detect low frequency (large scale) information, while shorter time windows can be used to detect high frequency (small scale) information.

Wavelet analysis offers a different approach for looking at signals. In effect, by using the Wavelet transform the signal is decomposed into components of different scales. This offers a method for a local outlook of the signal, a multi-scale outlook, and a time-scale analysis (Misiti et al., 1996). Wavelet analysis can be used as a tool for performing a wide range of tasks which include: detecting discontinuities in signals, detecting long-term evolution (trends), detecting self similarity (fractal signals), identifying pure frequencies (time-frequency analysis), and suppression, and compression of signals.

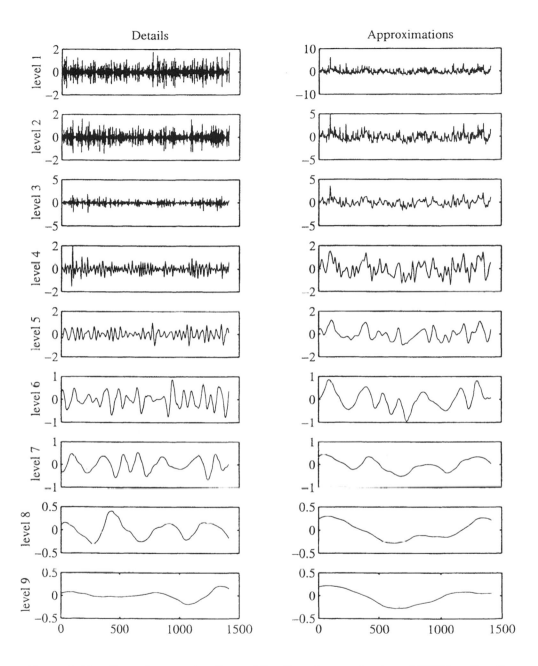

Figure 8. Details and approximations of the flow time series from USGS station 2.

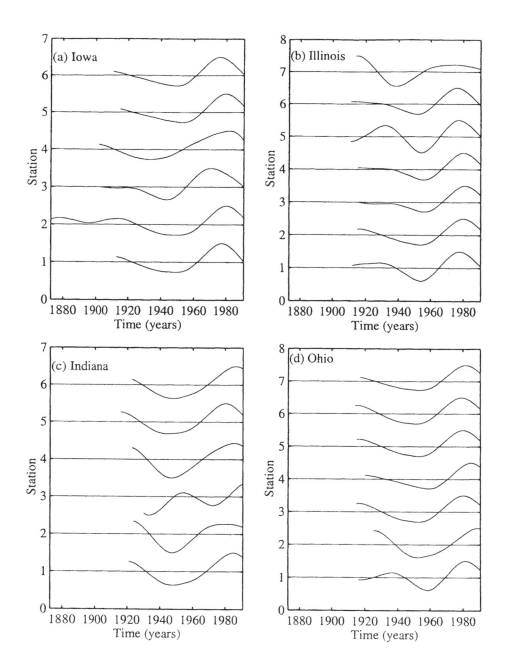

Figure 9. Details of the USGS flow time series at Level 9.

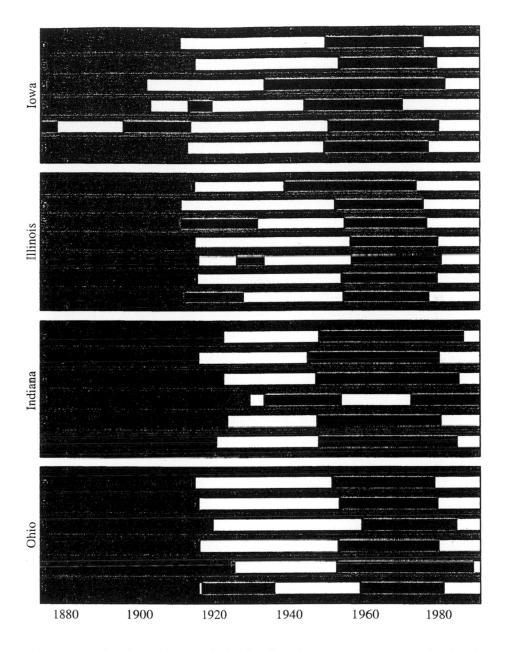

Figure 10. Directions of trends in USGS flow time series based on details at Level 9.

Wavelet Trend Analysis

By definition, a trend in a signal is the slowest part of that signal. Based on wavelet analysis, a signal can be decomposed into components at different scales. Trends can therefore be detected by inspecting details of the signals at large scales (Misiti et al., 1996). In this section, wavelet decomposition is used as a tool to identify trends in flow time series. For the purpose of decomposition a Daubechies wavelet of order 4 (db4) is used. For the data in this study, details of the signals at scales that are close to the length of the data are investigated for the existence of slow components representing trends in the time series. Figure 8 shows an example of the decomposition of a signal into its details and approximations at different levels by using the db4 wavelet. The decomposed signal in Figure 8 is the flow time series from USGS station 2. For trend analysis we focus on the large scales such as levels 8 and 9, which represent the slowest changing part of the signal.

Figure 9 shows the details at a scale of about 42 years for flow time series from 26 USGS stations. Figure 10 is a representation of trends in Figure 9, where increasing trends are depicted as black lines, while decreasing trends are depicted as white lines. It can be observed from these figures that there is a decreasing trend in the first part of the record up to about 1950 and an increasing trend afterwards.

Conclusions

Irrespective of the method used, whether it is spectral analysis, evolutionary spectral analysis or the time scale analysis, it is quite obvious that the river flow time series exhibit strong evidence of being nonstationary. The future research in this area should look into the methods of modeling these types of nonstationarity.

References

Burg, J.P., 1968. "A new analysis technique for time series data." *NATO Advanced Study Institute on Signal Processing with Emphasis on Underwater Acoustics*. Reprinted in Childers (1978).

Burg, J.P., 1975. *Maximum Entropy Spectral Analysis*. Ph.D. Dissertation, Dept. of Geophysics, Stanford University.

Chui, C.K., 1992. *An Introduction to Wavelets*. Academic Press.

Combes, J.M., Grossmann, A., Tchamitchian, Ph., eds. 1990. *Wavelets, Time-Frequency Methods, and Phase Space*. Springer.

Daubechies, I., 1992. *Ten Lectures on Wavelets*. SIAM, Philadelphia, PA.

Kay, S.M., 1988. *Modern Spectral Estimation, Theory and Application*. Prentice-Hall Signal processing Series. Alan V. Oppenheim, Series Editor. Prentice-Hall, N.J.

Mallat, S., 1996. *Wavelet Signal Processing*. Academic Press, New York, NY.

Meyer, Y., 1993. *Wavelets: Algorithms and Applications*. SIAM.

Misiti, M., Misiti, Y., Oppenheim, G., Poggi, J.M., 1996. *Wavelet Toolbox for use with MATLAB*. The Mathworks, Inc., Natick, Mass.

Percival, D.B. and Walden, A.T., 1993. *Spectral Analysis for Physical Applications: Multi-taper and Conventional Univariate Techniques*. Cambridge Univ. Press, Great Britain.

Priestley, M.B., 1965. "Evolutionary Spectra and non-stationary processes." *J. Roy. Statist. Soc.*, Ser. B, 27, pp. 204-237.

Priestley, M.B., 1984. *Spectral Analysis and Time Series*. Third Printing. Academic Press, London.

Priestley, M.B., 1988. *Non-linear and Non-stationary Time Series Analysis*. Academic Press, London.

Priestley, M.B. and Subba Rao, T., 1969. "A test for non-stationarity of time series." *J. Royal Statist. Soc.*, 31(1), pp. 140-149.

Strang, G. and Nguyen, T., 1996. *Wavelets and filter banks*. Wellesley. Cambridge Press, Wellesley, MA.

Thomson, D.J., 1982. "Spectrum estimation and harmonic analysis." *Proc. IEEE*, 70, pp. 1055-1096.

Thomson, D.J., 1990a. "Time series analysis of Holocene climate data." *Phil. Trans. Roy. Soc. London*, Series A, 330, pp. 601-616.

Thomson, D.J., 1990b. "Quadratic-inverse spectrum estimates: Applications to paleoclimatology." *Phil. Trans. Roy. Soc. London*, Series A, 332, pp. 593-597.

United States Geological Survey (USGS), 1992. *USGS daily values*. Data Files, Computer Laser Optical Disks, EarthInfo, Boulder, CO.

Surface Water Hydrology, Singh, Al-Rashed & Sherif (eds)
© 2002 Swets & Zeitlinger, Lisse, ISBN 90 5809 363 8

Effects of climate and soil moisture characteristics on the probability distribution of floods

Mauro Fiorentino
Department of Engineering and Environmental Physics
University of Basilicata, C/da Macchia Romana, 85100, Italy
e-mail: fiorentino@unibas.it

Vito Iacobellis
Dipartimento di Ingegneria Civile e Ambientale
Technical University of Bari, Via E. Orabona 4, 70125, Bari, Italy
e-mail: v.iacobellis@poliba.it

Abstract

The influence of climate and soil moisture characteristics on the flood generation process is here discussed with particular regard to the probability distribution of floods in the theoretical framework proposed by Iacobellis and Fiorentino (2000). The variability of the coefficient of variation of annual flood series with respect to basin characteristics is also investigated, and a link between these characteristics and the chance of a basin to yield floods is found. In addition, the probability distribution of partial areas contributing to the peak discharge is analyzed with regard to its link to prevailing runoff generation mechanisms in different climates and soils. All these issues are developed with particular attention to arid basins.

Introduction

The general context of this paper is to promote a deeper knowledge of the physical mechanisms involved in the flood generation process with particular regard to its use in the framework of flood frequency analysis. Within this field, important problems may be considered still open notwithstanding the great efforts produced in the last twenty years of hydrologic research. Among those, we like to point out that most procedures for estimating the mean annual flood, in the context of the index flood method (Dalrymple, 1960), are still empirical and they are often different from one region to another. Moreover, statistical regional analyses may lead to consider regions different for geology, morphology, climate, etc. as if they were homogeneous. The available methods for flood estimation, including regional analysis, usually make use of statistical procedures which essentially exploit the hydrometric and pluviometric information leaving only a marginal role to other information (geology, climate, vegetation and so on) which is crucial to the investigation of the base process whose maxima we are dealing with.

The influence of climate on the flood generation process is nowadays a topic of striking impact and particular remark due to the necessity of improving the available techniques and procedures for risk assessments and land protection. In particular great efforts of the recent research are provided to support the flood frequency analysis exploiting the amount of information available by the observation of the frequency of precipitation, vegetation coverage, soil permeability, etc in order to improve the performances of models for flood prediction. In facts, the climate influences the flood regime not only through the precipitation patterns but also by means of antecedent soil moisture condition and its most probable state which depends on the frequency and intensity of precipitation, vegetation coverage, soil permeability, etc. The antecedent soil moisture is then responsible of boundary and initial conditions. At the basin scale, these last quantity has to be considered as a result of a number of overlapping processes related to the mechanisms of runoff generation, and to the variability in time and space of precipitation, sorptivity and permeability of soil. In flood frequency analysis water losses processes are often taken into account in order to estimate position parameters like the mean annual flood. More rare is instead the case where these processes are invoked to infer variability characteristics of the flood distribution. However, we believe that more attention should be paid to this issue.

From a more general viewpoint, the estimation of flood frequency curves in ungauged basins is an important field of application of advanced research concerning the knowledge on physical processes and the development of statistical tools. Regional statistical analysis and physically-consistent derivation of probability density functions (*pdf*) are the key fields to rely on for providing robustness to estimation of flood quantiles, and for transferring hydrological information between basins. In this context, physically consistent reasoning applied to the regional statistical analysis can be closely connected with some basis of a geomorphoclimatic approach for derivation of the flood *pdf*. In facts, the characterization of peculiar features of the flood process finds a significant interpretative context within the geomorphoclimatic derivation of a flood probability distribution. The probability function of peak streamflow can be derived, either in an analytical or synthetic way, from the probability density function of rainfall, by using the functional relationships provided by the basin's hydrologic response. Following Eagleson (1972) several studies have been conducted in this field. Amongst them, the majority has been devoted to improving the interpretation of the hydrologic response and the modeling of rainfall (e.g. Hebson and Wood, 1982; Cordova and Rodriguez-Iturbe, 1983; Diaz-Granados et al., 1984; Wood and Hebson, 1986; Moughamian et al., 1987; Adom et al., 1989, Shen et al., 1990, Cadavid et al., 1991, Kurothe et al., 1997). In these works, the effect of partial contributing areas as well as that of different streamflow regimes, climate, vegetation and soil characteristics has been practically neglected. Sivapalan et al. (1990) suggested the need to take into account the role of different mechanisms of runoff generation, while Gottschalk and Weingartner (1998), pointed out the intrinsic randomness of the runoff coefficient. On the other hand, the impact of the climate on the probability distribution of floods has been clearly shown. For instance, Farquharson et al. (1992) demonstrated that the coefficient of variation of the flood series is systematically smaller in humid zones than in arid or semi-arid areas. Therefore, over the last twenty years, there has been a growing number of researchers who believe that, in order to improve the knowledge about the statistics of floods, more research needs to be developed with regard to the role played by runoff generating mechanisms. These are in fact strongly related to climate, vegetation and soil characteristics and exercise a significant control on the catchment areas that contribute to the flood peak.

A new rationale for deriving the probability distribution of floods and help in understanding the physical processes underlying the distribution itself was presented by Iacobellis and Fiorentino (2000) and Fiorentino and Iacobellis (2001). Based on this, a model that presents a number of new assumptions was developed, based on the following assumptions: (i) the peak direct streamflow Q can always be expressed as the product of two random variates, namely the average runoff per unit area u_a and the peak contributing area a; (ii) the distribution of u_a conditional on a can be related to that of the rainfall depth occurring in a duration equal to a characteristic response time τ_a of the contributing part of the basin; (iii) τ_a is assumed to vary with a according to a power law. Consequently, the probability density function of Q can be found as the integral, over the total basin area A, of that of a times the density function of u_a given a. In the model, it is suggested that u_a can be expressed as a fraction of the excess rainfall and that the annual flood distribution can be related to that of Q by the hypothesis that the flood occurrence process is Poissonian. In the proposed model it is assumed that a and u_a are gamma and Weibull distributed respectively. The model was applied to the annual flood series of 20 gauged basins in Basilicata and Puglia (Southern Italy) with catchment areas ranging from 40 to 1650 km^2. These regions are quite heterogeneous for climatic, geologic and land use characteristics. Thus, a stronger validation of the theoretical model was provided and more light was shed on the existing meaningful matches between the prevailing mechanisms which control the formation of partial contributing areas, the role of hydrologic losses within the rainfall-runoff process, the large scale permeability of the outcropping units, and important features of the probability distribution of extreme floods. The results showed strong physical consistence as the parameters tended to assume values in good agreement with well consolidated geomorphologic knowledge and suggested a new key to understanding the climatic control of the probability distribution of floods.

In this paper, the model is discussed with particular regard to its outcomes in arid basins. In particular, a significant interpretation to the observed relationships observed between characteristic water losses, climate and basin area A is provided. By means of results obtained on basins in Southern Italy, it is possible to suppose that in dry regions rain losses are mainly due to the initial abstraction phenomenon. Such findings are consistently matched to what one could expect from Hortonian behavior of soils. Conversely, in humid basins, the water losses show a strong relationship with climate represented by a climatic index. In this case, the basin's behavior is consistent with Dunne's model of runoff generation. Those results are supported by the theoretical framework provided by the classical analysis of infiltration in unsaturated porous media.

Theory

The quoted theoretical model (Iacobellis and Fiorentino, 2000) for analytical derivation of the flood frequency distribution is based on a simple rainfall-runoff model in which the peak flow Q_P is treated as a stochastic variable function of two other mutually dependent stochastic variables, namely the mean areal rainfall intensity $i_{a,\tau}$ and the peak runoff contributing area a:

$$Q_P = \xi \ (i_{a,\tau} - f_a) \ a + q_o \qquad (1)$$

where ξ is a constant routing factor, $i_{a,\tau}$ is the areal rainfall intensity in the duration equal to the lag-time τ_a within the source area a, f_a is the average water loss rate, within the same duration τ_a and area a, q_o is a constant base flow.

The time-space behavior of the involved quantities is basically controlled by the commonly observed geomorphologic power-type relationship between basin lag-time τ_A and basin area A which can be written as:

$$\tau_a = \tau_1 a^\nu \quad \text{with} \quad \tau_1 = \tau_A A^{-\nu} \qquad (2)$$

where τ_A is the lag-time of the basin and ν usually assumes values close to 0.5. The mean areal rainfall intensity $E \ [i_{a,\tau}]$ is usually found to scale with a according to the power law

$$E\left[i_{a,\tau}\right] = i_1 a^{-\varepsilon} \quad \text{with} \quad i_1 = E\left[i_A\right] A^\varepsilon \qquad (3)$$

where i_1 is rainfall intensity referred to the unit area. Also, in the model f_A is in general supposed to scale with the basin area A through a relationship of the type:

$$f_a = f_1 a^{-\varepsilon'} \quad \text{with} \quad f_1 = f_A A^{\varepsilon'} \qquad (4)$$

in which f_A represents the average water loss rate when the entire basin contributes to the flood peak. Indeed, τ_A, $E \ [i_A]$, f_A, ν, ε and ε' are characteristic features of basins.

Under the assumption of a rainfall process with poissonian occurrences and Weibull distributed intensity, the average areal water loss f_A may be related to the ratio between the average annual rates of rainfall and flood events, respectively Λ_p and Λ_q, by means of the equation (Fiorentino and Iacobellis, 2001):

$$f_A = \frac{E[i_A]}{\Gamma(1+1/k)} \left[\log\left(\frac{\Lambda_p}{\Lambda_q}\right) \right]^{1/k} \qquad (5)$$

where k is the exponent of the Weibull distribution of rainfall intensity and $\Gamma(\ )$ is the gamma function.

In the model, an important characteristic of the probability distribution of floods is the mean annual number Λ_q of independent floods, which, when this distribution is schematized as a compound Poisson process (e.g. EV1, GEV, TCEV), represents the distribution parameter that more strongly controls the coefficient of variation Cv. As shown by the above equation, Λ_q is related to the water loss parameter f_A through the formula:

$$\Lambda_q = \Lambda_p \exp\left(-\frac{f_A^k}{E[i_{A,\tau}^k]}\right)$$

(6)

Given a series of rainfall annual maxima, parameters p_1 and n of the at-site intensity-duration-frequency curve (idf) are derived according to equation:

$$E[p_t] = p_1 t^{n-1}$$

(7)

This is referred to the expected value of the annual maximum rainfall intensity p_t, in the duration t.

Assuming the hypotheses of Weibull distribution of rainfall intensity and poissonian occurrence of events, the distribution of annual maxima turns out to be a Power Extreme Value (PEV) type.

In the same framework, exploiting the relationship between the averages of *annual maxima* and of the *base process*, the average areal rainfall intensity may be estimated by means of the equation:

$$E[i_A] = \frac{p_1 \tau_A^{n-1}\left[1 - \exp\left(-1.1\tau_A^{0.25}\right) + \exp\left(-1.1\tau_A^{0.25} - 0.004A\right)\right]}{\Lambda_p \sum_{j=0}^{\infty} \frac{(-1)^j \Lambda_p^j}{j!(j+1)^{(1/k+1)}}}$$

(8)

This is valid for PEV distributed data, where p_1 is the mean annual rainfall depth in 1 hour. Equation (8) includes the well-known Weather Bureau areal reduction formula, here used replacing time with the basin lag-time τ_A.

Replacing relations (4) and (5) in Equation (6), after considering that

$$E[i_A^k] = \left(\frac{E[i_A]}{\Gamma(1+1/k)}\right)^k$$

(9)

we obtain:

$$\Lambda_q = \Lambda_p \exp\left[-\left(\frac{f_1 \Gamma(1+1/k) A^{\varepsilon'-\varepsilon}}{i_1}\right)^k\right]$$

(10)

From the above relation it appears that the scaling relationship Λ_q-area is clearly dependent on the values assumed by the parameters ε and ε'. The first parameter depends on the slope of the IDF curve and on the exponent of the geomorphological relationship τ-area and usually assumes values around 0.3-0.4. The second one is much more variable and its value, according to Fiorentino and Iacobellis (2001) and Fiorentino et al. (2001), may be characterized by the long term climate by way of the probable state of the basin in terms of its antecedent soil moisture conditions.

The *cdf* (cumulative density function) of the annual maximum values of Q_p was then obtained as:

$$
F_{Q_p}(q_p) = \exp\left\{-\Lambda_q \int_0^A \left[\left(\frac{1}{\alpha \Gamma(\beta)}\left(\frac{a}{\alpha}\right)^{\beta-1} \exp\left(-\frac{a}{\alpha}\right) + \delta(a-A)P_A\right)\right.\right.
$$
$$
\left.\left. \exp\left(-\frac{\left((q_p - q_o)/(\xi a) + f_i a^{-\varepsilon'}\right)^k - \left(f_i a^{-\varepsilon'}\right)^k}{\left(i_i a^{-\varepsilon}/\Gamma(1+1/k)\right)^k}\right)\right]da\right\}
$$

(11)

where α and β are position and shape parameters respectively of the a variate probability density function, q_o is the average base flow within the flood season, and P_A is the finite discrete probability of the total contributing area:

$$
P_A = prob[a = A] \tag{12}
$$

The impulsive function $\delta(\cdot)$ in Equation (11) is the Generalized Function which gives

$$
\int_{-\infty}^{\infty} P_A \delta(a-A)da = P_A \tag{13}
$$

Within the same theoretical approach a formulation of the index flood was suggested as:

$$
Q_I = E\left[u_{E[a]}\right]E[a]\Lambda_q S_{\Lambda_q} + q_o \tag{14}
$$

where $E[a]$ is the mean contributing area and $E[u_{E[a]}]$ is the expected peak runoff per unit area referred to $E[a]$. In the equation, S_{Λ_q} is a function of Λ_q and k, which stems from the PEV formulation:

$$
S_{\Lambda_q} = \sum_{j=0}^{\infty} \frac{(-1)^j \Lambda_q^j}{j!(j+1)^{(1/k+1)}} \tag{15}
$$

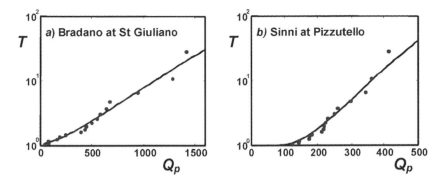

Figure 1. Peak flood derived distribution fitted to data; T is in *years* and Q_p in m³/s.

The probability distribution in Equation (11) has been shown to provide a reasonable good fit to observed annual flood series. An example of such a fit is given in Figure 1 for two basins whose main characteristics are listed in Table 1, where T is the return period and Q_p is the maximum annual peak discharge.

Study Area and Data Set

The investigated zone consists of 32 gauged basins of three administrative regions, namely Basilicata, Puglia, and Calabria, in Southern Italy (Figure 2).

These basins are listed in Table 1. In this area the climate is quite variable due to the morphological differences of lands. Within the northeastern sector (Puglia), characterized by low hills or flat lands, climate is of the hot-dry Mediterranean (semiarid or dry sub-humid) type, with mild, not very rainy winters and warm, dry summers. As one proceeds to the West-Southern sector (Basilicata and Calabria), climate turns cold and humid, of mountain type (Southern Appennine). The annual average rainfall depth goes from minimum values (about 600 mm per year) observed in Puglia and higher values (up to 1800 mm) in Basilicata and Calabria.

Figure 2. Investigated region.

Table 1. Basin Features in Puglia, Basilicata, and Calabria (Southern Italy).

Site	A (km^2)	τ_A (h)	Λ_p (years^{-1})	Λ_q (years^{-1})	Cv	f_A (mm/h)	I
Puglia Region							
St Maria at Lucera-Torrem.	58	2.6	44.6	2.6	0.92	6.48	-0.28
Triolo at Lucera-Torrem.	56	2.6	44.6	3.1	0.70	8.67	-0.25
Salsola at Foggia-San Sev.	455	7.3	44.6	5.0	0.54	1.89	-0.27
Casanova at Lucera Motta	57	2.6	44.6	3.7	0.81	5.62	-0.14
Celone at Foggia-San Sev.	233	5.2	44.6	6.6	0.72	2.09	-0.24
Celone at San Vincenzo	92	3.3	44.6	6.1	0.61	3.41	-0.06
Cervaro at Incoronata	539	8.0	44.6	5.2	0.58	1.80	-0.19
Carapelle at Carapelle	715	9.2	44.6	8.5	0.57	1.19	-0.23
Venosa at Sant'Angelo	263	5.6	44.6	4.2	1.18	2.64	-0.17
Arcidiaconata at Rapol.-Lav.	124	3.8	44.6	4.1	0.65	3.67	-0.04
Ofanto at Rocchetta	1111	11.5	21.0	4.7	0.58	1.13	0.16
Atella at Ponte sotto Atella	176	4.6	21.0	6.3	0.57	1.76	0.17
Basilicata Region							
Bradano at Ponte Colonna	462	4.3	21.0	4.0	0.76	2.36	-0.08
Bradano at San Giuliano	1657	7.1	21.0	2.9	0.79	2.17	-0.17
Basento at Menzena	1382	5.95	21.0	6.6	0.63	1.22	0.7
Basento at Pignola	42	2.9	21.0	19.6	0.43	0.07	0.28
Basento at Gallipoli	853	4.8	21.0	8.5	0.63	0.97	0.08
Agri at Tarangelo	511	8.9	21.0	16.8	0.49	0.14	0.47
Sinni at Valsinni	1140	5.6	21.0	19.1	0.56	0.07	0.57
Sinni at Pizzutello	232	2.4	32.0	31.0	0.51	0.03	1.26
Calabria Region							
Esaro at La Musica	520	4.7	20	3.0	0.82	3.5	0,77
Coscile at Camerata	285	3.7	20	3.2	0.74	4.5	0,65
Trionto at Difesa	32	2.8	20	10.7	1.09	1.0	0,90
Tacina at Rivioto	79	3.0	10	4.0	1.27	3.2	1,43
Alli at Orso	46	3.0	20	4.0	0.72	5.8	1,26
Melito at Olivella	41	3.0	20	4.8	0.62	4.5	0,72
Corace at Grascio	182	3.8	20	4.5	0.70	3.4	0,90
Ancinale at Razzona	116	3.9	10	3.3	0.73	4.4	1,34
Alaco at Mammone	15	1.3	10	3.5	0.75	7.4	1,66
Amato at Marino	113	4.6	20	5.0	1.18	2.6	0,86
Lao at Piè di Borgo	280	3.7	34	5.5	0.59	3.9	1,16
Noce at La Calda	42.5	1.3	34	13.7	0.41	2.9	1,58

Rainfall is distributed quite irregularly over the year, with the highest average shown in the October-March semester, more than twice that of the April-September period. July is the least rainy month, while the highest varies from October to January.

The climatic classification was performed by means of the climatic index (Thornthwaite, 1948):

$$I = \frac{h - E_p}{E_p} \tag{16}$$

In Equation (16), h is the mean annual rainfall depth and E_p the mean annual potential evapotranspiration calculated according to Turc's formula (Turc, 1961), dependent on the mean annual temperature only.

The rainfall annual maxima in about 450 raingauge stations (with record length not less than 20 years) were used in the relevant regional statistical analysis. In this analysis, the key parameter to estimate is the mean annual number, Λ_p, of independent rainfall events. Local and regional values of Λ_p were obtained by means of a regional model based on a Two Component Extreme Value distribution (Rossi et al., 1984). Parameters Λ_1, θ_1, and Λ_2, θ_2 of the TCEV were estimated using a Maximum Likelihood (TCEV-ML) procedure (Gabriele and Iiritano, 1994) with hierarchical estimation of parameters (Fiorentino et al., 1987), based on the homogeneous areas found in Versace et al. (1989). The estimated values of Λ_p, which can be related to the coefficient of variation of the rainfall annual maxima, are displayed in Table 1.

A regional estimation based on the Power Extreme Value-Maximum Likelihood procedure (Villani, 1993) was then applied to the same stations, providing the estimates of the shape factor k defined in Equation (9). Regional values of k were thus put to 0.8 in Puglia and Basilicata and to 0.53 in Calabria. The estimates of $E[i_A]$ obtained from Equation (8) are shown in Figure 3 (Puglia and Basilicata) and Figure 4 (Calabria). In these figures different plots were used at the aim to emphasize the goodness of fit of the regression line.

In each region (or sub-region) $E[i_A]$ estimates are in a good agreement with the power law relationships given in Equation (3). In particular, as regards Calabria, basins belonging to the Thyrrenian and Central zones lead to $i_1 = 11.5$ [mm h^{-1} km$^{-2\varepsilon}$] and $\varepsilon = 0.28$ ($R^2 = 0.94$), while basins of the Jonian Zone are characterized by higher rainfall intensities with $i_1 = 28.8$ [mm h^{-1} km$^{-2\varepsilon}$] and $\varepsilon = 0.32$ ($R^2 = 0.98$). With regard to the other regions, the following parameters were achieved: $i_1 = 10$ [mm h^{-1} km$^{-2\varepsilon}$] and $\varepsilon = 0.39$ ($R^2 = 0.90$) in Puglia, $i_1 = 13$ [mm h^{-1} km$^{-2\varepsilon}$] and $\varepsilon = 0.33$ ($R^2 = 0.78$) in Basilicata. It is noteworthy that the slopes obtained with regard to any regions are not far from each other, with a slight increase in the north-east part of the study area.

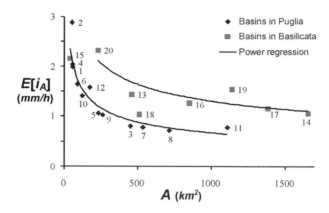

Figure 3. Expected value of the space-time averaged rainfall intensity vs catchment area.

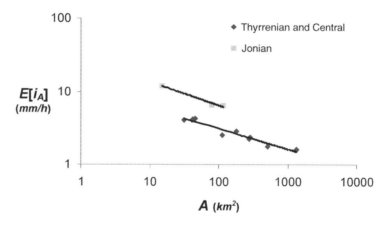

Figure 4. Expected value of the averaged space-time rainfall intensity vs catchment area.

The lithological and hydrological aspects of outcrops appear to be differentiated in the area of interest, with the highest large scale permeability in Puglia and the lowest in Basilicata.

To fulfill the aim of the study, a permeability index ψ was calculated for the basins in Puglia and Basilicata, where more detailed information were available:

$$\psi = \psi_h + 0.9\ \psi_m \tag{17}$$

In this equation, ψ_h and ψ_m are the fraction of the total area with outcrops belonging respectively to the high and medium class of permeability. The estimated values of ψ_h and ψ_m were drawn from Mongelli and Salvemini (1994) for basins in Puglia and

504

from DIFA (1998) for basins in Basilicata, and are reported in Fiorentino and Iacobellis (2001).

The morphology of the study area is overall characterized by mountainous units belonging to Southern Appenine with relief more pronounced and tall within the Center-Southern sector of the study area. Formations are less rugged and steep toward both the Adriatic (Northern) and Jonian (Southern) coast. Surface area and characteristic lag-times of basins are listed in Table 1. More details about other features of basins are found in Fiorentino and Iacobellis (2001).

The available hydrological information is completed by the mean annual number of independent events of rainfall and flood, Λ_p and Λ_q, taken from Iacobellis et al. (1998) and the coefficient of variation Cv of the observed annual flood series (see Table 1) (Claps et al., 1994; DIFA, 1998). In order to obtain estimates of the mean annual number Λ_q of floods shown in Table 1, the recorded series of annual flood maxima were analyzed by a regional GEV-PWM procedure (Hosking and Wallis, 1993). In particular, the Λ_q values were obtained using the regional estimate of L-skewness and the at-site estimates of the L-coefficient of variation.

Infiltration and Areal Water Losses

Infiltration Is the movement of water into the soil matrix. In particular, in the unsaturated zones of soil, the vertical flow is described by the Richards' equation which can be wrote in the form:

$$\frac{\partial \theta}{\partial t} = \frac{\partial}{\partial z}\left[D(\theta)\frac{\partial \theta}{\partial z} + K_z(\theta)\right] \qquad (18)$$

where D is the diffusion coefficient or diffusivity, K_z the effective permeability, θ the volumetric water content and z the coordinate direction, positive upward, with origin at the earth's surface.

Different analytical solution are achievable for particular assumptions, boundary conditions and initial conditions. An exact solution to the variable diffusivity problem of Equation (18), for initial and boundary conditions:

$$\theta = \begin{cases} \theta_i & z \leq 0, \quad t = 0; \\ \theta_o & z = 0, \quad t > 0; \end{cases} \qquad (19)$$

was given by Philip (1957) for θ_o correspondent to saturated surface, in the form of a series expansion whose first three terms give:

$$f(t) = \frac{1}{2}St^{-1/2} + c \qquad (20)$$

505

where the local infiltration process, $f(t)$, depends on time t with coefficients, sorptivity S and gravitational term c, equal to:

$$S = 2(\theta_o - \theta_i)\left[\overline{D}/\pi\right]^{1/2} \tag{21}$$

where $\overline{D}$ is an effective diffusivity over the range of possible moisture values $[\theta_o, \theta_i]$ (e.g. Eagleson, 1970).

$$c = \frac{1}{2}\left[K(\theta_o) - K(\theta_i)\right] \tag{22}$$

Integrating Equation (20) with respect to time, the cumulated infiltration volume $F(t)$ can be found as equal to:

$$F(t) = St^{1/2} + ct \tag{23}$$

Then, in the Equations (20) and (23) the infiltration process is assumed on condition of saturated terrain surface and sorptivity depending on the soil moisture at time zero. Therefore they do not take into account such phenomena as interception, depression storage, and initial moisturizing of vadose zone. In order to account for them, it is possible to introduce a third quantity, the volume of initial abstraction W. Let t_b be the time at which W reaches its maximum value W_b, thus, putting $t' = t - t_b$, we obtain

$$F'(t) = St'^{1/2} + ct' + W \; ; \; t = 0 \;\; \text{if} \; t' < t_b \tag{24}$$

Equations (23) and (24) can be presumed to hold at the basin scale if parameters S, c and W are assumed as the spatial averages of their respective local values.

In order to evaluate the significant rainfall abstraction, the unit runoff occurring on a basin with surface area A may be related, by Equation (1), to the maximum rainfall in the time interval τ equal to the basin lag time and to the time where this interval drops into the time history defined by Equation (24). Let t_i be the initial time of this time window and t_f its final time ($t_f - t_i = \tau$). According to the relative timing of the storm with respect to the $F'(t)$ history, we may distinguish three significant cases.

Case 1: $t_f < t_b$.

This case (Figure 5a) is representative of arid and semi-arid climates where, because of long lasting periods with no rains and high temperatures, the soil is likely to be very dry at the time of the storm. Consequently, at that time, W is close to zero. Therefore, in these climates, the majority of water losses during storms is due to the initial abstraction so that, according to Equation (24), we obtain

$$F_A = W_b \tag{25}$$

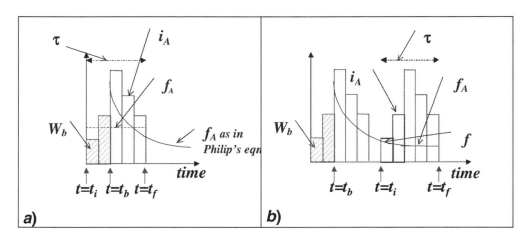

Figure 5. Precipitation and infiltration rate versus time; *a)* case 1; *b)* case 3.

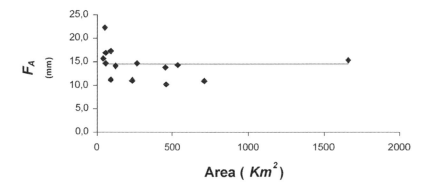

Figure 6. Total abstraction loss in semi-arid basins of Southern Italy.

Results found by Fiorentino and Iacobellis (2001) over arid and semi-arid basins in Southern Italy are in a significant agreement with the considerations presented above. In fact, the F_A values obtained in the investigated region show a quite constant pattern as in Figure 6. Data used in the figure can be reconstructed from Table 1 taking into account that $F_A = f_A \tau_A$.

Case 2: $t_i \cong t_b$, $t_f > t_b$.

This case is representative of dry-sub-humid and sub-humid climates where the water losses in the time window τ (defined as above) are usually due to sorptivity (S in Equations (5) and (6)) and, eventually, to the initial abstraction. Here, the behavior of water losses is also controlled by the quantity $t'_i = t_i - t_b$, that is in turn strongly affected by the rainfall antecedent to the storm. In fact, the greater the rainfall the higher the t'_i and the less, for a given area A, is F_A. One should note that the average antecedent rainfall depends on the yearly average number of storms and on their mean duration. In addition, the characteristic value of t'_i is also af-

507

fected by the average evaporation rate between storms. All this quantities depend on climate and on its average characteristic. Yet, since the majority of, or the entire, time interval responsible for the flood peak, falls on the rapidly declining limb of the $f(t)$ function, f_A depends on the area A too.

For instance, for $t'_i = 0$ we obtain

$$F_A = S\tau^{1/2} + c\tau \tag{26}$$

According to consolidated geomorphological theories that assume:

$$\tau = \tau_1 A^v \tag{27}$$

Equation (11) leads to the individuation of a dependence of F_A on basin area.

On the other hand, for a generic t'_i, we get

$$F_A = S[(t'_i + \tau)]^{1/2} - St_i'^{1/2} + c\tau \tag{28}$$

which shows that F_A also depends on the characteristic values t'_i.

As already stated, t'_i is dependent on climate and it is possible to observe that the individuation of the dominant factor between τ and t'_i seems to be sensitive to the way t'_i is correlated to climate.

In this case results provided by Fiorentino and Iacobellis (2001) for climate from dry sub-humid to humid show F_A estimates strongly correlated to climate represented by the climatic index (Thornthwaite, 1948) given in Equation (16).

Case 3: $t_i \gg t_b$.
This case, schematically depicted in (Figure 5b) is representative of humid and hyper-humid climates. Here, in fact, because of the very high number of rainy days, the rainfall responsible for the flood peak is likely to occur when the infiltration rate into the soil is close to the gravitational capacity. Therefore, neglecting the first term at the right-hand side of Equation (8) and, following Equation (12) we deduce

$$F_A = F_1 A^v \tag{29}$$

Looking at Equations (24), 26) and (29) one could suggest to account for different processes of the global hydrologic loss, by means of a global expression:

$$F_A = \alpha_1 a_1 + \alpha_2 a_2 A^{v/2} + \alpha_3 a_3 A^v \tag{30}$$

where a_1, a_2 and a_3 are coefficients mainly depending on the spatial averages of initial abstraction, characteristic sorptivity and gravitational infiltration rate. In Equa-

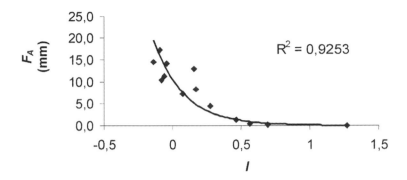

Figure 7. Total abstraction loss in dry-sub-humid and sub-humid basins of Southern Italy.

tion (30), α_1, α_2, and α_3 are weights, ranging from 0 to 1, whose relative value is a function of climate. In particular, the first of them tend to prevail in arid and semi-arid climate, while the last one is expected to be mainly active in hyper-humid climates. Thus, when the first term prevails, Equation (30) reduces to Equation (10), on a constant value. In the case of prevalence of the other terms, related to the infiltration rate through unsaturated soils, one would expect a stronger influence of climate, but also a scaling relationship with basin area particularly conditioned by the gravitational infiltration rate.

An analogous relationship was derived in Fiorentino and Iacobellis (2001) with regard to f_A:

$$f_A = \vartheta_1 c_1 A^{-v} + \vartheta_2 c_2 A^{-v/2} + \vartheta_3 c_3 \qquad (31)$$

where c_1, c_2 and c_3 are coefficients related to a_1, a_2 and a_3 respectively, and ϑ_1, ϑ_2, and ϑ_3 are weights, ranging from 0 to 1, playing the same role as α_1, α_2, and α_3 in Equation (30). When the first terms in Equations (30) and (31) prevail ($\alpha_1 = \vartheta_1 = 1$, all other weights = 0), the same equations give $F_A = a_1$ and $f_A = c_1 A^{-v}$. As according to the above hypotheses this case should be characteristic of arid basins, it is suggested that in arid basins total significant water losses (in mm) should remain constant in any sub-basin and that significant loss rates (in mm/h) tend to scale with the basin size raised to the power $-v$. It is noteworthy to remark that common values of v stay around 0.5. In the opposite case of hyper-humid zones ($\alpha_3 = \vartheta_3 = 1$, all other weights = 0), Equations (30) and (31) give $F_A = a_3 A^{v}$ and $f_A = c_3$. In this case the water loss rate does not vary at varying basin sizes while the total amount of significant loss tends to increase as the basin area increases.

A typical behavior of the case 3 above is shown by the majority of basins in Calabria, where estimated values of f_A are almost constant (Figure 8). Some consistent deviations from this behavior may be observed in small basins (say below 50 km^2) where local heterogeneity may lead to consistent oscillations of the runoff threshold. In the figure, the horizontal line represents a typical behavior of humid

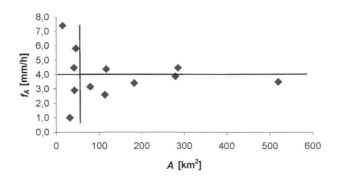

Figure 8. Average space-time water loss intensity versus basin area A in Calabria.

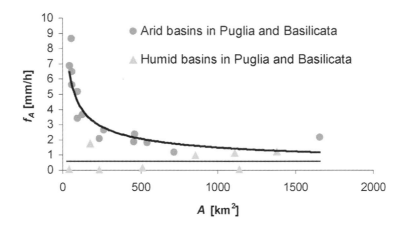

Figure 9. Average space-time water loss intensity versus basin area A in Puglia and Basilicata.

basins where water losses do not scale with area. The constant value represented by the horizontal line is quite higher with respect to those observed in humid basins of Basilicata and Puglia (Figure 9) where basins with similar climatic conditions in terms of mean annual precipitation and temperature are considered. This may reflect the fact that basins in Calabria are characterized by higher mean permeability compared to basins in Basilicata.

Particular support to the theoretical scheme depicted above is provided by Figure 9, also, where it is shown how different are the scaling properties of water losses with the basin size from arid to humid zones.

The contiguity between the three cases commented above is confirmed by Figure 10, where the three-dimensional representation of the f_A estimates versus climatic index and basin area shows such consistent pattern over the regions investigated. Data used in the figure are from Puglia and Basilicata regions. This figure also supports the conjecture of Equation (31). It seems worth remarking that

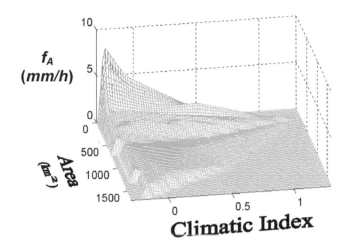

f_A
(mm/h)

10

5

0

0

Area
(km^2)

500

1000

1500

0

0.5

1

Climatic Index

Figure 10. Three-dimensional representation of total abstraction rate vs climatic index and catchment area A.

the f_A parameter accounts for the long-term average abstraction rate with specific regard to the time interval responsible for the flood peak only. In other words, it does not take into account rain losses that occur at a time insignificant for the flood peak. Moreover, f_A, rather than representing the most probable rain losses during a storm, is a position parameter of Equations (4), characterizing the way these losses tend to scale with the partial watershed area that contributes to peak runoff.

Average Flood Yield

The capability of a catchment to yield floods is here shown to depend upon the scaling properties derived above. In fact, due to Equations (10) and (31), the way f_A scales with the basin area A controls the flood number ratio Λ_q / Λ_p.

Therefore, with regard to the case 1 above (arid zones), due to the significant tendency of f_A to decrease as the basin size increases, one may expect this ratio to increase at increasing Λ. Yet, since f_A and $E[i_A]$ tend to scale with A by power laws with exponents not far from each other, this increase may be not very significant.

A support to this hypothesis is given by data achieved in the arid part of the study area (Figure 11), also in comparison with the opposite case of humid basins (basins of Calabria) where one can note a significant trend of this ratio to decrease as A increases, which means that rainfall events tend to reduce their capability to yield floods as the basin area becomes larger.

The intermediate case discussed in the section above is well represented by humid basins in Puglia and Basilicata, where the flood number ratio is more sensitive to the climatic index I than to catchment size.

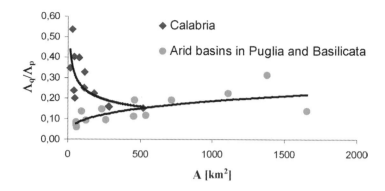

Figure 11. Mean annual number of flood and rainfall events ratio versus basin area A in Calabria and arid basins in Puglia and Basilicata.

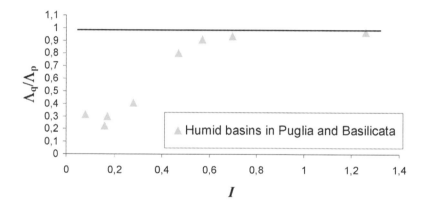

Figure 12. Mean annual number of flood and rainfall events ratio versus the climatic index in humid basins in Puglia and Basilicata.

Scaling Properties of the Coefficient of Variation

The behavior of the coefficient of variation Cv of annual floods was investigated with particular regard to the way it tends to scale with the basin area A. According to the probability distribution proposed by Iacobellis and Fiorentino (2001), Cv is mainly controlled by the mean annual number of floods, Λ_q, and by the shape parameter of the probability distribution of rainfall, k. In the model, Λ_q is in turn dependent on Λ_p, f_A and $E[i_A]$, whose definitions are provided earlier in this paper.

In the theoretical framework presented above, and in a region where k ad Λ_p are constant, parameters $E[i_A]$ and f_A scale with the basin area by a power law with exponents ε and ε' respectively. Therefore, following Equation (10), the relationship Cv–A depends on ε and ε'. This result adds new arguments to the controversial question whether Cv should theoretically increase or decrease as the basin size becomes larger (e.g. Smith, 1992; Gupta and Dawdy, 1995; Robinson and Sivapalan, 1997a, b; Blöschl and Sivapalan, 1997). In fact, many observed datasets, reported in literature, show a decrease of Cv with area, usually ascribed to the limited spatial extent of extreme events which leads to a decrease of Cv of areal rainfall intensity. An increase of Cv with the area at small scales is sometimes recognized as well. For instance, the observation that rainfall presents skewness practically constant for durations ranging between a few hours and one or two days while this statistics is usually lower at smaller durations could partly account for the above mentioned increase of Cv.

In particular our scheme suggests that a significant role on the control of the Cv behavior is played by the abstraction characteristics at the basin scale. In addition, as these features have been shown to be strongly related to the long term climate, it is also pointed out how the climate drives the scaling relationship Cv–A. In particular, a double control is recognized: the first one relates to the precipitation IDF and their scaling behavior (through the model parameter ε), and the latter is strongly influenced by the basin response in terms of the flood number ratio Λ_q / Λ_p as determined by the rainfall threshold for runoff generation f_A. In other words, the presented theoretical framework allows us to shed more light on the relationship between the coefficient of variation Cv of annual floods and the basin scale just looking at the dependence of the involved parameters with basin scale.

In the context of the presented model the Cv behavior could be accounted for the way parameters like Λ_p, k, ε and others scale with the basin area.

Let us now assume on a first approximation that $k = 1$, corresponding to the hypothesis of exponential distribution of rainfall intensity. In this case the model proposed by Iacobellis and Fiorentino leads to a distribution that under a very wide range of situations is not very far from a simple Gumbel distribution (EV1). Consequently, to the aim of this paper, the relationship between Cv and Λ_q can be confidently expressed in the closed form:

$$Cv = \frac{1.28255}{\log\Lambda_q + 0.5771} \tag{32}$$

Replacing Equation (10) into (9) we get:

$$Cv = \frac{1.28255}{\log\Lambda_p - \dfrac{f_1}{i_1}\ A^{\varepsilon-\varepsilon'} + 0.5771} \tag{33}$$

which represents a simple Cv–A relationship.

In Figure 13 we show, as an example, the functional relationship of Equation (33) for a given set of rainfall characteristic parameters, namely $\Lambda_p = 20$, $i_1 = 13$ mm/h and $\varepsilon = 0.33$. The patterns shown in Figure 13 for different values of ε' (0, 0.16, 0.33 and 0.5) highlight that the scaling relationship Cv–A, the other quantities constant, is significantly dependent on the values ε and ε' and their difference.

Commenting on Figure 13 and taking into account the typical values of ε' presented in the previous sections, we should point out that Cv is likely to decrease with the basin area A in arid basins where the prevailing runoff generation mechanism is of the Horton type and f_A tends to scale with A raised to the power $\varepsilon' = -0.5$ (for a more general comment on this topic, see also Fiorentino and Iacobellis, 2001). On the contrary, in humid and vegetated basins, where the prevailing runoff generation mechanism can be reconnected to soil saturation excess, and ε' tends to be zero, Cv may increase as the basin area increases. On the other hand, Figure 13 also points out that the achieved relationship Cv–A shows a more significant sensitivity of Cv to A when ε' is greater than ε, and that this may provide an explanation for the fact that in the real world a negative scaling of Cv with the basin area A seems to be prevailing.

Flood Peak Contributing Area

In the framework of the presented theory, floods are produced by runoff coming from an area that may represent only a portion of the entire basin. Therefore, the basin is thought as composed of some independent sub-basins, and is supposed that the storm may affect only one or a few of them. Furthermore, it is taken into account that within any of these sub-basins the conditions needed to produce surface runoff, and eventually significant throughflow, are reached not everywhere during the flood event. Therefore, the contributing area is composed by the sum of the partial areas belonging to the contributing sub-basins. In other words, the flood peak is given by the superposition of flows coming from a random number of sub-catchments interested by the storm according to the storm size and movement. This number depends on the basin size relative to two characteristic space scales,

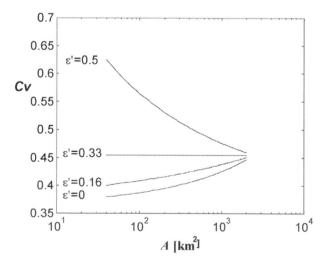

Figure 13. Coefficient of variation of floods vs basin area A according to Equation (33).

514

namely the storm and rain-cell sizes, hereafter indicated as S_s and S_c. In fact, for very small basins, whose area is comparable to or less than S_c, there is a high probability that the entire catchment contributes to the peak flood with space-time homogeneous rainfall. Whereas, for basins whose area ranges between S_c and S_s, it is much more likely that a is given by the sum of separate sub-catchment contributing areas. This is due to different timing and duration with respect to which separate portions of the basin are interested by the rain cells and to different response times of these portions.

In order to derive Equation (11), we assumed that the probability density function (*pdf*) of the peak contributing area is a gamma distribution for any a smaller then A. Thus, letting P_A be the finite discrete probability that the whole catchment is contributing to the flood peak, we achieved:

$$g(a) = \frac{1}{\alpha \Gamma(\beta)} \left(\frac{a}{\alpha}\right)^{\beta-1} \exp\left(-\frac{a}{\alpha}\right) + \delta(a-A)P_A \qquad (34)$$

where, indicating with $\gamma(\cdot,\cdot)$ the incomplete gamma function (e.g. Gradshteyn and Ryzhik, 1980, *p.* 940), P_A, given in Equation (12), specifies in

$$P_A - prob[a - A] = (A/\alpha, \beta) = \int_A^\infty \frac{1}{\alpha\Gamma(\beta)} \left(\frac{a}{\alpha}\right)^{\beta-1} \exp\left(-\frac{a}{\alpha}\right) da \qquad (35)$$

Basically, the choice of the gamma distribution is a working hypothesis. Nevertheless, it can be justified when it is reminded that this function arises as the distribution of the sum of β stochastic (independent) variables exponentially distributed with equal mean value α.

This justification helps in giving a meaning to the distribution parameters. In fact, without any loss of generality, we may think at any flood peak as being due to the superposition of flows coming from a number of sub-basins which can be differently interested by the storm. In addition, it is always possible to postulate that the sub-basins that may provide runoff have comparable sizes so that it is consequent to suppose, for modeling purposes, that their contributing areas have equal mean. They can also be thought of as being independent of each other, with the warning that in the eventuality of lack of independence attention should be paid to the meaning of β.

The number of these sub-basins is lower bounded by unity for very small basins. Instead, for basins whose area is comprised between S_c and S_s, it is possible to think at β as the number N_ω of sub-basins of Horton order immediately smaller than that of the whole basin. In fact, these sub-basins are in average of comparable sizes and may consequently be modeled by the same value of α. For larger catchments, β could be even greater than N_ω due to the probable superposition of discharges coming from different sub-basins invested by more than one storm. Incidentally, it may be interesting to remind that, according to a well consolidated geomorphologic knowledge, N_ω tends to be invariant at any scale and assumes values ranging between 3 and 5 in nearly all cases (Horton, 1945). According to Gupta and Waymire (1983), its expected value is close to 4.

However, the estimation of the gamma parameters could be left to direct observation and it is pointed out that even other assumptions for the probability distribution of a could be made, specially if local information is available to support them.

The α parameter in Equation (35) is related to the mean contributing area, also used in Equation (14), by means of:

$$E[a] = \alpha\,\beta\,\gamma\,(A/\alpha, \beta + 1) + AP_A \tag{36}$$

The ratio r, of the mean contributing area to the total basin area A, defined as:

$$r = E[a]/A \tag{37}$$

It is noteworthy that in the arid part of the study area r estimates match a meaningful correspondence with the values of pervious fraction calculated by Equation (17). In Figure 14 such a relationship is shown, and a trend line is drawn of equation:

$$r = -0.7\,\psi + 0.8 \qquad (R^2 = 0.84) \tag{38}$$

The only exception is represented by the basin of Ofanto at Rocchetta St Antonio that has been pooled together with the other basins of the arid region. This basin, despite the positive climatic index ($I = 0.16$), shows the highest mean contributing area ratio, which is likely to be due to the prevailing control of impermeable soils extensiveness. We should also notice that the scarcity of vegetation is peculiar for

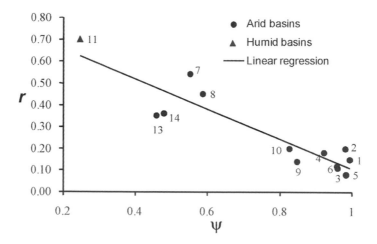

Figure 14. Ratios of the mean contributing area to the basin area A vs pervious fraction of A in the arid zone.

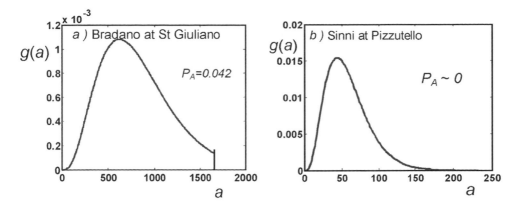

Figure 15. Estimated probability density functions of the contributing area (km^2).

this basin in comparison with the others (mainly located in Basilicata region) showing positive climatic index values.

Different behavior was recognized in humid basins of Basilicata, where the r estimates are quite low (always less the 0.3) independently of ψ.

An explanation for the different behavior of arid basins in comparison to humid ones can be related to the characteristics of runoff generation mechanisms. In fact, as pointed out by Dunne (1966, quoted by Leopold, 1974), in humid and semi-humid basins, where it is likely to find high vegetated zones, one can observe that, depending on conditions, only the part of a basin that is near the channels contribute to surface runoff and the rest of the basin area, further away, makes no contribution. This contributing area expands and contracts depending on the surface and subsurface conditions such as vegetation and soil moisture at the time prior to the flood event. Accordingly, in such basins, the area that contributes to overland runoff depends on the antecedent soil conditions and on the storm rainfall depth. The spatial extension of the storm plays a role too. A typical probability density function of partial areas contributing to flood peak for this kind of basins is given in Figure 15b, where it is shown that most probable contributing areas are much lower than the catchment area. In semi-arid zones, according to Horton runoff generation model, in presence of intense rainfall a wide part of the drainage area contributes to overland flow and the probability that the entire basin concurs to the peak discharge may not be neglected (e.g. Figure 15a). In such a case, the antecedent soil conditions is crucial to allow activation of overland flow but, once surface runoff has begun, the contributing area is rather controlled by the storm extension.

Conclusions

In this paper, a climatically consistent interpretation to observed relationships between water losses, climate and basin area is provided based on the use of the theoretical model proposed by Iacobellis and Fiorentino (2000). Particular attention was paid to the role played by the physical factors, their descriptive ability and the relative weight in characterizing the flood frequency distribution. By means of the investigations conducted, the following conclusions can be drawn.

In arid and semi-arid zones, dry and scarcely vegetated soils lead to usually consistent rainfall abstractions within the flood peak contributing areas. Moreover, the average rate of water loss in the time interval responsible for the peak runoff tends to be controlled by the total basin extension, quickly decreasing as the contributing area grows. This behavior can be accounted for by considering that in these regions rain losses are mainly due to the initial abstraction phenomenon and the estimated value of areal averages of total water losses are constant over basins of different area. Such findings are consistently matched to what one could expect from Hortonian mechanism of runoff generation (infiltration excess). Therefore, in these kind of basins one should expect that the greater the basin permeability, the lower is the mean contributing area, due to the more difficult activation of runoff.

Conversely, in humid zones, due to the high probability of saturated and intensely vegetated soils, runoff can be produced by rainfall of low intensity also; then the mean annual number of flood events approaches the mean annual number of rainfall events. Their ratio tends to one as far as the climatic index is high and the total abstraction rate goes to zero. In the same climates, the Dunne type runoff generation mechanism (saturation excess) prevails. Vegetation becomes the controlling factor, overcoming permeability. In fact, only small portions of the area contributes on average to runoff, independently of geology.

All that is reflected in the behavior of the coefficient of variation Cv of floods. In fact, according to the theoretical distribution of floods proposed by Iacobellis and Fiorentino (2001), Cv is mainly controlled by mean annual number of floods, Λ_q, and by the shape parameter of the probability distribution of rainfall, k. In the model, Λ_q is in turn dependent on Λ_p, f_A and $E[i_A]$. In particular, since $E[i_A]$ and f_A scale with the basin area by a power law with exponents ε and ε' respectively, the relationship that relates Cv to the catchment area A depends on ε and ε'. This result adds new arguments to the controversial question whether Cv should theoretically increase or decrease as the basin size becomes larger. In particular it suggests that a significant role on the control of the Cv behavior is played by the abstraction characteristics at the basin scale. In addition, as these features have been shown to be strongly related to the long term climate, it is also pointed out how the climate drives the scaling relationship Cv–A.

Acknowledgements

This work was supported by funds granted by the Italian CNR-GNDCI (under the research program RIVERS) and by the Italian Ministry of the University and Research (under the research program ECAPI).

References

Adom, D.N., B. Bacchi, A. Brath and R. Rosso (1989). On the geomorphoclimatic derivation of flood frequency (peak and volume) at the basin and regional scale, *New Directions for Surface Water Modelling,* edited by M.L. Kavvas, IAHS Publ. no. 181, 165-176.

Blöschl, G. and M. Sivapalan (1997). Process controls on regional flood frequency: Coefficient of variation and basin scale, *Water Resour. Res., 33*(12), 2967-2980.

Cadavid, L., J.T.B. Obeysekara and H.W. Shen (1991). Flood-frequency derivation from kinematic wave, *J. Hydraul. Eng, 117*(4), 489-510.

Claps, P., V.A. Copertino and M. Fiorentino (1994). Analisi Regionale dei Massimi Annuali delle Portate al Colmo di Piena (in Italian), in *Valutazione delle Piene in Puglia,* report, Dipartimento di Ingegneria e Fisica dell'Ambiente and Gruppo Nazionale per la Difesa dalle Catastrofi Idrogeologiche, Univ. della Basilicata, Potenza, Italy.

Cordova, J.R. and I. Rodriguez-Iturbe (1983). Geomorphoclimatic estimation of extreme flow probabilities, *Jour. Hydrol., 65,* 159-173.

Dalrymple, T. (1960). Flood frequency analysis, U.S. Geological Survey, *Water Supply Paper,* 1543-A.

Diaz-Granados, M.A., J.B. Valdes and R. L. Bras (1984). A physically based flood frequency distribution, *Water Resour. Res., 20*(7), 995-1002.

DIFA (1998). Valutazione delle Piene in Basilicata (in Italian), Report, Dipartimento di Ingegneria e Fisica dell'Ambiente and Gruppo Nazionale per la Difesa dalle Catastrofi Idrogeologiche, Univ. della Basilicata, Potenza, Italy.

Eagleson, P.S. (1972). Dynamics of flood frequency, *Water Resour. Res.,* 8(4), 878-898.

Eagleson, P.S. (1970). Dynamic Hydrology, McGraw-Hill Book Company.

Farquharson, F.A.K., J.R. Meigh and J.V. Sutcliffe (1992). Regional flood frequency analysis in arid and semi-arid areas, *Jour. Hydrol., 138,* 487-501.

Fiorentino M. and V. Iacobellis (2001). New insights about the climatic and geologic control on the probability distribution of floods, *Water Resour. Res., 37*(3), 721-730.

Fiorentino, M., Gabriele, S., Rossi, F. and P. Versace (1987). Hierarchical approach for regional flood frequency analysis, in V.P. Singh (eds), *Regional flood frequency analysis,* 35-49, D. Reidel, Norwell, Mass.

Fiorentino, M., Margiotta M.R. and Iacobellis V. (2001). Effects of Climate and Antecedent Soil Moisture on the Areal Average Abstraction Losses, accepted to XXIX IAHR Congress, September 16-21.

Gabriele, S. and G. Iiritano (1994). Alcuni aspetti teorici ed applicativi nella regionalizzazione delle piogge con il modello TCEV, (in Italian), GNDCI – Linea 1 U.O. 1.4, Pubblicazione N. 1089, Rende (Cs).

Gottschalk, L. and R. Weingartner (1998). Distribution of peak flow derived from a distribution of rainfall volume and runoff coefficient, and a unit hydrograph, *Jour. Hydrol., 208*, 148-162.

Gradshteyn, I.S. and I. M. Ryzhik (1980). *Table of integrals, series, and products*, Academic Press.

Gupta, V.K. and E. Waymire (1983). On the formulation of an analytical approach to hydrologic response and similarity at the basin scale, *Jour. of Hydrology, 65*, 95-123.

Gupta, V.K. and Dawdy D.R. (1995). Physical interpretations of regional variations in the scaling exponents of flood quantiles, *Hydrological Processes*, 9, 347-361.

Hebson, C. and E.F. Wood (1982). A derived flood frequency distribution using Horton order ratio, *Water Resour. Res., 18*(5), 1509-1518.

Horton, R.E. (1945). Erosional development of streams and their drainage basins: hydrophysical approach to quantitative geomorphology, *Geol. Soc. Amer. Bull., 56*, 275-370.

Hosking, J.R.M. and J.R. Wallis (1993). Some statistic usefol in regional frequency analysis, *Water Resources Research*, 29(2), 271-281.

Iacobellis, V. and M. Fiorentino (2000). Derived distribution of floods based on the concept of partial area coverage with a climatic appeal, *Water Resour. Res., 36*(2), 469-48.

Iacobellis, V., P. Claps and M. Fiorentino (1998). Sulla dipendenza dal clima dei parametri della distribuzione di probabilità delle piene, (in Italian). In: *Proceedings of the XIV Convegno di Idraulica e Costruzioni Idrauliche*, vol. II, 213-224, Univ. of Catania, Catania, Italy.

Kurothe, R.S., N.K. Goel and B.S. Mathur (1997). Derived flood frequency distribution for negativaly correlated rainfall intensity and duration, *Water Resour. Res., 33*(9), 2103-2107.

Leopold, L.B. (1974). Water a Primer, W.H. Freeman and company.

Mongelli, G. and A. Salvemini (1994). Caratterizzazione geolitologica e carta della permeabilità della Puglia settentrionale (in Italian). In: *Valutazione delle Piene in*

Puglia, report, Dipartimento di Ingegneria e Fisica dell'Ambiente and Gruppo Nazionale per la Difesa dalle Catastrofi Idrogeologiche, Univ. della Basilicata, Potenza, Italy.

Moughamian, M.S., D.B. McLaughlin and R.L. Bras (1987). Estimation of flood frequency: An evaluation of two derived distribution procedures, *Water Resour. Res.*, 23(7), 1309-1319.

Philip, J.R. (1957). The theory of infiltration - Sorptivity and algebraic infiltration equation, *Soil Sci.*, 84, 257-264.

Robinson, J.S. and M. Sivapalan (1997a). An investigation into the physical causes of scaling and heterogeneity of regional flood frequency, *Water Resour. Res.*, 33(5), 1045-1059.

Robinson, J.S. and M. Sivapalan (1997b). Temporal scales and hydrological regimes: Implications for flood frequency scaling, *Water Resour. Res.*, 33(12), 2981-2999.

Rossi, F., M. Fiorentino and P. Versace (1984). Two component extreme value distribution for flood frequency analysis, *Water Resources Research*, 20(7), 847-856.

Shen, H.W., G.J. Koch and J.T.B. Obeysekara (1990). Physically based flood features and frequencies, *J. Hydraul. Eng, 116*(4), 495-514.

Sivapalan, M., E.F. Wood and K.J. Beven (1990). On hydrologic similarity, 3, A dimensionless flood frequency model using a generalized geomorphologic unit hydrograph and partial area runoff generation, *Water Resour. Res.*, 26(1), 43-58.

Smith, J.A. (1992). Representation of basin scale in flood peak distributions, *Water Resour. Res.*, 28(11), 2993-2999.

Thornthwaite, C.W. (1948). An approach toward a rational classification of climate, *Am. Geograph. Rev.*, 38, 1, 55-94.

Turc, L. (1961). Estimation of irrigation water requirements, potential evapotranspiration: a simple climatic formula evolved up to date, *Annals of Agronomy*, 12, 13-14.

Versace, P., E. Ferrari, S. Gabriele and F. Rossi (1989). Valutazione delle piene in Calabria (in Italian), CNR-IRPI, Geodata.

Villani, P. (1993). Extreme flood estimation using Power Extreme Value (PEV) distribution, *Proceedings of the IASTED International Conference, Modeling and Simulation*, edited by M.H. Hamza, 470-476, Univ. of Pittsburgh, Pittsburgh, Pa.

Wood, E.F. and C.S. Hebson (1986). On hydrologic similarity, 1, Derivation of the dimensionless flood frequency curve, *Water Resour. Res.*, 22(11), 1549-1554.

Surface Water Hydrology, Singh, Al-Rashed & Sherif (eds)
© 2002 Swets & Zeitlinger, Lisse, ISBN 90 5809 363 8

Physically based model of discontinuous distribution for hydrological samples with zero values

Witold G. Strupczewski
Water Resources Department, Institute of Geophysics
Polish Academy of Sciences
Ksiecia Janusza 64, 01-452 Warsaw, Poland
e-mail: wgs@igf.edu.pl

Vijay P. Singh
Department of Civil and Environmental Engineering
Louisiana State University
Baton Rouge, Louisiana 70803-6405, USA
e-mail: cesing@lsu.edu

Stanislaw Węglarczyk
Institute of Water Engineering and Water Management
Cracow University of Technology, Warszawska 24, 31-155 Cracow, Poland
e-mail: sweglar@lajkonik.wis.pk.edu.pl

Abstract

It is hypothesized that the impulse response of a linearized kinematic diffusion (LK) model is a probability distribution suitable for frequency analysis of hydrological sample with zero values. Such samples are observed in arid and semiarid regions, and may include monthly precipitation in dry seasons, annual low flow and annual maximum peak discharge data. This model has two parameters, which are derived using the method of moments. The model is tested on real data.

Introduction

Hydrological statistics is dominated by flood frequency modeling and the developed techniques have been transferred to modeling of other hydrological and hydrometeorological variables. Mathematical models of flood frequency analysis (FFA) can be broadly classified into: (1) empirical, (2) phenomenological, and (3) physically-based. An excellent discussion of empirical models is given by Stedinger et al. (1993), Rao and Hamed (2000), among others. Till today these models continue to be most popular for doing FFA all over the world. Phenomenological models employ a set of probabilistic axioms which lead to a probabilistic model of one or more flood characteristics. Examples of this type of models are those based on the use of random number of random variables (Todorovic, 1982), the entropy theory (Singh, 1998), and the like. These models received a good deal of attention in the 1970s and the 1980s but did not become popular, partly because of higher mathematical demands. Physically-based models employ dynamical principles of flood generation. Eagleson (1972) was probably one of the first to employ such a model.

Another example of such a model is the use of watershed models, as for example, the stochastic flood model developed by Schaefer (1998).

Along the lines of physically based models and recognizing that channels are the dominant conduits for transmission of flood waters, it is plausible to develop a model that employs the physics of channel flow routing and in which no explicit consideration is given to the hydrologic processes occurring on the land areas of the watershed. It is well accepted that the complete linearized Saint Venant equation and its simplifications give a good representation of the physics of channel flow. It is then hypothesized that impulse response function (IRF) of such models can be considered as a probability density function (PDF) for FFA. Although the impulse response of a hydrologic system or the response of an initially relaxed linear deterministic system for the Dirac-δ impulse belongs to the class of purely deterministic functions, it is not difficult to find a stochastic interpretation of the impulse response. If one imagines that the unit volume of the Dirac-δ impulse consists of an infinite number of particles (or drops) then the integral of the impulse response $\int_0^T h(x,t)dt$ determines the probability that a single particle passes the outlet at x during time $(0,T)$, where $h(x, t)$ is the impulse response function at time t and position x. Apart from its stochastic interpretation, it should be noted that the impulse response function fulfills several requirements normally expected of the flood frequency models, namely, (1) semi-infinite lower bounded range with a non-negative value of the bound; (2) positive skewness and the unit integral over whole range $\int_0^\infty h(x,t)dt$; and (3) uni-modality, which is the property of all single component FF distributions. As an example, the gamma function is used both as the impulse response of a cascade of equal linear reservoirs and PDF in flood frequency analysis (FFA).

Because of the practice of applying existing probability distributions in FFA, emphasis in statistical hydrology has been on assessing the accuracy of parameter estimators using the Monte Carlo simulation techniques. As a result, not much attention has been paid to the development of physically based probability distributions taking into account peculiarities of hydrologic phenomena and the attendant statistical reasoning. To that end, this study espouses the use of IRFs as PDFs.

The objective of this paper is to hypothesize the linear kinematic diffusion (LK) model of flow routing as a probability distribution function for modeling hydrological sample with zero values and to evaluate the validity of this hypothesis.

Modeling Hydrological Samples with Zero Values

In the frequency analyses of hydrological data in arid and semiarid regions, one often encounters data series that contain several zero values while zero is the lower limit of the variability range. From the viewpoint of probability theory, the occurrence of zero events can be expressed by placing a nonzero probability mass on a zero value, i.e. $P(X=0) \neq 0$, where X is the random variable, and P is the probability mass. Therefore, the distribution functions from which such hydrological series were drawn would be discontinuous with discontinuity at the zero value having a form

$$f(x) = \beta \delta(x) + f_c(x;\mathbf{h}) \cdot 1(x) \qquad (1)$$

where β denotes the probability of the zero event, i.e. $\beta = P(X=0)$, $f_c(x;\mathbf{h})$ is the continuous function such that $\int_0^\infty f_c(x;\mathbf{h})dx = 1-\beta$, $\mathbf{h}$ is the vector of parameters, $\delta(x)$ is the Dirac delta function and $1(x)$ is a unit step function. Obviously, Equation (1) violates the basic assumption of continuity made in conventional frequency analyses.

The estimation procedures for a hydrological sample with zero events have been a subject of several publications. Omitting some misconceptions, it is the theorem of the total probability, which has been employed (Jennings and Benson, 1969; Woo and Wu, 1989; Wang and Singh, 1995) to model such series. Then, Equation (1) takes the form

$$f(x) = \beta \delta(x) + (1-\beta)f_1(x;\mathbf{g}) \cdot 1(x), \quad \beta \notin \mathbf{g} \qquad (1a)$$

where $f_1(x;\mathbf{g})$ is the conditional probability density function (CPDF), i.e. $f_1(x;\mathbf{g}) = f(x|X>0)$, which is continuous in the range $(0,+\infty)$ with a lower bound of zero value. Wang and Singh (1995) estimated β and the parameters of CPDF separately considering the positive values as a full sample for the purpose. Having estimated $\mathbf{g}$ and the β, the conditional distribution can be transformed to the marginal distribution, i.e. to $f(x)$, by Equation (1a). Among several PDFs with lower bound at zero recognized in FFA (e.g. Rao and Hamed, 2000), the Gamma distribution has been chosen by Wang and Singh (1995) as example of CPDF. Four estimation methods have been applied by them, i.e. the maximum likelihood method (MLM), method of moments (MOM), probability weighted moments (PWM) and ordinary least squares method.

The Kinematic Diffusion (KD) and the Rapid Flow (RF) Model

One of the most important problems in one-dimensional flood routing analysis is the downstream problem, i.e. the prediction of flood characteristics at a downstream section on the basis of the knowledge of flow characteristics at an upstream section. Using the linearization of the Saint Venant equation, the solution of the upstream boundary problem was derived by Deymie (1939), Masse (1939), Dooge and Harley (1967), Dooge et al. (1987a, b), among others; a discussion of this problem is presented in Singh (1996). The solution is a linear, physically based model with four parameters depending on hydraulic characteristics of the channel reach at the reference level of linearization. However, the complete linear solution is complex in form and is relatively difficult to compute (Singh, 1996). Two simpler forms of the linear channel downstream response are recognized in the hydrologic literature and are designated as linear diffusion analogy model (LD) and linear rapid flow model (LRF). These correspond to the limiting flow conditions of the linear channel response, i.e. where the Froude number is equal to zero (Hayami, 1951; Dooge, 1973) and where it is equal to one (Strupczewski and Napiorkowski, 1990).

The complete linearized Saint Venant equation is of hyperbolic type and may be written as

$$a\frac{\partial^2 Q}{\partial x^2} + b\frac{\partial^2 Q}{\partial x \partial t} + c\frac{\partial^2 Q}{\partial t^2} = d\frac{\partial Q}{\partial x} + e\frac{\partial Q}{\partial t} \tag{4}$$

where Q is the perturbation of flow about an initial condition of steady state uniform flow, x is the distance from the upstream boundary, t is the elapsed time, and a, b, c, d and e are the parameters being functions of channel and flow characteristics at the reference steady state condition. A number of models of simplified forms of the complete Saint Venant equation have been proposed in the hydrological literature.

If all three of the second-order terms on the left-hand side of that equation are neglected we obtain the linear kinematic wave model. Expressing the second and the third second-order term in terms of the first on the basis of the linear kinematic wave approximation leads to the advective diffusion equation, i.e. to the linear diffusion analogy model (LDA). If the diffusion term is expressed in terms of two other terms using the kinematic wave solution, one gets the rapid flow (RF) equation, which is of parabolic-like form (Strupczewski and Napiorkowski, 1990a). Therefore, it filters out the downstream boundary condition. It provides the exact solution for a Froude number equal to one and consequently can be used for great Froude numbers. If the alternative approach is taken of expressing all the second-order terms as cross-derivatives, one gets the equation representing the diffusion of kinematic waves (Lighthill and Witham, 1955; Strupczewski and Napiorkowski, 1989). The kinematic diffusion (KD) model, being of the parabolic-like form, fits satisfactorily the solution of the complete linearized Saint Venant equation only for small Froude number and slow rising waves.

Although RF and KD models correspond to quite different flow conditions, the structure of their impulse response is similar (Strupczewski et al., 1989; Strupczewski and Napiorkowski, 1990b). In both cases, the impulse response is:

$$h(x,t) = P_o(\lambda)\delta(t-\Delta) + \sum_{i=1}^{\infty} P_i(\lambda) \cdot h_i\left(\frac{t-\Delta}{\alpha}\right) \cdot 1(t-\Delta) \tag{5}$$

where

$$P_i(\lambda) = \frac{\lambda^i}{i!}\exp(-\lambda) \tag{6}$$

is the Poisson distribution and

$$h_i\left(\frac{t}{\alpha}\right) = \frac{1}{\alpha(i-1)!}\left(\frac{t}{\alpha}\right)^{i-1}\exp\left(-\frac{t}{\alpha}\right) \tag{7}$$

is the Gamma distribution and $1(x)$ is the unit step function. Parameters α, λ and Δ are functions of both channel geometry and flow conditions, which differs for the both models. Furthermore, there is no time lag (Δ) in the impulse response function of KD model.

Both models can be considered as hydrodynamic and conceptual. Note that the solution of both models can be represented in terms of basic conceptual elements used in hydrology, namely as cascade of linear reservoirs and a linear channel in case of the RF model. The upstream boundary condition is delayed in the RF model by a linear channel with time lag, Δ, divided according to a Poisson distribution with mean λ, and then transformed by parallel cascades of equal linear reservoirs (with time constant α) of varying lengths. Note that λ is the average number of reservoirs in a cascade. Strupczewski and Napiorkowski (1986, 1989) and Strupczewski et al. (1989) have derived the distributed Muskingum model from the multiple Muskingum model and then they have shown its identity to the KD model. Similarily, the RF model happens to be identical with the distributed delayed Muskingum model (Strupczewski and Napiorkowski, 1990b).

Einstein (1942) introduced the function given by (1) to hydrology as the mixed deterministic-stochastic model for the transportation of the bed load. For the second time, it has appeared in hydrology as the probability distribution function of the total rainfall depth got from the Partial Duration Series model built on the assumption of the Poisson process for storm arrivals and of the exponential distribution for storm depths (Eagleson, 1978).

The function in Equation (5) is considered to be the flood frequency model in this study.

KD Probability Density Function and its Properties

Since our interest is in frequency estimation for samples with zero values, KD model is more adequate then the RF model for the purpose, while the RF model can serve to model an incomplete sample censored by the Δ-value. Putting the delay (Δ) to equal zero in Equation (5) and renaming t as x, one gets a two-parameter probability distribution function of the form:

$$f(x) = P_0(\lambda)\delta(x) + \sum_{i=1}^{\infty}P_i(\lambda)\cdot h_i\left(\frac{x}{\alpha}\right)\cdot 1(x) \tag{5a}$$

where

$$P_i(\lambda) = \frac{\lambda^i}{i!}\exp(-\lambda) \qquad (6a)$$

is the Poisson distribution and

$$h_i\left(\frac{x}{\alpha}\right) = \frac{1}{\alpha(i-1)!}\left(\frac{x}{\alpha}\right)^{i-1}\exp\left(-\frac{x}{\alpha}\right) \qquad (7a)$$

is the Gamma distribution and $1(x)$ is the unit step function. Note that Equation (5a) differs from Equation (1a) since its second term can not be expressed as the product of probability of non-zero value, i.e. $(1 - P_o(\lambda))$, and the conditional probability density function (CPDF), i.e. $f_1(x;g)$.

The second term of the PDF, i.e.

$$f_c(x) = \sum_{i=1}^{\infty} P_i(\lambda) \cdot h_i\left(\frac{x}{a}\right) \qquad (8)$$

can be expressed by the 1st order modified Bessel function of the 1st type, $I_1()$:

$$f_c(x) = \exp\left(-\lambda - \frac{x}{\alpha}\right)\sqrt{\frac{\lambda}{\alpha x}}I_1\left(2\sqrt{\frac{\lambda x}{\alpha}}\right)\cdot 1(x) \qquad (8a)$$

So we, have

$$f(x) = P_o(\lambda)\delta(x) + \exp\left(-\lambda - \frac{x}{\alpha}\right)\sqrt{\frac{\lambda}{\alpha x}}I_1\left(2\sqrt{\frac{\lambda x}{\alpha}}\right)\cdot 1(x) \qquad (5b)$$

Modal Value

The modal value can be obtained from the solution of equation $\dfrac{\partial f_c(x)}{\partial x} = 0$, or equivalently of $\dfrac{\partial f_c(y)}{\partial y} = 0$, where $y = x/\alpha$. It gives:

$$\frac{I_0\left(2\sqrt{\lambda y}\right)}{I_1\left(2\sqrt{\lambda y}\right)} = \sqrt{\frac{y}{\lambda}} + \sqrt{\frac{1}{\lambda y}} \qquad (9)$$

for $\lambda \geq 2$. The modal value (y_{mod}) as the function of λ is shown in Fig.1. As one can see, the maximum of $f_c(y)$ and consequently of $f_c(x)$ is for $\lambda < 2$ at the origin of y-axis.

Cumulants and Moments

The *R-th* order cumulant of LD can be expressed as (Strupczewski & Napiorkowski, 1989):

$$k_R = R!\alpha^R\lambda \qquad (10)$$

Using the relations between moments and cumulants (Kendall and Stuart, 1969, p. 70) and Equation (10) the expression for the first four moments of Equation (1) are given below:

$$\mu_1' = \alpha\lambda \qquad (11)$$

$$\mu_2 = 2\alpha^2\lambda \qquad (12)$$

$$\mu_3 = 6\alpha^3\lambda = \frac{3\mu_2^2}{2\mu_1'} \qquad (13)$$

$$\mu_4 = k_4 + 3k_2^2 = 12\alpha^4\lambda(2+\lambda) = 3c_v^4(\mu_1')^4(c_v^2+1) \qquad (14)$$

Dimensionless Coefficients

The coefficient of variation is:

$$c_v = \sqrt{\frac{2}{\lambda}} \qquad (15)$$

The coefficient of skewness is:

$$c_s = \frac{\mu_3}{(\mu_2)^{\frac{3}{2}}} = \frac{3}{\sqrt{2\lambda}} = \frac{3}{2}c_v \qquad (16)$$

which for the gamma distribution is $c_s = 2c_v$.

The coefficient of kurtosis is:

$$Kurtosis = \frac{\mu_4}{(\mu_2)^2} = 3\left(\frac{2}{\lambda}+1\right) = 3(c_v^2+1) \qquad (17)$$

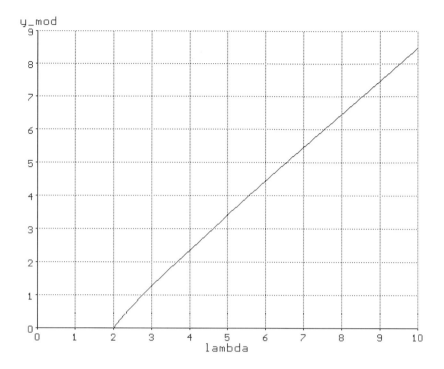

Figure 1. The modal value ($y_{mod} = x_{mod}/\alpha$) versus the parameter λ.

Shape of the Density Function

For $c_v \rightarrow 0$ the distribution tends to be symmetric like lognormal (LN), linear diffusion (LD) and gamma distributions. Typical graphs of the distribution for some selected values of λ versus $y = x/\alpha$ are presented in Fig. 1. Note that α is the scale parameter. For increasing λ-value the maximum of $f(x)$ decreases and shifted along the y-axis.

The value of λ defines from Equation (2) the probability of no-occurrence of the event, i.e. $P_o(\lambda)$.

Probability of Exceedance

A quantile corresponding to the probability of exceedance p, x_p is obtained by integrating Equation (5b) as

$$x_p = \alpha \cdot t_p(\lambda) \tag{18}$$

where $t_p(\lambda)$ is the lower limit of the integral

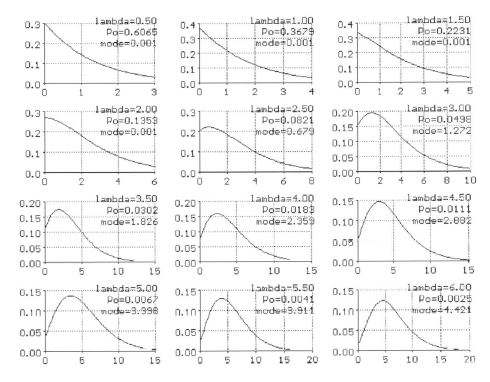

Figure 2. Graphs of $f_c(x)$ (Eq. (8a)) for $\alpha = 1$ and some values of λ P_0 denotes the probability of zero value, $P_0 = e^{-\lambda}$ (see Eq. (6)).

$$p = \int_{t_p}^{\infty} \exp(-\lambda - y)\sqrt{\frac{\lambda}{y}} I_1\left(2\sqrt{\lambda y}\right) dy \qquad (19)$$

and p must by less than $1 - e^{-\lambda}$, otherwise $t_p = 0$. Some values of t_p for given λ and p are listed in Table 1.

Table 1. KD quantile $t_p(\lambda)$ for given values of λ and probability of exceedance p.

$p(\%)$	λ										
	0.5	1	1.5	2	3	4	5	6	7	8	9	10
50	0.000	0.396	0.950	1.469	2.483	3.487	4.490	5.492	6.493	7.494	8.495	9.496
40	0.000	0.764	1.394	1.980	3.111	4.215	5.305	6.385	7.458	8.524	9.588	10.65
30	0.350	1.226	1.939	2.599	3.857	5.065	6.248	7.410	8.559	9.698	10.83	11.95
20	0.871	1.861	2.673	3.420	4.823	6.156	7.446	8.706	9.945	11.17	12.37	13.57
10	1.749	2.906	3.860	4.728	6.333	7.836	9.275	10.67	12.03	13.37	14.68	15.98
5	2.616	3.918	4.989	5.956	7.729	9.371	10.93	12.44	13.90	15.33	16.74	18.12
2	3.750	5.216	6.420	7.500	9.460	1.259	12.96	14.59	16.17	17.71	19.21	20.69
1	4.597	6.178	7.468	8.622	10.71	12.61	14.41	16.12	17.77	19.38	20.95	22.49
0.5	5.439	7.122	8.494	9.716	11.91	13.91	15.79	17.58	19.31	20.98	22.61	24.21
0.2	6.542	8.352	9.820	11.13	13.46	15.58	17.56	19.44	21.25	23.00	24.71	26.38
0.1	7.372	9.269	10.81	12.17	14.60	16.80	18.85	20.80	22.67	24.47	26.23	27.95

Parameter Estimation by the Method of Moments

Solving Equations (11) and (12) for parameters α and λ, one gets

$$\alpha = \frac{\mu_2}{2\mu_1'} = \frac{\mu_1'}{2}c_v^2 \tag{52}$$

$$\lambda = \frac{2(\mu_1')^2}{\mu_2} = \frac{2}{c_v^2} \tag{53}$$

Equations (52) and (53) are used in MOM to estimate parameters α and λ from sample moments.

Application of KD Model

Four flow gaging stations data were selected from the USGS data bank to evaluate the KD-model performance for annual peak flow discharges of arid zones. Some basic data are shown in Table 2 and the resulting KD cumulative distribution functions (CDF's) are presented in Figs. 3 a, b, c, d together with empirical CDF defined as $F_{emp}(x_{(i)})=i/(n+1)$, where $x_{(i)}$ is the i-th element of the sample arranged in ascending order. The MOM-estimated probability density functions are shown in Fig. 4. In general, MOM seems to perform reasonably well.

Table 2. Selected flow gauging stations and KD-model characteristics.

	Station			
USGS No.	8404000	06713000	13132500	10258500
River	Pecos	Cherry Creek	Big Lost River	Palm Cyn
Station	Avalon Dam, NM	Cherry Creek Lake, Co	Arco, Id	Palm Springs, Ca
Catchment area, km^2	46761.7	995.8	3646.8	240.8
n	48	50	51	66
n_2	39	42	46	64
α_{MOM}	304.251	6.227	8.966	27.919
λ_{MOM}	0.49	1.778	1.963	1.094
$x_{0.1\%,MOM}$ m^3/sec	2229.1	72.1	108.2	267.3

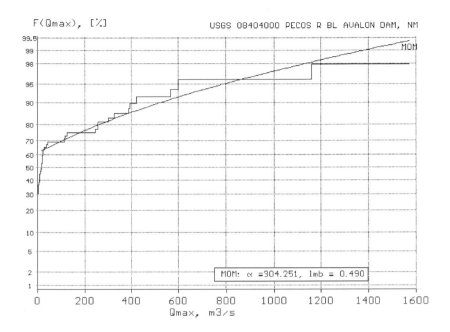

Figure 3. Empirical and theoretical KD CDF's.
a. The Pacos River, Avalon Dam, Nm.

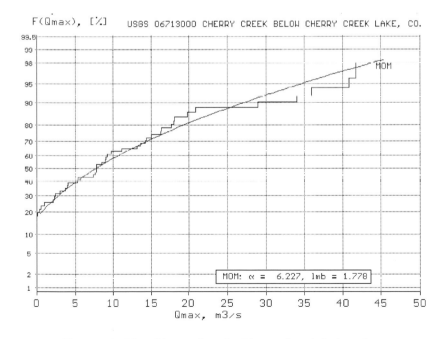

Figure 3b. The Cherry Creek, Cherry Creek Lake, Co.

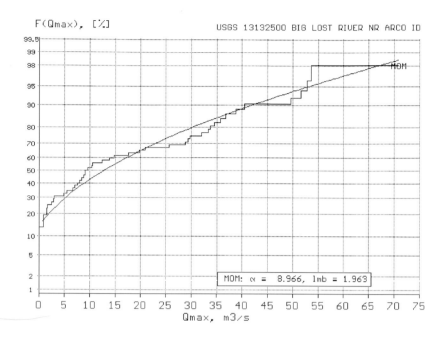

Figure 3c. The Big Lost River, Arco, Id.

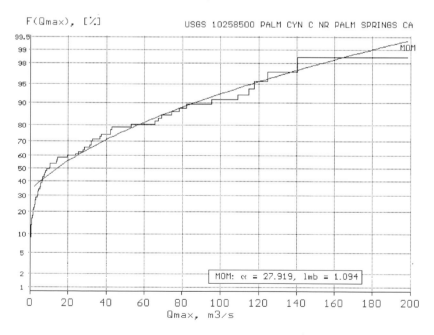

Figure 3d. The Palm Cyn River, Palm Springs, Ca.

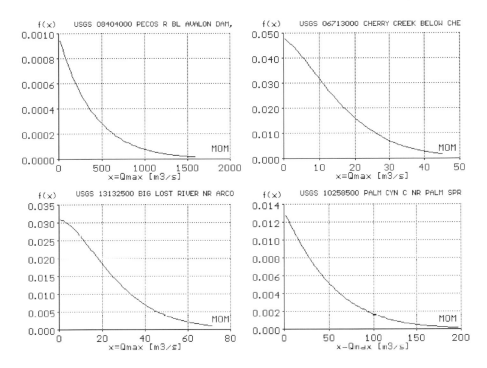

Figure 4. The theoretical KD probability density functions for the four selected flow gauging stations estimated with MOM.

Conclusions

The following conclusions are drawn from this study:

(a) The impulse response of the linear kinematic diffusion (KD) model is a promising model for frequency analysis of hydrological sample with zero values. It is easy from a computational point particularly if the method of moments is applied.

(b) Being a two-parameter model, KD is especially attractive in case of short samples which is very common in arid and semiarid regions.

(c) With its two parameters the KD model can reproduce a variety of probability density forms encountered in nature (Fig. 2).

(d) An examination of MOM (Figs 3 a,b,c,d) shows that MOM reproduces better the upper tail of the distribution.

(e) The KD model represents flood frequency characteristics of arid zones quite well, as shown using the USGS data.

Acknowledgment

This study has been partly financed by Polish Committee for Scientific Research, Grant No. 6 P 4D 056 17.

References

Deymie, P. (1939). "Propagation d'une intumescence allongee (Problem aval)" [Propagation of an elongated intumescence]. *Proc. 5th Internatl. Cong. Appl. Mech.,* 537-544, New York (in French).

Dooge, J.C.I., Harley, B.M. (1967). "Linear routing in uniform open channels". *Proc. Int. Hydrol. Symp. Fort Collins, Co.* Sept.1967, Paper No. 8, 1, 57-63.

Dooge, J.C.I. (1973). *Linear theory of hydrologic systems. Tech. Bull.,* 1468, Agricultural Research Service, Washington.

Dooge, J.C.I., Napiorkowski J.J., Strupczewski, W.G. (1987 a). "The linear downstream response of a generalized uniform channel". *Acta Geoph. Pol.* 35, 277-291.

Dooge, J.C.I., Napiorkowski, J.J., Strupczewski, W.G. (1987 b). " Properties of the general downstream channel response". *Acta Geoph. Pol.* 35, 405-418.

Eagleson, P.S. (1972). "Dynamics of flood frequency". *Water Resour. Res.*, 8, 878-898.

Eagleson, P.S. (1978). "Climate, soil and vegetation. 2. The distribution of annual precipitation derived from observed storm sequences". *Water Resour. Res.* 14, 5, 713-721.

Einstein, H.A. (1942). "Formulas for the transportation of bed load". *Transactions ASCE,* Paper No. 2140, pp. 561-597.

Hayami, S. (1951). "On the propagation of flood waves", *Kyoto Univ. Japan, Disaster Prevention Res. Inst. Bull.,* 1, 1-16.

Jennings, M.E. and Benson M.A. (1969). "Frequency curve for annual flood series with some zero events or incomplete data". *Water Resour. Res.,* 5(1), 276-280.

Kendall, M.G., Stuart, A. (1969). *The advanced theory of statistics. V.1. Distribution Theory.* Charles Griffin & Company Ltd., London.

Lighthill, M.H., Witham, G.B. (1955). "On kinematic waves. I. Flood movements in long rivers". *Proc. R. Soc., London,* Ser. A, 229, 281-316.

Masse, P. (1939). "Recherches sur la theorie des eaux courantes". [Researches on the theory of water currents], *Proc. 5th Internatl. Cong. Appl. Mech.*, 545-549, New York (in French).

Rao, A.R. and Hamed, K.H. (2000). *Flood Frequency Analysis,* CRC Press, Boca Raton, Florida.

Schaefer, M.G. (1998). "General storm stochastic event flood model: Technical support manual". *Technical Report,* MGS Engineering Consultants Inc., Olympia, Washington.

Singh, V.P. (1996). *Kinematic Wave Modeling in Water Resources: Surface Water Hydrology.* John Wiley & Sons, New York.

Singh, V.P. (1998). *Entropy-based parameter estimation in hydrology.* Kluwer Ac. Press, Dordrecht.

Stedinger, J.R., Vogel, M.V., Foufoula-Georgiou, E. (1993). Ch. 18. "Frequency analysis of extreme events". In: *Handbook of Hydrology.* Ed. D.R., Maidment, McGraw-Hill Inc.

Strupczewski, W.G. and Napiorkowski, J.J. (1986). "Asymptotic behaviour of physically based multiple Muskingum model". *Proc. Fourth Intern. Hydrol. Symp.,* H.W. Shen, J.T.B. Obeysekera, V. Yevjevich and D.G. Decoursey, eds., Eng. Res. Center, Colorado State Univ., Fort Collins, Colo., 372-381.

Strupczewski, W.G., Napiorkowski, J.J., Dooge, J.C.I. (1989). "The distributed Muskingum model". *J. of Hydrol.,* 111, 235-257.

Strupczewski, W.G. and Napiorkowski, J.J. (1989). "Properties of the distributed Muskingum model". *Acta Geoph. Pol.,* V.XXXVII, 3-4, 299-314.

Strupczewski, W.G. and Napiorkowski, J.J. (1990a). "Linear flood routing model for rapid flow". *Hydrol. Sc. J.,* 35(1, 2), 49 64.

Strupczewski, W.G., Napiorkowski, J.J. (1990b). "What is the distributed delayed Muskingum model ?". *Hydrol. Sc. J.,* 35, (1, 2), 65-78.

Todorovic, P. (1982). Stochastic modeling of floods. In: *Statistical Analysis of Rainfall and Runoff,* edited by V.P. Singh, Water Resour. Publications, Littleton, Colorado, pp. 597-636.

Wang, S.X. and Singh, V.P. (1995). "Frequency estimation for hydrological samples with zero value". *J. of Water Resour. Planning and Management,* ASCE, 121(1), 98-108.

Woo, M.K. and Wu, K. (1989). "Fitting annual floods with zero flows". *Can. Water Resour. J.,* 14(2), 10-16.

Surface Water Hydrology, Singh, Al-Rashed & Sherif (eds)
© *2002 Swets & Zeitlinger, Lisse, ISBN 90 5809 363 8*

Multivariate estimation of floods

Jose A. Raynal-Villasenor
Department of Civil Engineering
Universidad de las Americas-Puebla
72820 Cholula Puebla, Mexico
e-mail: jraynal@mail.udlap.mx

Abstract

Recently, the multivariate approach to extreme value distributions has started to be studied towards its application in the solution of hydrological problems. The theoretical foundations of such an approach has been proposed almost twenty years ago, but they have had very little impact in the field of hydrology. In the paper, a classification of such models is made and in the particular case of the differentiable models, characteristics, properties, and parameter estimation procedures for the method of maximum likelihood are given for the bivariate and trivariate distributions. Finally some applications of that distributions are described.

Introduction

The use of extreme value distributions has grown steadily in the last fifty years in the field of hydrology, mainly in performing frequency analyses for annual maxima or minima. Such univariate distributions have arisen as particular solutions to the Stability Postulate, which any extreme must satisfy. Jenkinson (1955), have found the general solution to the Stability Postulate, and some authors after him have called that solution the general extreme value (GEV) distribution. This distributions able to represent the extreme value distributions types II and III directly, and when one takes the limit of the shape parameter going to zero, the result comes to be the extreme value distribution type I. An excellent reference in this topic is NERC (1975).

In pioneer papers Finkelshtein (1953), Gumbel (1958) and Tiago de Oliveira (1958) gave the foundations for the multivariate approach to extreme value distributions. After this work, several models for bivariate extreme value distributions began to appear in the literature. Two classes of them are now known: the differentiable and the non-differentiable models. Among the latter class are, Tiago de Oliveira (1982): the biextremal, Gumbel and Naturals models.

Since this models do not have probability density functions, which makes more difficult the parameter estimation phase, and given that they have been developed for the particular case when both marginals are Gumbel distributions, they will not be described here any further. The interested reader is referred to Tiago de Oliveira (1982) for any additional details.

The only two known differentiable models for bivariate extreme value distributions are: the logistic and mixed models. Such a names for the models have been coined, Tiago de Oliveira (1982), from the fact that the reduced difference, when both marginals are Gumbel distributions, has the standard Logistic distribution for the first case. In the second case, the model has a dependence function coming for a mixture of such function for the cases of the dependence and independence, when the marginals are Gumbel distributions. General characteristics and multi-variate extensions of these models will be given in the following sections.

In the paper, general procedures for obtaining maximum likelihood estimators of the parameters of bivariate and trivariate extreme value distributions are given. Also, some identified fields of application of such distributions are contained in the paper.

Differentiable Models for Bivariate Extreme Value Distributions

As it said before, there are only two known models of the class of differentiable models: the logistic and the mixed models.

The general form of the logistic model for bivariate extreme value distributions is, Gumbel (1960):

$$F(x,y,m) = \exp\left\{-\left\{\left[-\ln F(x)\right]^m + \left[-\ln F(y)\right]^m\right\}^{\frac{1}{m}}\right\} \tag{1}$$

where m is the association parameter, F(u) is the marginal distribution function of u, $m \geq 1$ and $0 \leq \rho \leq 1$, where ρ is the population product – moment correlation co-efficient.

For m = 1, the independent case, the bivariate distribution function splits into the product of its marginals:

$$F(x,y,1) = F(x)F(y) \tag{2}$$

when m = ∞, the bivariate distribution function is:

$$F(x,y,\infty) = \min[F(x),F(y)] \tag{3}$$

as it was shown by Johnson and Kotz (1970). This is the diagonal case and for the case when both marginals are Gumbel distributions, this point is the only one that does not have a planar density. In this case, it is possible to obtain an analytical expression for the population product – moment correlation coefficient, Gumbel (1967):

$$\rho = 1 - \frac{1}{m^2} \qquad (4)$$

in the case of the mixed model, its general form is, Gumbel (1960):

$$F(x,y,a) = F(x)F(y)\exp\left\{a\left[\frac{1}{-\ln F(x)} + \frac{1}{-\ln F(y)}\right]^{-1}\right\} \qquad (5)$$

where a is the association parameter, $0 \le a \le 1$ and $0 \le \rho < 0.66$. For the case a = 0, the bivariate distribution splits into the product shown in Eq. (2), and this is the independent case. When a = 1, complete dependence is observed but this is not the diagonal case. This type of model has planar density in all points when is marginals are both Gumbel distributions. When the last condition is met, it is possible to obtain an analytical expression for the population product – moment correlation coefficient, this is, Tiago de Oliveira (1982):

$$\rho = \frac{6}{\pi^2} arc\ \cos\left(1 - \frac{a}{2}\right)^2 \qquad (6)$$

From the characteristic and properties of the differentiable models for bivariate extreme value distributions, Tiago de Oliveira (1982) started: "It is, then, intuitive that the distance between independence (m = 1) and the assumed model for m > 1 is small in general, and for small samples it will probably be impossible to distinguish them", in the case of logistic model. For the mixed model, he stated: "The smaller variation of the correlation coefficient and of the distance shows that the derivation from independence is smaller and most difficult to detect".

From these statements, the logistic model has been chosen, due to its greater flexibility and wider applicability, to be pursued further.

Multivariate Extension of the Logistic Model

For the case of logistic model for bivariate extreme value distributions, the following multivariate extension has been given by Gumbel (1960b):

$$F(x,y,..,m) = \exp\left\{-\left[(-\ln F(x))^m + (-\ln F(y))^m + ...\right]^{1/m}\right\} \qquad (7)$$

and Tiago de Oliveira (1975) has provided the following inequalities for the model contained in Eq. (7):

$$F(x_1)F(x_2)\ \ ...\ \ F(x_n) \le F(x_1,x_2,...,x_n) \le min\left[F(x_1),F(x_2),\ ...\ ,F(x_n)\right] \qquad (8)$$

$$\left[\prod_{i \ne j} F\left(x_i, x_j\right)\right]^{\frac{1}{2(n-1)}} \le F\left(x_1, x_2, \ldots, x_n\right) \le \frac{\left[\prod_{i \ne j} F\left(x_i, x_j\right)\right]^{\frac{1}{2}}}{\left[\prod_i F\left(x_i\right)\right]^{n-2}} \tag{9}$$

Maximum Likelihood Estimation of the Parameters

For the case of bivariate distributions functions, the samples usually do not have equal length of record, so it is necessary to have a formulation flexible enough to cover all possible combinations of arrangements of data. Such a formulation has been proposed by Raynal (1985), based on the generalization given by Anderson (1957):

$$L\left(x, y, \underline{\theta}\right) = \left[\prod_{i=1}^{n_1} f\left(s_i, \underline{\theta}_1\right)\right]^{l_1} \left[\prod_{i=1}^{n_2} f\left(x, y, \underline{\theta}_2\right)\right]^{l_2} \left[\prod_{i=1}^{n_3} f\left(t_i, \underline{\theta}_3\right)\right]^{l_3} \tag{10}$$

where $L(\bullet)$ is the likelihood function of $(\bullet)$, n_1, n_2, n_3 are the lengths of record before, on and after the common period n_2, respectively. S and t are the variables with records before and after the common period, respectively. l_i are indicator numbers such that $l_i = 1$ if $n_i > 0$ and $l_i = 1$ if $n_i = 0$. $\underline{\theta}_j$ is the vector of parameters.

Given the property that the maximum of a function and its logarithm occur in the same point and due to the fact the expressions provided by the logarithm of Eq. (10) are much easy to handle than those produced by such equation, the log-likelihood function will be used instead of its natural version. So, Eq. (10) it is transformed in, Raynal (1985):

$$LL\left(x, y, \underline{\theta}\right) = l_1 \left[\sum_{i=1}^{n_1} Ln \ f\left(s_i, \underline{\theta}_1\right)\right] + l_2 \left[\sum_{i=1}^{n_2} Ln \ f\left(x, y, \underline{\theta}_2\right)\right] + l_3 \left[\sum_{i=1}^{n_3} Ln \ f\left(t_i, \underline{\theta}_3\right)\right] \tag{11}$$

An extension of Eq. (10) for the case of trivariate distribution functions is:

$$L\left(x, y, z, \theta\right) = \left[\prod_{i=1}^{n_1} f\left(s_i, \underline{\theta}_1\right)\right]^{l_1} \left[\prod_{i=1}^{n_2} f\left(s_i, t_i, \underline{\theta}_2\right)\right]^{l_2} \left[\prod_{i=1}^{n_3} f\left(x, y, z, \underline{\theta}_3\right)\right]^{l_3}$$
$$\times \left[\prod_{i=1}^{n_4} f\left(u_i, v_i, \underline{\theta}_4\right)\right]^{l_4} \left[\prod_{i=1}^{n_5} f\left(u_i, \underline{\theta}_5\right)\right]^{l_5} \tag{12}$$

and the corresponding log-likelihood equation for this level is:

$$LL(x,y,z,\underline{\theta}) = I_1 \left[\sum_{i=1}^{n_1} Ln \; f(s_i,\underline{\theta}_1) \right] + I_2 \left[\sum_{i=1}^{n_2} Ln \; f(s_i,t_i,\underline{\theta}_2) \right] + \tag{13}$$

$$I_3 \left[\sum_{i=1}^{n_3} Ln \; f(x,y,z,\underline{\theta}_3) \right] + I_4 \left[\sum_{i=1}^{n_4} Ln \; f(u_i,v_i,\underline{\theta}_4) \right] +$$

$$I_5 \left[\sum_{i=1}^{n_5} Ln \; f(u_i,\underline{\theta}_5) \right]$$

Since those parameter values which maximize Eq. (11) or Eq. (13) are the maximum likelihood estimators of the parameters of such distribution functions and given that Equations (11) and (13) are very suitable to be solved through optimization procedures, like the Rosenbrock method multiple constrained variables, this is the proposed approach to estimate the parameters of bivariate and trivariate extreme value distributions.

Application of Multivariate Extreme Value Distributions to Hydrology

The main fields of application of multivariate extreme value distributions, are those related with improvement in parameter estimation and transfer of extreme value information.

It has been shown (Clarke, 1980; Rueda, 1981 and Raynal, 1985) that there exists an improvement in the parameter estimation phase, when the bivariate approach to distribution functions is used. Particularly, in the last two cases, related with flood information data, the improvement has been observed even in the case when both samples have the same length of record. As a sample, Table 1 taken from Raynal (1985) shows such condition. The measure of improvement is the relative information ratio, defined as, Raynal (1985):

$$I_a(\theta_i) = \frac{Var_a(\theta_i)}{Var_a(\theta_{i,B})} \tag{14}$$

Where $I_a(0_i)$ is the asymptotic relative information ratio, $Var_a(\theta_i)$ and $Var_a(\theta_{i,B})$ are the asymptotic variances of parameter θ obtained through univariate and bivariate maximum likelihood procedures, respectively.

Table 1. Relative asymptotic information rations for the parameters of the bivariate Gumbel distribution ($u_1 = 10.0$, $\alpha_1 = 1.0$, $u_2 = 15.0$, $\alpha_2 = 2.0$, and m = 2.0).

Parameter			$n_2 + n_3$	
	25	50	75	100
u_1	1.00	1.32	1.47	1.56
α_1	0.99	1.24	1.36	1.43
u_2	1.00	1.00	1.00	1.00
α_2	0.99	0.98	0.97	0.97

543

For the bivariate case, when the marginals are Gumbel distributions, such relative information ratios are given in Table 1, Raynal (1985).

For the trivariate case, when the marginals are Gumbel distributions, such relative information ratios are given in Table 2, Escalante-Sandoval and Raynal-Villasenor (1998).

For comparison purposes, in Tables 3 and 4 the design values for several return periods are shown between the univariate estimation against the bivariate values, and between the univariate estimation and the trivariate values.

Table 2. Relative asymptotic information rations for the parameters of the trivariate Gumbel distribution ($n_3 = 25$, $u_1 = 14.0$, $\alpha_1 = 1.4$, $u_2 = 12.0$, $\alpha_2 = 1.2$, $u_3 = 10.0$, $\alpha_3 = 1.0$, $m_b = 2.0$, and $m_t = 2.0$).

Parameter			n_5	
	n_4	25	50	75
u_1	25	1.37	1.54	1.64
	50	1.63	1.75	
	75	1.79		
α_1	25	1.32	1.44	1.51
	50	1.52	1.59	
	75	1.63		
u_2	25	1.37	1.54	1.64
	50	1.22	1.34	
	75	1.16		
α_2	25	1.32	1.44	1.51
	50	1.18	1.28	
	75	1.12		
u_3	25	1.02	1.02	1.01
	50	1.02	1.02	
	75	1.02		
α_3	25	1.02	1.00	0.99
	50	1.01	1.00	
	75	1.00		

Table 3. Design values for the univariate and bivariate estimation of parameters for the Gumbel distributions for Station Toahayana, Sin., Mexico.

Return Period (Years)	Univariate $Q_T(m^3/s)$	Bivariate $Q_T(m^3/s)$
5	1384	1439
10	1722	1810
20	2047	2166
50	2467	2627
100	2782	2973

Table 4. Design values for the univariate and trivariate estimation of parameters for the Gumbel distributions for Station Tuxtepec, Oax., Mexico.

Return Period (Years)	Univariate $Q_T(m^3/s)$	Trivariate $Q_T(m^3/s)$
5	3624	2764
10	3942	4119
20	4246	4460
50	4640	4900
100	4935	5231

Conclusions

The multivariate approach to extreme value distributions has been presented towards its application in the solution of hydrological problems.

It was shown that the use of multivariate extreme value distributions might improve the estimation of parameters of extreme value distributions. An observed application was the ability of such distribution to establish a process of transfer of information related with extreme value data.

The author considers this is a very promising field of research and would like to quote Tiago de Oliveira (1975b) to end this paper: "if the bivariate extremes area is largely open we can say that the field of multivariate extremes is almost completely open."

Acknowledgements

The author wish to express his gratitude to the Universidad de las Americas-Puebla, and the Engineering Graduate Studies Division, Universidad Nacional Autónoma de México, for the support provided to the realization of this paper.

References

Clarke, R.T. (1980). Bivariate gamma distributions for extending annual stream flow records from precipitation, Wat. Resour. Res., Vol. 16, No. 6, 863-870.

Finkelshtein, B.V. (1953). Limiting distributions of extremes terms of the variational series of a two dimensional random value. Dokl. Ac. Nauk, SSSR, 91.2.

Escalante-Sandoval, C.A. and Raynal-Villasenor, J.A. (1998). Multivariate Estimation of Floods: The Trivariate Gumbel Distribution. J. Statist. Comput. Simul., Vol. 61, no. 64 , pp. 313-341.

Gumbel, E.J. (1958). Fonctions de probabilites a deus variables extremales independantes. C.R. Acad. Sci. Paris, Vol. 246, 49-50.

Gumbel, E.J. (1960 a). Multivariate extremal distributions. Bulletin of the International Statistical Institute, Vol. 39, No. 2, 471-475.

Gumbel, E.J. (1960b). Distributions des valeurs extremes en plusieurs dimension, Publ. Inst. Statist, Paris, Vol. 9, 171-173.

Gumbel, E.J. (1967). Some analytical properties of bivariate extremal distributions, American Statistical Association Journal. Vol. 62, June, 569-588.

Johnson, N.L. and Kotz, S. (1972). Distribution is Statistics: Continous Multivariate Distributions, John Wiley and Sons, New York, N.Y., USA.

Linsley, R.K. and Franzini, J. (1979), Water Resources Engineering, 3rd ed., McGraw-Hill Book Co., New York, N.Y, USA.

Natural Environment Research Council, (1975). Flood Studies Report, Vol. 1 Hydrological Studies, Whitefriars LDT., London.

Raynal-Villaseñor, J.A. (1985). Bivariate Extreme Value Distributions Applied to Flood Frequency Analysis. Ph. D. Dissertation, Colorado State University, Fort Collins, Colorado, USA.

Raynal-Villaseñor, J.A. and Salas, J.D. (1986). A probabilistic model applied to the description of flooding downstream of a junction of two rivers. Proc. of the Int. Symp. On flood Freq. And Risk Analyses, Baton Rouge, Lousiana. USA.

Rueda, E. (1981), Transfer of Information for Flood Related Variables. M. Sc. Thesis, Colorado State University, Fort Collins, Colorado, USA.

Salas, J.D. (1980). Transfer of Information to improve estimates of flood frequencies. Hydrology for Transportation Engineers. (ed. By T.G. Sanders), U.S. Dept. of Transportation, 592–653.

Tiago de Oliveira, J. (1958). Extremal distributions. Faculdade de Ciencias de Lisboa, 2 Serie, A, Mat., Vol. VII.

Tiago de Oliveira, J. (1975a). Bivariate extremes: Decisions, Bulletin of the International Statistical Institute. (Proc. of the 40th Session,Warsaw), Vol. XLVI, Book 2, 241–251.

Tiago de Oliveira, J. (1975b). Bivariate and multivariate extreme distributions. In: Statistical Distributions in Scientific Work, Vol. 1 (ed. by G.P. Patil et al.) 355–361. D. Reidel Pub. Co.

Tiago de Oliveira, J. (1982). Bivariate extremes: Models and statistical decision. Tech. Report No. 14, Center for Stochastic Processes, Department of Statistics, University of North Carolina, Chapel Hill, North Carolina, U.S.A.

Woodroofe, M. (1975). Probability with Applications. McGraw-Hill Book Co. New York, N.Y, USA.

Surface Water Hydrology, Singh, Al-Rashed & Sherif (eds)
© 2002 Swets & Zeitlinger, Lisse, ISBN 90 5809 363 8

Effect of percent non-detects on estimation bias in censored distributions

Z. Zhang and U.S. Panu
Department of Civil Engineering
Lakehead University, Thunder Bay, Ontario, Canada
Department of Civil Engineering
University of Waterloo, Waterloo, Ontario, Canada
e-mail: uspanu@lakeheadu.ca

W.C. Lennox
Department of Civil Engineering
University of Waterloo, Waterloo, Ontario, Canada

Abstract

Uniqueness of the problem surrounding non-detects has been a concern alike to researchers and statisticians dealing with summary statistics while analyzing censored data. To incorporate non-detects in the estimation process, a simple substitution by the MDL (method detection limit) and the maximum likelihood estimation method are routinely implemented as standard methods by the US EPA laboratories.

In situations, where numerical standards are set at or near the MDL by regulatory agencies, it is not only prudent but also important to closely investigate both the variability in test measurements and the estimation bias. It is because an inference based on the biased estimates could entail significant liabilities. The variability is understood to be not only inevitable but also an inherent and integral part of any chemical analysis or test. In situations where regulatory agencies fail to account for the inherently present variability of test measurements, there is a need for regulated facilities to seek remedial action merely as a consequence of inadequate statistical procedure.

The paper utilizes the mathematical approach to derive the bias function for substitution method and furthermore bias curve are developed to investigate the censored samples from a variety of probability distributions such as the normal, lognormal, gamma, and Gumbel distributions. The bias functions and bias curves are also verified by using the Monte Carlo simulations.

To assess the influence of percent non-detects on the estimation bias, the censoring rate was used as an independent variable. Results show that half detection limit method yields very reasonable estimates of the mean and standard deviation for a censoring rate of 48% and 32% respectively for samples from censored Normal distribution. For a censoring rate of 73% and 61% respectively, the method yields reasonable estimates of the mean and standard deviation for samples from the Log

Normal distribution. For the gamma distribution, the limits of censoring rate are found to be 84% and 62% respectively for the mean and standard deviation. The half-detection limit was not found to be suitable for the Gumbel distribution because a significant bias was found to exist even at a considerable low censoring rate.

Introduction

Protection and safeguard of environment from disposal of hazardous materials is a major environmental concern throughout the world. In an effort to detect the earliest possible release of pollutants from disposal facilities, various governmental regulations in Canada have increased the requirement for extensive monitoring at potential contaminated sites. A common problem encountered in environmental data is that the sample often contains non-detects. The presence of non-detects in environmental data poses not only interesting and but unique statistical challenges, which considerably exceed beyond the norms of standard statistical tests. The existance of non-detects signifies that samples come from a truncated distribution.

The truncation of a distribution occurs in many situations. For instance, when an instrument has a detection limit, small values are not observable or cannot be measured accurately. As censored data arise in a sample, the sample is referred to as truncated or censored. In environmental data where the concentration of trace substances in soil, water, and air are investigated, left censored samples are always encountered. An example of left censoring is the low flow data. Seasonal rivers sometimes have very little flow or no flow during a drought season. Such small flow rates are often below the stage of measuring device and are recorded as either under-detection limit or zero. On the other hand, an example of right censored data are the water levels constrained by the upper limit stated as "danger level" on a river or an upper finite flow or water level before corrective measures can be initiated. Likewise, the surface water quality may be considered truncated by setting up of upper limits for undesirable water quality parameters. For samples that are drawn from a screened population with known truncation limit, the estimation of parameters of the original distribution becomes somewhat difficult. For samples with unknown truncation limit, the estimation of parameters becomes highly unreliable because of the existance of extraordinary bias.

The environmental monitoring data inherently contains some measurements that are below a detection limit. In such situations, the application of traditional statistical methods often leads to biased results and incorrect tests of hypotheses. Further, environmental monitoring at hazardous facilities presents even more complex statistical problems that could not be easily solved using standard statistical methods. Despite such statistical difficulties, the regulators need to set compliance standards; managers need to make decisions to meet the permit limitation. In environmental monitoring, there exist some statistical procedures that are practically simple to use and yet technically satisfactory.

It is in this vein that the focus of this paper is to (i) find the best estimation method for censored environmental data through analytical and simulation approaches,

(ii) test the suitability of the Half-detection limit method (MDL/2) in environmental applications, and (iii) suggest avenues for further investigations.

For conceptual illustration of the method of approach, consider that there is a population of true solute concentrations from which a sample of size n has been drawn, and that variation in the sampled population can be represented by a continuous distribution. Also, consider that the sample can be represented by a log-transformed sample of solute concentrations with mean 1 and standard deviation 0.5. Such an underlying Log Normal distribution LN (1, 0.5) is shown in Figure 1.

The actual underlying distribution (in this case a Log Normal distribution) of a censored sample can be visualized to consist of two parts, a *Dirac* function, and a part of the complete PDF to the right of the cut-off point. The vertical line crossing the detection limit bisects the PDF into two parts, a censoring proportion $\int_0^\xi f(x)dx$, and the remainder portion corresponding to the uncensored proportion of the sample. If the Half-detection limit method (MDL/2) is applied, the resulted PDF turns out to consist of a *Dirac* function at MDL/2, and the part on the left of the MDL. The mean and the standard deviation of the resulted PDF can be calculated. The difference

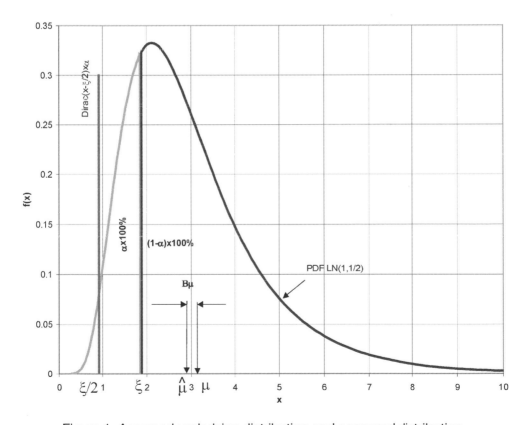

Figure 1. Assumed underlying distribution and censored distribution.

549

between the new parameter and the original parameter is the bias, which is derived as a function of the censoring proportion.

The estimation of parameter is the start point to conduct statistical inference. The estimation methods are compared on their ability to recover the true parameters. But in environmental data analysis, a more important task is to provide the confidence interval at a desired significance level and to predict future individual measurements. Therefore, the method should be evaluated not only by the ability to recover the true population parameter but also the ability to minimize overall false positive rate.

Another challenge in censored samples is the existence of multiple detection limits. Multiple censoring is caused by several reasons, such as multiple analysts, labs, experiments in multiple time, samples analyzed on-site and off-site, etc. Which will form the basis of further investigations.

Some Relevant Preliminaries of Environmental Data

Environmental data on some specified solute concentrations can be considered to be a random sample. For example, solute concentrations in groundwater are measured at a specific site that are the result of a potentially varying (space and time) source, precipitation history, travel through heterogeneous porous media, and random errors in measurement. It is in this sense that actual measurements are considered to constitute a random sample.

Variability and Liability

The variability in environmental measurements arises in several ways. Firstly, some degree of variability is inherently present in any chemical test analysis. Secondly, the chance of random error may causes measurements of a sample, even from the same laboratory, to be different. Lastly, as the concentration of the chemical decreases, so does our ability to measure and consequently, the measurement will contain more and more uncertainty, until eventually the analytical method is no longer being able to detect the substance of interest. The un-detected measurements that are below the pre-specified detection limit are scientifically unreliable because of the unknown probability distribution function associated with such un-detected measurements. On the other hand, the second kind of variability arises in environmental data, when tests are conducted at a different laboratory, using a different detection instrument, or by a different technician. Therefore, the variability is inherent and integral part of a test procedure. Should the permit limitation fails to account for the variability that is inherent in the test procedure, regulated facilities may require remedial action merely as a consequence of false detection of pollutants. Uncertainty near the detection limit can be as large as the reported value itself and thus it is virtually impossible to confirm the identity of the chemical substance detected. Therefore, there are legal arguments against using any regulatory standards that expose a party to liability without an objective standard.

Sample Size	N	10	20	50	100	200
Probability of at Least one of n Measurements being Significant	$1-0.99^n$	0.0956	0.1821	0.3950	0.6340	0.8660

A typical groundwater-monitoring problem is more complex and as a results require multiple comparisons of the key issues. Consequently, the groundwater quality is evaluated on measurements of many wells that are often analyzed for multiple constituents. For example, consider that n measurements are independent of each other and further that the probability of a false positive rate is very small, say 0.01, then the probability of at least one of the n measurements being significant by chance alone is large, as shown below.

Censoring and Censored Samples

Censored data emerges frequently in many areas. Standard environmental laboratory procedures have limitations in their measurement capabilities. Environmental monitoring and sampling programs usually involve a number of constituents that are stipulated as a standard set for which measurements are mandatory, and low concentrations below the MDL have to be reported as non-detects. For example, to prevent groundwater pollution, monitoring of groundwater involves collection of data on concentration of contaminants. Environmental data analyses usually encounter censored-data sets, which contain data value(s) below a predetermined detection limit or too low to be quantifiable. Therefore, a data set containing measurements that are non-detectable is said to be a censored data set. The limit, to which measurements can be observed, is called the censoring point i.e., the censoring point is the maximum detect limit. For example, if a device cannot detect the concentration that is lower than 1 µg/L, i.e. the MDL is said to be 1 µg/L, then the censoring point is 1 µg/L.

Singly or Doubly Censoring

In a left censored sample, consider that the observed measurements represent an independent and identically distributed (IID) random variable. These values are recorded in increasing order of magnitude; that is, the data appear as the vector of order statistics in a natural way. In such situations, one would have a censored sample in which order statistics play an important role. In cases, when large values on the right of the distribution are censored, one obtains a right-censored sample. One can have left censoring wherein smaller values are censored. In other situations, one can have censoring of left or right regions.

Type-I and Type-II Censoring

There are different classifications of censored samples, which are being used to distinguish different ways of censoring. If the censoring point is fixed and known, the censoring is called Type-I censoring. If the censoring point is unknown but the censoring proportion is determined, and then the censoring is called Type-II censoring. In general, Type-I censoring refers to the situation where a sample of size n is drawn, and the observations below a predetermined limit are censored. It is noted that the limits are fixed for Type-I censoring while the number of censored observations is random variable. For Type-II censoring, the number or the percentage of censored observations is fixed while the censoring limits can be considered random. Type-I censoring lead to a truncated sample, while Type-II censoring leads to a censored sample.

Besides, Type-I and Type-II censoring, another prominent classification of censored samples includes random censoring. Taking clinical trials, as an example, in which lifetime of the specimen is censored by an independent time. Such a scheme is known as a random censoring scheme.

Practical problems arise, when one has to estimate the parameters. It is important to understand the degree of bias as the results are used in making decisions about compliance. In environmental data analysis, depending on the method used to deal with non-detects (NDs), the estimates are somewhat biased from the true mean or true standard deviation. If all non-detect entries reported as "less than" the MDL are neglected, the resulting average from the remaining censored data will be upwardly biased. If all non-detects are substituted by MDL, the resulting sample will yield positively biased estimates.

Method Detection Limit

In environmental monitoring, a threshold limit below which measurements can not be numerically quantified is referred to as the method detection limit (MDL). Literally, it is the limitation of detection capability of laboratory instrumentation for a constituent. For example, the concentrations of many inorganic and organic contaminants in ambient waters are so low that they are below limits of detection established by analytical laboratories. Kaiser (1965) defined it as the minimum concentration of a substance that can be identified, measured, and reported with 99% confidence that the analytic concentration is greater than zero.

The measure is suitable only for detecting a pollutant, which is actually present. The MDL is estimated on the basis of ideal laboratory conditions with ideal analyte (e.g., nitrogen) samples and does not account for matrix or other inferences encountered during analysis of specific and/or actual field samples. Traditionally, detection limit estimators have been concerned only with providing protection against Type-I errors, or false positive conclusions. The corresponding need for similar protection against Type-II errors, or false negative values has not yet been considered. The stringency of environmental quality standards has increased to the point that the MDL is very near the maximum concentration limit (MCL) allowable by environmental regulations.

Substitution Methods

Substitution methods have always played an important role in tackling the issues and concerns related to non-detects. They involve the replacement of the missing (i.e., non-detect) data by a constant or a variate (such as expected value of order statistics). There are infinite values that can be used for the constant. Zero is one of the choices and yields the smallest estimation of mean along with the highest variation. On the other hand, if the censored values were replaced by the MDL, one would obtain the highest estimation of mean and lowest estimation of variance. Any value between zero and the MDL can be used as proxy for the data below the detection limit. Clearly, the most conservative estimator for the mean is the one with non-detects replaced by the MDL. Though the estimation is heavily biased, it has often been suggested by the US EPA (1989) as its accepted procedure. The most liberal policy is the one that substitutes zero for the censored data. The half-MDL (i.e., MDL/2) method may be used as a balance to the two extreme cases.

In environmental monitoring, the values smaller than the predefined detection limit are not required to be reported. To obtain reliable estimate of the populations mean and standard deviation, US EPA (1989) has advocated that non-detected measurements to be replaced by one-half the detection limit and data is analyzed as if all measurements were observable with equal precision. Other substitution values for non-detects required by the US EPA (1989) includes zero for drinking water contaminants. The simple substitution method is computationally attractive and is often adequate for most practical purposes if the non-detects proportion is below a certain censoring level. However, the US EPA (1989) does not give the theoretical basis of this particular percentage value.

There are a number of methods for handling non-detects in environmental data. Historically, the estimators have been compared on the basis of their ability to recover the true population parameters. And there have been several studies comparing the statistical properties of the various estimators, both analytically and via Monte Carlo simulation. The maximum likelihood method by Cohen (1959, 1961, 1991) performs well and in some cases the linear estimators can outperform the MLE estimator.

The efficiency of the methods was found to be dependent on the sample size and the censoring level. Haas and Scheff (1990) found that the bias became large for samples of size 10. Gleit (1985) pointed out that the environmental data are characterized not only by detection limits but also by small sample size and the MLE method is inefficient and highly biased for small sample size. The effect of censoring level is more pronounced than the sample size. Results of study by Haas and Scheff (1990) revealed that the estimation of mean was negatively biased for low degrees of censoring but positively biased for high degrees of censoring.

Probability Distributions for Environmental Data

All contaminants of concern are usually subject to regulatory restrictions denoted by threshold values or maximum contaminant level (MCL), and these are often specified in terms of µg/L or ppm. For a given contaminant, the MCL is specified as x_0. Samples are collected and different statistics of the sample are calculated such as mean $\bar{x}$, standard deviation s, skewness μ_3, kurtosis μ_4, and higher moment (order r) $\frac{1}{n}\sum_{i=1}^{n}x_i^r$, etc. Depending on the censoring schemes, the estimates are somewhat biased from the true values. It has long been recognized that the sample parameters suffer from an extreme sensitivity to normality violations of the distribution. Data sets from a number of distributions such as Normal, Log Normal, Gamma, and Gumbel are the focus of investigations in this paper.

The random variables x_1, x_2,..., x_n are associated with a probability density function $f(x)$. The form of $f(x)$ may be known from physical arguments or can be inferred from histogram of large sample data set. For environmental data involving chemical concentrations (which are known to vary from $0 \le x \ge \infty$) tend to follow Log Normal distribution. The shape of this distribution has been found to have considerable potentials in describing environmental quality data (McBean and Rovers, 1992).

Parameter Estimation Methods and Hypothesis Testing

It is noted that since the upper limit is theoretically infinite, there is always a finite probability that $x > x_0$. For cases of known (or assumed to be known) probability distributions, the estimation of population parameters (mean μ, standard deviation σ, skewness μ_3, kurtosis μ_4, etc.), are evaluated using parametric methods. For cases of unknown probability distributions, the parameters are estimated using non-parametric methods.

For obtaining estimates of the population parameters, the simplest approach is to use the method of moments (MOM). The confidence about these estimates in terms of a probability statement can be expressed as Prob $(a_1 < \mu < a_2) = 1 - \alpha$. Where, α is referred to as the level of significance (often $\alpha = 0.05$ or 0.01). For the normal variate Y, ($a_2 = \bar{y} + \frac{s_y}{\sqrt{n}}t_{\alpha/2,n-1}$), where, n is the sample size; $\bar{y}$ is the mean of variate Y; s_y is the standard deviation of variate Y; $t_{\alpha/2,n-1}$ is an appropriate student t variable. Of particular concern is whether or not $\mu_y < a_2 < y_0$. By increasing α, one can assure that the above inequality is true but it is done at the expense of decreasing the probability or confidence level.

The hypothesis testing is achieved as follows.

H_0: $\mu_y \le y_0$ no contamination
H_1: $\mu_y > y_0$ contamination present

The probability of accepting H_0 when, in fact, H_1 is true is referred to as Type-I error and occurs with probability α. The probability of accepting H_1 when, in fact, H_0 is true, is referred to as a Type-II error. The probability used to explain this event is described by the power curve. In most monitoring literature the occurrence of a Type II error is referred to as a "false positive" rate, i.e., concluding that there is contamination when none is present, or below the MCL. The alternative is referred to as a "false negative" rate.

The methodology of conducting numerical calculations and Monte Carlo simulations, manipulation of algebraic equations, and plotting of graphs was achieved with the use of the Waterloo Maple 6® (hereafter referred to Maple®).

Experimental Investigations on Probability Distributions

It is often impossible to make observations on all members of the population; therefore, estimates of population parameters are obtained through investigations of samples. It is therefore, the statistical analysis of samples obtained through environment monitoring of groundwater and/or soil and air helps the process of decision-making in the face of uncertainty and statistical inference provides a tool for the process of estimation and hypothesis testing and ultimately, decisions regarding risk.

To a significant extent, statistical estimates of a data set are obtained based on the underlying distribution. In cases, where the assumption of normality is untenable, one would seek out other distributions. For example, When a data exhibits a few elevated values that are particularly larger than the remaining values in the data set, one would consider that the data set might be coming from a skewed distribution with a long tail. In such situations, one could inspect the distribution by plotting the histogram of the original data. If the distribution is skewed, Log Normal, Gamma, Gumbel, and other skewed distributions may be an obvious choice for better describing the underlying distribution.

Some typical probability distributions commonly considered in environmental data analyses are: Normal, Log Normal, Gamma, Beta, and Gumbel. The probability plots are widely used in statistical analyses of environmental data because the probability paper provides subjective analyses and is easy to understand. In environmental context, a relevant detail on these probability distributions such as probability plotting positions and to facilitate the computer plotting, the ranked observations are plotted against the expected values of the order statistics, and the approaches to estimate the bias are provided by Zhang et al. (2001). They further observed that Blom's plotting position is the best to fit a hypothetical distribution. The fitted straight line on the probability graph paper can best recover the statistical parameters of assumed distribution. In addition, it is shown that the approach also affects the preciseness of the estimation bias. They used maximum likelihood approach and Monte Carlo simulation approach to calculate the estimation bias and also showed that similar results are achieved for either approach. Specifically, bias of estimation is calculated using linear estimator and order statistical methods.

Methodology for the Bias of Estimation in Censored Samples

The estimation bias is based on the distribution analysis and the censoring proportion. First, an appropriate distribution model is chosen, and the censored distribution function is mathematically represented with recourse to two mathematical functions. Further, the parameters for censored distribution are derived and the parameter estimation bias is plotted against the censoring proportion. The bias curves provide a graphical method to investigate the half detection limit method, which then is applied to various data samples. Bias estimation procedures for the mean and variance were developed under four probability distributions (Normal, Log Normal, Gamma, and Gumbel) using substitution methods of MDL and MDL/2. The procedural detail for censored samples from Normal probability distribution is presented in Appendix-A. The procedural detail for other probability distributions is provided by Zhang et al. (2001).

Censored Samples from Normal Distribution

Normal probability distribution does not often perfectly fit sequences of environmental data, however, is widely used, for example, in dealing with transformed data that do follow the Normal distribution and in estimation of sample reliability by virtue of the central limit theorem. It describes many processes that are subject to random and independent variations. The Normal distribution is completely determined by two parameters, i.e., the mean μ and the standard deviation σ, and is described as $N(\mu,\sigma)$. By a simple transformation, the general form of Normal distribution can be written as a univariate, single-parameter function. If we define

$$z = \frac{x - \mu}{\sigma} \tag{1}$$

Where, z is called the standardized variate, then accordingly $dx = \sigma dz$. The PDF becomes

$$f(z) = \frac{1}{\sqrt{2\pi}} \exp\left(-\frac{1}{2}z^2\right) \tag{2}$$

Which is standard Normal distribution. The random variable z is normally distributed with zero mean and unit standard deviation represented as $N(0, 1)$.

Integrate with respect to z from $-\infty$ to t, and one obtains a function with an independent variable t as

$$F(t) = \int_{-\infty}^{t} f(z)dz \tag{3}$$

Which is the cumulative density function (CDF) of standard Normal distribution and represents the area under the PDF. The CDF is tabulated to serve all Normal distributions after standardization. Given a cumulative probability, the *z-score* is found in the table and the variate x is computed from the inverse transformation $x = \mu + z\sigma$.

Probability Paper

Probability paper is a specially designed graph paper. Probability papers for Normal, Log Normal, Gamma, and Gumbel distributions have been developed on Ma-

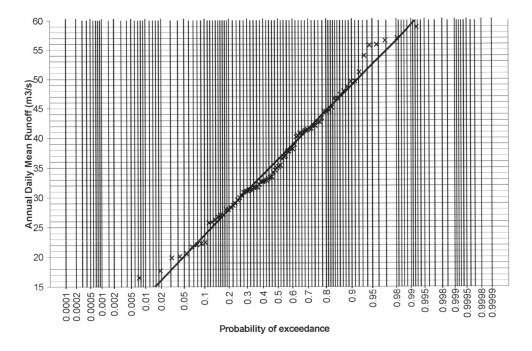

Figure 2. Probability paper for the Normal distribution.

ple® and Microsoft Excel® in this paper. The advantage of computer-based probability papers is that one can plot the empirical points directly on the desired probability paper. These programs fit a straight line to plotted points and also provide the goodness of fit in terms of RMSE (i.e., root mean square error). For example, Figure 2 exhibits plotted points and a straight line fit to annual daily flows of the Grand River at Galt (Ontario) on the Normal probability paper.

Bias of Estimation by Substitution Method

Consider that the underlying distribution is standard normal, i.e., the PDF is

$$f(x) = \frac{1}{\sqrt{2\pi}} \exp\left(-\frac{1}{2}x^2\right) \tag{4}$$

The mean is 0 and the standard deviation is 1. Suppose a censoring point ξ divides the PDF into two parts. The left part constitutes the censoring proportion, which is of an area of

$$a = \int_{-\infty}^{x} f(x)dx = \frac{1}{2}erf\left(-\frac{\sqrt{2}}{2}\xi\right) + \frac{1}{2} \tag{5}$$

557

Then, one can write the censored PDF with recourse to the Dirac and Heaviside functions. A description of the two functions is provided in Appendix-C.

$$g(x) = a \ \ Dirac(x - \xi - \ln 2) + Heaviside(x - \xi)\frac{1}{\sqrt{2p}}\exp\left(-\frac{1}{2}x^2\right) \tag{6}$$

In the Half-detection limit method, all values that are below detection limits are assigned a numerical value of the detection limit ξ divided by 2. Since the motivation for this work assumes that the actual measurements are lognormally distributed, and that a logarithmic transformation is performed to achieve normality, all non-detectable values are replaced by ($\xi - \ln 2$).

Integrate Equation 6 with respect to x from $-\infty$ to ∞

$$m = \int_{-\infty}^{\infty} g(x)x\,dx \tag{7}$$

and eliminate the independent variable x, to obtain the formula for calculation of the mean μ of the new PDF:

$$\mu = \frac{1}{2}(\xi - \ln(2))\,\mathrm{erf}\left(\frac{1}{2}\sqrt{2}\,\xi\right) + \frac{1}{2}\xi - \frac{1}{2}\ln(2) + \frac{\frac{1}{2}e^{(-1/2\,\xi^2)}\sqrt{2}}{\sqrt{\pi}} \tag{8}$$

To calculate the standard deviation σ for the new PDF, integrate $g(x)(x - \mu)^2$ from $-\infty$ to ∞

$$s^2 = \int_{-\infty}^{\infty} g(x)(x-m)^2\,dx = \int_{-\infty}^{\infty} g(x)x^2\,dx - m^2 \tag{9}$$

Eliminating x in Equation 9 by using Equations 6 and 8, the standard deviation σ is obtained as follows,

$$\sigma^2 = \frac{1}{2}(\xi - \ln(2))^2\,\mathrm{erf}\left(\frac{1}{2}\sqrt{2}\,\xi\right) + \frac{1}{2}(\xi - \ln(2))^2$$
$$+ \frac{\frac{1}{2}\left(-\sqrt{\pi}\,\mathrm{erf}\left(\frac{1}{2}\sqrt{2}\,\xi\right) + \sqrt{\pi} + \xi\,e^{(-1/2\,\xi^2)}\sqrt{2}\right)}{\sqrt{\pi}}$$
$$- \left(\frac{1}{2}(\xi - \ln(2))\,\mathrm{erf}\left(\frac{1}{2}\sqrt{2}\,\xi\right) + \frac{1}{2}\xi - \frac{1}{2}\ln(2) + \frac{\frac{1}{2}e^{(-1/2\,\xi^2)}\sqrt{2}}{\sqrt{\pi}}\right)^2 \tag{10}$$

Which is solely a function of independent variable ξ.

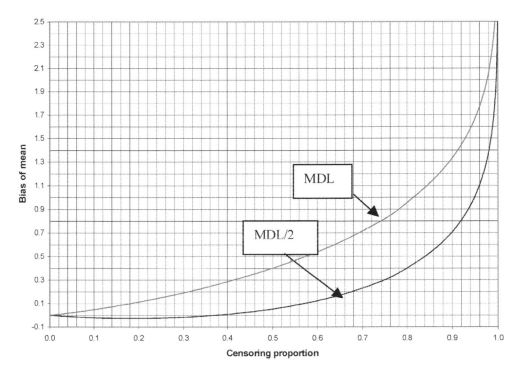

Figure 3. Bias of the estimation of mean for censored Normal distribution.

Obviously, when the censoring point ξ is small enough (e.g., −4 in standard Normal distribution), the censoring rate is about zero and there is no bias for the estimation of both mean and standard deviation. For a large censoring point (e.g., 4 in a standard Normal distribution), the censoring rate approaches 1 (that is, there is almost no quantified measurement). The bias for the estimation of mean approaches its maximum, and the bias for the estimation of standard deviation approaches −1 (that is, the resulted PDF is a Dirac function with the variance zero).

The bias calculation results for the substitution method (i.e., when censored values are replaced with half of the censoring point $\xi/2$ or censoring point ξ) are presented in Table A-1. The plots of the bias of mean and standard deviation against the censoring proportion are given in Figures 3 and 4.

In these figures, the vertical axes represent the difference of resulted parameter (computed from Equations 8 and 10) and the original parameter. As shown in these figures, the half detection limit method yields a slightly biased estimation of mean for censoring rate up to 50%. When the censoring rate is less than 36%, the estimator is negatively biased. Bias of mean becomes positively biased when the censoring rate is greater than 36%. The method leads to slightly biased estimation of standard deviation for censoring rate up to 30%.

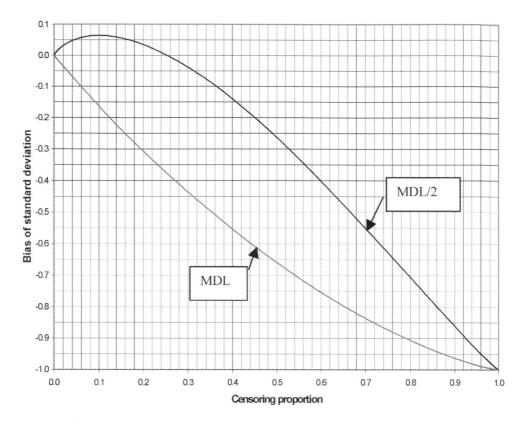

Figure 4. Bias of the estimation of standard deviation for censored
Normal distribution.

Variance of Maximum Likelihood Estimation

Cohen (1961 and 1991) gave formulas to compute the confidence interval for estimations of the mean and standard deviation. The asymptotic variance of the maximum likelihood estimation of the mean $\hat{\mu}$ and standard deviation $\hat{\sigma}$ is given as:

$$V(\hat{\mu}) = \frac{\sigma^2}{N}\mu_{11} \qquad (12)$$

$$V(\hat{\sigma}) = \frac{\sigma^2}{N}\mu_{22} \qquad (13)$$

For singly left truncated samples, with $N = n$, one has

$$\mu_{11} = \frac{\phi_{22}}{\phi_{11}\phi_{22} - \phi_{12}^2} \qquad (14)$$

560

$$\mu_{22} = \frac{\phi_{11}}{\phi_{11}\phi_{22} - \phi_{12}^2} \tag{15}$$

Where

$$\phi_{11}(\xi) = 1 - Q(\xi)[Q(\xi) - \xi]$$
$$\phi_{12}(\xi) = Q(\xi)\{1 - \xi[Q(\xi) - \xi]\} \tag{16}$$
$$\phi_{11}(\xi) = 2 + \xi\phi_{12}(\xi)$$

and

$$Q(\xi) = \frac{\phi(\xi)}{1 - \Phi(\xi)} \tag{17}$$

For singly left censored samples, with the expectation $E(n) = N[1 - \Phi(\eta)]$,

$$\mu_{11} = \frac{1}{1 - \Phi(\eta)} \frac{\phi_{22}}{\phi_{11}\phi_{22} - \phi_{12}^2} \tag{18}$$

$$\mu_{22} = \frac{1}{1 - \Phi(\eta)} \frac{\phi_{11}}{\phi_{11}\phi_{22} - \phi_{12}^2} \tag{19}$$

Where

$$\phi_{11}(\eta) = 1 + Q(\eta)[Q(-\eta) + \eta]$$
$$\phi_{12}(\eta) = Q(\eta)\{1 + \eta[Q(-\eta) + \eta]\} \tag{20}$$
$$\phi_{22}(\eta) = 2 + \eta\phi_{12}(\eta)$$

and

$$Q(\eta) = \frac{\phi(\eta)}{1 - \Phi(\eta)} \tag{21}$$

A computational aid for the calculation of asymptotic variances of the mean $\bar{\mu}$ and standard deviation $\bar{\sigma}$ is provided in Table A-2. Cohen (1959, 1961, 1991) provided a table for η up to 2.5 ($\Phi = 99.38\%$). This paper expands the table for η up to 4 ($\Phi = 99.996\%$).

Similar to the Normal distribution, other distributions such as Log Normal, Gamma, and Gumbel were also used for estimation of parameters and obtaining bias of estimation.

Bias of the Estimation curve of standard deviation for Log Normal distribution suggests that the estimation of standard deviation is unbiased when the censoring rate is less than 60%. Comparing with the upper limit-censoring rate of 30% in the case of normally distribution data, the MDL/2 method yields reasonable estimation of standard deviation up to a high censoring percentage.

For Gamma probability distribution, the bias of mean is zero at censoring rate 74%, and the bias of standard deviation is zero at censoring rate 54%. The qualitative performance of the method in the case of the Gamma distribution is even better than the Log Normal distribution. Further, the bias curve is almost parallel to the horizontal axis and represents quite low bias value up to a quite large censoring proportion.

It is apparent that due to the significant bias even for the low censoring degree, the half detection limit method gives the poorest performance and is not suitable to estimate parameters for the samples coming from Gumbel distribution.

Case Study: An Example

During monitoring of soil, air, and water quality, sampling is performed on a regular basis in response to government regulations, which are established to assist in meeting the legislated requirements. All samples are analyzed with the methods approved by the provincial or federal government. There are government regulations, which prescribe the maximum concentrations of contaminants in all media (soil, air, and water). In Ontario, for example, the Provincial Water Quality Objectives give direction on how to manage the quality and quantity of both surface and ground waters. The drinking water quality is protected by the Ontario Drinking Water Standards.

Surface water and groundwater quality refers to not only chemical quality, but also radiological and microbiological quality. Generally, water quality refers to both natural and man-made compounds found in water. Examples of natural chemicals include fluoride and arsenic. Man-made compounds include industrial, automotive, and agricultural chemicals. Although several groundwater quality data including the concentration of trichloroethylene (TCE), pentachloroethylene (PCE), cis-1, 2 dichloroethane (cis-1, 2-DCE), and vinyl chloride (VC) were investigated. However, investigations on the concentration of trichloroethylene (TCE) are only reported in this paper. These substances are regulated by the legislation and if a substance is not detected in a water quality sample, it is either not present (and is indicated as "ND"), or present in a level below the detection limit of the standard analysis method (and is indicated as "< MDL"). The full data sets of above chemical compounds are provided by Zhang et al. (2001). From 1997 – 1998, Conestoga Rovers and Associates Ltd. measured these data sets during the plume remediation and source area investigations. Although, the results of both on-site and off-site laboratory analyses of groundwater samples were available but only the relatively more accurate off-site data sets were analysed in this paper. The results of data analyses on parameter basis are presented as follows.

Analysis of Trichloroethylene (TCE) Data Set

Trichloroethylene (TCE) is an organic chemical mostly used in dry cleaning, while some is used in metal degreasing operations and in tetrachloroethylene production. TCE may be introduced into surface and ground waters through industrial spills and illegal disposal of effluents. The detection limit for TCE is 1 m/L, while the government guideline (Ontario Drinking Water Standards established by Ministry of Environment under the Ontario Water Resources Act) for the maximum acceptable concentration in drinking water is 50 μg/L.

The TCE data was plotted on various probability papers to investigate the underlying probability distribution. For example, the TCE data is plotted on a Normal probability paper (Figure 5). From this figure, it is apparent that the data does not follow a Normal distribution. However, plot of the TCE data on the Log Normal probability paper (Figure 6) indicates that the Log Normal distribution is a potential candidate for the underlying probability distribution. For such probability plots, the mean square error (MSE) is used to test the goodness of fit. A majority of the solute concentration data sets are found to best fit a Log Normal distribution (McBean and Rovers, 1992).

Based on above observation, the TCE sample is split into two components, uncensored and censored. The total sample size including both components is N = 368. The number of observed measurements as uncensored is n = 306. The number of censored measurement c is 62. Therefore, the censoring rate h is c/N = 16.85%.

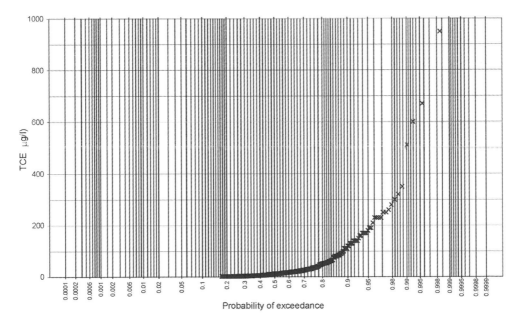

Figure 5. Fitting TCE data points to Normal probability paper.

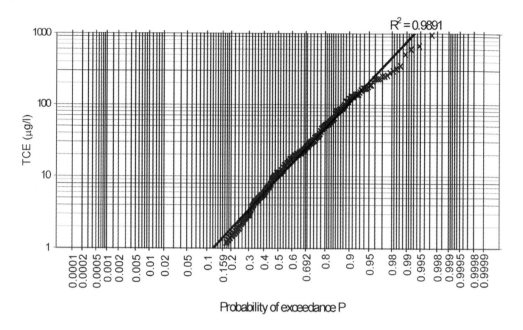

Figure 6. Fitting TCE data points to Log Normal probability paper.

Other statistics of the TCE sample data are summarized as:

$$\bar{X}_n = 48.69$$

$$\bar{X}_{n,\,log} = 1.2121$$

$$\bar{S}_n = 96.94$$

$$\bar{S}_{n,\,log} = 0.6474$$

$$T = 1$$

$$T_{log} = 0$$

$$g = \frac{0.6474^2}{(1.2121 - 0)^2} = 0.2853$$

The values of h (= c/N) = 16.85% and g = 0.2853 are entered in Table B-1 (Appendix B) and through interpolation, the value of $\lambda(g,h) = 0.25$ is obtained for substitution into Cohen's formulae for MLE:

$$\hat{\mu} = \bar{X}_{n,log} - \lambda(g,h)(\bar{X}_{n,log} - T_{log}) = 1.2121 - 0.25 \times (1.2121 - 0) = 0.91$$

$$\hat{\sigma}^2 = \bar{S}^2 + \lambda(g,h)(\bar{X} - T)^2 = 0.6474^2 + 0.25 \times (1.2121 - 0)^2 = 0.7864$$

or
$$\hat{\sigma} = 0.887$$

The standardized truncation point $\hat{\xi} = \dfrac{T - \hat{\mu}}{\hat{\sigma}} = \dfrac{0 - 0.91}{0.887} = -1.03$

With truncation point $\hat{\xi} = -1.03$, one obtains $\mu_{11} = 3.26$, $\mu_{22} = 1.53$ from Table A-2. One then proceeds to calculate the asymptotic variance of the maximum likelihood estimation of the mean $\hat{\mu}$ and standard deviation $\hat{\sigma}$ from Equations 12 and 13 as follows.

$$V(\hat{\mu}) = \frac{\sigma^2}{N}\mu_{11} = \frac{0.7864}{368} \times 3.26 = 0.006966$$

$$V(\hat{\sigma}) = \frac{\sigma^2}{N}\mu_{22} = \frac{0.7864}{368} \times 1.53 = 0.003269$$

Accordingly,

$$\sigma_{\hat{\mu}} = 0.0818$$

and

$$\sigma_{\hat{\sigma}} = 0.0572$$

The 95% confidence intervals are calculated as

$$\hat{\mu} - 1.96 \times \sigma_{\hat{\mu}} < \mu < \hat{\mu} + 1.96 \times \sigma_{\hat{\mu}}$$
$$\hat{\sigma} - 1.96 \times \sigma_{\hat{\sigma}} < \sigma < \hat{\sigma} + 1.96 \times \sigma_{\hat{\sigma}}$$

Thus,

$$0.750 < \mu < 1.070$$
$$0.775 < \sigma < 0.999$$

In the original unit, the mean concentration of TCE is

$$5.62 < \mu_{TCE} < 11.75$$
$$5.96 < \sigma_{TCE} < 9.98$$

With the half detection limit method (MDL/2), one obtains the detection limit $T = 1$. Substitute $T/2$ with all the non-detects (NDs) and take logarithm of the data to calculate the mean and standard deviation of the sample as follows:

$$\mu = 0.96$$

$$\sigma = 0.815$$

In the original unit, the mean and standard deviation of the concentration of TCE are

$$\mu_{TCE} = 9.12$$

$$\sigma_{TCE} = 6.61$$

The critical value corresponding to the significance level $\alpha = 95\%$ is $t_{\alpha/2} = 1.96$. Therefore, the 95% confidence interval of the mean is

$$0.877 < \mu < 1.043$$

In the original unit, the confidence interval for the mean TCE is

$$7.53 < \mu_{TCE} < 11.05$$

Comparing the results of the half detection limit method (MDL/2) and the maximum likelihood estimation (MLE) method, it is found that the two methods yield very similar results.

To provide additional information on the applicability of the methodology to different sample sizes, several sub-sets were created through random selection. The homogeneity of data set was ensured and statistical similarity was found to hold from one to another sub-data sets. Calculated results for various sample sizes of TCE data are summarized in Table 1 as follows.

First, samples with sizes 25, 50 are 100 are randomly selected. They represent small, medium, and large samples. Together with the complete sample, four sample sizes are considered for the TCE data set. Second, the censoring rates for the four samples turn out to be quite similar. Such an observation reinforces the fact that the random selection process performs well.

In this summary table, the results in rows corresponding to $\hat{\mu}$ and $\hat{\sigma}$ show the estimations of mean and standard deviation using the half detection limit method (MDL/2) are very close to the maximum likelihood estimators (MLE). The estimation of the mean ranges from 0.90~1.04. For the MLE, the estimated standard deviation ranges from 0.76~0.92. On the other hand, for the MDL/2 method, the estimated standard deviation ranges from 0.74~0.84. For both estimation methods, smaller sample yields larger standard deviation.

The MDL/2 method gives similar confidence interval, i.e. the upper limit (UL) and lower limit (LL), to the inference compared to the maximum likelihood estimation method. For example, the 95% confidence interval is 5.55~11.77 and 7.53~11.05 respectively for the MLE and MDL/2 methods.

Table 1. Summary of estimation of mean, standard deviation, and confidence interval of TCE samples.

	MLE	MDL/2	MLE	MDL/2	MLE	MDL/2	MLE	MDL/2
N	25	25	50	50	100	100	368	368
c	3	-	8	-	17	-	62	-
n	22	-	42	-	83	-	306	-
h	12%	-	16%	-	17%	-	16.85%	-
$\bar{X}_{n,\log}$	1.22	-	1.28	-	1.26	-	1.21	-
$\bar{S}_{n,\log}$	0.587	-	0.658	-	0.634	-	0.647	-
T	1	1	1	1	1	1	1	1
$T/2$	-	0.5	-	0.5	-	0.5	-	0.5
$T_{\log}$	0	-	0	-	0	-	0	-
g	0.232	-	0.264	-	0.253	-	0.286	-
$\lambda(g,h)$	0.160	-	0.200	-	0.285	-	0.250	-
$\hat{\mu}$	1.025	1.04	1.024	1.03	0.901	1	0.908	0.96
$\hat{\sigma}$	0.763	0.74	0.872	0.838	0.924	0.829	0.886	0.815
$\hat{\xi}$	-1.342	-	-1.174	-	-0.975	-	-1.025	-
μ_{11}	2.86	-	3.00	-	2.90	-	3.26	-
μ_{22}	1.40	-	1.46	-	1.42	-	1.53	-
$\sigma_{\hat{\mu}}$	0.258	-	0.214	-	0.157	-	0.083	-
$\sigma_{\hat{\sigma}}$	0.917	-	0.502	-	0.282	-	0.161	-
$t_{2.5\%}$	1.96	1.96	1.96	1.96	1.96	1.96	1.96	1.96
$UL_{\log}$	1.531	1.330	1.443	1.262	1.209	1.162	1.071	1.043
$LL_{\log}$	0.519	0.750	0.605	0.798	0.592	0.838	0.744	0.877
UL	33.95	21.38	27.72	18.29	16.20	14.54	11.77	11.05
LL	3.30	5.62	4.03	6.28	3.91	6.88	5.55	7.53

Results and Discussion

The results provide support to the half detection limit (MDL/2) method that has been advocated by US EPA (1989) to be effective and efficient. Because, the method outperformed many other methods and thus appeared to be an appropriate method for estimation of the mean and standard deviation up to a significant censoring proportion. For the Gamma distribution, such a censoring rate was found to be up to 75% of data samples. The MDL/2 method is practicable because of its simplicity and ease of operation.

For small sample size (n < 10), the maximum likelihood estimators are positively biased for Normal, Log Normal, and other distributions, although this bias can be reduced by using the restricted maximum likelihood estimators. Among other methods, the substitution method with expected values of order statistics, always yields positively biased estimation of the mean over the entire range of censoring level (percentage from 0% to 100%).

For censored samples coming from the Normal distribution, the estimation of mean based on MDL/2 method is slightly biased up to the censoring rate of 50%. This means that the parameter estimation is accountable when less than half of the data is censored. When the censoring level is near 36%, the method leads to the zero biased estimate of the mean. The MDL/2 method is negatively biased for censoring rate of less than 36%, and is positively biased for censoring rate of higher than 36%.

For estimation of the standard deviation of a censored sample from Normal distribution, the MDL/2 method is less effective than for estimation of the mean, but it still gives asymptotically unbiased estimation up to censoring rate of 30%. The method performs even better for samples coming from the Log Normal or the Gamma distribution. Environmental samples often come from heavy tailed or skewed distributions; such an observation of the MDL/2 is desirable. On the other hand, the maximum likelihood method is sensitive to these types of distributions, and the method become less functional when the underlying distribution of samples is unknown.

Estimation of the Standard Deviation

It is often desirable to estimate the mean and standard deviation for the censored sample, especially for statistical hypotheses testing. In this paper, the bias function of both the mean and standard deviation is derived using similar approach and is also investigated in a parallel manner. In general, the estimate of standard deviation is more biased than that of the mean for any kind of probability distribution. It is also more sensitive to the censoring percentage.

With respect to the estimation of standard deviation for censored distribution, it is easy to understand that there is always a limit to the resultant bias. For the substitution method, censored PDF is represented by the Dirac function. Therefore, as censoring rate approaches to its highest, i.e. one, the standard deviation of resulted PDF is zero due to the feature of the Dirac function.

Statistical Inference

Besides the estimation of the population parameters, the confidence interval helps the engineers and scientists to draw conclusions and/or to make decisions. The procedures that lead to the acceptance or rejection of statistical hypotheses comprise a major area in environmental data analysis. Confidence intervals set bounds on the estimator and provide venues to assess the reliability of the estimators. Once the estimations of mean and standard deviation are obtained, the subsequent statistical inference is to estimate the confidence interval at a desired significance level. Results of the analyses presented in this paper show that the confidence intervals based on the MDL/2 method lead to the same conclusion as the maximum likelihood method.

Recommendations

Further investigations related to the effect of changing detection limits (multiple detection limits) on the estimation are recommended. Other estimation parameters such as percentile should be explored and also, statistical inference and the estimation of confidence intervals should be further studied.

References

Cohen, A.C. (1959). Simplified estimators for the Normal distribution when samples are singly censored or truncated. Technometrics (A Journal of Statistics for the Physical, Chemical and Engineering Sciences, published by the American Statistical Association and the American Society for Quality Control), Vol. 1, p. 217-237.

Cohen, A.C. (1961). Tables for maximum likelihood estimates: Singly truncated and singly censored samples. Technometrics, Vol. 3, p. 535-541, Chapter 4.

Cohen, A.C. (1991). Truncated and censored samples: Theory and applications. Marcel Dekker Inc, New York.

Gleit, A. (1985). Estimation for small normal data sets with detection limits. Environmental Science and Technology, 19, p. 1201-1206.

Haas, C.N. and Scheff, P.A. (1990). Estimation of averages in truncated samples. Environmental Science and Technology, 24, p. 912-919.

McBean, E.A. and Rovers, F.A. (1998). Statistical procedures for analysis of environmental monitoring data and risk assessment. Prentice Hall PTR Environment Management & Engineering Series, Volume 3.

US EPA (1989). Statistical analysis of groundwater monitoring data at RCRA facilities: Interim final guidance. Office of Solid Waste, US Environmental Protection Agency, Washington, D.C.

Zhang, Z., Panu, U.S. and W.C. Lennox (2001). Assessment of the Effect of Percent Non-Detects on Estimation Bias for Censored Distributions in Environmental Data. Civil Engineering Technical Report, Department of Civil Engineering, Lakehead University, Thunder Bay, Ontario, Canada.

APPENDIX-A

The bias calculation results for the substitution method (censored values are re-placed with half of the censoring point $\xi/2$ or censoring point ξ) are presented in Table A-1.

Table A-1. Bias of mean (μ) and variance (σ^2) for the Normal distribution.

MDL	Censoring level	Bias of mean		Bias of standard deviation	
	*100%	MDL/2	MDL	MDL/2	MDL
ξ	$\Phi(\xi)$	B_μ	B_μ	B_σ	B_σ
-4.0	0.000032	-0.000015	0.000007	0.000131	-0.000060
-3.9	0.000048	-0.000022	0.000011	0.000192	-0.000091
-3.8	0.000072	-0.000033	0.000017	0.000279	-0.000137
-3.7	0.000108	-0.000049	0.000026	0.000401	-0.000204
-3.6	0.000159	-0.000071	0.000039	0.000571	-0.000300
-3.5	0.000233	-0.000103	0.000058	0.000803	-0.000437
-3.4	0.000337	-0.000147	0.000087	0.001118	-0.000632
-3.3	0.000483	-0.000208	0.000127	0.001540	-0.000903
-3.2	0.000687	-0.000291	0.000185	0.002098	-0.001280
-3.1	0.000968	-0.000403	0.000267	0.002827	-0.001796
-3.0	0.001350	-0.000554	0.000382	0.003766	-0.002497
-2.9	0.001866	-0.000752	0.000542	0.004960	-0.003437
-2.8	0.002555	-0.001010	0.000761	0.006458	-0.004687
-2.7	0.003467	-0.001343	0.001060	0.008312	-0.006330
-2.6	0.004661	-0.001767	0.001464	0.010570	-0.008469
-2.5	0.006210	-0.002300	0.002004	0.013279	-0.011224
-2.4	0.008198	-0.002962	0.002720	0.016477	-0.014734
-2.3	0.010724	-0.003772	0.003662	0.020186	-0.019159
-2.2	0.013903	-0.004750	0.004887	0.024406	-0.024679
-2.1	0.017864	-0.005914	0.006468	0.029107	-0.031490
-2.0	0.022750	-0.007278	0.008491	0.034223	-0.039804
-1.9	0.028717	-0.008850	0.011054	0.039637	-0.049842
-1.8	0.035930	-0.010629	0.014276	0.045181	-0.061830
-1.7	0.044565	-0.012603	0.018288	0.050626	-0.075989
-1.6	0.054799	-0.014742	0.023242	0.055673	-0.092527
-1.5	0.066807	-0.017000	0.029307	0.059963	-0.111626
-1.4	0.080757	-0.019308	0.036668	0.063068	-0.133437
-1.3	0.096800	-0.021569	0.045528	0.064508	-0.158060
-1.2	0.115070	-0.023658	0.056102	0.063758	-0.185540
-1.1	0.135666	-0.025417	0.068620	0.060268	-0.215856

Table A-1. (continued).

-1.0	0.158655	-0.026656	0.083315	0.053488	-0.248912
-0.9	0.184060	-0.027150	0.100431	0.042892	-0.284535
-0.8	0.211855	-0.026640	0.120207	0.028011	-0.322471
-0.7	0.241964	-0.024837	0.142879	0.008459	-0.362394
-0.6	0.274253	-0.021425	0.168673	-0.016033	-0.403907
-0.5	0.308538	-0.016065	0.197797	-0.045594	-0.446559
-0.4	0.344578	-0.008405	0.230439	-0.080196	-0.489856
-0.3	0.382089	0.001918	0.266761	-0.119639	-0.533279
-0.2	0.420740	0.015260	0.306895	-0.163552	-0.576304
-0.1	0.460172	0.031968	0.350935	-0.211403	-0.618421
0.0	0.500000	0.052369	0.398942	-0.262516	-0.659155
0.1	0.539828	0.076755	0.450935	-0.316100	-0.698077
0.2	0.579260	0.105382	0.506895	-0.371284	-0.734823
0.3	0.617911	0.138458	0.566761	-0.427158	-0.769101
0.4	0.655422	0.176135	0.630439	-0.482813	-0.800699
0.5	0.691462	0.218511	0.697797	-0.537381	-0.829484
0.6	0.725747	0.265623	0.768673	-0.590071	-0.855401
0.7	0.758036	0.317449	0.842879	-0.640197	-0.878466
0.8	0.788145	0.373907	0.920207	-0.687199	-0.898760
0.9	0.815940	0.434865	1.000431	-0.730658	-0.916414
1.0	0.841345	0.500140	1.083315	-0.770294	-0.931602
1.1	0.864334	0.569509	1.168620	-0.805964	-0.944524
1.2	0.884930	0.642715	1.256102	-0.837652	-0.955401
1.3	0.903200	0.719478	1.345528	-0.865447	-0.964459
1.4	0.919243	0.799497	1.436668	-0.889529	-0.971923
1.5	0.933193	0.882467	1.529307	-0.910145	-0.978012
1.6	0.945201	0.968079	1.623242	-0.927588	-0.982928
1.7	0.955435	1.056031	1.718288	-0.942178	-0.986858
1.8	0.964070	1.146033	1.814276	-0.954248	-0.989970
1.9	0.971283	1.237812	1.911054	-0.964124	-0.992409
2.0	0.977250	1.331113	2.008491	-0.972119	-0.994303
2.1	0.982136	1.425704	2.106468	-0.978524	-0.995761
2.2	0.986097	1.521377	2.204887	-0.983604	-0.996872
2.3	0.989276	1.617948	2.303662	-0.987592	-0.997711
2.4	0.991802	1.715255	2.402720	-0.990692	-0.998339
2.5	0.993790	1.813161	2.502004	-0.993079	-0.998805
2.6	0.995339	1.911548	2.601464	-0.994898	-0.999147
2.7	0.996533	2.010316	2.701060	-0.996272	-0.999396
2.8	0.997445	2.109385	2.800761	-0.997300	-0.999576
2.9	0.998134	2.208688	2.900542	-0.998061	-0.999705

3.0	0.998650	2.308171	3.000382	-0.998620	-0.999797
3.1	0.999032	2.407791	3.100267	-0.999026	-0.999861
3.2	0.999313	2.507514	3.200185	-0.999319	-0.999906
3.3	0.999517	2.607315	3.300127	-0.999528	-0.999937
3.4	0.999663	2.707173	3.400087	-0.999676	-0.999958
3.5	0.999767	2.807073	3.500058	-0.999779	-0.999972
3.6	0.999841	2.907002	3.600039	-0.999851	-0.999982
3.7	0.999892	3.006953	3.700026	-0.999900	-0.999988
3.8	0.999928	3.106920	3.800017	-0.999934	-0.999992
3.9	0.999952	3.206897	3.900011	-0.999957	-0.999995
4.0	0.999968	3.306882	4.000007	-0.999972	-0.999997

A computational aid to calculate asymptotic variances of the mean $\hat{\mu}$ and standard deviation $\hat{\sigma}$ is provided in Table A-2. Cohen provided a table for η up to 2.5 (Φ = 99.38%). This paper expands the table for η up to 4 (Φ = 99.996%).

Table A-2. Variances of estimators for singly left truncated and singly left censored samples from Normal distribution.

H	Truncated samples		Censored samples		
	μ_{11}	μ_{22}	$\Phi\eta$	μ_{11}	μ_{22}
-4	1.0005382	0.5022870	0.0000317	1.0000015	0.5000289
-3.9	1.0007807	0.5031625	0.0000481	1.0000023	0.5000437
-3.8	1.0011211	0.5043251	0.0000723	1.0000036	0.5000655
-3.7	1.0015943	0.5058507	0.0001078	1.0000057	0.5000973
-3.6	1.0022457	0.5078294	0.0001591	1.0000087	0.5001430
-3.5	1.0031340	0.5103665	0.0002326	1.0000133	0.5002081
-3.4	1.0043343	0.5135835	0.0003369	1.0000200	0.5003001
-3.3	1.0059428	0.5176188	0.0004834	1.0000300	0.5004286
-3.2	1.0080810	0.5226282	0.0006871	1.0000445	0.5006062
-3.1	1.0109025	0.5287855	0.0009676	1.0000654	0.5008493
-3	1.0146012	0.5362833	0.0013499	1.0000955	0.5011788
-2.9	1.0194208	0.5453330	0.0018658	1.0001382	0.5016206
-2.8	1.0256685	0.5561666	0.0025551	1.0001983	0.5022073
-2.7	1.0337310	0.5690378	0.0034670	1.0002823	0.5029788
-2.6	1.0440964	0.5842240	0.0046612	1.0003986	0.5039833
-2.5	1.0573811	0.6020290	0.0062097	1.0005586	0.5052787
-2.4	1.0743655	0.6227860	0.0081975	1.0007768	0.5069336
-2.3	1.0960380	0.6468620	0.0107241	1.0010722	0.5090285

Table A-2. (continued).

-2.2	1.1236526	0.6746625	0.0139034	1.0014695	0.5116568
-2.1	1.1587999	0.7066365	0.0178644	1.0020001	0.5149262
-2	1.2034978	0.7432830	0.0227501	1.0027044	0.5189600
-1.9	1.2603044	0.7851575	0.0287166	1.0036338	0.5238979
-1.8	1.3324577	0.8328796	0.0359303	1.0048537	0.5298987
-1.7	1.4240492	0.8871409	0.0445655	1.0064472	0.5371412
-1.6	1.5402370	0.9487133	0.0547993	1.0085197	0.5458272
-1.5	1.6875055	1.0184585	0.0668072	1.0112049	0.5561845
-1.4	1.8739815	1.0973373	0.0807567	1.0146726	0.5684711
-1.3	2.1098154	1.1864200	0.0968005	1.0191387	0.5829797
-1.2	2.4076392	1.2868970	0.1150697	1.0248780	0.6000445
-1.1	2.7831127	1.4000901	0.1356661	1.0322412	0.6200493
-1	3.2555715	1.5274641	0.1586553	1.0416766	0.6434374
-0.9	3.8487917	1.6706390	0.1840601	1.0537592	0.6707242
-0.8	4.5918876	1.8314027	0.2118554	1.0692281	0.7025132
-0.7	5.5203582	2.0117237	0.2419637	1.0890360	0.7395151
-0.6	6.6773018	2.2137650	0.2742531	1.1144150	0.7825727
-0.5	8.1148188	2.4398975	0.3085375	1.1469634	0.8326911
-0.4	9.8956209	2.6927142	0.3445783	1.1887607	0.8910755
-0.3	12.0948710	2.9750443	0.3820886	1.2425212	0.9591797
-0.2	14.8022732	3.2899679	0.4207403	1.3118004	1.0387657
-0.1	18.1244371	3.6408302	0.4601722	1.4012706	1.1319805
0	22.1875396	4.0312566	0.5000000	1.5170940	1.2414530
0.1	27.1403067	4.4651670	0.5398278	1.6674265	1.3704187
0.2	33.1573415	4.9467908	0.5792597	1.8631032	1.5228788
0.3	40.4428151	5.4806806	0.6179114	2.1185735	1.7038069
0.4	49.2345560	6.0717275	0.6554217	2.4531823	1.9194159
0.5	59.8085532	6.7251751	0.6914625	2.8929335	2.1775077
0.6	72.4838709	7.4466310	0.7257469	3.4729289	2.4879302
0.7	87.6281016	8.2420864	0.7580363	4.2407538	2.8631808
0.8	105.6631583	9.1179207	0.7881446	5.2612027	3.3192058
0.9	127.0717064	10.0809198	0.8159399	6.6229042	3.8764656
1	152.4040774	11.1382898	0.8413447	8.4476551	4.5613620
1.1	182.2856318	12.2976630	0.8643339	10.9036375	5.4081608
1.2	217.4248406	13.5671160	0.8849303	14.2242298	6.4615981
1.3	258.6216866	14.9551705	0.9031995	18.7349178	7.7804317
1.4	306.7771806	16.4708251	0.9192433	24.8920002	9.4423105
1.5	362.9021379	18.1235130	0.9331928	33.3385579	11.5504899
1.6	428.1296698	19.9232203	0.9452007	44.9858344	14.2431512
1.7	503.7225433	21.8803180	0.9554345	61.1322266	17.7064115

573

Table A-2. (continued).

1.8	591.0914411	24.0058663	0.9640697	83.6382690	22.1926036
1.9	691.7963504	26.3111595	0.9712834	115.1854404	28.0461186
2	807.5740552	28.8083224	0.9772499	159.6612298	35.7401829
2.1	940.3207648	31.5092070	0.9821356	222.7354561	45.9295272
2.2	1092.1843160	34.4280423	0.9860966	312.7281525	59.5263307
2.3	1265.4606640	37.5774632	0.9892759	441.9245139	77.8104404
2.4	1462.7123840	40.9719251	0.9918025	628.5796985	102.5904270
2.5	1686.7267580	44.6258561	0.9937903	899.9950266	136.4405722
2.6	1940.5603910	48.5544893	0.9953388	1297.2682930	183.0519606
2.7	2227.4527060	52.7713629	0.9965330	1882.6771080	247.7562743
2.8	2551.5018760	57.3035117	0.9974449	2751.2318050	338.3129547
2.9	2915.6280170	62.1448599	0.9981342	4048.8695150	466.1001324
3	3326.2586300	67.3579833	0.9986501	6001.3133890	647.9309715
3.1	3785.0643910	72.8997746	0.9990324	8960.1319140	908.8407114
3.2	4302.8699000	78.8986881	0.9993129	13476.8242000	1286.4018400
3.3	4882.2362520	85.3132784	0.9995166	20422.7093500	1837.4438430
3.4	5495.0703230	91.5986700	0.9996631	31184.3628800	2648.6027840
3.5	6204.2190070	98.7425902	0.9997674	47984.9205600	3853.0446670
3.6	6996.7995230	106.4149800	0.9998409	74414.7467800	5657.0633550
3.7	8004.7119860	116.4313525	0.9998922	116316.7708000	8382.9136890
3.8	9156.7097100	127.4818570	0.9999277	183270.0686000	12537.9964100
3.9	10527.2093800	140.3941568	0.9999519	291101.3508000	18927.9949800
4	12270.6497200	156.8730111	0.9999683	466161.5063000	28842.8426300

APPENDIX-B

Table B–1. Auxiliary estimation function λ(g, h) for singly censored samples from the Normal probability distribution.

h=0.1		h=0.2		h=0.3		h=0.4		h=0.5		h=0.6	
g	λ	g	λ	G	λ	g	λ	g	λ	G	λ
-5.499E-02	1.050E-01	-1.698E-01	2.089E-01	-2.820E-01	3.116E-01	-3.917E-01	4.132E-01	-4.989E-01	5.137E-01	-6.037E-01	6.131E-01
-5.476E-02	1.051E-01	-1.697E-01	2.090E-01	-2.819E-01	3.117E-01	-3.917E-01	4.133E-01	-4.989E-01	5.138E-01	-6.038E-01	6.132E-01
-5.454E-02	1.051E-01	-1.695E-01	2.090E-01	-2.818E-01	3.118E-01	-3.916E-01	4.134E-01	-4.989E-01	5.138E-01	-6.038E-01	6.132E-01
-5.431E-02	1.051E-01	-1.694E-01	2.090E-01	-2.818E-01	3.118E-01	-3.916E-01	4.134E-01	-4.989E-01	5.139E-01	-6.038E-01	6.133E-01
-5.408E-02	1.051E-01	-1.692E-01	2.091E-01	-2.817E-01	3.119E-01	-3.915E-01	4.135E-01	-4.989E-01	5.140E-01	-6.038E-01	6.133E-01
-5.384E-02	1.052E-01	-1.690E-01	2.091E-01	-2.816E-01	3.119E-01	-3.915E-01	4.135E-01	-4.989E-01	5.140E-01	-6.038E-01	6.134E-01
-5.361E-02	1.052E-01	-1.689E-01	2.092E-01	-2.815E-01	3.120E-01	-3.914E-01	4.136E-01	-4.989E-01	5.141E-01	-6.038E-01	6.134E-01
-5.337E-02	1.052E-01	-1.687E-01	2.092E-01	-2.814E-01	3.120E-01	-3.914E-01	4.137E-01	-4.989E-01	5.142E-01	-6.038E-01	6.135E-01
....	....	....	....	....	....	....	....	....	....	....	....
2.662E-02	1.124E-01	-1.149E-01	2.218E-01	-2.486E-01	3.283E-01	-3.749E-01	4.319E-01	-4.942E-01	5.328E-01	-6.070E-01	6.310E-01
2.769E-02	1.125E-01	-1.142E-01	2.220E-01	-2.482E-01	3.285E-01	-3.747E-01	4.321E-01	-4.941E-01	5.330E-01	-6.070E-01	6.312E-01
2.876E-02	1.126E-01	-1.135E-01	2.221E-01	-2.477E-01	3.286E-01	-3.744E-01	4.323E-01	-4.941E-01	5.332E-01	-6.071E-01	6.314E-01
2.985E-02	1.127E-01	-1.128E-01	2.223E-01	-2.473E-01	3.288E-01	-3.742E-01	4.325E-01	-4.940E-01	5.334E-01	-6.071E-01	6.317E-01
3.095E-02	1.128E-01	-1.120E-01	2.224E-01	-2.468E-01	3.290E-01	-3.740E-01	4.327E-01	-4.939E-01	5.336E-01	-6.071E-01	6.319E-01
3.207E-02	1.129E-01	-1.113E-01	2.226E-01	-2.464E-01	3.292E-01	-3.737E-01	4.330E-01	-4.938E-01	5.339E-01	-6.071E-01	6.321E-01
3.320E-02	1.130E-01	-1.105E-01	2.227E 01	-2.459E-01	3.294E-01	-3.735E-01	4.332E-01	-4.938E-01	5.341E-01	-6.072E-01	6.323E-01
....	....	....	....	....	....	....	....	....	....	....	....
3.027E-01	-4.605E-03	3.120E-01	-1.042E-02	3.241E-01	-1.800E-02	3.406E-01	-2.828E-02	3.644E-01	-4.303E-02	4.016E-01	-6.596E-02
2.993E-01	-4.492E-03	3.083E-01	-1.016E-02	3.201E-01	-1.755E-02	3.361E-01	-2.757E-02	3.591E-01	-4.194E-02	3.952E-01	-6.425E-02
2.960E-01	-4.382E-03	3.047E-01	-9.914E-03	3.161E-01	-1.712E-02	3.317E-01	-2.688E-02	3.540E-01	-4.087E-02	3.890E-01	-6.259E-02
2.927E-01	-4.275E-03	3.012E-01	-9.670E-03	3.123E-01	-1.669E-02	3.274E-01	-2.621E-02	3.490E-01	-3.983E-02	3.829E-01	-6.096E-02
2.895E-01	-4.170E-03	2.977E-01	-9.431E-03	3.085E-01	-1.628E-02	3.231E-01	-2.555E-02	3.441E-01	-3.882E-02	3.769E-01	-5.938E-02
2.863E-01	-4.067E-03	2.943E-01	-9.197E-03	3.048E-01	-1.587E-02	3.190E-01	-2.491E-02	3.394E-01	-3.783E-02	3.711E-01	-5.784E-02
2.832E-01	-3.966E-03	2.910E-01	-8.969E-03	3.011E-01	-1.547E-02	3.149E-01	-2.428E-02	3.347E-01	-3.687E-02	3.654E-01	-5.634E-02
2.801E-01	-3.868E-03	2.877E-01	-8.746E-03	2.975E-01	-1.509E-02	3.109E-01	-2.367E-02	3.301E-01	-3.592E-02	3.599E-01	-5.487E-02
2.771E-01	-3.772E-03	2.844E-01	-8.528E-03	2.940E-01	-1.471E-02	3.070E-01	-2.307E-02	3.256E-01	-3.501E-02	3.545E-01	-5.344E-02
2.741E-01	-3.678E-03	2.813E-01	-8.315E-03	2.906E-01	-1.434E-02	3.032E-01	-2.248E-02	3.212E-01	-3.411E-02	3.492E-01	-5.205E-02
....	....	....	....	....	....	....	....	....	....	....	....
9.962E-02	-9.202E-05	9.976E-02	-2.071E-04	9.994E-02	-3.550E-04	1.002E-01	-5.524E-04	1.005E-01	-8.288E-04	1.010E-01	-1.244E-03
9.899E-02	-8.886E-05	9.913E-02	-2.000E-04	9.930E-02	-3.428E-04	9.953E-02	-5.334E-04	9.985E-02	-8.003E-04	1.003E-01	-1.201E-03
7.074E-02	-1.004E-05	7.076E-02	-2.258E-05	7.078E-02	-3.871E-05	7.080E-02	-6.022E-05	7.084E-02	-9.034E-05	7.089E-02	-1.355E-04
7.037E-02	-9.640E-06	7.038E-02	-2.169E-05	7.040E-02	-3.718E-05	7.042E-02	-5.784E-05	7.046E-02	-8.677E-05	7.051E-02	-1.302E-04
7.000E-02	-9.258E-06	7.001E-02	-2.083E-05	7.003E-02	-3.571E-05	7.005E-02	-5.555E-05	7.008E-02	-8.333E-05	7.013E-02	-1.250E-04
6.963E-02	-8.891E-06	6.964E-02	-2.000E-05	6.966E-02	-3.429E-05	6.968E-02	-5.335E-05	6.971E-02	-8.002E-05	6.975E-02	-1.200E-04
6.926E-02	-8.537E-06	6.927E-02	-1.921E-05	6.929E-02	-3.293E-05	6.931E-02	-5.123E-05	6.934E-02	-7.684E-05	6.938E-02	-1.153E-04
6.890E-02	-0.197E-06	6.891E-02	1.844E 06	6.802E 02	3.162E 06	6.804E 02	-4.918E-06	6.897E-02	-7.378E-05	6.901E-02	-1.107E-04
6.854E-02	-7.869E-06	6.855E-02	-1.771E-05	6.856E-02	-3.035E-05	6.858E-02	-4.722E-05	6.861E-02	-7.083E-05	6.865E-02	-1.062E-04
6.818E-02	-7.554E-06	6.819E-02	-1.700E-05	6.820E-02	-2.914E-05	6.822E-02	-4.533E-05	6.825E-02	-6.799E-05	6.829E-02	-1.020E-04
6.783E-02	-7.251E-06	6.784E-02	-1.631E-05	6.785E-02	-2.797E-05	6.787E-02	-4.351E-05	6.789E-02	-6.526E-05	6.793E-02	-9.790E-05
6.747E-02	-6.959E-06	6.748E-02	-1.566E-05	6.750E-02	-2.684E-05	6.751E-02	-4.176E-05	6.754E-02	-6.264E-05	6.757E-02	-9.396E-05
6.712E-02	-6.679E-06	6.713E-02	-1.503E-05	6.715E-02	-2.576E-05	6.716E-02	-4.007E-05	6.718E-02	-6.011E-05	6.722E-02	-9.017E-05
6.678E-02	-6.409E-06	6.679E-02	-1.442E-05	6.680E-02	-2.472E-05	6.681E-02	-3.846E-05	6.683E-02	-5.768E-05	6.687E-02	-8.653E-05
6.643E-02	-6.149E-06	6.644E-02	-1.384E-05	6.645E-02	-2.372E-05	6.647E-02	-3.690E-05	6.649E-02	-5.535E-05	6.652E-02	-8.302E-05
6.609E-02	-5.900E-06	6.610E-02	-1.327E-05	6.611E-02	-2.276E-05	6.612E-02	-3.540E-05	6.614E-02	-5.310E-05	6.617E-02	-7.966E-05
6.575E-02	-5.660E-06	6.576E-02	-1.274E-05	6.577E-02	-2.183E-05	6.578E-02	-3.396E-05	6.580E-02	-5.094E-05	6.583E-02	-7.642E-05
6.542E-02	-5.429E-06	6.542E-02	-1.222E-05	6.543E-02	-2.094E-05	6.545E-02	-3.258E-05	6.547E-02	-4.887E-05	6.549E-02	-7.330E-05
6.508E-02	-5.208E-06	6.509E-02	-1.172E-05	6.510E-02	-2.009E-05	6.511E-02	-3.125E-05	6.513E-02	-4.687E-05	6.516E-02	-7.031E-05
6.475E-02	-4.994E-06	6.476E-02	-1.124E-05	6.477E-02	-1.926E-05	6.478E-02	-2.997E-05	6.480E-02	-4.495E-05	6.482E-02	-6.743E-05
6.442E-02	-4.789E-06	6.443E-02	-1.078E-05	6.444E-02	-1.847E-05	6.445E-02	-2.874E-05	6.447E-02	-4.311E-05	6.449E-02	-6.466E-05
6.410E-02	-4.593E-06	6.410E-02	-1.033E-05	6.411E-02	-1.771E-05	6.412E-02	-2.756E-05	6.414E-02	-4.133E-05	6.416E-02	-6.200E-05
6.377E-02	-4.403E-06	6.378E-02	-9.907E-06	6.379E-02	-1.698E-05	6.380E-02	-2.642E-05	6.381E-02	-3.963E-05	6.384E-02	-5.945E-05
6.345E-02	-4.221E-06	6.346E-02	-9.498E-06	6.347E-02	-1.628E-05	6.348E-02	-2.533E-05	6.349E-02	-3.799E-05	6.351E-02	-5.699E-05
6.313E-02	-4.047E-06	6.314E-02	-9.105E-06	6.315E-02	-1.561E-05	6.316E-02	-2.428E-05	6.317E-02	-3.642E-05	6.319E-02	-5.463E-05
6.282E-02	-3.879E-06	6.282E-02	-8.727E-06	6.283E-02	-1.496E-05	6.284E-02	-2.327E-05	6.285E-02	-3.491E-05	6.287E-02	-5.237E-05
6.250E-02	-3.718E-06	6.251E-02	-8.365E-06	6.252E-02	-1.434E-05	6.253E-02	-2.231E-05	6.254E-02	-3.346E-05	6.256E-02	-5.019E-05

Maple Help File: Dirac and Heaviside - the Delta and Step Function

Calling Sequence: Dirac(t), Dirac(n, t), and Heaviside(t)

Parameters: t - algebraic expression and n – nonnegative integer

Description:

- The Dirac and Heaviside functions are distributions. The Dirac(t) delta function is defined as zero everywhere except at t = 0 where it has a singularity. It also has the property that

$$\int_{-\infty}^{\infty} \mathrm{Dirac}(t)dt = 1$$

- Derivatives of the Dirac function are denoted by the two-argument Dirac function. The first argument denotes the order of the derivative. For example, diff(Dirac(t), t$n) will be automatically simplified to Dirac(n, t) for any integer n.
- The Heaviside(t) function is defined as zero for t < 0, 1 for t > 0 and is not defined at 0. It is related to the Dirac function by

$$\frac{\partial}{\partial t} Heaviside(t) = Dirac(t)$$

- These functions are typically used in the context of integral transforms such as laplace(), mellin() or fourier() or the solving of differential equations. They are also used to represent piecewise continuous functions with conversion routines available.

➢ plot(Dirac(t),t = -infinity..infinity);

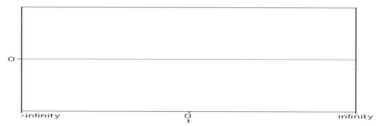

➢ plot(Heaviside(t),t = -infinity..infinity);

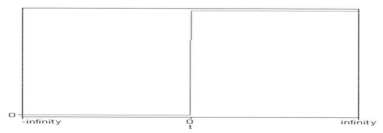

See Also: <u>inttrans[laplace]</u>, <u>inttrans[mellin]</u>, <u>inttrans[fourier]</u>, <u>convert[piecewise].</u>

Surface Water Hydrology, Singh, Al-Rashed & Sherif (eds)
© 2002 Swets & Zeitlinger, Lisse, ISBN 90 5809 363 8

Some recommendations for hydrological use of the two-parameter Weibull model

Fahim Ashkar
Department of Mathematics and Statistics, Université de Moncton
Moncton, N.B., E1A 3E9, Canada
e-mail: ashkarf@umoncton.ca

Smail Mahdi
Department of Computer Science, Mathematics & Physics
University of the West Indies, Cave Hill Campus, Barbados
e-mail: smahdi@uwichill.edu.bb

Nassir El-Jabi
School of Engineering, Université de Moncton
Moncton, N.B., E1A 3E9, Canada
e-mail: eljabin@umoncton.ca

Abstract

The *Weibull (WEI)* distribution is suitable for studying a wide variety of hydrological phenomena. We discuss some properties of this model that make it appropriate for hydrological applications. We then present some estimation techniques for this distribution, based on generalized probability weighted moments *(GPWM)*, generalized moments *(GM)*, and maximum likelihood *(ML)*. Some comparisons are given of the different estimators. Although the *GPWM* method may in some situations lead to a slight gain in quantile estimation accuracy, the *ML* method is the one to be generally recommended. Constructing confidence intervals (CI's) for quantiles of a hydrological variable is a means of assessing the precision of quantile estimates. Inaccurate methods for calculating such CI's have been used, although more accurate techniques may be available. We review and recommend specific methods for calculating such CI's for the two-parameter *WEI* model.

Introduction

In hydrological activities, such as the design of a flood-control structure, the estimation of flood damage, or the modeling of low stream-flow, decisions are often made on the basis of a small sample of historical observations. A statistical distribution is chosen to fit the sample values, and then used through extrapolation to estimate events corresponding to small probabilities of exceedance (extreme quantiles). The *Weibull (WEI)* distribution is a model that has been found suitable for studying hydrological phenomena such as low stream-flow. In fact it is a model adequate for representing both high stream-flow (e.g., annual flow maxima) as well as low stream-flow, such as Annual flow minima. It is also a model applicable in both the *'Peaks-Over-Threshold' (POT)* as well as the *'Deficit-Below-Threshold' (DBT) methods* for modeling hydrological extremes. These threshold-based methods are based on truncation of an observed hydrological time series.

We shall first discuss some of the properties of the two-parameter *WEI* model that make it appropriate for a wide variety of hydrological applications. We shall then discuss some estimation techniques for this distribution, based on the methods of generalized probability weighted moments *(GPWM),* generalized moments *(GM),* and maximum likelihood *(ML).*

Constructing confidence intervals (CI's) for a quantile, x_p of a hydrological variable X, is a usual means of assessing the precision of estimates of x_p. Recall that the quantile x_p is defined as the value of X exceeded with probability equal to $1 - p$, or corresponding to a return period $T = 1/(1- p)$. Inaccurate methods for calculating such CI's have often been used in engineering practice, although more accurate small-sample techniques may be available. We shall review and recommend for hydrological applications specific small-sample methods for use when fitting a two-parameter *WEI* model.

Weibull Model for Hydrological Applications

The probability density function (pdf) for the two-parameter *WEI* distribution is given by

$$f(x) = ase^{-(ax)^s}(ax)^{s-1} \qquad 0 \le x < \infty; \qquad s > 0, \alpha > 0 \qquad (1)$$

where α is the scale parameter and s the shape parameter. It is noted that if X follows the *WEI(s,α)* distribution, then $Y = log\,X$ follows an Extreme Value type I *(EVI)* distribution for the minima; see, e.g., Mahdi and Ashkar (2001). Figure 1 depicts some shapes of the *WEI* pdf.

Based on density function (1), the two-parameter *WEI(s,α)* cumulative distribution function (cdf) is given by

$$F(x) = 1 - e^{-(\alpha x)^s} \qquad (2)$$

The quantile X_T is the x-value with return period T, or which is exceeded with probability $1 - p$. It is obtained by solving the equation

$$T = \frac{1}{1-p} = \frac{1}{1 - F(x_T)} \qquad (3)$$

which, from Eq. (2), admits the solution $X_T = \alpha^{-1}(\ln(T))^{1/s}$.

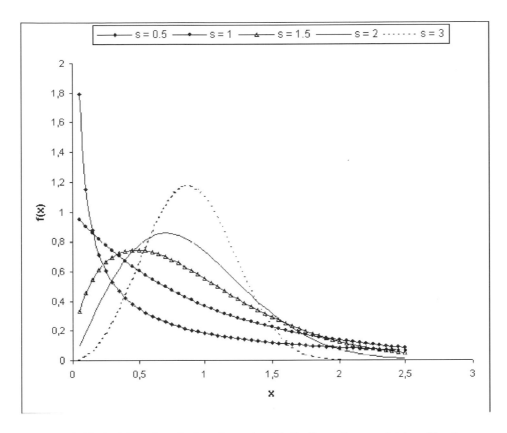

Figure 1. Probability density function of a Weibull random variable with shape
parameter s.

Hydrologists have used the *WEI* model in many applications. As a matter of fact,
the *WEI* distribution plays an important role not only in hydrology, but also in the
more general field of applied statistics. In lifetime testing, for instance, it is used to
model the lifetime of objects with linear failure rates. It also models the situation of
constant failure rates, since it reduces to the Exponential distribution for s = 1
(Fig. 1). In Hydrology, the 3-parameter *WEI* distribution was one of the main models used by Matalas (1963) and Loganathan and al. (1985), among others, to study
low stream-flow. In the *Peaks-Over-Threshold (POT)* method for modeling hydrological extremes (e.g., flood analysis; Fig. 2), the generalized Pareto, 2-parameter
WEI, and Exponential distributions, are widely used for fitting peaks over the
threshold, or 'exceedances' (Rasmussen et al., 1993; Ouarda and Ashkar, 1995).
These three distributions were applied by Ouarda and Ashkar (1994) to model
flood-exceedance magnitudes at a large number of stream-gauging stations in
Canada. Different distribution types were found to be adequate for different flood
sites within this Canadian flood database.

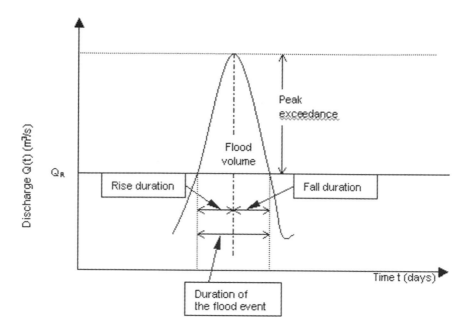

Figure 2. Peaks-Over-Threshold *(POT)*: Application to flood flow.

The *WEI* model is useful not only with the *POT* method (modeling high extremes) but with the *'Deficit-Below-Threshold' (DBT)* method as well (modeling low extremes). *POT* modeling gives rise to a series of *exceedances of the threshold*, whereas *DBT* modeling gives rise to *deficits below the threshold* (Fig. 3). The adequacy of the *WEI* distribution for such threshold-based applications is based on some desirable features of the *WEI* pdf. In fact, when the hydrologist is using *DBT* modeling for studying low stream-flow, for example, and is interested in estimating design events with high return period (i.e. very low flows, or very severe water deficits: Fig. 3), a low threshold for the variable X must be chosen. In this case a monotonic decrease of the density function of deficits is encountered. In Fig. 1 it is shown that the *WEI* distribution exhibits this feature of monotonic decrease in its pdf (i.e., a pdf with no mode) when its shape parameter is less than or equal to one. On the other hand, when interest is with design events of *small* return period, a less extreme threshold must be used, and water deficits below the threshold may in this case exhibit modality. Therefore, it may be necessary to consider modal distributions in this kind of situation. The *WEI* model is again found to be useful in this case, because when the shape parameter is greater than one (Fig. 1), the Weibull distribution is seen to exhibit a modal density.

Some Estimation Techniques for the Weibull Distribution

The maximum likelihood (*ML*) method is the most important distribution-fitting method used in statistics because it often leads to efficient parameter and quantile estimators with Gaussian asymptotic distribution. However, the *ML* method is often

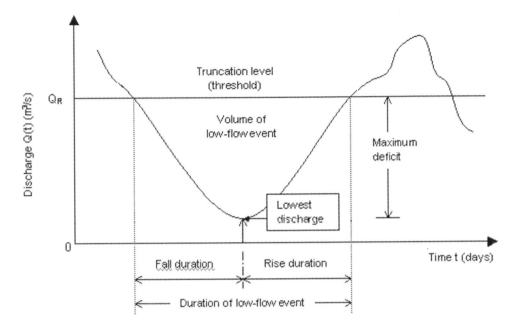

Figure 3. Deficit-Below-Threshold *(DBT)*: Application to low flow.

highly computational, and may not work well in small samples. On the other hand, the method of moments *(MM)* is mostly used because of its relative ease of application, and furthermore it can help to obtain starting values for numerical procedures involved in *ML* estimations. Therefore, other estimating methods have recently been developed as alternatives to these two classical estimation methods. The most serious competitor to the *ML* and *MM* methods, according to Davison and Smith (1990), is the one based on probability-weighted moments *(PWM's)*. This method is strongly advocated in Hosking (1986). Specifically, the *PWM* of order *(i, j, k)* for a random variable *X* with distribution function *F* is defined as

$$M_{i,j,k} = E\left(X^i\left[F(x)\right]^j\left[1 - F(x)\right]^k\right) \tag{4}$$

The quantities $M_{i,j,k}$ exist if and only if $E(x^i)$ exists since $|[F(x)]^j[1-F(x)]^k| \le 1$ for any positive integers j and k.

PWM's can therefore be expressed in the form:

$$M_{i,j,k} = \int_{-\infty}^{\infty} x^i \left(F(x)\right)^j \left(1-F(x)\right)^k dF(x) \tag{5}$$

or in the alternative form:

581

$$M_{i,\,j,k} = \int_0^1 \left(x(F)\right)^i F^j \left(1-F\right)^k dF \tag{6}$$

where $x(F)$ is solution to the equation $F(x) = F$.

The order $i = 1$, in Eqs. (4–6), is the most commonly used power for X. The choice of $i = 1$ has the double advantage of not over-weighting sample values unduly, and also leads to a class of linear L-moments (Hosking, 1986; 1990) with asymptotic normality. The two other orders, j and k, are often restricted to small positive integer values, but according to Rasmussen (2001) a lot may be gained by using real numbers for j and k, that are not necessarily small. The extended class of *PWM's* with real orders was referred to in Rasmussen (2001) as the class of Generalized probability weighted moments *(GPWM's)*. Mahdi and Ashkar (2001) recently studied *GPWM's* for the 2-parameter *WEI* model.

Prior to the Rasmussen's generalization of *PWM's* to *GPWM's*, Ashkar and Bobée (1987) had generalized the classical *MM* method into one that they called method of "generalized moments" *(GM)*. The *GM* method, like *GPWM*, estimates distribution parameters using moment orders that are not necessarily small positive integers. The *GM* method has been successfully applied for fitting the generalized Pareto distribution in Ashkar and Ouarda (1996).

Mahdi and Ashkar (2001) have derived point estimators for the parameters and quantiles of the 2-parameter *WEI* model using *GPWM*, *GM*, and *ML* estimation methods. They also gave analytical expressions for the asymptotic variances of the parameter and quantile estimators. In the following section we compare the accuracy of estimates of the *WEI* shape parameter and quantiles.

Comparing the Accuracy of Estimates of the *WEI* Shape Parameter and Quantiles

To check the performance of the three estimating methods *GPWM*, *GM* and *ML*, Mahdi and Ashkar (2001) performed a simulation study for several values of sample size n, shape parameter s for the *WEI* random variable, and return period T. *GPWM's* were considered with orders $j = k_1$ and $k = k_2$ taking all values between 0 and 4, with pace 0.1, for a total of 1600 (k_1, k_2) pairs. The best parameter estimates obtained with the *GPWM* method were not always reached at integer orders j and k, which agrees with the findings of Rasmussen (2001) for the Pareto distribution. In the *GM* method, in which the classical moments $E(X^{k_1})$ and $E(X^{k_2})$ are used in the estimation, the same 1600 order couples (k_1, k_2) were used as with the *GPWM* method.

Mahdi and Ashkar (2001) generated 1000 samples for each combination of n, s, and pair (k_1, k_2). In Tables 1–4, we present some brief comparisons between the three estimation methods based on the simulations done by Mahdi and Ashkar (2001). In these comparisons, the root mean square errors *(RMSE's)* of parameter/quantile estimates, which have been rounded to 2 decimal places, are used as performance index. The overall conclusions that can be drawn from Tables 1–4 are as follows:

- The *GM* method does not perform well in comparison to the *ML* and *GPWM* methods (Tables 1 and 2). Therefore, in Tables 3 and 4 no results for the *GM* method are presented.

- The *ML* estimator of s outperforms the other estimators (Tables 1 and 2). However, it is overall only slightly better than the *GPWM* estimator.

- For estimating quantiles, the *ML* method almost systematically gives better results than the *GPWM* method (Tables 3 and 4).

- Although the *GPWM* method in some situations leads to a slight gain in quantile estimation accuracy, the *ML* method is the one to be recommended because it generally performs either better or almost as good as the *GPWM* method.

Due to the advantages that have just been shown for the *ML* method, we restrict attention in the remainder of the paper to *ML* estimation.

Table 1. *RMSE* for estimates of s obtained with the *GPWM*, *GM* and *ML* methods in the cases of s = 0.5(0.5)3 and n = 15 (small sample size).

N = 15	GPWM(s)	GM(s)	ML(s)
s = 0.5	0.15	0.33	0.14
s = 1.0	0.31	0.34	0.27
s = 1.5	0.46	0.47	0.41
s = 2.0	0.61	0.59	0.55
s = 2.5	0.74	0.72	0.68
s = 3.0	0.86	0.86	0.82

Table 2. *RMSE* for estimates of s obtained with the *GPWM*, *GM* and *ML* methods in the cases s = 0.5(0.5)3, for n = 30, 60, 120 combined (relatively large sample size).

n = 30, 60,120	GPWM(s)	GM(s)	ML(s)
s = 0.5	0.07	0.19	0.06
s = 1.0	0.14	0.19	0.12
s = 1.5	0.21	0.24	0.17
s = 2.0	0.29	0.29	0.23
s = 2.5	0.37	0.36	0.29
s = 3.0	0.44	0.42	0.35

Table 3. *RMSE* of X_T for T = 10, 50, 100 and 200 obtained with the *GPWM* and *ML* methods in the case n = 10 (small sample size).

S	GPWM T = 10	GPWM T = 50	GPWM T = 100	GPWM T = 200	ML T = 10	ML T = 50	ML T = 100	ML T = 200
0.50	3.12	10.05	14.52	19.90	3.06	9.59	13.90	19.17
1.00	1.49	2.60	3.14	3.70	1.47	2.60	3.14	3.71
1.50	1.08	1.57	1.78	2.00	1.06	1.55	1.77	1.99
2.00	0.89	1.17	1.29	1.41	0.86	1.15	1.28	1.40
2.50	0.77	0.96	1.04	1.12	0.74	0.94	1.02	1.10
3.00	0.69	0.83	0.88	0.94	0.66	0.80	0.86	0.92

Table 4. *RMSE* of X_T for T=10, 50, 100 and 200 obtained with the *GPWM* and *ML* methods in the cases n = 30, 60 and 120 combined (relatively large sample size).

S	GPWM T = 10	GPWM T = 50	GPWM T = 100	GPWM T = 200	ML T = 10	ML T = 50	ML T = 100	ML T = 200
0.50	2.62	8.17	12.20	17.25	2.62	8.18	12.23	17.32
1.00	1.47	2.56	3.09	3.65	1.43	2.56	3.09	3.65
1.50	1.07	1.55	1.76	1.97	1.07	1.55	1.76	1.97
2.00	0.87	1.16	1.28	1.39	0.87	1.15	1.27	1.39
2.50	0.75	0.94	1.03	1.10	0.75	0.94	1.02	1.10
3.00	0.67	0.81	0.87	0.93	0.67	0.81	0.86	0.92

Some *ML* Results for the *WEI(s,α)* Distribution

The *ML* estimator of the parameter s of a *WEI*(s,α) distribution is obtained through a numerical solution of the following equation (Mahdi and Ashkar, 2001):

$$\hat{s} = [(\sum_{i=1}^{n} x_i^{\hat{s}} \ln(x_i))(\sum_{i=1}^{n} x_i^{\hat{s}})^{-1} - \frac{1}{n}\sum_{i=1}^{n}\ln(x_i)]^{-1} \tag{7}$$

The *ML* estimator of the parameter α is then given by

$$\hat{\alpha} = \left[\frac{n}{\sum_{i=1}^{n} x_i^{\hat{s}}}\right]^{1/\hat{s}}. \tag{8}$$

The covariance matrix Δ of estimators $\hat{\alpha}$ and $\hat{s}$ is obtained as the inverse of the Fisher-information matrix I, since $\sqrt{n}(\hat{\alpha} - \alpha, \hat{s} - s)$ converges in distribution as $n \rightarrow \infty$ to a Gaussian vector with mean 0 and covariance matrix I^{-1}. The terms of the matrix I (Mahdi and Ashkar, 2001) are given by

$$I_{11} = ns^2\alpha^{-2} \tag{9}$$

584

$$I_{21} = I_{12} = \frac{n}{\alpha} J \tag{10}$$

$$I_{22} = \frac{n}{s^2} [1 + J'] \tag{11}$$

where $J = 0.42784$ and $J' = 0.823683$.

The asymptotic variance of the quantile estimator $\hat{X}_T$ is evaluated using the variances and covariance of $\hat{\alpha}$ and $\hat{s}$ as follows:

$$Var(\hat{X}_T) = (\frac{\partial X_T}{\partial \alpha})^2 Var(\hat{\alpha}) + (\frac{\partial X_T}{\partial s})^2 Var(\hat{s})$$
$$+ 2(\frac{\partial X_T}{\partial \alpha}) \; (\frac{\partial X_T}{\partial s}) \; Cov(\hat{\alpha}, \hat{s}) \tag{12}$$

where X_T and the partial derivatives involved in this formula are as follows:

$$X_T = \alpha^{-1}(\ln(T))^{1/s} , \tag{13}$$

$$\frac{\partial X_T}{\partial \alpha} = -\alpha^{-2} \; (\ln(T))^{\frac{1}{s}} \tag{14}$$

and

$$\frac{\partial X_T}{\partial s} = -\alpha^{-1} s^{-2} \ln(\ln(I)) \; (\ln(I))^{\frac{1}{s}} . \tag{15}$$

To obtain the variance of $\hat{X}_I$ we only need to substitute the *ML* parameter estimates into Formula (12).

Small-Sample Confidence Intervals for *WEI* Quantiles Based on the *ML* Method

The accuracy with which a quantile x_p is estimated depends on the accuracy with which the parameters of $f(x)$ have been obtained. In hydrology, it is important not only to estimate x_p, but also to give a measure of the error of estimation by constructing CI's for x_p. This problem of constructing confidence intervals for the quantiles *(CIQ's)* has been given considerable attention in the statistical literature, mainly due to its interest in Reliability analysis. For certain types of distributions, small-sample procedures have been successfully developed for calculating *CIQ's*. One notes, however, that the majority of these procedures have been based on the *ML* method, and much research remains to be done for developing *CIQ's* for other estimation methods. It is to be noted that small-sample results have been relatively easier to obtain for distributions whose parameters are of the location-scale type, such as the Normal, Gumbel, and Exponential, or for distributions related to these, such as the Lognormal, Weibull and Pareto.

Several authors have studied the problem of constructing approximate small-sample *CIQ's* for the *WEI* and *EV1* distributions (which are related to each other by a logarithmic relationship, as noted earlier). These many studies include ones by Lawless (1975; 1978), Thoman and al. (1970), Mann and Fertig (1977), Bain and Engelhardt (1981), and others. However, the method proposed by Bain and Engelhardt (1981) is particularly simple, and accurate, for most hydrological applications. This method is applied as follows:

Assume that a hydrological sample of size n is given from a 2-parameter *WEI* population, *X*, and that we wish to obtain a 100(1−2α)% CI for the pth quantile x_p, of *X*. This *CIQ* is calculated by the following procedure:

- Let: $$\lambda_p = \ln(-\ln(1-p)) \tag{16}$$

$$a = 1 - (z_{1-a})^2/2(n-1) \tag{17}$$

$$b = l_p^2 - (z_{1-a})^2/n \tag{18}$$

where $z_{1-\alpha}$ is the (1−α)th quantile of the standard normal distribution;

- Calculate: $$d = [n/(n-1)]^{1/2} [\lambda_p + ((\lambda_p)^2 - a.b)^{1/2}]/a \tag{19}$$

$$c = [n/(n-1)]^{1/2} [-\lambda_p + ((\lambda_p)^2 - a.b)^{1/2}]/a \tag{20}$$

$$p_1 = 1 - \exp(-\exp(-c)), \tag{21}$$

$$p_2 = 1 - \exp(-\exp(d)). \tag{22}$$

Then [x_{p1}, x_{p2}] will be the desired 100(1 - 2α)% CI for x_p, where x_{p1} and x_{p2} are the distribution's p_1th and p_2th quantile estimates, obtained by maximum likelihood, from the sample of size *n*. A numerical application of this method is now given.

Numerical Application

The knowledge of various characteristics of low-flow events (duration, volume, etc.) in stream channels is essential for water management and the design of hydraulics structures. In this numerical example we consider the volume of low-flow events for hydrometric Station BP001 on the Little Southwest Miramichi River at Lyttleton in New Brunswick (Canada). The drainage area for this station is 1340 km². The series from which low-flow volumes were taken is composed of 45 years (1952–1996) of daily river-flows. We were interested in low-flow discharges below a threshold,

Q_R (Fig. 3). A truncation level of Q_R = 4.75 m^3/s was chosen, on the flow hydrograph, and volumes of low-flow events below this level were calculated in m^3/s x days. During the period of record, there were n = 53 low-flow events. The chosen Q_R value represented approximately the 2nd percentile in the daily-flow series classified in ascending order.

Among the recommended models to represent the various characteristics of low-flow events is the *WEI* model. In fact, the *WEI* model was found to be flexible enough to provide a good fit to the Little Sowthwest Miramichi low-flow volume data. Some statistics relevant to this data set are: *sample size* = 53; *mean* = 5.80 m^3/s x days; *median* = 0.92 m^3/s x days; *standard deviation* = 11.12 m^3/s x days.

The *ML* method was used to fit the *WEI* distributions to the data set. The numerical solution of Eq. (7) was done using the widely available computer software *Excel*. The *ML* parameter estimates for α and s are $\hat{\alpha}$ = 0.343 and $\hat{s}$ = 0.525.

The steps involved in calculating small-sample *CIQ*'s based on the Little Southwest Miramichi low-flow volume data, by Eqs. (16–22), will now be given.

Suppose that we are interested in calculating a 95% CI (i.e., 2α = 0.05; α = 0.025) for the 100-year low-flow volume x_p (T = 100; p = 0.99). The calculation is done as follows:

- Calculate $\lambda_p = \ln(-\ln(1-p))$. Here we have p = 0.99; this gives λ_p = 1.527.

- Calculate $a = 1 - (z_{1-\alpha})^2 / 2(n-1)$, Here α – 0.025, $z_{1-\alpha}$ = 1.96, n = 53; this gives:
  $a = 1 - (1.96)^2/2(52) = 0.9631$

- Calculate $b = (\lambda_p)^2 - (z_{1-\alpha})^2/n$
  $= (1.527)^2 - (1.96)^2/53 = 2.260$

- Calculate $d = [n/(n-1)]^{1/2} [\lambda_p + ((\lambda_p)^2 - a.b)^{1/2}]/a$
  $= [53/(53-1)]^{1/2} [1.527 + ((1.527)^2 - 0.9631*2.260)^{1/2}]/0.9631$
  $= 2.015$

- Calculate $c = [n/(n-1)]^{1/2} [-\lambda_p + ((\lambda_p)^2 - a.b)^{1/2}]/a$
  $= [53/(53-1)]^{1/2} [-1.527 + ((1.527)^2 - 0.9631*2.260)^{1/2}]/0.9631$
  $= -1.187$

- Now applying formulas (21–22) yields p_1 = 0.9623 and p_2 = 0.9994.

- From Eq. (2), we find $x(F) = (-\ln(1-F))^{1/s}/\alpha$. Substituting F = p_1 = 0.9623, and F = p_2 = 0.9994 into this equation, and replacing α and s by their *ML* estimates $\hat{\alpha}$ = 0.343 and $\hat{s}$ = 0.525, gives the required 95% CI for the 100-year low-flow volume.

The calculated 95% CI for the 100-year low-flow volume turns out to be [27.92; 135.04] m^3/sec x days, which is shown in Fig. 4. This figure also shows CI's for other events with return periods of up to 1000 years (T = 1000, log(T) = 3).
On the other hand, the *asymptotic* CI is calculated as follows:

- From Eqs. (9–11) we obtain the matrix

$$I = \begin{bmatrix} I_{11} & I_{12} \\ I_{21} & I_{22} \end{bmatrix} = \begin{bmatrix} 124.339 & 66.114 \\ 66.114 & 350.235 \end{bmatrix}.$$

- It's inverse is given by

$$I^{-1} = \begin{bmatrix} 0.0089 & -0.0017 \\ -0.0017 & 0.0032 \end{bmatrix}.$$

- Therefore Var($\hat{\alpha}$) = 0.0089, Var($\hat{s}$) = 0.0032, and Cov($\hat{\alpha}$, $\hat{s}$) = –0.0017.
- From Eq. (12), and with the help of Eqs. (13–15), we find $Var(\hat{X}_T) = 648.32$.

- The asymptotic CI is therefore given by $\hat{X}_T = 1.96 \pm \sqrt{Var(\hat{X}_T)}$, which yields [3.46; 103.27] m^3/sec x days.

It is seen that the asymptotic *CIQ* differs significantly from the one based on small-sample theory. It is recommended that the later one be used in hydrological applications. Figure 4 depicts asymptotic *CIQ's* for various return periods.

Summary

The *Weibull (WEI)* distribution has been presented as a candidate for representing a variety of hydrological phenomena including both high and low stream-flows. This model should be particularly useful in arid regions for modeling extreme hydrological observations that are mathematically expressible as deficits *below a threshold'* (DBT's). The maximum likelihood *(ML)* method is the one to be recommended for fitting this distribution to hydrological data, because:

- It is efficient relative to the other popular estimation methods discussed;

- Small-sample procedures are available for calculating confidence intervals for the quantiles (*CIQ's*) of this distribution, when the ML method is used.

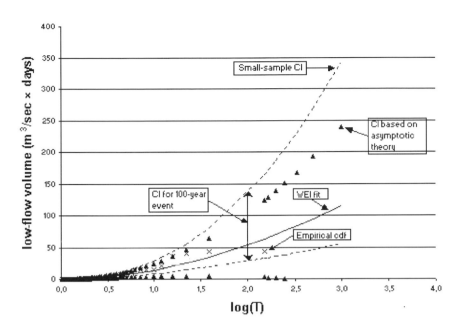

Figure 4. Observed and fitted *WEI* cdf, with 95% CI's for quantiles.

Acknowledgments

The financial support of the Natural Sciences and Engineering Research Council of Canada and of UWI is gratefully acknowledged. The authors wish also to thank Miss Lampouguin Bayentin for her assistance in preparing this manuscript.

References

Ashkar, F. and Bobée, B. (1987). "The generalized method of moments as applied to problems of flood frequency analysis: Some practical results for the log-Pearson type 3 distribution". *J. Hydrol., 90, 199-217.*

Ashkar, F. and Ouarda, T.B.M.J. (1996). "On some methods of fitting the generalized Pareto distribution". *J. Hydrol., 177, 117-141.*

Bain, L.J. and Engelhardt, M. (1981). "Simple approximate distributional results for confidence and tolerance limits for the Weibull distribution based on maximum likelihood estimators". *Technometrics, 23(1), 15-20.*

Davison, A.C. and Smith, R.L. (1990). "Models for exceedances over high threshold". *J. R. Stat. Soc. B, 52 (3), 393-442.*

Hosking, J.R.M., Wallis, J.R. and Wood, E.F. (1985). "Estimation of the generalized extreme-value distribution by the method of probability weighted moments". *Technometrics, 27, 251-261.*

Hosking, J.R.M. (1986). "The theory of probability weighted moments". *Research Report RC12210*, IBM Thomas J. Watson Research Center, New York.

Hosking, J.R.M. (1990). "L-Moments: analysis and estimation of distributions using linear combinations of order statistics". *J. R. Stat. Soc. B, 52(1), 105-124.*

Lawless, J.F. (1975). "Construction of tolerance bounds for the extreme value and Weibull distributions". *Technometrics, 17, 225-261.*

Lawless, J.F. (1978). "Confidence interval estimation for the Weibull and extreme value distributions". *Technometrics, 20, 225-261.*

Loganathan, G.V., Kuo, C.Y., and McCormick, T.C. (1985). "Frequency analysis of low-flows". *Nordic Hydrology, Vol. 16, 105-128.*

Mahdi, S. and Ashkar, F. (2001). "Exploring generalized probability weighted moments, generalized moments and maximum likelihood estimating methods in two-parameter Weibull model". Submitted for publication in *Water Resources Research.*

Mann, N.R. and Fertig, K.W. (1977). "Efficient unbiased quantile estimators for moderate-size complete samples from extreme value and Weibull distributions: confidence bounds and tolerance and prediction intervals". *Techonmetrics, 19, 87-93.*

Matalas, N.C. (1963). "Probability distributions of low flows". *U.S. Geological Survey. Prof. Rep. 437-A, 27 p.*

Ouarda, T.B.M.J. and Ashkar, F. (1994). "Regional multiple regression flood frequency estimation by the peaks-over-threshold method". *Internal Report*, Department of Mathematics and Statistics, Université de Moncton, Moncton, N.B., Canada.

Ouarda, T.B.M.J. and Ashkar, F. (1995). "The peaks-over-threshold method for regional flood frequency estimation". *48th Canadian Water Resources Association Annual Conference*, Fredericton, New Brunswick, June 20-23. Proceedings, 641–659.

Rasmussen, P., Ashkar, F., Bobée, B. and Rosbjerg, D. (1993). "The POT method for flood estimation: a review". *International Conference on Stochastic and Statistical Methods in Hydrology and Environmental Engineering*, Waterloo, Ont., Canada, June 21-23.

Rasmussen P. (2001). "Generalized probability weighted moments: Application to the generalized Pareto distribution". *Water Resour. Res.*, 37(6), 1745-1751.

Thoman, D.R., Antle C.E. and Bain, L.J. (1970). "Reliability and tolerance limits in the Weibull distribution". *Technometrics, 12, 363-371.*

Surface Water Hydrology, Singh, Al-Rashed & Sherif (eds)
© 2002 Swets & Zeitlinger, Lisse, ISBN 90 5809 363 8

The bigeneral extreme value distribution applied to drought frequency analysis

Jose A. Raynal-Villasenor
Department of Civil Engineering
Universidad de las Americas-Puebla
72820 Cholula Puebla, Mexico
e-mail: jraynal@mail.udlap.mx

Antonio Acosta
Water Management under Direction
National Water Commission
Insurgentes Centro # 30-32
06600 Mexico, D.F., Mexico
e-mail: gasir1@sgt.cna.gob.mx

Abstract

Drought frequency analysis has been carried out by using wide-known univariate probability distributions for the minima. Multivariate extreme value models for flood and drought frequency analyses have been explored towards their application in hydrological problems. The selected method of estimation of parameters is the method of maximum likelihood. From the study reported here, it has been shown the suitability of using multivariate modeling of extremes when it is needed to improve the estimation of parameters and confidence limits of design values of a gauging station with short record when gauging stations with longer records are available in its neighborhood. In the paper an example of application is given to show the suitability of the model in engineering practice.

Introduction

During the last sixty years the use of the extreme value distributions types I (Gumbel), II (Frechet), and III (Weibull) in the field of frequency analyses for the maxima and minima has grown steadily. Such distributions, as three particular solutions to the Stability Postulate which any extreme must satisfy, they have been integrated in one form after Jenkinson (1955) found the general solution to that postulate giving birth to the so-called general extreme value (GEV) distribution.

The GEV distribution has been used widely in flood frequency analysis and in other fields of the geophysical sciences, but such a distribution is applied to drought frequency analysis very rarely.

Recently, the multivariate models for extreme value distributions have been started to be explored for application in flood and drought frequency analyses (Raynal-Villasenor, 1985 and 1986; Raynal-Villasenor and Salas, 1987; Escalante-

Sandoval, 1991, and Acosta, 1993). The results obtained so far shown the suitability of the use of these models for flood and drought frequency analyses. They have been explored up to the trivariate level for the case of the maxima and up to the bivariate level for the case of the minima.

The General Extreme Value Distribution for the Minima

The probability distribution function of the GEV distribution for the maxima is, NERC (1975):

$$F(x)_{max} = \left\{ -\left[1 - \left(\frac{x - x_0}{\alpha} \right) \beta \right]^{1/\beta} \right\} \tag{1}$$

where α, β, and x_0 are the scale, shape and location parameters.

The probability distribution function of the GEV distribution for the minima is obtained by using the symmetry principle, Gumbel (1958):

$$F(x)_{min} = 1 - F(-x)_{max} \tag{2}$$

and:

$$F(x)_{min} = 1 - exp\left\{ -\left[1 - \beta(w - x)/a \right]^{1/\beta} \right\} \tag{3}$$

where α, β, and ω are the scale, shape and location parameters.

Using the following conventions:

$$\Phi(x) = F(x)_{min} = Pr(X \le x) \tag{4}$$

$$\Pi(x) = 1 - \Phi(x) = Pr(X > x) \tag{5}$$

the corresponding exceedance probability, $Pr(X > x)$, is:

$$\Pi(x) = exp\left\{ -\left[1 - \beta(\omega - x)/\alpha \right]^{1/\beta} \right\} \tag{6}$$

and its probability density function (pdf) is given by:

$$\pi(x) = \frac{1}{\alpha} exp\left\{ -\left[1 - \beta(\omega - x)/\alpha \right]^{1/\beta} \right\} \left[1 - \beta(\omega - x)/\alpha \right]^{1/\beta - 1} \tag{7}$$

Bivariate Extreme Value Distribution for the Minima

The general form of the Logistic model for bivariate extreme value distributions for the minima is, Gumbel (1962):

$$\Pi(x,y,m) = exp\left\{-\left[\left(-Ln\Pi(x)\right)^m + \left(-Ln\Pi(y)\right)^m\right]^{\frac{1}{m}}\right\} \qquad (8)$$

where $\Pi(x,y,m)$ is the bivariate extreme value distribution, $\Pi(.)$ is the univariate extreme value distribution of $(.)$, m is the association parameter for the Logistic model, and Ln is the natural logarithm.

The particular form of Equation (8), when both marginals are GEV distributions for the minima is, Raynal (1986):

$$\Pi(x,y,\theta) = exp\left\{-\left[\left(1-\left(\frac{\omega_1-x}{\alpha_1}\right)\beta_1\right)^{m7b1} + \left(1-\left(\frac{\omega_2-x}{\alpha_2}\right)\beta_2\right)^{\frac{m}{\beta_2}}\right]\right\} \qquad (9)$$

where ω, α and β are the location, scale and shape parameters of the marginal GEV distributions for the minima.

The corresponding probability density function is, Raynal (1986):

$$\pi(x,y,m) = \frac{1}{\alpha_1\alpha_2}\left(1-\left(\frac{\omega_1-x}{\alpha_1}\right)\beta_1\right)^{m7\beta_1-1}\left(1-\left(\frac{\omega_2-x}{\alpha_2}\right)\beta_2\right)^{m7\beta_2-1}$$

$$exp\left\{-\left[\left(1-\left(\frac{\omega_1-x}{\alpha_1}\right)\beta_1\right)^{m7\beta_1} + \left(1-\left(\frac{\omega_2-x}{\alpha_2}\right)\beta_2\right)^{m7\beta_2}\right]^{1/m}\right\}$$

$$+\left\{\left[\left(1-\left(\frac{\omega_1-x}{\alpha_1}\right)\beta_1\right)^{m7\beta_1} + \left(1-\left(\frac{\omega_2-x}{\alpha_2}\right)\beta_2\right)^{m7\beta_2}\right]^{1/m} + m-1\right\} \qquad (10)$$

Bivariate Maximum Likelihood Estimation of Parameters

Using the following generalized log-likelihood function for bivariate distributions, when the sample sizes of the marginals are not equal, Raynal-Villasenor (1985), the corresponding Log-likelihood function for the bivariate general extreme value (BGEV) distribution function based in Equation (11) is, Raynal (1986):

$$L(x,y,\theta) = \left[\prod_{i=1}^{N_1}\pi(r_1,\theta_1)\right]^{l_1}\left[\prod_{i=1}^{N_2}\pi(r_2,\theta_2)\right]^{l_2}\left[\prod_{i=1}^{N_3}\pi(r_3,\theta_3)\right]^{l_3} \qquad (11)$$

where I_1, I_2, I_3 are indicator numbers with value equal to one only if $N_i > 1$ and zero otherwise.

Computational Procedure for Maximum Likelihood Estimation of the Parameters

Given the complexity of the mathematical expressions in Equation (11), and their partial derivatives with respect to the parameters, the constrained Rosenbrock method, Kuester and Mize (1973), was applied to obtain the maximum likelihood estimators for the parameters, by the direct maximization of Equation (11).

The required initial values of the parameters to start the optimization of Equation (11) were provided by the univariate maximium likelihood estimators of the parameters for the case of the location, scale and shape parameters. The initial value of the association parameter was set equal to 2, following the procedure developed by Raynal-Villasenor (1985) for the case of the maxima.

Reliability of Estimated Parameters

The indicator selected to detect the reliability of estimated parameters when using the bivariate distribution as compared with to the univariate counterpart was the asymptotic relative information ratio. Table 1 shows a sample of relative information ratios obtained by using the following set of parameters:

$$\omega_1 = 10; \ \alpha_1 = 4; \ \beta_1 = 0.15; \ \omega_2 = 2; \ \beta_2 = 0.10; \ m = 2$$

Table 1. Asymptotic relative information ratios of the parameters of the bivariate extreme value distribution for the minima.

Parameter		$N_1 + N_2$		
	25	50	75	100
ω_1	1.05	1.32	1.45	1.53
α_1	0.97	1.10	1.16	1.19
β_1	1.14	1.24	1.19	1.31
ω_2	1.05	1.02	1.01	1.01
α_2	0.97	0.96	0.96	0.95
β_2	1.16	1.07	1.04	1.02

Case Study

To apply the proposed methodology, five gauging stations were selected within the Fuerte River basin in the state of Sinaloa in Northern Mexico. Tables 2–4 shows the results of the application of the BEV distribution for the minima to the data recorded in such gauging stations.

Table 2. Pairs, sample product-moment correlation coefficient and association parameter for gauging stations in case study.

Station	Univariate		Values		Bivariate		Values	
	ω	α	β	EE	ω	α	β	EE
Huites	4.10	2.23	1.05	1.54	4.10	2.23	1.05	1.54
Choix	0.07	0.11	1.48	0.15	0.08	0.13	1.60	0.14
Palo Dulce	0.73	0.48	0.71	0.30	0.78	0.52	0.70	0.32
Chinipas	0.59	0.33	0.73	0.21	0.60	0.38	0.80	0.17
Urique	2.66	2.50	1.45	2.02	2.78	2.70	1.45	2.10

Table 3. Univariate maximum likelihood estimates of the parameters of the GEV distributions defined by the data of the gauging stations of case study.

Station	ω	α	β
Huites	4.10	2.23	1.05
Choix	0.07	0.11	1.48
Palo Dulce	0.73	0.48	0.71
Chinipas	0.59	0.33	0.73
Urique	2.66	2.50	1.45

Table 4. Bivariate maximum likelihood estimates of the parameters of the GEV distributions defined by the data of the gauging stations of case study.

Station	ω	α	β
Huites	4.10	2.23	1.05
Choix	0.08	0.13	1.60
Palo Dulce	0.78	0.52	0.70
Chinipas	0.60	0.38	0.80
Urique	2.78	2.70	1.45

In order to compare the goodness of fit between the univariate and bivariate maximum likelihood estimates of the parameters of the GEV distributions defined by the data of gauging stations considered in the case study the standard error of fit, as defined in Kite (1977), was obtained and is displayed in Table 5.

Table 5. Standard errors of fit for gauging stations of case study.

Stations	Standard	Error of Fit
	Univariate	Bivariate
Huites	1.54	1.54
Choix	0.15	0.14
Palo Dulce	0.30	0.32
Chinipas	0.21	0.17
Urique	2.02	2.10

Conclusions

The Logistic model for bivariate general extreme value distribution for the minima has been presented for its application to drought frequency analysis. From the obtained values, both asymptotic and data based, the authors suggest the proposed model as a suitable option to be considered when performing drought frequency analysis.

Acknowledgments

The authors express their deepest gratitude to Universidad de las Americas-Puebla, Comision Nacional del Agua, and Universidad Nacional Autonoma de Mexico for the support given for the publication of this paper. Funding for this study was provided by the Comision Nacional del Agua through the Agreement SARH-CNA-UNAM.

References

Acosta, A. (1993). "Application of a Bivariate Extreme Value Distribution for Low-Flow Frequency Analysis". Ph. D. Dissertation, Engineering Graduate Studies Division, Universidad Nacional Autonoma de Mexico, p. 215 (in Spanish).

Escalante-Sandoval, C.A. (1991). "Trivariate Extreme value Distributions and its Applications to Flood Frequency Analysis". Ph. D. Dissertation, Engineering Graduate Studies Division, Universidad Nacional Autonoma de Mexico, p. 315 (in Spanish).

Gumbel, E.J. (1958). "Statistics of Extremes", Columbia University Press, p. 375.

Gumbel, E.J. (1962). "Statistical Theory of Extreme Values (Main Results)", Chapter 6. In: Contributions to Order Statistics, Sarhan, A.S. and Greenberg, B.G., editors, John Wiley & Sons, pp. 59-63.

Jenkinson, A.F. (1955). "The Frequency Distribution of the Annual Maximum (or Minimum) Values of Meteorological Elements". Quart. J. of the Roy. Met. Soc., Vol. 87, pp. 158-171.

Kuester, J.L. and Mize, J.H. (1973). "Optimization with FORTRAN", McGraw-Hill Book Co., pp. 386-398.

Raynal-Villasenor, J.A. (1985). "Bivariate Extreme Value Distributions Applied to Flood Frequency Analysis". Ph. D. Dissertation, Civil Engineering Department, Colorado State University, 237 pp.

Raynal, J.A. (1986). "A Bivariate Extreme Value Model Applied to Drought Frequency Analysis". In: Multivariate Analysis of Hydrologic Processes, H.W. Shen, J.T.B. Obeysekera, V. Yevjevich and D.G. De Coursey, editors. H.W. Shen, pp. 717-7.

Raynal-Villasenor, J.A. and Salas, J.D. (1987). "Multivariate Extreme Value Distributions in Hydrologic Analyses". Water for the Future, IAHS Publication No. 164, pp. 111-119.

Surface Water Hydrology, Singh, Al-Rashed & Sherif (eds)
© 2002 Swets & Zeitlinger, Lisse, ISBN 90 5809 363 8

FLODRO 3.0: A user friendly personal computer package for flood and drought frequency analyses

Jose A. Raynal-Villasenor
Department of Civil Engineering, Universidad de las Americas-Puebla
72820 Cholula, Puebla, Mexico
e-mail: jraynal@mail.udlap.mx

Abstract

User friendly personal computer package FLODRO 3.0 was developed under the programming environment of Visual Basic (Visual Basic is a trademark of Microsoft, Inc.). FLODRO 3.0 has two independent programs: FLOOD and DROUGHT. On program FLOOD, flood frequency analysis is performed by using eight typical distribution functions: Normal, Two and Three Parameter Log-Normal, Two and Three Parameter Gamma, Log-Pearson Type III, Extreme Value Type I and General Extreme Value. On program DROUGHT, drought frequency analysis is performed by using four typical distribution functions: Three Parameter Log-Normal, Extreme Value Types I and III and General Extreme Value. The characteristics, properties and construction of the programs FLOOD and DROUGHT of computer package FLODRO 3.0 are displayed in the paper towards its application in flood and drought frequency analyses. The selected methods of estimation of parameters are those of moments, maximum likelihood, sextiles and probability weighted moments.

Introduction

A subject of paramount interest in planning and design of water works is that related with the analysis of flood and drought frequencies. Due to the characteristic that design values have, they are linked to a return period or to an exceedance or non-exceedance probabilities, and the use of mathematical models known as probability distribution functions is a must. Among the most widely used probability distribution functions for hydrological analyses are the following, Kite (1988), Matalas (1963), Salas and Smith (1980), and Raynal-Villasenor (1998).

For flood frequency analysis: Normal, Two and Three parameters Log-Normal, Two and Three parameters Gamma, Log-Pearson Type III, Extreme Value Type I and General Extreme Value distributions. For drought frequency analysis: Three parameters LogNormal, Extreme Value types I and III, and General Extreme Value distributions.

In the light of the personal computer applications in education and training in all the fields of science, a personal computer program was designed to take care of the processes of flood and drought frequency analyses, in particular in engineering hydrology but easily extended to other fields related with frequency analyses dealing with maxima and minima, providing a wide number of options in the models to be used as in the analyses that can be done with such a tool as well. The resulting code has been named FLODRO 3.0, as it will be referred herein. The paper contains the key features of FLODRO 3.0 and a few examples for flood and drought frequency analyses are included to show the main results that FLODRO 3.0 can supply to the user.

Framework of FLODRO 3.0

FLODRO has been developed under Visual Basic 6.0, (Visual Basic 6.0 is a registered trademark of Microsoft Corporation), a BASIC compiler compatible with IBM (IBM is a registered trademark of International Business Machines) personal computers. The interactive mode in which FLODRO 3.0 is written makes it to have a high user friendly component. In any step, the user has the control on the processes that the program executes, from data input to printing of results of the analysis. The personal computer package FLODRO 3.0 has the structure shown in Figure 1, Raynal-Villasenor (1998).

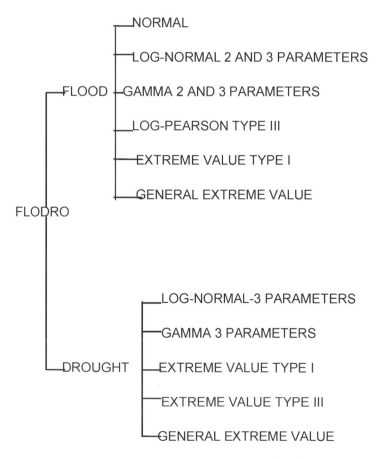

Figure 1. Framework of computer package FLODRO 3.0.

All the probability distribution functions mentioned in the previous section are contained in FLODRO 3.0 and it is divided in two independent computer programs: FLOOD and DROUGHT.

In the FLOOD computer program, the flood frequency analysis is performed by the use of eight probability distribution functions, as it is shown in Fig. 1.

In the DROUGHT computer program, the drought frequency analysis is performed by the use of five probability distribution functions, as it is shown in Fig. 1.

Both programs can perform the required computations to obtain, as shown in Fig. 2, Raynal-Villasenor (1998):

a) Estimation of parameters, by the methods of moments in DROUGHT, and moments, maximum likelihood and probability weighted moments in FLOOD, where applicable.

b) Computation of probability distribution function (CDF) for sample values or for any other values provided by the user.

c) Computation of probability density function (PDF) for sample values or for any other values provided by the user.

d) Inverse of the CDF for a fixed number of values or for any other values provided by the user.

e) Confidence limits for design events, by the methods of moments in DROUGHT and moments and maximum likelihood in FLOOD.

f) Goodness of fit tests based in the standard error of fit, Kite (1988), and based in a graphical comparison between the empirical and theoretical CDF and PDF's.

Personal computer program FLODRO has been designed to use minimum of memory and computer peripherals. Each computer program FLOOD or DROUGHT have less than 1.4 MB so there is no need to have a hard disk to run any of such programs. The graphs provided by FLOOD or DROUGHT are printed in a common printer there is no need to use costly plotters to get in paper these graphs. These features make FLODRO 3.0 very suitable in programs of hydrology education and training particularly in developing countries and in continuing education as well.

Numerical Examples

Gauging stations Jaina and Villalba in Northwestern Mexico have been selected to analyze the annual floods and the one-day low flow, repectively using the General

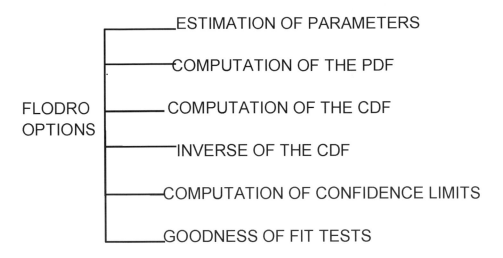

FLODRO
OPTIONS

ESTIMATION OF PARAMETERS

COMPUTATION OF THE PDF

COMPUTATION OF THE CDF

INVERSE OF THE CDF

COMPUTATION OF CONFIDENCE LIMITS

GOODNESS OF FIT TESTS

Figure 2. Options of analysis in computer package FLODRO.

Extreme Value probability distribution function and the method of maximum likelihood. The parameters obtained through the use of FLODRO 3.0 are:

a) For floods, gauging station Jaina:

> Location parameter = 651.25
> Scale parameter = 328.93
> Shape parameter= –0.5489
> Mean = 1135.41
> Standard Deviation = 1130.93
> Skewness = 4.01

b) For droughts, gauging station Villalba:

> Location parameter = 0.37
> Scale parameter = 0. 17
> Shape parameter = 0.53
> Mean = 0.34
> Standard Deviation = 0.15
> Skewness = 0.64

A sample of the screens and graphic displays provided by FLODRO 3.0 are contained in Figs. 3 to 12, for the analyzed gauging stations.

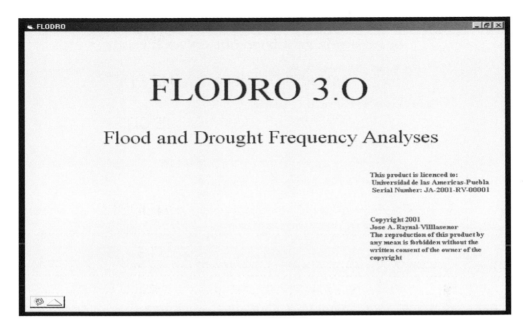

Figure 3. Screen of presentation of FLODRO 3.0.

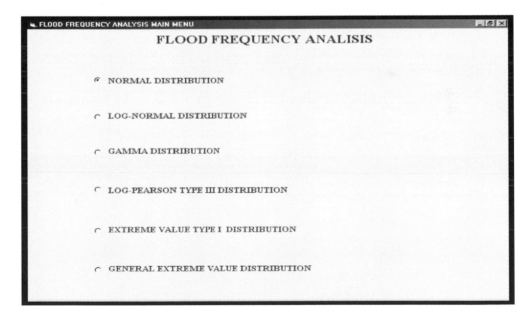

Figure 4. Flood frequency analysis Main Menu.

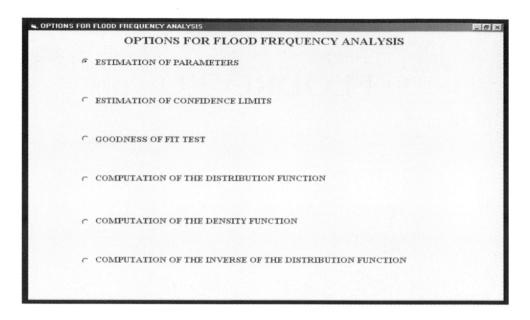

Figure 5. Flood frequency analysis options Menu.

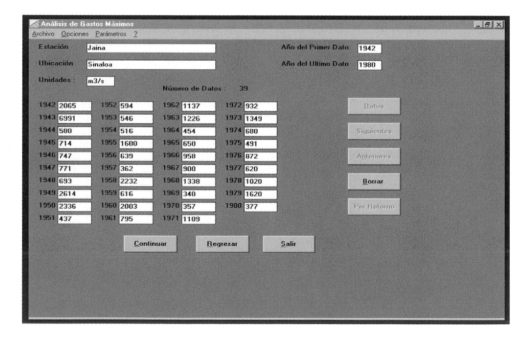

Figure 6. Flood frequency analysis access of data screen.

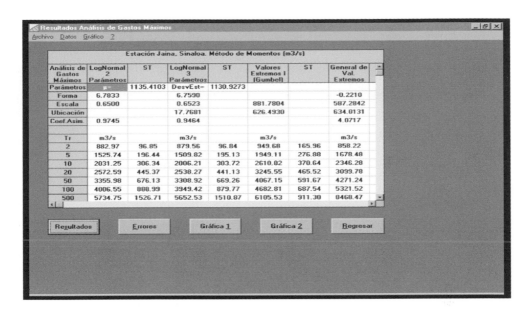

Figure 7. Flood frequency analysis screen of results.

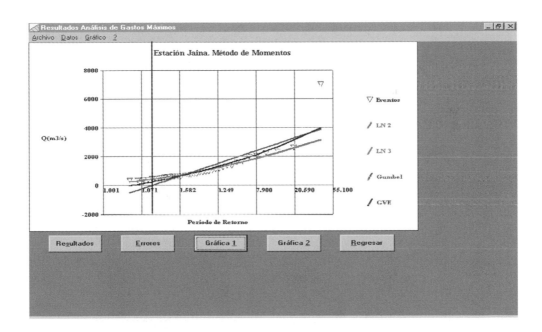

Figure 8. Empirical and theoretical frequency curves for gauging station Jaina
(Flood frequency analysis).

605

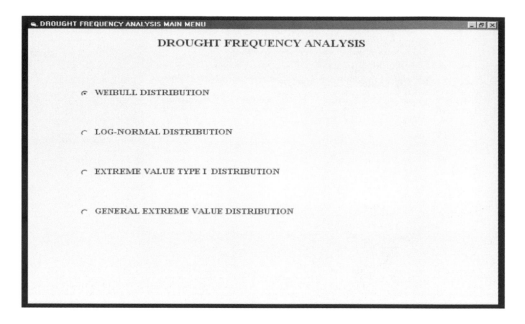

Figure 9. Drought frequency analysis Main Menu.

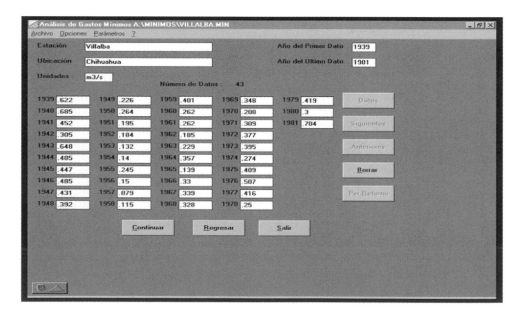

Figure 10. Drought frequency analysis access of data screen.

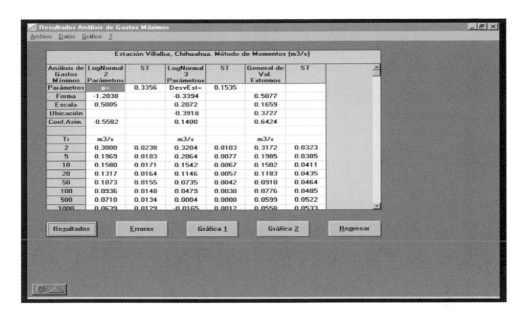

Figure 11. Drought frequency analysis screen of results.

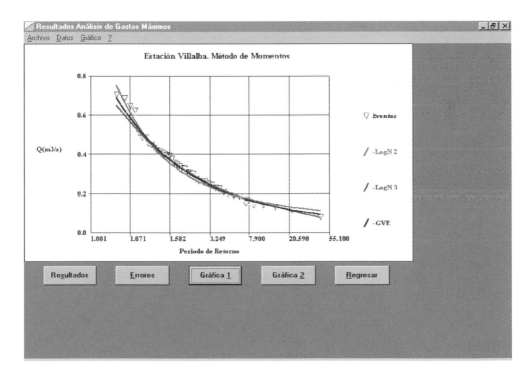

Figure 12. Empirical and theoretical distribution functions for gauging station
Villalba (Drought frequency analysis).

Conclusions

A personal computer program has been presented for flood and drought frequency analyses education and training. The computer code has been applied successfully to train students coming from Latin American, Asian and African countries, showing the user friendly component of such computer code, given that most of the students have not have any previous computer experience. Due to the minimum requirements of central memory and computer peripherals that the personal computer program has, as it has been shown in the paper, makes it a versatile tool to train students or technical personnel in the field with a personal computer without a plotter.

Acknowledgments

The author wishes to express his deepest gratitude to the Universidad de las Americas-Puebla for the support given to produce this paper.

References

Kite, G.W. (1988). "Flood and Risk Analyses in Hydrology". Water Resources Publications, Littleton, Colorado, USA, pp. 187.

Matalas, N.C. (1963). "Probability Distribution of Low Flows, Statistical Studies in Hydrology". U.S. Geological Survey Professional Paper 434-4, USGS, pp. A1-A27.

Raynal-Villasenor, J.A. (1998). "FLODRO 2: A User-Friendly Personal Computer Package for Flood and Drought Frequency Analyses". Computer Techniques to Environmental Studies, WIT Press, Southampton, UK, pp.181–187.

Salas, J.D. and Smith, R. (1980). "Computer Programs of Distribution Functions in Hydrology". Colorado State University, Fort Collins, Colorado, USA, 108 p.

Section 8: Hydrologic applications

Surface Water Hydrology, Singh, Al-Rashed & Sherif (eds)
© *2002 Swets & Zeitlinger, Lisse, ISBN 90 5809 363 8*

Downstream effects of dams:
The Lower Sakarya River

Necati Ağiralioğlu
Istanbul Technical University, Department of Civil Engineering
80626, Maslak, Istanbul, Turkey
e-mail: necati@itu.edu.tr

Sabahattin Işık, Mustafa Şaşal and Lütfi Saltabaş
Sakarya University, Department of Civil Engineering
54187 Esentepe, Sakarya, Turkey
e-mail: sisik@sakarya.edu.tr

Abstract

Dams and some other hydraulic structures cause some changes in the riverbed downstream after their construction. After constructing the Gökçekaya Dam which started to operate on the Middle Sakarya River in 1975, some changes occurred on the Lower Sakarya River such as river regime, sediment transport rates, and morphological changes. The measurements of sediment rate and the water flow are evaluated graphically. Results show that flood peak discharges in the Lower Sakarya River decreases as 30–50% after 1975. Furthermore, sediment transport rates decrease 50–53% after constructing of Gökçekaya Dam. There is also some degradation in the riverbed. Thalweg elevations of the river for selected Cross-sections were determined in 1965 and 1998. It was observed that the level of degradation varies 0–10 m during last 35 years. These changes investigated in this study affect water the table, the irrigation and drainage system, the gravel mining, and the navigability of the river.

Introduction

It is known that dams cause some changes in river characteristics at the downstream after their constructions. Many researchers investigated these changes, affecting services and structures. The downstream effects of dams can be classified according to their locations as (1) effects on below dams, (2) downstream effects, and (3) effects on river mouth. The third one is not related only with river characteristics, but also with sea conditions. Although first one has been studied widely (Komura and Simons, 1967; Stevens et al., 1975; Chen and Simons, 1979), downstream effects have not so much (Kondolf, 1997; Abam and Omuso, 2000). On the other hand, there is no any investigation on the changes in Sakarya River, although large dams have been taken to operation since 1956. Some dikes also constructed on the lower Sakarya River during 1960–1970. Furthermore, gravels have been mining from the Lower Sakarya River especially within 5 decades in order to meet construction materials for the surrounding large cities. In this study, the effects of the man activities, such as construction of dams and dikes, and gravel mining, on the Lower part, are to be investigated.

The Characteristics of Sakarya River

The Sakarya River is located in the northwest of Turkey, and its length is nearly 416 km, totally. The drainage area of the Sakarya River is about 55 312 km². The area of Lower Sakarya River includes all the drainage basin of the Sakarya River between the Doğançay gage station and the Black Sea (Fig. 1). Bottom elevation change from Black Sea to downtown Adapazari is about 24 m, i.e., the bottom slope of the river bed is around 0.3 m per km. River has the top width of 50 to 150 m, depth of 1.5 to 8.0 m, and moderately meandering alignment downstream portion.

In 1950's, some institutions start to investigate and to improve the water resources in Sakarya River. The first large dam, Sarıyar Dam, was constructed, and was started to operate in 1956 on the Middle Sakarya basin. Below its reservoir another large dam was built and start to operate in 1975. These two large dams and some other small dams have affected the characteristics of the Lower Sakarya River. Furthermore, between 1960–1970, some dikes were built on the Lower Sakarya river. Apart from these, subtraction of gravels by mining has increased with time because of rapid grow of the surrounding cities in order to supply construction materials needed. Beside, some irrigation and drainage projects where realized for agricultural purposes in these area.

Effects on River Hydrology

Discharges of the Lower Sakarya River were measured at two stations, Doğançay and Botbaşı (EPRSD, 1995). Mean flows of the Lower Sakarya River before and after 1975 were calculated and presented for Doğançay and Botbaşi gauging stations in Figs. 2 and 3, respectively. Flow regime is regulated at both stations. The annual mean flows in Doğançay are 137 and 115 m³/s before and after 1975, respectively. As seen from Fig. 2, the river flow for peak months decreases and the river flow for drought months slightly increases after dam construction upstream. The annual mean flows in Botbaşi are 196 and 160 m³/s before and after 1975, respectively.

After 1975, the river was regulated decreasing the peak discharges and increasing the minimum discharges. For observation of two stations, "Frequency Analysis of Maximum Discharge" was made by using "Gumbel Distribution Function" [SHW, 1994]. The computed values are shown in Table 1. Before 1975, probable maximum floods for 25-, 50-, and 100-year frequencies were expected as 1137, 1293, and 1448 m³/s, respectively. Probable maximum floods in Doğançay with data after 1975 for 25-, 50-, and 100-year frequencies were expected as 618, 680, and 741 m³/s, respectively.

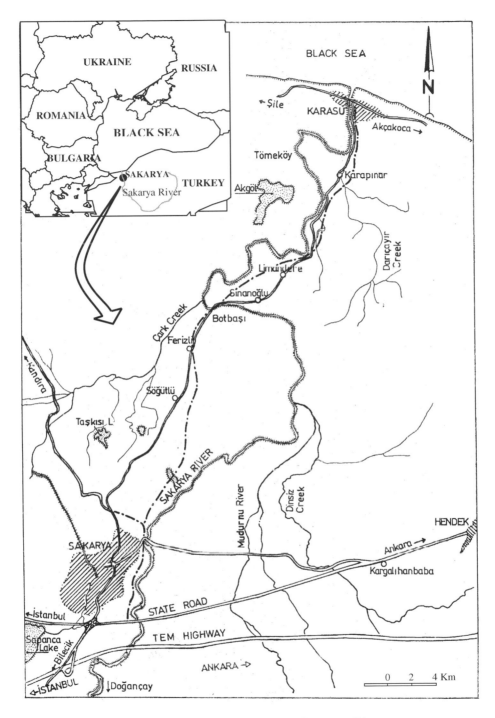

Figure 1. Location of the Lower Sakarya River.

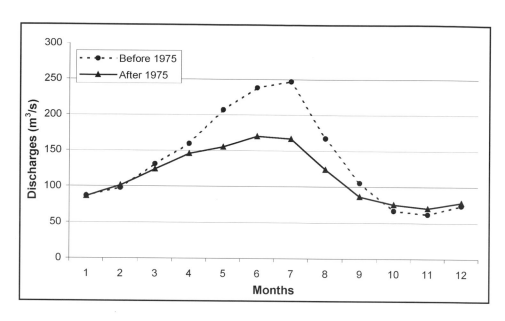

Figure 2. Mean discharges for the Doğançay gauging station.

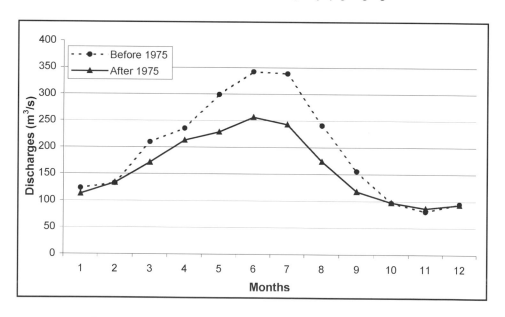

Figure 3. Mean discharges for Botbaşi gauging station.

Probable maximum floods decreases between 46–49% in Doğançay after 1975 as seen from the table. Comparison of expected maximum discharges with before and after is shown in Fig. 4 for different recurrence intervals.

Table 1. Probable maximum floods for the Doğançay gauging station.

INTERVALS	Expected Maximum Floods (m³/s)			
(Years)	1953-1994	After 75	Before 75	Decrease (%)
25	947	618	1137	46
50	1075	680	1293	47
100	1201	741	1448	49

Table 2. Probable maximum floods for the Botbaşi gauging station.

INTERVALS	Expected Maximum Floods (m³/s)			
(Years)	1960-1994	After 75	Before 75	Decrease (%)
25	900	734	1065	31
50	996	798	1187	33
100	1091	861	1309	34

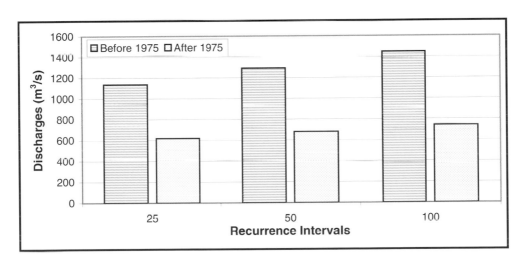

Figure 4. Maximum expected discharges for different recurrence intervals for the Doğançay gauging station.

Similarly, probable maximum floods in Botbaşi with data before 1975 for 25-, 50-, and 100-year frequencies were predicted as 1065, 1187, and 1309 m³/s, respectively. With data after 1975 probable maximum floods for 25-, 50-, and 100-year frequencies were predicted as 734, 798, and 861 m³/s, respectively. Probable Maximum Floods decreases between 31–34% in Botbaşi after 1975 as seen from Table 2. Comparison of expected maximum discharges with before and after is shown in Fig. 5 for different recurrence intervals. The expected maximum floods between 1953–1994 and 1960–1994 in Doğançay and Botbaşi, respectively, vary between values of previous two periods.

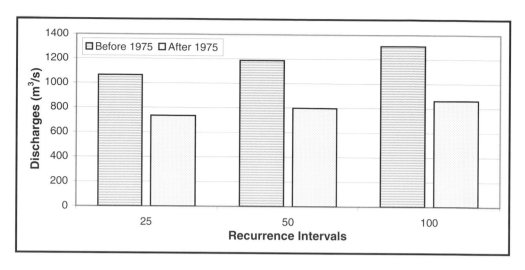

Figure 5. Maximum expected discharges for different recurrence intervals for the Botbaşi gauging station.

Effects on Sediment Transport Rate

Measurements of sediment transport rate have been made in the Botbaşi station since 1964 (EPRSD, 1993). Suspended sediment sampling of the Sakarya River at Botbaşi during 1964–1985 indicates that the average flow of river carries approximately 1200 ppm sediment. The sand size carried by flow varies from 30 to 60 percent of the sediment total load. During high flows the concentration of the sediment jumps over 5000 ppm.

Recorded sediment values were evaluated and shown in Fig. 6 for with data measured before 1975. The figure shows that the relationship between discharge and sediment transport rate is obtained as follows;

$$Q_s = 2.5866Q^{1.6247}$$

(1)

where Q_s in tons/day is sediment transport rate and Q in m^3/s is the water discharge. Correlation coefficient for these data is found as 0.705. As seen from Eq. 1 the rate of sediment transport increases as a power function of flow.

The relationship between Q_s versus Q is plotted in Fig. 7 with data after 1975. This figure indicates that a relationship can be written as:

$$Q_s = 0.9578Q^{1.6716}$$

(2)

In this case, the correlation coefficient for these data is found as 0.5744.

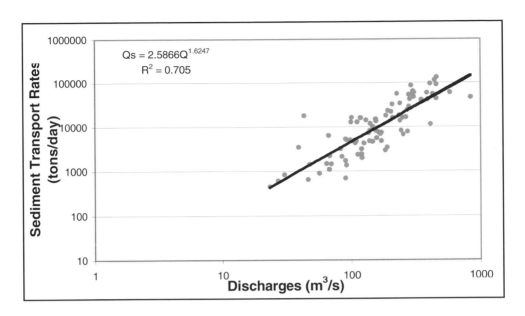

Figure 6. Relationship between sediment transport rates-discharges before 1975.

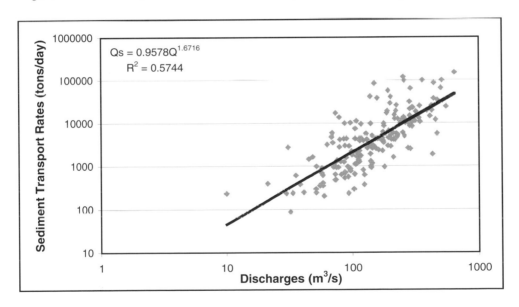

Figure 7. Relationship between sediment transport rates-discharges after 1975.

The response to dam construction is shown in Fig. 8 for comparison. As seen from this figure, sediment transport rates decrease about 50–53% after construction of Gökçekaya Dam. For instance, sediment transport rates for the mean annual discharge 160 m^3/s are found as 9587 and 4631 tons/day before and after 1975, respectively.

617

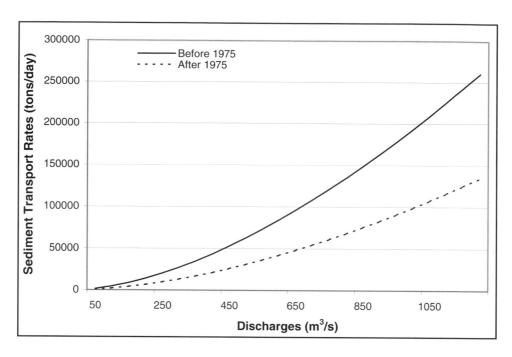

Figure 8. Comparison of sediment transport rates before and after 1975.

Effects on River Morphology

Riverbed elevations were measured in 1965 (SHW, 1965), and also measured in 1998 (Ağıralioğlu et al., 1998). The profile of the river depending on the related measurements is plotted in Fig. 9. In general, the riverbed degraded due to trapping of sediment at dam reservoirs. For example, at Sinanoğlu, the thalweg elevation is 8.90 m in 1965 and 2.78 m, in 1998, respectively. The decrease is about 6.12 m at this section of the river. At the distance from river mouth of 83 km, the decrease of riverbed elevation is being maximal as 10.14 m as seen from Fig. 9.

Results of Effects

All these impacts are not only caused by construction of dams. There are some other factors, which influence the decreasing of the riverbed elevation as gravel mining and construction of dikes on the Lower Sakarya River. Especially, gravel mining, which is very effective in this area directly, alters the channel geometry and bed elevation.

Degradation of the bed elevation causes some practical results. The ground water table that was previously very high is in dangerous condition for earthquake and construction stability in the area. When the bed elevation decreases, the river surface elevation also decreases, and consequently so does ground water table.

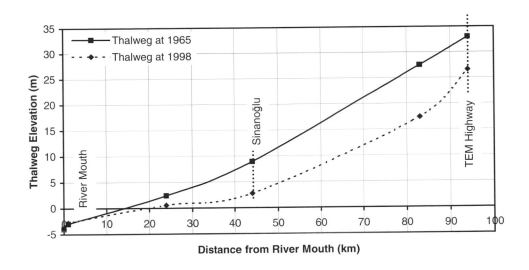

Figure 9. Changes of riverbed elevations.

As water table decreases, the possibility of irrigation condition becomes more difficult, while the agricultural drainage becomes easier. The navigability of the river is higher than the previous condition because of increasing of water depth and decreasing of bottom slope of the river. On the other hand, possibility of flood occurrence also decreases in the Lower Sakarya River. In order to investigate all these changes a lot of measurements and studies should be carry out in this area.

Conclusions

The following conclusions can be drawn from this study.

1. For the Lower Sakarya River, monthly mean river flows decreased during 1975–1994.
2. Expected peak flows decreased as 30–50% after 1975 on the Lower Sakarya River.
3. Sediment transport rates also decreased as about 50% for last 35 years.
4. The decrease of riverbed elevation varies 0–10.14 m depending on the location between 1965 and 1998.
5. The changes of river characteristics on the Lower Sakarya River yield many practical results. For instance, the river water surface level decreases, the very large peak flows do not occur in the area as frequently now as in the past.

References

Abam, T.K.S. and Osumo, W.O. (2000). "Technical Note on River Cross-Sectional Change in the Niger Delta". Geomorphology, Elsevier Science, 34, pp.111-126.

Ağıralioğlu, N., I. Çallı, L. Saltabaş, B. Eryılmaz, Z. Gündüz, M. Şaşal, S. Karpuz and İ.H. Demir. (1998). "Feasibility Study on Navigation Project for the Lower Sakarya River". Final Report, Sakarya University in Turkish.

Chen, Y.H. and Simmons, D.B. (1979). "Geomorphic Study of Upper Mississippi River". J. of the Waterway Port and Coastal and Ocean Division, ASCE, 105(3), pp. 313-328.

SHW, (1965). "Feasibility Report for Lower Sakarya Project". State Hydraulics Works Press, Ankara.

SHW, (1994). "Frequency Analysis of Maximum Discharges for River Basin of Turkey". State Hydraulics Works, Ankara.

EPRSD, (1993). "Sediment Data and Sediment Transport Amount for Surface Water in Turkey". Electrical Power Resources Survey and Development Administration, Ankara.

EPRSD, (1995). "Monthly Average Discharges (1935-1990)". Electrical Power Resources Survey and Development Administration, Ankara.

Komura, S. and Simons, D.B. (1967). "River-Bed Degradation Below Dams". J. of the Hydraulics Division, ASCE. 93(4), pp. 1-14.

Kondolf, G.M. (1997). "Hungry Water: Effects of Dams and Gravel Mining on River Channels". Environmental Management, Springer-Verlag New York Inc., 21(4), pp. 553-551.

Stevens, M.A., Simons, D.B., and Schumm, S.A. (1975). "Man-Induced Changes of Middle Mississippi River". J. of the Waterways Harbors and Coastal Engineering, ASCE, 101(2), pp. 119-133.

Surface Water Hydrology, Singh, Al-Rashed & Sherif (eds)
© 2002 Swets & Zeitlinger, Lisse, ISBN 90 5809 363 8

Flash flood mitigation using watershed management tools: Petra Area (Jordan)

Radwan Al-Weshah
University of Jordan (on leave), Regional Hydrologist
UNESCO Cairo Office, Cairo, Egypt
e-mail: weshah11@yahoo.com

Fouad El-Khoury
Resources and Environment Dept., Dar Al-Handasah
(Shair and Partners) Beirut, Lebanon

Abstract

Petra is located in the southwest region of Jordan about 200 km south of Amman between the Dead Sea and the Gulf of Aqaba. Petra was carved in sandstone canyons by the Nabatean Arabs over 2,000 years ago. It is a major tourism attraction, as its monuments are considered the jewels of Jordan. Floods pose a serious threat to the tourist activities as well as to the monuments themselves. To alleviate the impact of flood risk in Petra, different flood mitigation measures are investigated. The impact of these measures on flood-peak flow and volume is evaluated. They include afforestation, terracing, and the construction of check and storage dams, as well as various combinations of these measures. A flood simulation model depicts reductions of up to 70% in flood-peak flows and volumes due to these flood mitigation strategies.

Introduction

The Petra region is located in southwest Jordan, between the Dead Sea and the Gulf of Aqaba. It lies in the Sherah Mountains overseeing Wadi Araba in the Jordan Rift Valley at latitude 30°20' North and longitude 35°27' East, and at an average elevation of 950 m above mean sea level. It is about 200 km south of Amman. The monuments of Petra are concentrated in the lowest part of the area where many streams (*wadis*) meet. The main wadi is Wadi Musa (River of Moses) after which the town adjacent to Petra took its name. These monuments were built and carved by the Nabatean Arabs over 2000 years ago in sandstone canyons that are protected by limited and very narrow gorge-like accesses. The main gorge, a narrow passage bound by high cliffs, is called the *Siq,* Petra is a major tourism attraction in Jordan and is commonly referred to as the "rosy rock city."

Historical records and events concerning flash floods threatening Petra show that flood protection and mitigation measures are urgently needed to protect lives as well as the existing monuments. These measures shall be implemented on the basis of a detailed hydrological analysis and assessment of the Petra watershed to quantify the level of required flood mitigation and protection works. Due to Petra's unique nature, all measures will be evaluated to be sure that they would not affect the existing monuments negatively.

As documented in most of the available studies – e.g., Electricité de France (EDF, 1995) – the flood that occurred in 1963 was an extreme event. The intense and sudden rainfall caused flood water to flow from all wadis into the main wadi upstream of the *Siq*. The dam at the entrance of the *Siq* filled up with sediment and consequently, flood water overtopped the dam and entered the *Siq* instead of being diverted through the tunnel of Wadi Al-Mudhlim. Eyewitnesses stated that the floodwater depth was about 10 m in some areas of the *Siq* passage. Despite great emergency efforts by different Jordanian authorities, it was not possible to rescue all the tourists trapped in the *Siq*, and twenty lost their lives due to the flood event.

Recent floods occurred in 1991, 1995 and 1996. Although the floodwater did not enter the *Siq* during these events, traces of high water within Wadi Al-Matahah (into which the diversion tunnel of Wadi Al-Mudhlim discharges) indicated that the water level reached an elevation of more than 12 m above the wadi bed. During these events, the *Siq* entrance area was flooded and tourists had to be rescued.

Climate and Rainfall

The Petra region belongs to the Mediterranean climatic zone. The average annual precipitation is around 200 mm. Most rainfall is concentrated between October and April and is mainly of orographic origin.

There are three rainfall gauging stations in the Wadi Musa watershed. Annual total rainfall records are available for these stations since 1937, and daily total rainfall records are available since 1980.

Description of Catchments

The overall Petra catchment has an area of about 50 square kilometers. It can be divided into nine sub-catchments as shown schematically in Figure 1. The physical characteristics of each of the nine sub-catchments in the Petra catchment are described in Table 1 below.

Flood Analysis and Prediction

The Watershed Modeling System (WMS) developed by Brigham Young University (WMS, 1996) was used for hydrologic analysis. This model was calibrated and then used to estimate the peak flow and flood volume at the *Siq* entrance for return periods ranging from 10 to 100 years. The Intensity-Duration-Frequency (IDF) curves developed for Wadi Musa meteorological station by the Water Authority of Jordan (WAJ, open files) were used to estimate the 24-hour design storm for these return periods.

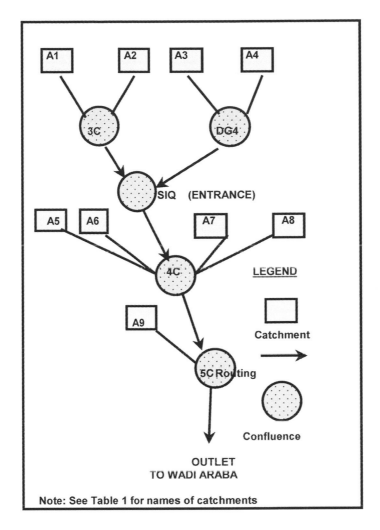

Figure 1. Schematic diagram of the hydrologic model for Petra catchment.

Flood Mitigation Measures and Management Schemes

Mitigation means activities and measures which actually eliminate or reduce the effect and the probability of occurrence of an extreme event. In this paper, the focus is on mitigation measures at the upstream side of the catchment areas, which are physically distant from the site of the monuments.

Four possible watershed management scenarios or measures were analyzed hydrologically to estimate their effectiveness in reducing floods at critical Petra sites such as the *Siq* entrance. The four scenarios are:

Table 1. Summary of catchment characteristics
for the Petra watershed.

Catchment Number*	Catchment Name	Area (km^2)	Average Slope (%)	Weighted CN	Lag Time (hrs)
A1	Kafr Isham	4.30	5.7	80	0.76
A2	Qurnat Bin Sa'd	21.11	6.2	81	1.50
A3	Al Hai	10.75	9.0	81	1.00
A4	Jebel Zubaira	13.90	5.3	75	1.19
A5	Al Madras	3.90	6.7	91	0.66
A6	Al Mataha	4.20	8.7	91	0.60
A7	Wadi Kharubit	6.90	10.0	91	0.96
A8	Wadi Al Ullyqa	7.05	9.1	91	1.02
A9	Wadi Siagh	7.90	11.8	91	0.67

1. Afforestation of selected parts of the watershed;
2. Contour terracing and construction of check dams with afforestation;
3. Construction of storage/detention dams; and
4. Combination of storage/detention dams and afforestation.

Most of these measures were originally used by the ancient Nabateans of Petra. Our aim is to restore parts of the old Nabatean hydraulic system where possible. Their system was unique, well-maintained, and integrated within their life style and agricultural practices. Our approach was careful to avoid any intrusion on the aesthetics of the historical setting of Petra.

First Scenario: Afforestation

Afforestation could be undertaken at selected areas in the two upstream watersheds of Jebel Zubaira and Qurnat Bin Sa'd. The afforestation scheme would cover an area of 1,000 ha (500 ha in Jebel Zubaira and 500 ha in Qurnat Bin Sa'd).

The analysis focuses on good afforestation practices according to the definition of the Soil Conservation Service (SCS, 1986). Good afforestation conditions are applicable when the total vegetated area is greater than 75% of the arable land in the catchment. The afforestation conditions affect the infiltration-runoff process in the catchment as reflected hydrologically by the Curve Number (CN) value (Chow et al., 1988). The flood-peak flow and volume are estimated at the bridge culvert near the entrance of the *Siq*. This analysis covers storm events with return periods of 10 and 100 years. Results of this analysis showed that this measure reduces the flood-peak flow by 28% for the 100-year flood and 39% for the 10-year flood. Similar reductions of 25 and 35% were noted in flood volumes for the 100- and 10-year events, respectively.

Details and plans of afforestation schemes including the vegetation and tree cover that best fit the area, and the socioeconomic aspects, need to be addressed in other studies with the help of the Ministry of Agriculture and other concerned agencies. Good examples of soil conservation and afforestation projects in Jordan are the Highland Development Project and the Zarqa River Basin Development Project. Both projects encourage local farmers to undertake soil conservation, afforestation (mainly with olive trees), and water harvesting. This approach can be immediately implemented in the Petra watershed.

Second Scenario: Terracing, Check Dams, and Afforestation

Terracing and check dams involve the construction of dry-stone walls and box gabions which follow the contour lines and intercept overland flow lines. Check dams are gabion structures a few meters high built across the course of the wadi. Terracing activities are usually accompanied by afforestation schemes. Terraces and check dams have three-fold interconnected effects, as follows:

Increasing the time of concentration (T_c) for the sub-basin. This will decrease the sensitivity of the watershed to short-duration high-intensity storms, making it less vulnerable to more frequent flood events. The value of T_c depends on the spacing, configuration, and height of terracing (Chow et al., 1988). From a practical point of view, it was found that T_c values (and thus lag-time values) can be increased by about 30% for the terraced catchments compared to the existing T_c. The increased T_c results from the effect of terracing, which increases the length of flow lines and decreases the slopes of overland and channel flow paths.

Decreasing values of the Curve Number (CN) for afforested terraced catchments. Values of CN are reduced by land conservation practices such as terraced catchments. Terracing allows for greater infiltration and abstraction during a storm event. Available literature suggests that conservation measures would reduce the CN value by 5 to 10%. However, in the present analysis it was considered that a reduction in CN value of 5% is a reasonable estimate for all terraced catchments upstream from the *Siq* entrance.

Providing additional shallow storage, which in turn increases the T_c. Storage provided by terraces and check dams may attenuate the hydrograph and delay the time to peak. The effect of terracing and check dam storage is already considered in the reduced values for CN and increased values for T_c. Therefore, the storage effect of check dams and terracing was not considered as an additional potential benefit.

Results of this analysis showed that this measure reduces the flood-peak flow by 52% for the 100-year flood and 65% for the 10-year flood. Similar reductions of 38 and 52% were noted in flood volumes for the 100- and 10-year events, respectively.

Third Scenario: Storage/Detention Dams

Storing flood water in reservoirs and detention basins behind dams is another alternative for flood mitigation and control. The topography of the catchments seems to allow the construction of these reservoirs. Based on available topographic maps, a number of locations are possible sites for these dams and reservoirs. However, a detailed field survey to check the exact location, storage volume, and outlet type of these dams would be required to confirm the feasibility and costs. Approximate elevation-storage volume data were obtained from the available topographic maps. As these dams are flood-control structures, their outlets can be designed similar to those of regular detention basins.

The reservoir storage system was assumed to consist of seven storage dams, constructed in various parts of the catchment, with a total capacity of 214,000 m^3. Results of this analysis showed that this measure reduces the flood-peak flow by 45% for the 100-year flood and 39% for the 10-year flood. Similar reductions of 6 and 9% were noted in flood volumes for the 100- and 10-year events, respectively.

The effect of this scenario on flood volume should be negligible. Any effect on flood volume can be attributed to additional infiltration due to the increased detention time.

Fourth Scenario: Combined Effect of Storage Dams and Afforestation

The combined effect of storage dams and afforestation was evaluated according to the schemes outlined in scenarios one and three above. Results of this analysis showed that this combined measure reduces the flood-peak flow by 62% for both the 100-year and the 10-year event. Similar reductions of 38 and 52% were noted in flood volumes for the 100-and 10-year events, respectively.

Analysis of Results and Discussion

The analysis of the above results focused on evaluating the relative percentage change in the peak flow, time to peak, and flood volumes for storm events of return periods of 10 and 100 years. Improvements are compared to existing conditions (do-nothing option). The station at the entrance of the *Siq* was selected because it is a very critical point for tourist activities.

Figures 2 and 3 represent the relative effectiveness of each of the four measures by comparing them with existing conditions (do-nothing option) for the ratios of peak flow and flood volume at the entrance of the *Siq*. Figure 2 shows the relative impact on peak flow. From this figure, it can be seen that afforestation and terracing as well as afforestation and dam storage are the most effective measures for all return periods. Terracing/check-dams and afforestation are slightly superior to afforestation and storage action for more frequent (small) events (e.g., return periods less than 10 years). The reservoir storage option starts to be slightly more effective than terracing for less frequent (larger) events. Figure 3 shows the relative impact on flood volumes for the four flood-mitigation scenarios. It can be seen that affore-

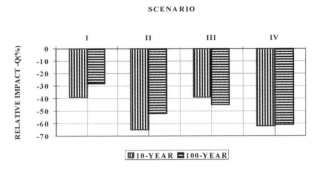

Figure 2. Relative mitigation impact on flood peak for different scenarios.

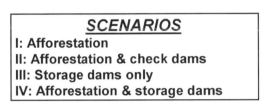

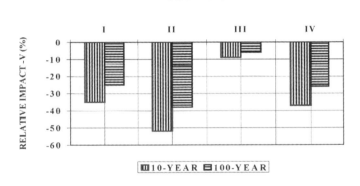

Figure 3. Relative mitigation impact on flood volume for different scenarios.

station and terracing as well as afforestation and storage are the most effective volume-control measures for all return periods.

A comprehensive evaluation of each alternative would consider economical, social, legal, and environmental aspects. From a flood-mitigation perspective, terracing and afforestation provide more than 60% reduction in the flood-peak flow and about 30% reduction in the flood volume at the *Siq* entrance. On the other hand, the terraces/check-dams are usually necessary to provide suitable land for afforestation. Terraces/check-dams can be used for water harvesting to sustain small amounts of water needed for afforestation. This was the ancient Nabatean practice for desert agriculture. Furthermore, compared to construction of dams, the effect of terraces/check-dams on the environment will be less pronounced.

Terraces/check-dams can be constructed by local people without any sophisticated engineering design, contrary to the storage option where skilled technicians and complete engineering designs are needed for dam construction. The dam storage option also is more costly compared to the terracing/check dam option. The risk involved in the construction of dams is usually higher due to a possible dam break, compared to the risk posed by failure of the terracing/check-dams option. Among the major advantages of reservoir storage is the use of stored water for different purposes (domestic use and irrigation). However, the amount stored is relatively small, with a seasonal storage volume of about 2 to 3 million cubic meters. This is quite small compared to the needs, and much of the stored volume can be lost by evaporation during dry hot seasons.

Conclusions

Based on the above analysis and discussion, the following conclusions and recommendations can be stated:

1. Flooding poses a serious risk to lives and property in the *Siq* area under the present conditions, and immediate actions should be taken to mitigate the flood risk. Structural measures that intrude on the integrity and aesthetics of the historical setting of Petra are not encouraged. In this case, restoring some watershed management practices used by the ancient Nabateans of Petra is more attractive and sound.

2. Afforestation only, of 1000 ha in both Jebel Zubaira and Qurnat Bin Sa'd sub-basins, provides about 30 to 50% reduction in flood-peak flow. However, it has a limited impact on flood volume (about 20%) for most storm events.

3. Afforestation combined with terracing and check dams in the watershed upstream from the *Siq* entrance produced about 50 to 80% reduction in the flood-peak flow, and about 40 to 70% reduction in flood volumes for most storm events.

4. Afforestation combined with dam storage of 214,000 m^3 produced about 60% reduction in the flood-peak flow, and about 30 to 50% reduction in flood volumes for most storm events. This scenario would require a chain of seven dams over major wadis in the watershed upstream of the *Siq* entrance.

5. From the socioeconomic, environmental, and risk perspectives, a terracing-with-check-dams measure would be more suitable, less expensive, and thus more desirable compared to dam storage. However, a further detailed evaluation of these two options must consider all relevant parameters, including economical, and other factors.

Acknowledgments

This research was partially done under the Petra Planning Project for the Ministry of Tourism and Antiquities (MOTA) of Jordan. The cooperation of the Ministry of Water and Irrigation in providing some of the raw rainfall and flow data is highly appreciated. A comprehensive paper upon which this article is based appeared in the ASCE's *Journal of Water Resources, Planning, and Management,* Vol.125, No. 2, May/June, 1999.

References

Al-Weshah R.A. and F. Khoury (1999). "Flood Analysis and Mitigation for Petra Area in Jordan." ASCE J. Water Resources Planning and Management, Volume 125 No, 2, May/June.

Chow, V.T., D.R. Maidment and L.W. Mays (1988). *Applied Hydrology.* New York, McGraw-Hill.

EDF, *Petra Site Hydraulic Project* (1995). Main report submitted to the Dept. of Antiquities, Electricité de France, Amman, Jordan.

SCS, *Urban Hydrology for Small Watersheds* (1986).Technical Release No. 55, Soil Conservation Service, US Dept. of Agriculture, Washington, DC.

WAJ, open files (1998). Water Authority of Jordan, Amman, Jordan.

WMS, *Watershed Modeling System (1998).* Brigham Young University, Provo, Utah.

Surface Water Hydrology, Singh, Al-Rashed & Sherif (eds)
© 2002 Swets & Zeitlinger, Lisse, ISBN 90 5809 363 8

Opportunities of dry seeding for drought alleviation in rainfed lowland rice systems

Abul Fazal M. Saleh

Institute of Flood Control and Drainage Research
Bangladesh University of Engineering and Technology
Dhaka 1000, Bangladesh
e-mail: saleh@ifcdr.buet.edu

M.A. Kashem Khan and M. Abdul Mazid

Bangladesh Rice Research Institute
Joydebpur, Gazipur 1701, Bangladesh
e-mail: brrihq@bdmail.net

Abstract

In South Asia, 15%–66% of the arable land is under rainfed agriculture, where rice is the predominant crop. Uncertainty and variability of rainfall, in time, amount and frequency have significantly affected the rice production. Studies on rainfed rice systems in Northwest Bangladesh have shown that with the traditional technology of crop establishment by transplantation, droughts are expected at the beginning of the crop season during transplanting and near the end of the season during flowering and ripening of the grains. Opportunities for drought alleviation through dry seeding technology of crop establishment have been studied during the 1994-2000 wet seasons, by carrying out field experiments and farmers' surveys. From comparative analysis of water availability from rainfall and relative water supply, the dry seeding technology has been affirmed as technically more feasible for drought alleviation, but the socio-economic constraints limiting its adoption by the farmers need further research.

Introduction

Rice is a staple food in South Asian countries and is grown in abundance in India, Bangladesh, Sri Lanka and Nepal. About 15% of the rice lands in Sri Lanka, 35% in India, 50% in Bangladesh and 66% in Nepal, are rainfed (IRRI, 1993). Due to economic and resource constraints, irrigation development in the region has almost stagnated in the recent past and a vast majority of these areas would therefore continue to remain under rainfed agriculture in the foreseeable future. In the South Asian rainfed systems, rice is the predominant crop and being a water loving plant it requires about 1000 mm of water for a decent yield. Although the rice growing regions of South Asia are in the monsoon belt and receive an average rainfall of more than 1000 mm, the uncertainty and variability of rainfall in time, amount, duration and frequency affect water availability and the rice production.

In Bangladesh, the Northwest region receives the lowest rainfall of about 1000 mm during the five monsoon months (June–October) and can be classified as low for a rainfed rice system (Garrity et al., 1986). The rainfed areas of the region are predominantly mono-cropped, with transplanted rice (TPR) as the main crop, grown during the monsoon months. Some upland crops, such as wheat, chickpea, lentils, oil seeds etc. are grown during the post-monsoon season depending upon the availability of the residual soil moisture. TPR establishment on puddled soil facilitates weed control and water retention, but requires large amount of water for crop establishment and is labour intensive. All of the farming activities leading to transplanted rice, such as seedbed preparation, sowing of seeds, culture of seedlings, land preparation for puddling and transplanting of seedlings, are all dependent upon water availability from rainfall. Droughts at the beginning of the season delay transplantation and lengthen the growing period beyond the monsoon and hence the crop again becomes vulnerable to droughts during its maturity. Even when timely transplanted, droughts are expected at the recession stage of the monsoon, if the crop has not yet been harvested. Droughts during the flowering and grain filling stages reduce yield due to sterility and unfilled grains.

Construction of on-farm reservoirs (OFRs) to harvest excess rainfall and runoff generated from the farm has played a crucial role in alleviating drought in rainfed systems. Studies in the Philippines, Indonesia and India have shown that OFRs are being used by the farmers for supplementary irrigation of monsoon rice, and the technology has been successful in drought alleviation and making a breakthrough towards increased productivity (Bhuiyan and Zeigler, 1994). But, although the OFR technology was found to be both technically and economically feasible in Bangladesh (Islam et al., 1998), because of mainly social constraints (high demand for land and high man-land ratio), the technology was not acceptable to the farmers.

Rainfall dependency associated with puddling and transplanting could be largely overcome through dry seeding, in which rice seeds are sown on dry tilled unsaturated fields early in the monsoon season, which subsequently may become flooded. Studies conducted in the drought prone rainfed systems of the Philippines and Indonesia have shown that the dry seeded rice (DSR) uses rainfall more efficiently, suffers less drought risks, produces similar yields and is more profitable than TPR (Saleh and Bhuiyan, 1995; Lantican et al., 1999).

To complement earlier research findings on the advantages of DSR compared to TPR in rainfed systems, this study was conducted in Northwest Bangladesh during the 1994–2000 period, with the objectives of understanding and evaluating the role of dry seeded rice in alleviating drought in rainfed rice systems.

Materials and Methods

Study Area

The study was conducted in Rajabari Union of Godagari Thana of Rajshahi District of Nothwest Bangladesh. The soil texture of the area varies from silt loam to silt clay loam and is poorly drained with 6–8 cm thick plow pan at 9–11 cm depth. The soil is low in organic matter (0.8%–1.2%) and is acidic in nature (pH varies from 5.5 to 6.5).

The average annual rainfall at Godagari is 1300 mm and 80% of it is confined during the monsoon months of June to September. The average daily evaporation at Rajshahi varies from about 2.3 mm in January to 6.3 mm in April. The four months when the rainfall exceeds evaporation are June to September. The averages of the maximum and minimum temperatures at Rajshahi are about 39 °C and 10 °C respectively, and occur in the months of April and January.

Analytical Procedure

The extent of droughts was studied by using the water balance method for finding out the number of 5-day water deficient periods during the monsoon season. Since rainfall during one day may be adequate to meet the crop water requirement for the entire 5-day period, the chances of a drought within the period are higher if the period is longer than five days. As soil-water may be available to the rice roots two to three days after standing water has disappeared, a shorter period was not considered. The methodology followed is similar to Thornthwaite's method of water balance (Steenhuis and Van Der Mollen, 1986). But, in this analysis the seepage and percolation loss (S&P) was also taken into account along with evapotranspiration (ET), as S&P is an integral part of crop water requirement for lowland rice (Saleh and Bhuiyan, 1996). The 5-day period water balance is written as follows:

$$H_t = R_t + H_{t-1} - ET_t - (S\&P)_t \tag{1}$$

where, H and is the bund storage, R is the rainfall and subscripts t and t – 1 denote time in 5-day time steps (present and previous 5-day, respectively). R, ET and (S&P) are all expressed in mm/day (in each 5-day time step). The maximum height of the bunds is 20 cm. Thus, the incidence of either drought or adequacy of water supply was determined by the following criteria:

$$\text{if } (R_t + H_{t-1} - ET_t) < 0; \text{ then } (S\&P)_t = 0; H_t = 0; \text{ there is drought;} \tag{2}$$

$$\text{if } ET_t < (R_t + H_{t-1}) < ET_t + (S\&P)_t; \text{ then } H_t = 0; \text{ there is no drought;} \tag{3}$$

$$\text{if } R_t + H_{t-1} - ET_t - (S\&P)_t > 0; H_t > 0; \text{ there is no drought; and} \tag{4}$$

$$\text{if } R_t + H_{t-1} - ET_t - (S\&P)_t > H_{max}; H_t = H_{max} = 20 \text{ cm;} \tag{5}$$

there is no drought.

The water availability during the crop growth season was determined by using the concept of relative water supply (RWS), which is defined as the ratio of supply of water to the demand by the crop (Levine, 1982). Thus, RWS for a given period t can be written as:

$$RWS_t = (R/(ET + S\&P))_t \tag{6}$$

A RWS value greater than 1 indicates that the water supply is abundant, where as, a value less than 1 indicates drought. For lowland rice, the pan evaporation data are good indicators of crop evapotranspiration (Tomar and O'Toole, 1980). The rainfall probability and the occurrences of droughts were analyzed by using the Gamma distribution.

For drought analysis, 31 years of daily rainfall data (1963–1993) of Godagari station situated about 15 km north-west of the study area were collected from Bangladesh Water Development Board, as long-term rainfall data were not available for the study site. Daily rainfall and US Class A Pan evaporation data for the study seasons were collected from a nearby temporary weather station.

Field Experiment and Farmers' Survey

The field study was carried out during 1994–2000 monsoon (July–November) seasons. The crop production practices of 50 randomly selected farmers were surveyed using a pre-designed questionnaire in order to gain qualitative and quantitative information on farmers' practices in crop establishment and crop management.

Field water status and water use of 30 farms were closely monitored by installing 100 cm long PVC tubes (50 cm perforated), 20 cm above and the rest below the ground surface. The water level readings in the PVC tubes (standing or perched water level) were taken every alternate day. Changes in standing water levels during rainless days were used in order to estimate S&P loss after deducting pan evaporation.

Field experiments on effect of crop establishment method (DSR and TPR) and time of seeding/transplanting date on crop yield, were carried out in a split-plot design. For TPR, the seeding in the wet bed for raising seedlings was done on the same day of seeding for DSR and then 30-day old seedlings were transplanted in puddled soil.

Results and Discussion

From the farmers' survey and field observation, it was ascertained that during the monsoon season growing of TPR is the traditional technology, and the preferred times of transplanting and harvesting are mid July and mid November. Earlier transplanting is generally not possible because of inadequate rainfall for land preparation. Similarly, later harvesting is not acceptable because of expected high yield loss due to drought. Thus, the 120 days between 15 July to 15 November was considered as the field duration of TPR. The duration was divided into three growth stages: vegetative, reproductive and grain filling, each stage being 40 days long. From field measurements, the average pan evaporation during the season was determined as 3 mm/day. The average S&P loss was determined from the PVC tube readings of rainless days as 7 mm/day. During the two months required from first plowing to transplanting (15 May to 15 July), the farmers required at least 400 mm of cumulative rainfall to complete the land preparation activities prior to transplantation. In a similar study in the Philippines, the cumulative rainfall required for land preparation was found to be about 600 mm (Saleh and Bhuiyan, 1995).

Drought Analysis of TPR

Drought was characterized through probability analysis of rainfall during each of the crop growth stages of the crop field duration (15 July to 15 November). Water balance with 5-day time steps was carried out for the crop field duration using Equations 1–5 and with 31 years (1963–1993) of rainfall data. As no water is required during the two weeks prior to harvesting, the duration of the grain filling stage with crop water requirement was 25 days (till the end of October). From the number of 5-day droughts that occurred during each of the crop growth stages, the probability of 5-day droughts during each stage was determined and is shown in Fig. 1. The figure shows that compared to other growth stages, droughts are more frequently expected during the grain filling stage. The probability of two and three 5-day droughts during the grain filling stage are more than 80% and 45%, respectively. But, as these droughts are not continuous (5-day drought followed by another 5-day drought) the crop is expected to recover during the in between drought less period and the crop yield would not be seriously affected. Hence, the probability of consecutive 5-day droughts (juxtaposed) during the crop field duration was analyzed and the results are shown in Fig. 2. Again, the frequent occurrence of droughts during the grain filling stage compared to other stages is evident from the figure. The probability of two and three consecutive 5-day droughts (10-day and 15-day duration), during the grain filling stage are 73% and 53%, respectively. Thus, a two week period without rain is expected once in two years during the grain filling stage and can be detrimental to crop yield.

The 50% and 80% probable 5-day rainfall for the May-November period were determined and are shown in Fig. 3. The figure shows that in the study area very little or no rainfall is expected after early October. Since early October is the beginning of the grain filling stage, the crop is vulnerable to damage by droughts.

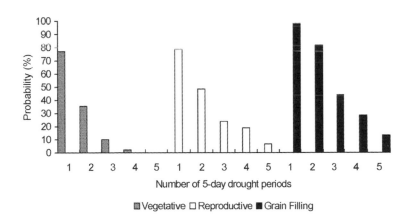

Figure 1. Probability of 5-day drought during crop field duration.

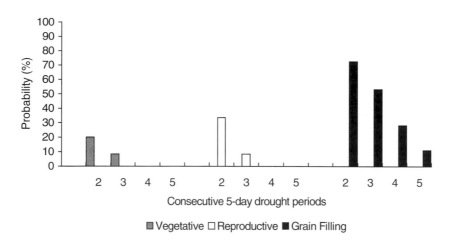

Figure 2. Probability of consecutive 5-day drought during crop field duration.

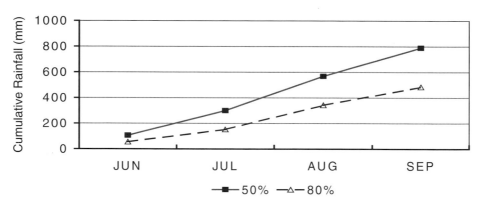

Figure 3. 5-day rainfall at 50% and 80% probabilities at Godagari, Rajshahi.

The probability of getting 400 mm of cumulative rainfall required for transplanting at different times during the beginning of the crop season is shown in Fig. 4. In an average year (50% probability), 400 mm of rainfall required for transplanting can be expected by 15 July. But, twice in ten years (80% probability), this amount may not be available before 15 August and transplanting may be delayed by one month. The delay in transplantation would lengthen the crop field duration till 15 December. Such delay would adversely affect the crop yield, as no rainfall is expected in the study area after early October (Fig. 3), to meet the crop water requirement.

From the preceding analysis it can be concluded that with the present TPR technology the study area is drought prone at two stages. Firstly, at the beginning of the crop season (which can cause delay in transplantation) and secondly, at the grain filling stage, both of which can drastically reduce the crop yield.

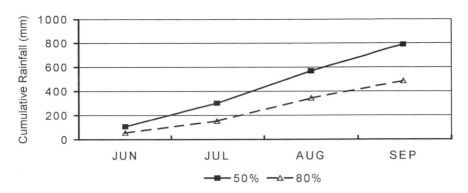

Figure 4. Cumulative rainfall at 50% and 80% probabilities at Godagari, Rajshahi.

Drought Alleviation Potentials of DSR

Hence, the strategy for drought alleviation would be earlier harvesting of the crop by avoiding the crop field duration during October. This can be done either by establishing the crop timely with shorter field duration (change of rice variety) or by establishing the crop early for avoiding the drought in October. Although varieties with shorter crop field duration are now available, they have lower yield and as such are not acceptable to the farmers. It has already been mentioned that DSR can be established early and can be very effective in drought alleviation. If dry seeded, drought at the end of the season is not expected to affect the yield because the earlier established crop would be near the harvesting stage by the time drought sets in during October. Moreover, earlier harvested dry seeded rice would leave a favorable soil-water regime for any subsequent non-rice crop.

Field experiments on early crop establishment showed that during the 1994–2000 period it was possible to establish DSR by 15 June in all the years except in 2000, when heavy rainfall (> 300 mm) washed away the seeds. As DSR establishment does not require puddling of the soil, water requirement for land preparation was much also less. The rainfall prior to seeding (May–15 June) during the 1994–2000 period is shown in Fig. 5. It is evident from the figure that DSR establishment would require about 150 mm of rainfall. Studies in the Philippines have also shown a similar requirement of about 150 mm of rainfall for crop establishment through DSR (IRRI, 1995).

Results of three-year experimental study (1994–1996) on comparative productivity of DSR and TPR are shown in Table 1. It can be seen from the table that the DSR yields are similar or slightly better than that of TPR yields for all the seeding/transplanting dates. The experiments also showed that DSR matured about 1–2 weeks earlier than TPR. The late maturity of TPR was due to transplantation shock, which is the additional time required by the uprooted seedlings to re-establish their roots and recover after transplantation.

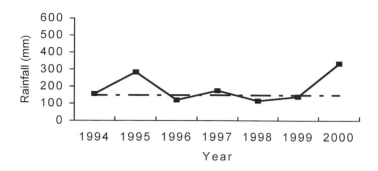

Figure 5. Cumulative rainfall prior to dry seeding (May–15 June) at Rajabari, Rajshahi.

Table 1. Effect of time of seeding (DSR) and transplanting (TPR) on grain yield (t/ha) of rice at Rajabari, Rajshahi, 1994–1996.

Method of Establishment	Time of Seeding/Transplanting					
	June 01	June 16	July 01	July 16	August 01	August 15
1994						
DSR	-	3.11	3.50	3.45	2.78	-
TPR	-	-	-	-	2.44	2.77
1995						
DSR	2.60	2.75	2.85	2.46	-	-
TPR	-	-	2.52	2.45	2.93	2.37
1996						
DSR	3.43	3.93	-	-	-	-
TPR	-	-	3.40	3.81	-	-

Rainfall probability analysis has shown that, if a cumulative rainfall of 150 mm is required for establishment of DSR, then once in two years (50% probability) crop establishment may be possible by the first week of June (Fig. 4). But, twice in ten years (80%) probability, the rainfall may be inadequate for crop establishment before the end of June. It is evident from Table 1 that even when seeded late, the DSR yields are not significantly affected by drought. This inherent drought resistance quality of DSR is probably because of deeper and wider penetration of the root system, which needs further research.

Comparative analyses of water adequacy during the crop field duration for DSR and TPR in an average and a dry year were made and the results are shown in Figs. 6 and 7. Water adequacy during each 5-day period of the crop field duration of was determined by calculating the relative water supply (RWS) using Equation 6. In an average year, when dry seeding is completed by 10 June and transplantation is completed by 15 July, the average RWS at 50% probable rainfall during the crop field duration for DSR and TPR are 0.94 and 0.79, respectively (Fig. 6). This means that, even in an average year, the water supply from rainfall is inadequate to meet the crop water requirement for TPR. Even if dry seeding is completed by 10 June and transplantation is completed by 15 July (average year), the RWS with 80% probable rainfall (dry year) during the crop field duration for DSR and TPR are 0.69 and 0.56, respectively (Fig. 6). Thus, the drought scenario for TPR becomes even worse during the dry years.

Because of inadequate rainfall at the beginning of the season, if dry seeding and transplanting are delayed (completed by June and 15 August, respectively), which may happen twice in ten years, then with 50% probable rainfall during the crop field duration, the RWS values drop to 0.78 for DSR and 0.50 for TPR (Fig. 7). With such delayed seeding and transplantation and with 80% probable rainfall during the crop field duration, the RWS values for DSR and TPR drop to 0.56 and 0.34, respectively. Thus, even during the dry years, the water availability from rainfall to meet the crop water requirement is relatively higher for DSR compared to TPR.

From the drought analyses presented in the preceding paragraphs, it is evident that DSR is better able to alleviate the drought compared to TPR. The low water requirement for crop establishment and earlier maturity are the two traits that make

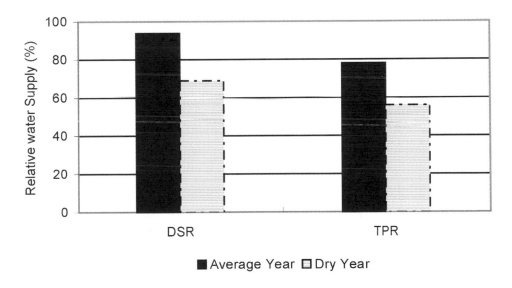

Figure 6. RWS during average and dry years with normal seeding and transplanting.

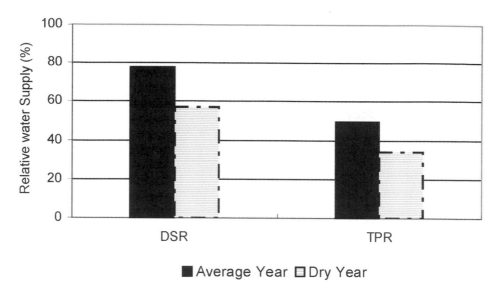

Figure 7. RWS during average and dry years with late seeding and transplanting.

the DSR technology more favorable over the traditional TPR technology for drought alleviation. Because of earlier establishment, DSR is in the field early and is more effective in utilizing the early seasonal monsoon rainfall. Because of higher water requirement for land preparation, this early seasonal rainfall is required for establishment of TPR and not for rearing of the crop. Because of earlier maturity, DSR would be harvested by October. Even though the study area is drought prone during October, because of the low crop water requirement during the two weeks prior to harvesting, the two-week drought would have very little impact on the yield of DSR.

Constraints to DSR Adoption

From the field experiments and rainfall probability analyses, it has been ascertained that the DSR technology for crop established has better potentials than the traditional TPR technology for drought alleviation. Yet, the rate of adoption of the DSR technology by the farmers during the study period has been far below the expectation. Farmers' survey revealed that the main reason behind the low adoption is the high infestation of weeds in the DSR farms compared to TPR farms. Because of puddling during land preparation, the suppression of weeds is more effective with the TPR technology. Moreover, during the vegetative growth stage, standing water (which is very effective in weed suppression) is maintained for a longer duration in TPR farms compared to DSR farms.

Although herbicides for weed management are now available, the technology is new to the farmers and is also expensive. Moreover, there are also concerns about the possible contamination of groundwater by herbicides. Therefore, although the DSR technology is expected to have definite positive impact in drought alleviation over TPR technology, the socio-economic constraints to wider adoption of the DSR technology need further research.

Conclusions

Experimental studies in rainfed systems of Northwest Bangladesh have shown that the traditional technology of transplanted rice is drought prone during transplantation and at the grain filling stage. Water available from rainfall in an average year is unable to meet the crop water requirement of transplanted rice. Drought alleviation is possible through dry seeding, which can be established early and hence can also be harvested early. Although the production trials did not show a significant difference between the yields of dry seeded and transplanted rice, the scenario can change in a dry year when transplantation would be late and yield would be affected by late season drought. From the rainfall probability analyses, it has been ascertained that, both during average and dry years, the dry seeded rice has much better potential for drought alleviation compared to transplanted rice. But, yet the adoption of the technology by the farmers has been far from expectation. The socio-economic constraints limiting the adoption of the technology by the farmers need further research.

References

Bhuiyan, S.I. and Zeigler, R.S (1994). "On-farm rainwater storage and conservation system for drought alleviation: Issues and challenges". *On-farm Reservoir Systems for Rainfed Ricelands.* S.I. Bhuiyan, ed., Int. Rice Res. Inst., Manila, Philippiners.

Garitty, D.P., Oldeman, L.R., Morris, R.A. and Lenka, D. (1986). "Rainfed lowland rice ecosystems: Characterization and distribution". *Progress in Rainfed Lowland Rice,* Int. Rice Res. Inst., Manila, Philippines.

IRRI (1993). *IRRI rice facts.* Int. Rice Res. Inst., Manila, Philippines.

IRRI (1995). *Program Report for 1994.* Int. Rice Res. Inst., Manila, Philippines.

Islam, M.T., Saleh, A.F.M. and Bhuiyan, S.I. (1998). "Agro-hydrologic and economic analyses of on-farm reservoirs for drought alleviation in rainfed ricelands of Northwest Bangladesh". *Rural and Env. Engrg.,* No.35, 15-26.

Lantican, M.A., Lampayan, R.M., Bhuiyan, S.I. and Yadav, M.K. (1999). "Determinants of improving productivity of dry seeded rice in rainfed lowlands". *Experimental Agriculture,* Cambridge University Press, 35, 127-140.

Levine, G. (1982). "Relative water supply: an explanatory variable for irrigation systems". *Determinants of Developing Country Irrigation Project Problems,* Technical Rep. No.6, Cornell University, New York.

Saleh, A.F.M. and Bhuiyan, S.I. (1995). "Crop and rain water management strategies for increasing productivity of rainfed lowland rice systems". *Agricultural Systems*, Elsevier Science Limited, 49, 259-276.

Saleh, A.F.M. and Bhuiyan, S.I. (1996). "Analysis of agricultural drought in rainfed lowland rice system: A case study in North-west Bangladesh". *J. Indian Water Resources Soc.*, 2(1), 51-56.

Steenhuis, T.M. and Van der Mollen, W.H. (1986). "The Thornwaite-Mather procedure as a simple engineering method to predict recharge". *J. of Hydro.*, 84, 221-229.

Tomar, V.S. and O'Toole, J.C. (1980). "Water use in lowland rice cultivation in Asia: a review of evapotranspiration". *Ag. Water Mgmt.*, 3, 83-106.

Surface Water Hydrology, Singh, Al-Rashed & Sherif (eds)
© *2002 Swets & Zeitlinger, Lisse, ISBN 90 5809 363 8*

Sand dams: Source of water in arid and semi-arid lands of Kenya

Joel K. Kibiiy
Department of Land Reclamation
P.O. Box 7070, Eldoret, Kenya

Tribeni C. Sharma
Department of Agricultural Engineering, University of Nairobi
P.O. Box 30197, Nairobi, Kenya
e-mail: tcsharma@clubinternetk.com

Abstract

Rural people in arid and semi-arid lands (ASALs) in Kenya collect their water from small non-perennial streams when they carry water in the wet season and from holes dug in the sand in the shallow riverbed during the dry periods. To ensure sustainable supply of water during the entire season, a weir called SAND DAM is constructed by erecting a concrete/stone masonry wall on a natural rock outcrop across a riverbed at a suitable site. The dam is constructed with the assumption that there will be sufficient amount of sediment eroded from the watershed, to be trapped by the dam. Water will be impounded in the pores of the accumulated sediments for use during the dry season.

This paper presents an assessment carried out in 1999 of the water resources available in some existing sand dams in the West Pokot district and their potential adequacy to match the demand. The amount of drainable water in the sand dams ranged from just over 44 m^3 to over 5500 m^3. The highest specific storage (defined as the volume of drainable water per unit area of dammed channel cross section, m^3/m^2) in sand dams is apparently obtained from river slopes between 1 and 2%. Other channel characteristics varied greatly from one river channel to another and even along a specific channel. Average channel – side slopes of 2:1 and 3:1 were adopted. The lowest dam had a height of 0.5 m while the highest was 3 m. The reach lengths of the reservoirs ranged from 40 to 500 m while channel widths ranged from 6.5 to over 50 m.

The paper also highlights the acute shortage of hydro-geological data that form the basis for planning and development of the sand dams and points to the urgent need to install a satisfactory gauging network in ASALs for monitoring rainfall and runoff. In particular, bedload transport is crucial in relation to the design of sand dams and therefore requires exhaustive research in the arid and semiarid regions.

Introduction

Over 80% of the total land area of Kenya is occupied by arid and semiarid lands (ASALs) characterised by an annual rainfall/potential evaporation ratio between 25 and 50% (Government of Kenya (GoK), 1992). These lands support about 20% of the country's population and approximately 50% of the livestock. A major proportion of the country's wildlife, the backbone of the nations tourist industry, also lies in ASALs.

Scarcity of water in the ASALs is perceived as the foremost constraint to development. Perennial water sources such as rivers and lakes are few and far apart. Most rivers are ephemeral seasonal sandy bed streams running only for short periods after rains. These streams are full of sand beds, which can be exploited for abstracting the water for meeting needs. One source of water supply in the ASALs has been recognised in the form of sand dams (Mburu, 1989; Waweru and Ngigi, 1999; Senete, 1999). A number of sand dams have been constructed to good effect in Kitui, Machakos and West Pokot districts in Kenya (Fig. 1). They store sufficient quantities of water for livestock and domestic use and the cost per unit of water stored is much less than for a rainwater tank (Thomas, 1999). However, the quality of water can be poor unless there is an adjacent well from which filtered water can be drawn, and the burden of transporting the water can be heavy.

Traditionally, water resources development in the ASALs of Kenya has centred on the construction of ground water and temporary surface water sources – in the form of pans and small earth dams. Ground water resources are unevenly distrib-

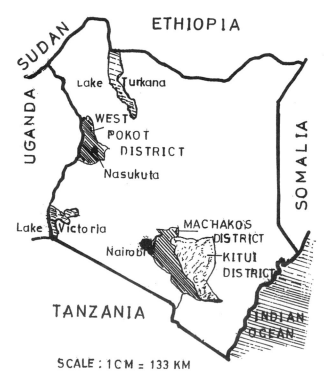

Figure 1. Location of West Pokot, Kitui and Machakos districts in Kenya.

644

uted and yields are low due to the Precambrian basement system of rocks that underlies large parts of the country (Republic of Kenya, 1992). Maintenance of borehole pumps and equipment has been difficult making them unsustainable in the long run. Soil erosion studies in the ASALs (Ministry of Land Reclamation, Regional and Water Development (MLRRWD), 1994b) have shown high net soil losses of up to 3740 tons/km^2/year. The high rates of soil loss are an indication of land degradation caused by overgrazing and repeated droughts that reduce vegetative ground cover. Charcoal burning being done to meet fuel demand at the urban centres contributes to the problem. Strong torrential rains and the resultant high overland flow together with high erodible soils in some areas combine to reduce the useful lives of pans and earth dams through siltation. Furthermore, siltation enhances the already high evaporation losses from the pans due to the diminishing depth/surface area ratios. This discourages further development of these surface water structures.

During dry periods, people in ASALs can be seen digging for water at certain points on the sand beds of the seasonal river channels. These water points are usually at the upstream sides of rock ledges cutting across the channels and thus preventing the downstream flow of the water contained in the sand. Sand dams are an enhancement of these natural aquifers. A concrete/masonry weir is built on the rock bars across the river channels so that it can trap and hold back the sand brought by the river during floods. The trapped sand will store water in its pores. This system is what is commonly called a sand dam. This is not to be confused with as sub-surface dam which could be made of clay or other material constructed below ground level in a sandy river or low lying area to prevent the flow of subsurface water and maintain or raise the water table. In contrast, sand dams are concrete/masonry weirs, which like retaining walls, have their upstream faces covered by sand while the downstream faces are fully exposed (Fig. 2). At either end of the dam, where the valley sides are flat, wing walls may be added at an angle to the main dam for the same purpose. A well constructed sand dam requires little or no maintenance. However, it is necessary to check after floods and repair any damage that is found.the wall is raised to prevent the river cutting round during a flood. Sand dams are thus a form of water harvesting structures in which flood water is stored in an artificially created perched sand aquifer. Abstracting water from sand dams can be done using scoop holes, but to improve the quality, water is best taken from a shallow well. Outlet pipes passing through the dam wall are sometimes used for drawing water from sand dams, but it is generally observed that tap is left open or gets broken and the water is lost.

Sand dams are not unique to Kenya. The literature indicates that they have been used in one form or another in a number of countries in various parts of the world including the USA (Sivils and Brock, 1981), India, Ethoipia, Namibia and Thailand (Wolde, 1986).

The first series of sand dams in West Pokot district were constructed in the pre-independence, colonial period. There are no records, but judging from observation, the few colonial era sand dams seem to have been built only as part of road works across the sandy riverbeds. There was no further construction of sand dams until

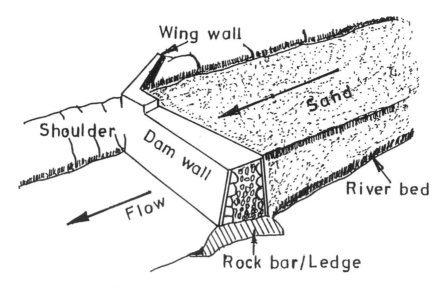

Figure 2. Sectional view of a sand dam.

the 1980s. Even then, by 1990 there were only nine sand dams in the district. Most of the existing sand dams in West Pokot district were constructed in the last decade. There are now about twenty five of them. In Kitui district, the earliest sand dams were also constructed during the colonial period and most of them are still in existence (Thomas, 1999). By 1999 there were over 30 sand dams in Kitui.

The idea behind sand dams is not new and the principle of the method is, seemingly, easy to understand. Nevertheless, most of the sand dams now in existence are huge structures that are clearly indicative of engineering over design. The design manual for water Development in Kenya (MWD, 1986) does not give any guidance on design procedure for sand dams. Some workers (e.g. Wolde, 1986) have treated sand dams as water-logged retaining walls. Consequently load analysis has been directed only to attaining stability against sliding and overturning. A sand dam, however, is a river structure that also passes floods of various magnitudes from time to time.

This fact is apparently recognised by builders of dams who attempt to cater for the unknown by investing in size resulting in over design. Although Sand dams are increasingly coming into use in the ASALs of Kenya, they still fall under the realm of unconventional water supply methods that is only slowly finding recognition by the Water Department – the lead water development agency in the country. Practically all the sand dams in Kenya have been constructed by non-governmental organisations or donor funded quasi-governmental development agencies. Design criteria and construction methods have not been established and standardised. The difference between sand dams in West Pokot and those in Kitui gives an ample demonstration of the difference in approach. In West Pokot sand dams are built gradually, one step on top of another each dry season until the desired height is attained. The aquifer is thus built up in stages by the sediments that accumulate in the subsequent wet seasons. In Kitui, however, the entire wall is raised at a go during a dry season ready to trap sediments and build its aquifer in the subsequent wet season. The different approaches result in qualitatively different aquifers.

The natural sorting and deposition of sediments in streams is a function of channel slope and so is the shape of channel cross section. All these in turn affect the storage capacities of sand dams. The apparent importance of channel slope in the siting of sand dams was noted by Hanson and Nilsson (1986) who state that an optimum composition of riverbed material is generally found in transition zones between mountains and plains. Around that time most of the sand dams in existence in Kenya had been constructed on slopes of 1–5% but there were some on slopes of 10–15%. It was thus necessary to investigate the relationship between channel slope and storage of sand dams.

The Study Area

The sand dams studied are in the ASALs of West Pokot district in Kenya. The district is located in the north western corner of the country over 400 km from Nairobi. Its western border is marked by the international boundary with the Republic of Uganda (Fig.1).

The district may be divided into two agro-ecological zones. The highland region covering the southern part of the district rising to an altitude of about 3000 m is characterised by a cool, humid tropical climate with high rainfall. The area to the north of these highlands constitutes the lowland ASAL plains with altitudes up to 900 m with characteristically high ambient temperatures, relatively low rainfalls and high evaporation rates. The area is inhabited by a semi nomadic pastoralist people (known as the Pokot), whose continuous movement in search of water and pasture often results in conflicts with people in the neighbouring districts and across the border in Uganda. The vegetation is dominated by bushes and shrubs of the *acacia* species and a variety of grasses.

Methods

Nine different sand dams in West Pokot district were visited and various measurements taken for the assessment of their water storage capacities. Experiments were conducted to determine the porosities, specific yields (available water) and the hydraulic conductivity of the residual sediments (sand) in the sand dam at Nasukuta. A textural analysis of the sand was also carried out.

Measurements were taken of dam heights and their widths across the rivers. Measurements were also taken for determination of channel slopes. The measurements consisted in taking the elevations of channel surface at regular intervals using an engineers level. The width of the dam across the channel was also measured. The widths of the channel reach forming the reservoir were measured using steel tape at regular intervals coinciding with the levelling stations. The surface of the water table was similarly established after first scooping away the sand. The heights of the dams up to the spillways were determined using an engineers level.

The extent of the reservoir upstream of the dams was determined from both visual observation at the site and from the longitudinal profiles of the water table and sand surface. The combined profile for Losam sand dam is given in Fig. 3. From the profile, it is seen that the water table slopes, follows the sand surface quite closely and emerges to the surface at one point (along a bend). A notable departure of the water surface profile from the sand surface is observed starting at distance 200 m where it begins to deepen through 360 m. From the shape of the water table profile and the fact that a rock ledge across the channel was encountered at distance 360 m, it is a good indication that the sand dam reservoir ended at this point.

Reservoir volume was calculated for this stretch of channel using Simpson's rule (Bannister et al., 1992). For this purpose it was assumed that the channel was prismatic and of trapezoidal cross section. The angles of channel sides were estimated from measurements and observations made at the sites. The slopes of channel sides were observed to vary greatly from one river channel to another and even along specific channels. They range from steep, near vertical sides in some places to open flat banks in others. Average values of 2:1 and 3:1 were invariably adopted for the calculation of the sectional areas of the sand dams depending on observations made at site. The reach lengths of channel forming the sand dam reservoir varied from 40 m to 500 m while the channel widths ranged from 6.5 to 25 m. The lowest dam had a height of 0.5 m while the highest was 3 m. Available water was calculated based on the specific yield determined for the material forming the reservoir at Nasukuta.

In order to assess the impact of channel slope on the storage capacities of the sand dams of different heights on different sizes of streams, use was made of the

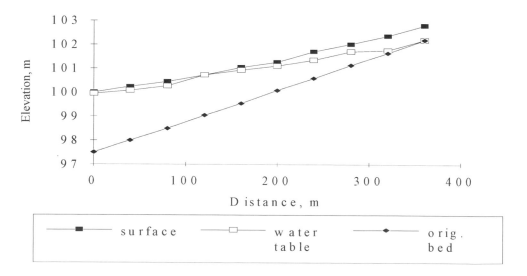

Figure 3. Longitudinal profile of Losam sand dam showing the original bed, the water table and the sand surface.

concept of specific storage – defined as the volume of available water per unit area of dammed channel cross section (m^3/m^2).

Results and Discussion

Site Selection and Dam Design

Note was taken of the siting of the dams with respect to the material forming the foundations and the banks. All the sand dams were found to be built at points with a rock outcrop across the river channel. The sites are presumed to provide for firmer foundations for the dams while at the same time minimising chances of seepage losses. However, geologic conditions do not seem to receive any further attention. The rock outcrops are more often sedimentary formations containing a number of individual sedimentary layers. The degree of weathering of these ex-posed rocks as well as their angle of dip relative to the direction of flow are impor-tant aspects that call for further work.

Capacities of Sand Dams

The volumes of available water in the nine sand dams ranged from 44 m^3 to 5522 m^3. These are shown in column 2 of Table 1. These were, however, calculated based on the charcteristics of residual sediments at the Nasukuta sand dam. The sediments had a median particle diameter of 0.64 mm, mean porosity of 43.7%, a specific yield of 23.3% and hydraulic conductivity of 28 m per day.

Table 1. Channel slope, available water and specific storage of some sand dams in West Pokot.

Sand Dam	Original Channel Slope (%)	Available Water (m^3)	Specific Storage (m^3/m^2)
Priro-lower	0.1	147	13
Kases	0.7	105	12
Chepkobegh	1.1	5522	44
Loうam	1.3	1421	52
Chemulunje	1.4	1118	37
Nasukuta	1.6	1095	44
Samor	2.2	1887	20
Kasaghon	4.9	65	2
Priro upper	11.8	1044	5

Effect of Channel Slope on Storage

Natural channel slopes at the sites of the nine dams ranged from 0.1 to 11.8%. Contrary to expectation the water table in the sand dams was observed not to be horizontal but sloping towards the dam approximately parallel to the channel sand surface. The reason for this is not clear. But it is suspected that this is the transi-tional water table constantly adjusting itself as the water is lost through leakage and drawn out for use. The effect of unequal evaporation losses along the length of the reservoir surface cannot be discounted.

The specific storage (amount of water stored per unit cross sectional area of dammed channel) was found to vary with slope. The specific storage seemed to increase with channel slope at the lower slopes (< 2%), while it seemed to decrease at higher slopes (> 2%). The relationship, depicted in graph (Fig. 4) indicates that sites on channel slopes between 1 and 2% would be expected to give the highest specific storage. Based on this limited data (n = 9), it seems that outside this range of channel slopes, specific storage decreases.

Whereas 2% may be taken as the limiting slope beyond which the construction of sand dams may be considered with reservation, the 1% slope may be viewed as the delineating slope below which the construction of sand dams ought to be replaced by subsurface dams.

Strategic Value of Sand Dams

Sand dams are small storage reservoirs of great strategic value. Their strategic significance lies in their resourcefulness during the dry season, a fact that is only fully appreciated upon a site visit during the height of a dry season. Sand dams in West Pokot are not generally resorted to until all other convenient surface water sources have been exhausted. Another important aspect of sand dams is that they, unlike the pan reservoirs, are recharged to capacity by the first flood on the river.

These unique aspects of sand dams were observed at Nasukuta in 1999. Nasukuta Farm is a government livestock farm provided with water from a pan and a sand dam located only a few dozen metres apart. The year 1999 was a dry year in West Pokot and Kenya in general. When the onset of the rainy season in April delayed and the drought persisted, the pan dried up from sustained exploitation and evaporation. The sand dam was then resorted to and exploited to near exhaustion. The rains eventually came in July, but they were below normal. In spite of this, the

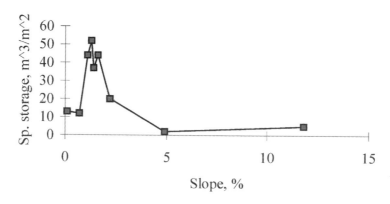

Figure 4. Relationship between channel slope and specific storage in sand dams.

sand dam was fully recharged by the first flood on the river. The pan, however, failed to make an effective recovery until further rains in September and October. Additional water collecting after a storm was subsequently reduced by evaporation and livestock use. This pan remained partially dry for the rest of the year.

Sand dams are thus valuable as standby arrangements and supplement other water supply systems that may be provided in the ASALs.

Some Aspects of Water Supply and Demand on Sand Dams

The Pokot are a pastoral people who keep large herds of cattle, sheep, goats and in some cases, camels. Livestock watering is the major demand on sand dams. Water demand for cooking and other domestic requirements are comparatively small albeit of great importance.

The Water Department (Ministry of Water Development (MWD), 1986) used the concept of livestock unit (LU) to estimate the water requirements of various domestic animals. One LU is equivalent to three indigenous cattle and is assigned 50 litres of water per day. This means that in West Pokot the water demand is about 17 litres per head of cattle per day.

While water requirements for livestock are thus easily estimated from knowledge of animal species and climatic conditions the per capita, human requirements of water varies greatly depending on culture and life styles of a community. Gleck (1996) puts the basic daily water requirement for human survival at 5 litres per head per day (5 l/h/d). The Water Department (MLRRWD, 1994a) gives a figure of 10 l/h/d for West Pokot. But it is worth noting that these rates are not static. Water consumption rates rise with time as peoples' knowledge and standard of living rise. Detailed studies would be required for an accurate estimate for any specific community. A consumption rate of 20 l/h/d would be reasonable estimate for planning purposes in West Pokot.

Distances to the water points also contribute to the water consumption rates. According to the Water Department (MLRRWD, 1994a) livestock do no graze over a distance of more than 10 km in a day. This means that water supplies for livestock should ideally be located within a 5 km radius of establish kraals. This also means that water supplies for livestock should be spaced at most 10 km apart. For domestic water supply the Department (MWD, 1986) stipulates that water points should be so distributed that the maximum walking distance for 90% of the consumers will be about 1.5 km for ASALs such as West Pokot. In the face of harsh realities where the hope of meeting the stated distance targets is receding, this position has been watered down so that the present policy guideline on distances to water points is "a reasonable distance".

Sand dams store reasonable amounts of water and serve both people and livestock. In view of the need to reduce distances to water points and increase the amount of available water, it is necessary to develop more water points in the ASALs, possibly by constructing more sand dams.

Data Requirements for Design of Sand Dams

The construction of any height of sand dam is contingent on the accumation of good quality coarse sand. Knowledge of the rainfall-runoff characteristics of the catchment and sediment transport regime of the river is a prerequisite.

Unfortunately, there is a dearth of hydrological data not only in the study area but in the ASALs of Kenya in general. It may be consoling that river flow records can be stretched statistically by correlating them with longer rainfall series based on studies of rainfall and runoff and used for the estimation of sediment yields. However, studies in rainfall-runoff relationships in West Pokot are few and have been based on experimental runoff plots (e.g., Katana, 1998; Cheruiyot, 2001). Even rainfall data are scarce and characterised by discontinuities.

Channel geometry at the prospective dam sites and in the channel reach immediately upstream are important aspects to be considered along with the final height of the dam and the shape of the spillway. After the construction of a sand dam a new channel cross section is created together with a new gentler channel slope immediately upstream of the dam. The modified channel must safely pass the highest expected flood without overflowing the banks and threatening the dam abutments. Thus, in addition to rock outcrops for firm foundations, high banks are another desirable feature for sand dam sites. Sites with a combination of these features are not always easily found. Where the riverbanks are low, the dam has to be raised on either or both sides and wing walls extended beyond the banks in order to direct flood water and prevent it from cutting around the dam. The importance of hydrological data in this respect cannot be over emphasized. Simulated river flow series can be used to estimate size of flows likely to occur over a specified period. Yet river flow records are not available for any of the ephemeral streams in the area. There is a real need to monitor rainfall and to intall river gauging structures in these areas. These would serve as a basis for further work including studies on sediment transport.

References

Bannister, A. and Raymond, S. 1984. Surveying. 5th ed. English Language Book Society/Longman, Essex, England. 510 p.

Cheruiyot, H.K. 2001. Green-Ampt based modelling of infiltration using micro-catchments in an ASAL soil. Msc thesis, Department of Civil and Structural Engineering, Moi University, Eldoret, Kenya.

Gleck, M.P. 1996. Basic water requirements for human activities: Meeting basic needs. *Water International*, 21(2), 83–92.

Government of Kenya 1992. Development Policy for Arid and Semi arid Lands (ASAL). Government Press, Nairobi, Kenya.

Hanson, G. and Nilsson, A. 1986. Ground water dams for rural water supplies in developing countries, *Ground Water*, 24(4), 497–506.

Katana, S. 1998. A hydrological study of water harvesting micro catchments in a semiarid area. MSc thesis, School of Environmental Studies, Moi University, Eldoret, Kenya.

Mburu, D.M. 1989. The Role of Sand Dams in Water Supply in Arid Areas (Machakos District - Kenya). MSc thesis, University of Nairobi, Kenya.

Ministry of Land Reclamation, Regional and Water Development (MLRRWD) 1994a. District Water Development Plan, West Pokot. Water Resources Assessment Planning Project (WRAP) Phase IV, Nairobi, Kenya.

Ministry of Land Reclamation, Regional and Water. Development (MLRRWD) 1994b. Soil erossion studies in West Pokot District. Water Resources Assessment and Planning Project, Nairobi, Kenya.

Ministry of Water Development (MWD) 1986. Design Manual for Water Supply in Kenya. Government Press, Nairobi, Kenya.

Republic of Kenya 1992. Guidelines for the design constructiona and rehabilitation of small dams and pans in Kenya. Kenya–Belgium Water Development Programme, Ministry of water development Nairobi, Kenya.

Senete, J.P. 1999. Availability and utilization of water from sand dams in West Pokot District, Kenya. MSc thesis, School of Environmental Studies, Moi University, Eldoret, Kenya.

Thomas, D.B. 1999. Where there is no water. A story of community water development and sand dams in Kitui District, Kenya. SASOL and Maji na Ufanisi, Nairobi, Kenya.

Sivils, B.E. and Brock, J.H. 1981. Sand dams as a feasible water development for arid regions. *Journal of Range Management,* 34(3), 238–239.

Waweru, D.W. and Ngigi, S.N. (eds.) 1999. Sand dams and subsurface dams. Proceedings of a workshop held in Machakos Kenya, Rainwater Harvesting Association, Nairobi, Kenya.

Wolde, G. 1986. Subsurface flow dams for rural water supply in arid and semi-arid regions of developing countries. MSC thesis, Tampere University of Technology, Department of Civil Engineering, Water Supply and Sanitation, Tampere, Finland.

Surface Water Hydrology, Singh, Al-Rashed & Sherif (eds)
© 2002 Swets & Zeitlinger, Lisse, ISBN 90 5809 363 8

Some hydrological investigations
toward the design of sand dams in Kenya

Joel K. Kibiiy
Department of Land Reclamation
P.O. Box 7070, Eldoret, Kenya

Tribeni C. Sharma
Department of Agricultural Engineering, University of Nairobi
P.O. Box 30197, Nairobi, Kenya
e-mail: tcsharma@clubinternetk.com

Abstract

Sand dams are unconventional, low technology option for water supply used in arid and semiarid lands (ASALs) in Kenya. The main feature of a sand dam is a concrete weir constructed across an ephemeral stream.

Design approaches vary from place to place with equally varied results. However, it is apparent that storage capacities of sand dams are dependent on hydrologic and topographical factors of the catchment.

This paper outlines the results of a field study carried out in 1997–1999 in West Pokot district of Kenya to investigate the role of hydrological factors in the design of sand dams. Rainfall and stage of flow at an ephemeral stream in a typical ASAL area were monitored. Stream discharge ranged from zero to 29 m^3/s. Suspended sediment concentrations were found to be as high as 24200 mg/l and consisted of around 30% clay. A maximum bedload flux of 2.02 kg/m/s was observed at a modest discharge of 2.55 m^3/s. Gradual construction of the weir in small step heights was shown to result in the deposition of sand that has a better grading than would be attained in single stage construction.

Introduction

Most rivers in the ASALs of Kenya are ephemeral sandy bed channels that are dry for most part of the year. At some points these rivers cross resistant rock outcrops. These rock outcrops form natural controls determining the slopes of riverbed upstream. Large quantities of sand are deposited on the channel upstream of the outcrops. If they are unfructured the outcrops also form impermeable barriers preventing the downstream subsurface flow of water in the sand bed. It is at such points along the dry rivers that people dig for water during dry periods. In an effort to increase the amount of water at such points, concrete or masonry walls (dams) are built atop the rock outcrops to raise the height of the impermeable barrier. The constructions result in more sediment deposits upstream of the dams and more water, where the work is successful. The system consisting of the constructed wall and the water holding sediment reservoir created is what is referred to as a SAND

DAM. A sand dam is, therefore, an artificially created perched aquifer on a dry sandy bed river channel.

One of the approaches aimed at achieving improved sediment characteristics to attain higher water storage capacities in these aquifers has been to construct the dams in stages: a small rise of the dam every season instead of raising it to full height at once. Among other things, the present study was set to experimentally investigate the effect of stepped construction on the quality of sediment deposits at an existing sand dam at Nasukuta in West Pokot district.

The main challenge in the design of sand dams is the determination of the size of dam step height: it should not be so high as to act as a silt trap, yet it should be sufficiently high to create a reservoir capable of holding all coarse sand sediment yield of the designated period.

Moving water is the prime factor in the sedimentation process, being the causative and transporting agent. The quantification of runoff in ephemeral ASAL streams is thus a precondition for the estimation of sediment yields necessary in sand dam design. The starting point, therefore, is the number of runoff events expected at the outlet of the catchment per season. In the absence of river flow records, this may be conveniently equated to the number of days with runoff producing events. This latter statistic may be deduced from daily rainfall data that is usually more readily available. The next step is to determine the magnitude of the runoff in the events. In the absence of records, they too are represented by the depths of the causative agent – the rainfall. Sediment yields may then be predicted by coupling stochastic hydrological models to deterministic models of sediment transport.

Study Area

Location

The study was centred at Nasukuta sand dam located on an ephemeral stream in West Pokot District of Kenya (Fig. 1).

Relief, Vegetation and Land Use

The Nasukuta dam is located at the outlet of an upland catchment with an area of about 8 km^2 and an elevation ranging from 1620 to 1960 m.

The vegetation is characteristic of the ASALs of Kenya. Trees and shrubs of the *acacia* species (*Hockii, Brevispica, Nilotica, Senegal, Tortilis* and *Etbaica*) are predominant. Several other tree species are found in the catchment. Towering high *Euphorbia obovolifolia* intersperse the catchment as is the *Terminalia brownii* which offers a magnificent sight when in bloom with its dark brown pods. Others are *Balanites aegypca, Pappea capensis* and *Fagara chalybea*. Grasses of different types grow among the shrubs and in the open land. There are over twelve grass species but a few are prevalent: *Cymbopogon sp., Erogrotis superba, Erogrotis ciliansis, Andropogon sp.* and *Themeda triandra*.

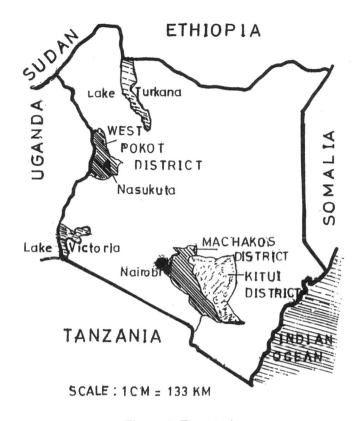

Figure 1. The study area.

Rainfall and Climate

There are no weather data specific to the catchment. The only reliable data is from Kacheliba weather station (elevation 1280 m) located about 30 km to the north west of Nasukuta. The area receives about 880 mm of annual rainfall and has a mean pan evaporation of about 2390 mm per annum. The rains start in April and end in August, with a peak in July. The months of September to March are characterised by low rainfall and generally form the dry season.

Soils and Geology

The soils in the Nasukuta catchment range from sandy loam on the surface (0-70 mm) and sandy clay in deeper (> 110 mm) horizons (Ministry of Land Reclamation, Regional and Water Development (MLRRWD), 1994). The District is underlain by Precambrian basement system of rocks which cover about 75% of the District (Ministry of Planning and National Development (MPND), 1985). At Nasukuta the geology is predominated by metamorphic sediments of quartz–muscovite gneisses (MLRRWD, 1994).

Channel Geometry and Flow Regime

The dam at Nasukuta forms a section control at which sediments have been deposited upto spillway crest level. The dam is 2.5 m high. The stream upstream of the dam has an active channel width of about 11.5 m and a whole-channel width of about 15.5 m. The side slopes of the channel varied. An average side slope of 3:1 was adopted. The channel section at the gauging station is characterised by a steep left bank and an open gentle slope on the right bank. The channel has a slope of 0.97% over a 400 m reach upstream of the dam. A resistant rock ledge at this point forms a natural control that determines the geometry of the channel at a higher elevation upstream. The bed material in this channel reach is poorly graded clean quartz sand of median diameter of 0.64 mm. The river is normally dry and only suddenly comes to life following storms that occur mostly in the afternoons. The flows are characterised by rapid changes in stage both in the rising and falling stages of flow.

Methods

Measurement of Rainfall

Rainfall was measured using an autographic raingauge and five standard rain-gauges distributed over the catchment.

Flow Gauging

A staff gauge was installed on the left bank of the channel about 23 m upstream of the dam. Flow stage was observed and recorded by a person on the right bank positioned directly opposite the staff gauge. A water level recorder was installed on the right bank. Attempts at discharge rating measurements using a mini Ott current meter were made during the floods of 22/5/98 and 27/5/98. Observations made during these two attempts clearly demonstrated that current metering was inappropriate for the stream because of the fast rate of change in stage. On 22/5/98 current metering took 35 minutes while it took only 4 minutes for the flood to peak. On 27/5/98 the exercise took 38 minutes while it took only 2 minutes for the flood to peak. During these attempts, high turbidity rendered the propeller invisible once lowered below the water surface. Floating debris, especially leaves, repeatedly stuck to the propeller blades occasioning stoppages and inevitable repeats. During high stages the sand bed became mobile and the flows were fearfully turbulent and occasionally had in transport the remains of trees and other vegetation. This made it hazardous to attempt wading for current metering.

Because of the above mentioned reasons the velocities of flow were estimated using the float gauging method (Herschy, 1985). Low density packaging material was dropped into the flow and the time taken by the floats to cover a straight stretch of channel reach were measured using a stopwatch. Notwithstanding the problems cited above, the current meter was used to measure surface velocities at various low flow stages. Corresponding surface float velocities were measured.

Sediment Sampling

Grab sampling (Sharma and Gikonyo, 1999; Makhoalibe, 1984) was used to collect water samples for the determination of suspended sediment loads. The samples were then transferred into 1 litre plastic containers and labelled ready for laboratory analysis. The samples were subjected to standard concentration analysis by evaporation (Okalebo et al., 1993). A suspended sediment rating curve was then developed. Textural analysis for the composition of the suspended sediment was carried out using the hydrometer method (Okalebo et al., 1993).

Bed load samples were collected using a variation of the movable pit sampler (Kinori and Mevorach, 1984). An open mouthed container was lowered and instantaneously placed against the downstream edge of the weir at low stages of flow. This acted as a trap into which all the water and its sediment load fell. It was lifted off immediately it filled and the sample removed for analysis. Bedload transport rates at various stages of flow were computed from the weights of dried bedload samples. Total observed sediment transport rates were obtained by adding the suspended load component to the bedload at the corresponding stages of flow.

Analysis of Hydrological Data and Modelling for Design of Sand Dams

Sand dams are built in the dry season in order to avoid possible damage or interference by floods and in order that they may be ready to harvest sediments and water in the subsequent wet season. An estimate of the magnitudes of expected floods and sediment yields may be made through modelling based on availability of relevant data.

Collection of daily rainfall data at Kacheliba seems to have started in 1957. However, daily records for the period up to 1965 would not be traced; only monthly summaries were available. Data for the years 1979–1981 were missing from record. Data for six other years had significant lacuna and was discarded. There was finally only 25 years of data for analysis.

Statistical models were fitted to historical rainfall data from Kacheliba. The binomial approximation to the normal was used for the number of runoff producing storms (R > 10 mm) per season. The depths of rainfall per event were fitted to the exponential distribution function. The SCS Curve Number (CN) method (Chow et al., 1988) was applied for the prediction of runoff volumes from observed rainfall data while the SCS dimensionless hydrograph method of unit hydrograph synthesis (Singh, 1988) and Bernard's distribution graph (DG) were used for the time distribution of flow, i.e., to generate runoff hydrographs. Observed sediment data was fitted to the sediment transport models of Engelund and Hansen (1967) and of Young and Stall (1976). The former model was selected and was then superimposed onto the hydrographs to predict sediment yields for individual events. The results of these computations were compared to observed data. Subsequently, the total sediment yield over any chosen period can be computed as the sum of the yields of individual events at any desired probability level in the period.

Stepped Construction

The effect of stepped construction on the potential of sand dams was studied by evaluating the textural composition of residual sediments after the erection of additional steps on the dam at Nasukuta. Three step heights of 0.27 m each were constructed one after the other on top of the existing dam. The volumes of sediment deposited behind the dam after every flow event were estimated from engineering survey measurements. Each new step was built as soon as the preceding one was filled flush with sediments. The texture of the sediments corresponding to each step was determined from a sieve analysis of samples collected before raising the new step.

Results and Discussion

Rainfall Characteristics

Rains in the study area fall in concentrated, short duration storms occurring mainly in the afternoons. Triangular rainfall hyetographs (Singh, 1988) developed from autographic raingauge data are given in Table 1. These storms give a mean maximum intensity I = 48.7 mm/hr with peak time T_p = 31 minutes, a recession time T_r = 87 minutes and a mean storm advancement coefficient (Chow et al., 1988), r_a = 0.30.

Table 1. Event rainfall characteristics for Nasukuta.

Date	I, mm/hr	T_p, min	T_r, min	T, min	$r_a = T_p/T$
16/5/98	64.8	7.5	232.5	240	0.03
23/5/98	12	30	90	120	0.25
10/6/98	36.4	15	105	120	0.13
1/7/98	87.2	22.5	67.5	90	0.25
20/7/98	28.4	60	30	90	0.67
20/7/98	24	22.5	15	37.5	0.60
2/7/98	144	7.5	22.5	30	0.25
25/7/98	32	30	217.5	247.5	0.12
26/7/98	32.8	135	60	195	0.69
1/8/98	72.2	45	22.5	67.5	0.67
10/8/98	64.8	7.5	22.5	67.5	0.67
15/8/98	41.6	22.5	52.5	75	0.30
16/8/98	80	22.5	105	127.5	0.18
16/8/98	41.6	15	7.5	22.5	0.67
26/8/98	66.4	15	45	60	0.25
17/9/98	42.4	37.5	52.5	90	0.42
4/10/98	47.2	22.5	90	112.5	0.20
10/10/98	30.4	22.5	157.5	180	0.13
12/11/98	29.6	75	180	255	0.29
12/7/99	12	22.5	165	187.5	0.12
15/7/99	32	7.5	82.5	90	0.08
Average	48.7	30.7	86.8	117.5	0.3

Although the study catchment is small, a marked variability in the areal distribution of rainfall was observed. This may be associated with the orographic effect of the mountain range bordering the catchment. But uneven areal distribution of rainfall in ASAL is not uncommon as evidenced by observations in South–Western United states (Osborn et al., 1979), Saudi Arabia (Wheater et al., 1991) and in Negev desert in Israel (Reid et al., 1995).

Return periods of the storms at Nasukuta could not be definitively determined because there are no historical records for the site. Estimates of recurrence intervals were however estimated from rainfall data from Kacheliba. Most of the storms during the study period were estimated to have recurrence intervals of less than two years.

Stochastic Features of Rainfall Data

The monthly distribution of annual rainfall at Kacheliba for the wet season months of April–August were studied. It was found that the means of the total rainfalls of these months were statistically the same (analysis of variance) at a significance level of $\alpha = 0.05$ ($F_{computed} = 1.72 < F_{cr} = 2.37$). Therefore, $R_4 = R_5 = R_6 = R_7 = R_8 = R_m$, where the subscripts 4,5,6,7 and 8 denote the months April to August and m denotes the mean. This five month period of April–August thus represents the wet season within which dam design related activity is to be considered. The number and depth of runoff producing events in this period is of great importance. There are about 30 runoff-producing events per year out of which 21 fall in this period.

Although the Poisson distribution is widely used in modelling the occurrence of wet and dry spells (Bogardi et al., 1988; Gupta and Duckstein, 1975), the normal approximation to the binomial was used in preference because of the Poisson parameter λ being greater than 0.05 (Freund, 1992) and due to 'n' the number of events being small.

The statistical requirement for the independence of rainfall events was assessed by the test for randomness based on the theory of runs (Freund, 1992). The results show that for all but one of the 25 years, the randomness of the sequences of wet days ($R \geq 10$ mm) and dry days ($R < 10$ mm) could not be rejected at 5% level of significance since the calculated standard normal variates Z fall between -1.96 and 1.96. In other words, the hypothesis of the independence of wet days above the threshold of 10 mm is acceptable. It is known that wet and dry spells tend to persist and may follow the Markov law of dependence (Sharma, 1996). Nevertheless, the high threshold adopted ($R > 10$ mm) resulted in the breaking of the wet spells into small isolated units rendering the occurrences of the spells to be nearly random.

Rainfall depth per event was modelled using the exponential distribution function (Tsagiris and Agrafiotis, 1988). A satisfactory fit was indicated by the χ^2 goodness of fit test.

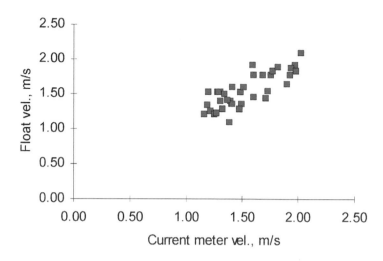

Figure 2. Comparison of float velocities and surface current meter velocities.

Runoff Volumes and Hydrographs

River flows were characterised by rapid changes in stage starting from zero to peak within minutes. The duration of flows were also short. The floods had an average time to peak t_p = 28 minutes, a maximum t_p = 73 minutes and a minimum t_p = 2 minutes. A mean flow duration t_b = 129 minutes was observed (n = 14).

The float velocities were analysed against the surface current meter velocities to assess their suitability for velocity rating. The velocities showed a close linear relationship (r = 0.99, n = 38). The t-test at 5% significance level showed that there was no significant difference between the means of the two arrays, which may thus be taken as samples from the same population (Fig. 2).

The catchment had a CN of 82.4. A round figure of 10 mm was determined as the threshold rainfall for the production of measurable channel flow at the outlet of the catchment. The CN method used for the prediction of runoff volumes from rainfall gave good results as indicated by a high coefficient of efficiency (R^2) between observed and predicted runoff volumes (R^2 = 0.93). R^2 in the present paper is termed as coefficient of efficiency in line with its usage ih the hydrologic literature (Nash and Sutcliffe, 1970; Onyando and Sharma, 1993). The relationship is depicted in Fig. 3.

The triangular hydrographs obtained were then transformed into dimensionless curvilinear hydrographs based on the tabulated ratios given by Singh (1988). The peak discharges of the predicted hydrographs were compared to the observed peak rates. The SCS hydrograph showed a higher predictive efficiency (R^2 = 0.67) than the Bernard's distribution graph (DG) with R^2 = 0.48. Fig. 4 shows the observed runoff hydrograph and those predicted by the DG and SCS methods.

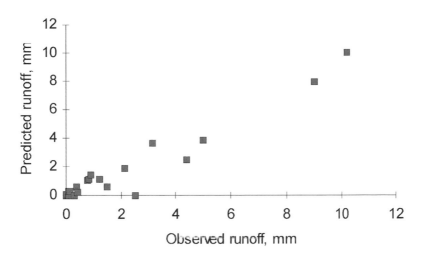

Figure 3. Observed and CN predicted runoff for Nasukuta.

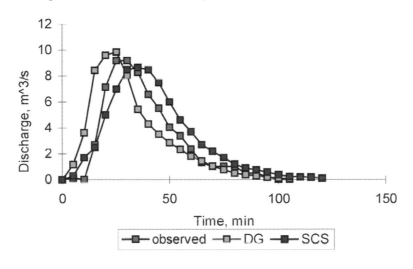

Figure 4. Observed and predicted hydrographs for the event of 2/6/98.

Sediment Transport and Sediment Yields

Instantaneous sediment concentrations ranged from a minimum of 3100 mg/l to 24200 mg/l. The observed suspended sediment transport rates are shown in Fig. 5.

Textural analysis of the suspended sediments loads indicated that its composition corresponds to the parent clay loam soils of the catchment with sand – 23%, silt – 47% and clay – 30%. This high clay content of the suspended sediment would have adverse effects on the hydraulic characteristic on the sand dam if allowed to settle therein.

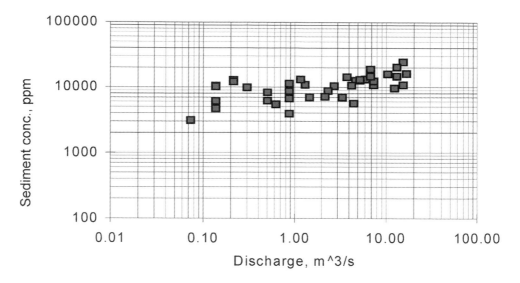

Figure 5. Suspended sediment transport rates at Nasukuta.

Observed sediment transport rates were found to be more accurately modelled by the Engelund–Hansen procedure (R^2 = 0.75) compared with the stream power approach of Yang and Stall (R^2 = 0.68).

The construction of three additional steps to the sand dam in 1999 gave an opportunity to compare the yields predicted by theoretical modelling and the actual deposits behind the new steps. There was however difficulty in determining the quantity of sediments behind the new steps. The difficulty lies in establishing the extent behind the dam to which the deposits may be ascribed to the new dam step heights. In the study, the point was determined from elevation plots of survey levels as the point where the monotonic change in elevation of the new surface begins to suffer persistent disruption. This point was designated 'the point of cessation'. A comparison of the observed and predicted sediment yields is shown in Table 2.

Table 2. Predicted and observed sediment yields at Nasukuta.

Date	Rainfall (mm)	Cessation Point, (m) Upstream of Dam	Observed Bedload Yield, (m^3)	Predicted Bedload Yield, (m^3)
31/8/99	24.8	300	107	127
5/9/99	14.1	100	22	8
12/9/99	13.5	35	12	4
8/10/99(1)	18.1	Not available	Not available	52
8/10/99(2)	31.2	300	Not available	349
8/10/99 (1+2)		300	305	401
13/10/99	17.8	80	79	48

Effects of Stepped Construction

The choice to construct in single rise or to construct in stages has an effect on the storage potential of sand dams.

Following the construction of addition steps raising the dam weir gradually in three equal heights of 0.27 m each at Nasukuta, the channel slope changed from the initial 0.97% to 0.89% at the 1st step, 0.77% at the 2nd and 0.73% at the final step. The changes in channel slope had a profound effect on the texture of the residual sediments. The median diameter of the residual sediments reduced consistently from 0.50 mm at the initial channel slope of 0.97% to 0.27 mm at the final slope of 0.73%. These results are depicted in Table 3 and the graph in Fig. 6.

Table 3. Change in median particle size with channel slope at Nasukuta.

Date	Channel Slope, %	Median Size (d_{50}), mm
14/8/99	0.97	0.50
5/9/99	0.87	0.42
4/10/99	0.77	0.33
11/10/99	0.73	0.27

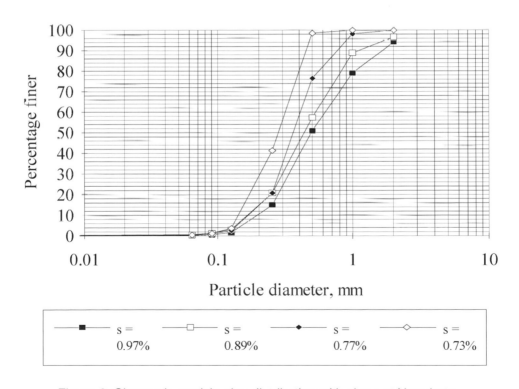

Figure 6. Change in particle size distribution with slope at Nasukuta.

665

The median particle diameter (d_{50}) at each step expressed as a percentage of (d_{50}) at the preceding step was 84, 78.6 and 81.8% for the first second and third steps respectively. However, the median diameter of 0.27 mm at the final step is only 54% of initial diameter of 0.50 mm.

This experiment, conducted at Nasukuta, showed that by raising the dam weir gradually in three equal stages of 0.27 m each, the sediment deposits behind the dam gradually became finer finally attaining a median diameter of about half the initial deposit. A single rise construction would have forestalled the possibility of trapping the sediment of the intervening intermediate diameters. This points to the need for the construction of sand dams in small step heights that will result in small percentage changes in median particle diameter from one step to the next.

Size of Dam Step Heights

With appropriate data, rainfall-runoff models can be developed and sediment yields over any desired period can be estimated. The next challenge is to determine the size of dam step height needed to store the expected sediments yields. This lies in the ability to theoretically determine the point behind the dam which marks the start of deposition induced by the height increment and the profile of the sediment deposits (Fig. 7).

The point at which dam induced deposition begins has been designated point of cessation corresponding to point C in Fig. 7. The amount of sediments deposited behind the new height increment is enclosed between points CD and the old surface profile (Figure 7). There is however, no evidence in literature of the nature of the new profile. It is thus proposed to approximate it using surface water profiles. It is known from open channel hydraulics that under the ideal conditions of steady uniform flow, the energy gradeline, the water surface and the channel bottom are all parallel. If a barrier is introduced on the bed of the channel, the flow regime is interfered with resulting in back water curve. The task therefore lies in determining the extent behind the dam of the backwater curve that also marks the start of dam-

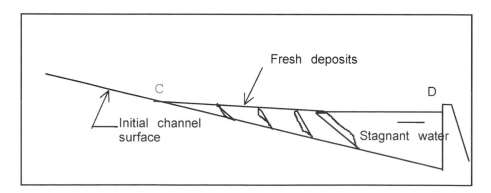

Figure 7. Scene behind a new step height following a small flood: formation of a new channel profile.

induced deposition of sediments. When the reservoir is filled flush to weir level with fresh sediments, the new profile of the channel surface is marked by a straight line joining the top of the weir and the point of convergence (points CD in Fig. 7).

There are two main groups of backwater curves, designated M and S (French, 1985) that are determined as follows:

Group M includes all cases in which the critical depth of flow Y_c is less than the normal depth of flow Y_n,

$$Y_n > Y_c \tag{1}$$

or

$$S_o < g/c^2 \tag{2}$$

Where g – acceleration due to gravity and C – Chezy's coefficient (m/s^2) given by

$$C = R^{1/6}/n_M \tag{3}$$

and R – hydraulic radius, n_M – Manning's roughness coefficient (s/m$^{1/3}$).

Group S includes all cases in which the critical depth of flow Y_c is greater than the normal depth of flow Y_n.

$$Y_n < Y_c \tag{4}$$

or

$$S_o > g/c^2 \tag{5}$$

The depths of critical and normal flow respectively are given by

$$Y_c = \{q/(g^{1/2}\, w)\}^{2/3} \tag{6}$$

and

$$Y_n = \{n_M q/(\, S_o^{1/2}\, w)\}^{3/5} \tag{7}$$

Where q – rate of flow and w – channel width.

Applying Eqs. (6) and (7) on the design flood of q = 1.05 m^3/s at Nasukuta and generally on the range of observed flows, it was found that $Y_n > Y_c$ at the initial channel slope of 0.97%. The surface water profile is thus of type M. A check using Eq. (2) confirms this position. The depth Y of the water just upstream of the dam (Fig. 8) is equal to the height of the dam increment dh and the depth of flow over the weir, $Y = Y_c + dh$, which is greater than Y_c. Thus, $Y > Y_n > Y_c$. Under this condition the curve is of the type M1.

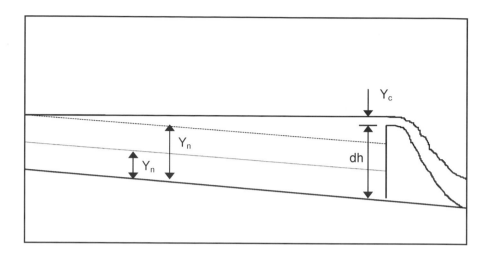

Figure 8. An M1 water profile on a channel behind a sand dam.

The upstream boundary of the backwater curve, and extent of sediment deposits behind a new dam increment, is marked by the attainment of the normal depth of flow Y_n. But since the depth of flow at the upstream boundary of the backwater curve asymptotically approaches the normal depth, a reasonable definition of convergence must be established (French, 1985). For example computation will proceed until the depth is 1.1 of the normal depth.

The water surface profile was computed for the peak discharge of the design flood. For the purpose of the computations, it was assumed that the channel was prismatic and of uniform slope. With the construction of a dam a control is introduced in the channel. Because the flow becomes critical at the control, the critical depth at the point was found by Eq. (6) and used as the starting point of the computations. By adopting various sizes of dam height increments corresponding surface profiles were drawn and their upstream boundaries determined. The volumes enclosed between the new and previous profiles of the channel were then easily calculated.

Using the above procedure at a design probability level of 0.5 (rainfall depth per event R = 15.1 mm), a sediment yield of 27 m^3 is expected. From the observed data it was found that the coarse sediment (d > 0.075 mm) to be trapped behind the dam constituted on average about 57% of the total sediment load – that is 15 m^3 per event or about 324 m^3 per season, the sediment yield in a season or any chosen period being the sum of the individual flood events in the period. In this case, there are 21 events in the wet season period of April–August.

Computation of water surface profiles and sediment volumes for Nasukuta using the step method for the design flood of R = 15.1 mm were carried out based on a channel slope of 0.89%, bed levels taken on 5/10/99 and Manning's $n_M = 0.03$ (French, 1985). The computations, give a sediment storage capacity of 190 m^3 for the first 0.3 m increment. Constructing another step of similar height on top of the preceding one gives an additional storage of 263 m^3. The corresponding upstream points of cessation are about 50 m and 100 m respectively. Therefore, if field and other conditions allow, two incremental steps of 0.3 m each would be constructed at Nasukuta in a season. Their combined storage of 453 m^3 would be higher than the design annual coarse sediment yield of 324 m^3 and part or whole of the extra 129 m^3 would be occupied by the finer sediment component. The provision of small holes suggested in the preceding section would drain the otherwise stagnant sediment-laden water behind the dam thus avoiding such an undesirable eventuality.

It is seen from the above results that by progressively superimposing new step heights on top of the preceding ones, the storage capacities for a constant size of step height increases on account of the trapezoidal channel section and the decreasing channel slope. A fixed step height increment cannot thus be recommended for any particular site or sand dams in general. The size of step heights are therefore, to be determined for each specific case based on the physio-hydrological factors affecting the site.

The size of step height is however, subject to practical demands of construction practice. The minimum height may be dictated, for example, by the sizes of the masonry stones, blocks or shuttering material normally supplied for construction. In addition there is usually the public driven urge for the quickest completion and commissioning of such high demand projects. Ways have to be devised to address these conflicting demands. One approach would be to construct dams to full height in single rise and then provide openings in the wall. The essence of this being the creation of the structure but an effectively lower slope and an escape route for sediment laden water — leaving no pools of stagnant water. This approach obviously requires more work.

Conclusions

Construction in stages is, perhaps, the single most important aspect of sand dams. Construction of small step heights every season means that the work will have to be done annually over a number of years with the attendant costs of remobilisation. This repeated remobilisation for stepped construction is an important consideration with a two-fold implication where the beneficiaries are required to make substantial contributions towards the project. Firstly, the project has to help the community to attain an element of control over the environment and in creating the perception that their environment is not irrevocably harsh but manageable even with simple methods. The lack of immediate dramatic results from stepped seasonal constructions may lead to a loss of vision from the beneficiary community. Secondly, it must be noted that the project is a competitor for labour and financial resources in the beneficiary communities' seasonal/annual calendar of activities. Most beneficiaries of sand dams are rural peasant or pastoralist communities who engage in subsistence cultivation and must promptly respond to seasonal weather changes in order to maintain their food security to stave off famine.

Construction in stages was however, shown to have a positive effect on the texture of the trapped sediments. Small steps result in the harvesting of the coarsest possible sediments for given channel slopes. The size of dam step height at any stage should be so determined that it will cause the deposition behind it of an amount of sediments equal or less than the expected yield. The study also showed suspended sediment concentrations to as high 24200 mg/l and to consist of upto 30% clay. The high clay content is detrimental to the water storage potential of sand dams and efforts must be made to exclude suspended sediments from sand dams.

The development of sand dams obviously needs more work on rainfall-runoff and sediment transport. There are also other aspects of sand dams – e. g. water quality, that need further study.

References

Bogardi, J.J., Duckstein, L. and Rumambo, O.H. 1988. Practical generation of synthetic rainfall event time series in a semi-arid climatic zone, *J. Hydrol*, 103, 357–373.

Chow, C.T., Maidment, D.R. and Mays, L.W. 1988. *Applied Hydrology*. McGrow - Hill Book Co., Singapore.

Engelund, F. and Hansen, E. 1967. *A monograph on sediment transport in alluvial streams.* Teknisk Forlag, Denmark.

French, R.H. 1985. *Open Channel hydraulics.* McGrow-Hill Book Co., New York.

Freund, J.E. 1992. *Mathematical Statistics.* Prentice-Hall Inc., Eaglewood Cliffs, New Jersey, USA.

Gupta, V.K. and Duckstein, L. 1975. A stochastic analysis of extreme droughts, *Water Resour. Res.*, 11(2), 221–228.

Herschy, R.W. 1985. *Stream flow measurement.* Elsevier Applied Science Publishers, UK.

Kinori, B.Z. and Mevorach, J. 1984. *Manual of surface drainage engineering, Vol.2, Streamflow engineering and flood protection.* Elsevier Publishers, B.V., Amsterdam.

Makhoalibe, S. 1984. Suspended sediment transport in Lesotho, Challenges in African hydrology and water resources, Proc. of the Harare Symposium, *IAHS Publ.* 144 , 313–321.

Ministry of Land Reclamation, Regional and Water Development, 1994. *Soil studies in West Pokot district.* Water Resources Assessment and Planning Project, Phase IV, Nairobi, Kenya.

Ministry of Planning and National Development, 1985. *District Atlas, West Pokot,* Kapenguria, Kenya.

Nash, J.E. and Sutcliffe, J.V. 1970. River flow forecasting through conceptual models, Part 1 – A discussion principles. *J. Hydrol.* 10(3), 282–290.

Okalebo, J.R., Gathua, J.W. and Woomer, P.L. 1993. *Laboratory methods of soil and plant analysis: A working manual.* Tropical soil Biology and Fertility Programme, Nairobi, Kenya.

Onyando, J.O. and Sharma, T.C. 1995. Simulation of direct runoff volumes and peak rates for rural catchments in Kenya, East Africa. *Hydrological Sciences Journal*, 40(3), 367–380.

Osborn, H.G., Renard, K.G. and Simanton, J.R. 1979. Dense networks to measure convective rainfalls in the South Western United States. *Water Resour. Res.* 15(6), 1701–1711.

Reid, I., Laronne, J.B. and Powell, D.M. 1995. The Nahal Yatir bedload data base: Sediment dynamics in a gravel-bed ephemeral stream. Earth Surf. Process. Landf., 20, 845–857.

Sharma, T.C. 1996. A Markov-Weibull model of rainsum for designing rainwater catchment systems. *Water Resources Management* 10(2), 147–162.

Sharma, T.C. and Gikonyo, J.K. 1999. In: Jayawardena, Lee and Wang (eds). An instantaneous unit sediment graph study of a small agricultural catchment in Kenya, East Africa. *Proc. of the Seventh International Symposium on River Sedimentation*, Hong Kong, China, Balkema, Rotterdam.

Singh, V.P. 1988. *Hydrologic systems. Vol.1, Rainfall-runoff modelling.* Prentice Hall, Eaglewood Cliffs, NJ, USA.

Tsagiris, G. and Agrafiotis, G. 1988. Aggregated runoff from small watersheds based on stochastic representation of storm events, *Water Resourc. Management* (2), 77–86.

Wheater, H.S., Butter, A.P. and Stewart, E.J. 1991. A multivariate spatial-temporal model of rainfall in Southern Saudi Arabia 1. Spatial rainfall characteristics and model formulation, *J. Hydrol., 125, 175–1099.*

Yang, C.T. and Stall, J.B. 1976. Applicability of stream power equation, Proc. ASCE, *J. Hydrol. Div.,* 102(5), 559–568.

Surface Water Hydrology, Singh, Al-Rashed & Sherif (eds)
© 2002 Swets & Zeitlinger, Lisse, ISBN 90 5809 363 8

Evaluation of environmental loads
on marine structures at the Marmara Sea

Murat Küçük and Necati Ağiralioğlu
Istanbul Technical University, Civil Engineering Department
80626, Maslak / Istanbul
e-mail: necati@itu.edu.tr

Abstract

In August 1999, an effective earthquake occurred on the East Coast of the Marmara Sea, and some marine structures destroyed. After destroying of these structures, environmental loads affecting on those are taken into consideration again by some investigators. In this study, the environmental loads that include the current, wave, wind and earthquake loads are compared with each other. A computer model for designing of marine structures is developed, and applied to a special structure as a pier in this region. SAP 2000, structures analyzing computer program, depends on finite element method is used for analyzing the marine structure. The findings show that the earthquake load is more effective than current and wave loads in the Marmara Sea coastal.

Introduction

Design of marine structures, especially is important for industries and economy in Marmara Region, since these structures are the best alternative way in transportation for products (Kucuk, 2000). For this region, waves, current, wind and earthquake provide the principal environmental loads on marine structures. In this study, the environmental loads that include the principal forces of environmental loads are evaluated respect to the design values of those. Forces of each environmental load are also compared with each of them.

Environmental Loads

Wave Forces

One of the most important aspects in the design of a coastal structure is the determination of the wave forces to which it is to be subjected during its lifetime. Design wave forces should be derived from the design wave parameters that include significant wave height, wave length and its period. The magnitude of wave forces depends not only on the wave height, wave period and dimensions of the structure, but also on the resulting hydrodynamic regime. The magnitude of wave force is controlled by the relationship between the width or diameter of the submerged part of the structure or member, D, and wavelength, L, as follow:

- For $D/L > 1$ situation, the reflection applies. The wave may be assumed to be non-breaking and to form a clapotis or standing wave.

- For 1 > D/L > 0.2 situation, the diffraction theory applies. Diffractive conditions are valid.
- For 0.2 > D/L situation, the Morison's equation applies. The total force imposed may be calculated from Morison's equation as the phase difference between of two components (BS 6349, 1984).

Linear wave theory may generally be assumed to be valid, but in shallow water, where the depth to deep water wave length ratio, d/Lo, is less than 0.1, as the wave form starts to deviate significantly from sinusoidal. Piles which used at the model structure as pier, satisfy 0.2 > D/L condition for using Morison's equations in the Marmara region and drag force and inertial force are considered as wave forces.

Drag Force

The flow with constant velocity (steady flow) of a fluid around an obstacle exerts resultant force that called drag force on the obstacle. It can be obtain from Eq. 1.

$$f_D = C_D \frac{g}{2g} D u^2 \tag{1}$$

where f_D is the unit drag force, C_D is the drag coefficient, γ is unit weight of fluid, u is the instantaneous acceleration of the water particles both measured normal to the axis of the member at the elemental length. The total drag force (F_{DM}) on the pile at any moment can be determined by integrating the below formula over the whole height of the pile (Quinn, 1961).

$$F_{DM} = C_D \frac{g}{2g} K_{DM} D H^2 \tag{2}$$

$$K_{DM} = \frac{1}{H^2} \int_{-d}^{h} \pm u|u| dz \tag{3}$$

where η is the height of the water surface above still water level, d is the still water depth, z is the vertical coordinate, K_{DM} is a coefficient as function of x/h for maximum solitary wave in shallow water, x is the distance from crest, h is the height of through above bottom and H is the wave height.

Figure 1 indicates maximum drag force and wave crest. The maximum drag force occurs where wave crest pass over pile. In the figure s represents the relative lever arm above the bottom, η_c is the height of the water surface from wave crest.

Inertial Force

Another force, in addition to the drag force, acting normally and tangentially on the surface of the pile is termed inertial force, which is combined in the term drag force.

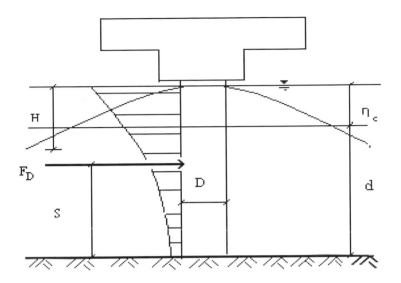

Figure 1. The wave crest position and the maximum drag force.

The constantly accelerating or decelerating masses of water exert also mass force, a kind of Impact force, on the pile. Unit inertial force can be written as

$$f_I = C_M \frac{g p}{4g} D^2 \frac{du}{dt} \tag{4}$$

where C_M is the coefficient of mass, du/dt is the horizontal fluid acceleration. The total inertial force (F_{IM}) on the pile at any moment can be found as:

$$F_{IM} = C_M \frac{g}{2g} K_{IM} D^2 H \tag{5}$$

$$K_{IM} = \frac{p}{2H} \int_{-d}^{h} \frac{du}{dt} dz \tag{6}$$

where K_{IM} is coefficient as function of x / h for maximum solitary wave in shallow water. This force is zero at crest and trough positions with the maximum values occurring between these two positions. Therefore, the total maximum force is not the summation of the maximum values of F_{DM} and F_{IM}, as they occur different times within the wave cycle (Quinn, 1961).

Total Force For Pile Group

The coefficients K_I and K_D vary with the phase angle, θ, K_D obtaining its maximum value K_{DM} at the crest (θ = 0°) and trough (θ = 180°) positions while the maximum value K_{IM} occurs at the phase angles, which depend on the type of wave. The total maximum force on a vertical cylindrical pile is, according to Morison (Abbott, 1993).

$$F_{TM} = C_D \frac{g}{2g} K_{DM} D H^2 \pm C_M \frac{g}{2g} K_{IM} D^2 H \qquad (7)$$

Another method to estimate the maximum total force (F_m) by Dean may be used

$$F_m = \Phi_m \gamma C_D H^2 D \qquad (8)$$

$$W = C_M D / C_D H \qquad (9)$$

where W is dimensionless coefficient, determined by Ref. U.S. Army Corp., 1984. Φ_m is the coefficient read from figures which depends on W (for 0,05 –0,1 –0,5 –1 values), H/gT^2 and d/gT^2. Similarly, the maximum moment M_m can be determined from figures, which are based on Dean's stream function theory as,

$$M_m = \alpha_m \gamma C_D H^2 D d \qquad (10)$$

in which α_m is the coefficient read from figures, which depends on W (for 0,05–0,1–0,5–1 values), H/gT^2 and d/gT^2 (U.S. Army Corp., 1984).

Current Forces

Currents may be sources of significant loads on marine structures, especially upon moored vessels and when additive to wave loads on offshore structures. Moreover, they generally present such as scouring, deposition, increased corrosion rates and modification of wave effects. Also they can cause impact of ice and flotsam (Gaytwaite, 1990). Drag, lift and possible dynamic amplification of forces should be considered.

Current Drag Force

Firstly, design current velocity and appropriate vertical distribution have been selected after the current loads can be evaluated using the drag force equation, which calculates from wave drag force. The drag force is acting in direction of flow. For exposed structures where wave forces are important, the current velocities (U) should be added vectorially to the wave partical velocities (u) to yield the total drag on the structure. Figure 2 indicates current pressure height in front of a pile.

Total drag force can be found as fellow,

$$F_{TD} = C_D \frac{g}{2g} A_p (U+u)^2 \qquad (11)$$

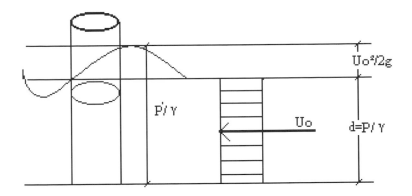

Figure 2. The current passes over the pile.

Table 1. Variety of C_D to d/D.

d/D	C_D
1.1	4.6
1.5	2.2
>7	1.0

in which γ is the unit weight of fluid, A_p cross section area (flow direction), C_D is the drag coefficient. Drag coefficient of current which depends on water depth, d, is defined in Table 1.

Lift Force

Hydrodynamic lift force which acting perpendicular to the direction of flow may be of critical importance in the case of long cylindrical components, such as pipelines with long unsupported lengths. Lift force per unit length is found as fellow,

$$f_L = \frac{\gamma}{2g} C_L D U^2 \tag{12}$$

where C_L is the lift force coefficient which can be taken as approximately equal to $C_D/3$ for circular cylinders and steady flow velocity (Sumer, 1997). This force will be neglected for the model structure in this study.

Wind Force

The magnitude and the strength of wind load acting on the structure depend on the velocity of prevailing wind, the structures orientation, and the amount of structure area exposed to wind. Two components of wind forces are usually considered. One of them acts perpendicularly to the dock structure, and the other one acts parallel

Table 2. Shape coefficients (C_s).

Type of Structures	Shape Coefficients
For beams	$C_s = 1.5$
For sides of buildings	$C_s = 1.5$
For cylindrical sections	$C_s = 0.5$
For overall projected area of platform	$C_s = 1.0$

to it. The maximum wind force is usually obtained when the wind is blowing at 90°. The wind force varies in regions from 50 kg/m² to 100 kg/m² for 100–145 km/h wind velocity. The wind force on an object should be calculated by using an appropriate method such as

$$F = 0.0473 \, C_s \, A \, V^2 \tag{16}$$

where F is the wind force, V is the wind velocity (km/h), C_s is the shape coefficient, A is the area of the object. In Table 2 shape coefficients, C_s, are recommended for perpendicular wind approach angles with respect to each projected area (Gaytwaite, 1990).

Earthquake Force

During an earthquake, a structure is subjected to a forced vibration imposed upon it by the movements of its foundation. The inertia of the structure tends to resist the movements of the foundation. Therefore, at the foundation there will be a sharing force, base share, during the period of the earthquake (Ordemir, 1984). In general more rigid structures such as batter-pile-supported piers in shallower water will have higher natural frequencies and hence shorter natural periods (T_n), 0.25 to 0.75 second, thus making them less likely to experience large deflections but more likely to develop high reactions forces from impact. By contrast, more flexible platform structures in deep water, often with $T_n > 1.0$ second, may absorb impact well (Gaytwaite, 1990). In deeper water, the structures may be more susceptible seismic loads.

In areas of low seismic activity, storm, wave or other environmental loading rather than earthquake would normally control the structure design. For areas where horizontal ground acceleration is more than 0.05 g, such as the Marmara Region, the design for environmental loading other than earthquake will not provide sufficient resistant against potential effects from seismically active zone. Thus, earthquake analysis is required. The earthquake analysis can be calculated with two methods, such as response spectrum analysis, and time history analysis in time domain (API, 1989).

Response Spectrum Method

The response spectrum for given earthquake is plot showing the variation in the maximum response of single degree of freedom oscillator versus the natural true

Table 3. Modal periods and frequencies.

MODE	PERIOD (TIME)	FREQUENCY (CYC/TIME)	FREQUENCY (RAD/TIME)	EIGENVALUE (RAD/TIME)**2
1	0.572143	1.747816	10.981849	120.601010
2	0.451465	2.215012	13.917333	193.692154
3	0.399717	2.501767	15.719067	247.089056
4	0.072524	13.788518	86.635816	7505.765
5	0.071141	14.056675	88.320691	7800.545
6	0.070693	14.145606	88.879465	7899.559
7	0.069210	14.448769	90.784293	8241.788

period of vibration, when subjected to the base acceleration. The response spectrum of a particular earthquake acceleration record is in fact a property of that ground motion, stated in terms of the maximum response of simple structures (single degree of freedom). The term "maximum response" refers to peak value of acceleration, velocity, and displacement of the oscillator (Hsu, 1984).

In this method, maximum internal forces and displacements are determined by the statistical combination of maximum contributions obtained from each of the sufficient number of natural vibration modes considered. For the response spectrum method, as many modes should be considered as required for an adequate representation of the response. At last four modes having the higher overall response should be included for each of the three principal directions plus significant torsion modes. For calculating the statistical combination of maximum contributions of response quantities, each vibration mode is shown in Table 3 (Aydinoglu, 1998).
When the response spectrum method is used and one design spectrum is applying equally in horizontal directions, the complete quadratic combination (CQC) method may be used. The combining modal responses and the square root of the sum of the squares (SRSS) may be used for combining the directional responses (Sahin, 2001).

Analysis Methods in the Time Domain

Time history is a continuous record over time of ground motion or response. Where time history method is used, the design response should be calculated as average of the maximum values for each of the time history considered.

Previously recorded ground motions may be used for the linear or nonlinear analysis of structures in the time domain. The most unfavorable response quantities shall be considered in the design. The duration of strong motion part of the acceleration record in which acceleration envelopes are not to be less than ±0.05 g, shall neither be less than 5 times the first natural vibration period of the building nor less than 15 seconds.

The most complete description of ground motion at a site would be there orthogonal components of ground acceleration. These could be used as input to the computer program, or physical model of the structure and/or its foundation. The use of physical models is expensive, but is common in research. Computer models of the structures and foundations are more common in practice. When time histories are used in design, it is common to select from a database a number of accelerograms caused by the design earthquake whose characteristics are magnitude and distance (Tsinker, 1996). Data of the time history for use in the design was obtained from the Yarimca-PETKİM data record station. The station positions are 40 45 51 latitude, 29 45 35.7 longitude, 13 m attitude. Event start time is on 08/17/99, and at 03:01:39. Record duration is 135.81 second. Number of records is 27163. Peaks values are +230 mg in the longitudinal direction, East-West, +322 mg in transverse direction, North-South, −241 mg in vertical, up, direction. Where, g is called ground acceleration.The magnitude of the Kocaeli Earthquake is 7.4.

Figure 3 shows that Kocaeli (Izmit) Earthquake accelerations record from the Yarımca-PETKİM record station.

From the time history analysis, base share and base moment of the structure can be found during the earthquake was occurred. Figure 4 shows that base share in X direction depended on the time history of the Earthquake.

Minimum and maximum base shares and base moments, which are calculated from the time history analysis of the Marmara Earthquake, are shown in Table 4. Table 4 also shows that displacements respect to each directions of the structure.

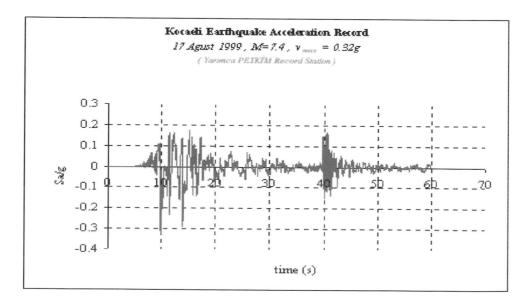

Figure 3. Kocaeli earthquake accelerations records.

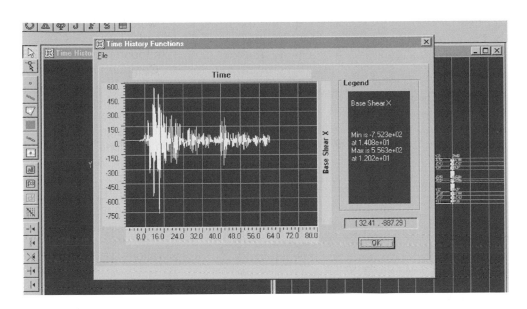

Figure 4. Base share X from the time history analysis.

Table 4. Minimum and maximum base shares and moments.

	DISPLACEMENT			BASE SHARE		BASE MOMENT	
	MIN/MAX (m)	TIME (s)	PILES END NO.	MIN/MAX (ton)	TIME (s)	MIN/MAX (ton-m)	TIME (s)
X direction	0,54	14,08	6-12-18-24-	556	12,02	2236	12,68
	-0,40	12,02	30-36-48	-752	14,08	-1835	14,08
Y direction	0,05	12,70	1-2-3-4-5-6	561	12,44	1674	12,02
	-0,04	12,42		-691	12,68	-2256	14,08
Z direction	0,01	12,44	3	7	12,9	9031	12,44
	-0,01	12,68		-7	12,7	-10053	12,66

Application

The structure which is selected for applying environmental forces in Istanbul / Kartal, and is to be designed to sand to be loaded on from vessels has the 2000 deadweight tonnage (DWT). The piers dimensions are 70 m x 12 m. It has 8 pile caps which designed as a concrete beam, 190 cm x 130 cm, and 35 bents which also designed as a concrete beams, 80 cm x 30 cm. It has also 48 piles. Figure 5 shows that plane of piles location, pile ends (KUC) and joint (DN) of the structure, which defined on computer model (CSM, 1997).

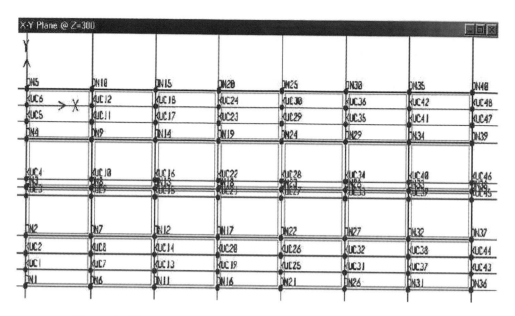

Figure 5. Plane of piles location, pile ends (KUC) and joint (DN).

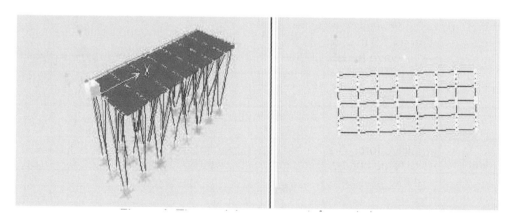

Figure 6. The model structures deformed shape.

Effects of the forces on the structure and comparison of the forces are considered based on an instance structure, and calculated by aid of a computer program. SAP 2000, Structural Analyzing Program, used in computations and calculations, is based on finite elements method. The normal force and the moments, which are used in design of the structures, can be found at the end of the calculations. Found axial forces of piles can allow the comparison of the forces, which called environmental loads on the structure. Figure 6 shows the model structures deformed shape from SAP 2000 computer programs while environmental loads were exposed on the structure.

682

Calculations of Forces

In order to calculate forces, the data of the Marmara Region are as follow

Significant wave height	(H): 3.0 m,
Wave period	(T): 6.5 s,
Water depth	(d): 10 m,
Wave length	(L): 65 m,
Gravity acceleration	(g): 9.81 m/sn^2,
Water density	(γ): 1031 kg/m^3,
Drag and mass coefficient	$C_D = 0.7$, $C_M = 2.0$,
Diameter of pile	(D): 0.6096 m.

Wave, current, and wind forces are considered at the deck elevation. Wave height, its period, current velocity profile and storm water depths are defined from the Marmara region. For static calculation, base share forces of the earthquake, found from time history analysis, are considered at the mess gravity center.

Wave Force

From Eq. 6, for d/T^2 = 0.0467 ft/sec^2, and for H/T^2 = 0.233 ft/sec^2,
K_{IM} = 13.5 ft/sec^2, K_{DM} = 28 ft/sec^2, selected from Ref. (U.S. Army Corp., 1984).
The forces can be calculated as F_D = 3795.62 ton, F_I = 1062.73 ton.
F_T can be found as 1,8 ton per pile from Ref. (Gaytwaite, 1981).
For all piles, total wave force can be found multiplying by 48 and found 86.4 ton.

In the other method, by using Dean's, total wave force can be calculated as follow. Dimensionless value of W is found 0.58 from Eq. 9. If H/gT^2 is 0.007, H/gT^2 is 0.024, Φ_m can be taken as 0.35. Ref. (Abbott, 1993).

The maximum wave force can be calculated from Eq. 8 as F_m = 13438 N per pile.
For all piles total force can be found multiplying with 48 and found as 64.7 ton.

Current Force

For d/D > 7, C_D drag coefficient may be taken 1.0.
Current direction surface area for overall piles, A_p, is 918 m^2.
Total current force which can be calculated from Eq. 12 is F_{TD} = 50 ton.

Wind pressure

Wind shape coefficient (C_s) is 1.3 and effected area of structure, A, is 292 m^2, wind velocity is 40 m/s then F can be found from Eq. 16 as 30.40 ton.

Earthquake force

It was found from the time history analysis as in Table 4.

Force Application Directions

Wave force is 64 ton for whole structure length and uniform disturbed load is 1.83 ton/m in −Y direction. Wind force is 32 ton for whole structure length and uniform disturbed load is 0.91 ton/m, in −Y direction. Current force is 50 ton for whole structure length and uniform disturbed load is 1.43 ton/m in −Y direction. Earthquake force is, maximum, 556 ton or, minimum, −752 ton. It applies at gravity center of the structures and in X directions.

Evaluations

In order to computerize the earthquake force, the natural period of the structure was calculated from the dynamic analysis of SAP 2000 computer program. As seen from Table 3, the model structure has natural period with 0.57 second. Since it is smaller than 1.0 second, the structure must have high resistance to movement of foundations of the structure. However, as seen from Table 4, the pile ends of the structure moved 0.50 m in X direction due to the Kocaeli Earthquake. From the Yarimca-PETKİM data record station, the maximum earthquake acceleration, multiplied with g, was found as 322 mg in transverse direction, North-South, at 9.88th second of the earthquake. On the other hand, the total base share force of the structure is occured as 750 tons in X direction, North-South, at 14.08th second of the earthquake.

During the application of the enviromental loads to the structure, current, wave and wind forces were applied at same directions as static loads and dynamic effect of these forces were neglected. After analyzing, as seen from Table 5, the maximum axial force for the pile 7 was occurred 16 tons by the wave force, although axial forces with 8 tons and 13 tons respectively by the wind force and the current force were occurred. If the wind, wave and current forces are compared with each oth-

Table 5. Maximum and minimum axial forces of piles of the structure due to the environmental and dead loads.

LOAD TYPE	MIN/MAX (TON)	PILE NO
CURRENT	11	12
	-13	7
WAVE	15	12
	-16	7
WIND	7	12
	-8	7
EARTH QUAKE	49	24
	-64	24
DEAD	0	1
	-51	13

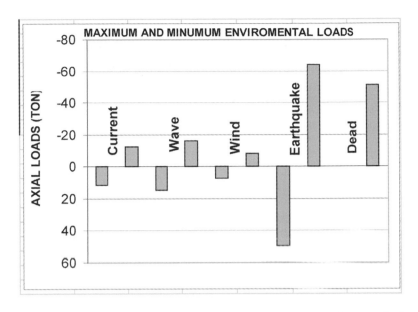

Figure 7. The comparison of the maximum and minimum axial forces of piles of the structure.

ers, it can be infered from the results that maximum or minimum axial forces of piles will be occured by the wave force. However, the maximum axial force of the pile 24 occured by the earthquake was found as 64 tons from the time history analysis of SAP 2000 computer program. It shows that earthquake force is more effective than the environmental forces which include wind, wave and current forces for the structure in the Marmara Region. The axial force of the piles occured from the weight of the structure, called as dead load, is about 51 tons. It is almost equal to earthquake force which was occurred in vertical direction during the earthquake.

Figuro 7 chows that the comparison of the maximum and minimum axial forces of piles occurred from both the environmental forces and the earthquake force. Maximum axial force of the piles due to the earthquake, is higher than axial force of the piles due to the dead load of the structure, except other fixed loads above of the structure, while occurring the earthquake.

Conclusions

If the environmental loads is compared with each others, the earthquake force is almost three times more higher than the other environmental loads which include wind, wave and current forces in the Marmara Region. The earthquake force is also higher than the dead load of the structures. The base share force, occurred due to the earthquake, in X direction is higher than base share forces both in Y and Z directions, since the earthquake acceleration is the highest in X direction. Thus, the direction of base share force in an earthquake must be considered in design of structures, when the earth acceleration is the highest. As a result, North-South is effective direction for the Marmara Region.

References

Abbott, M.B. and Price, W.A. (1993). "Coastal, Estuarial and Harbor Engineer's Reference Book". Chapman & Hall, London.

American Petroleum Institute, API, (1989). "Draft Recommended Practice for Planning, Designing and Constructing Fixed Offshore Platforms-load and Resistance". American Petroleum Institute, Northwest Washington, DC 20005.

Aydinoglu, M.N. (1998). "Specification for Structures to be Built in Disaster Areas. Part III – Earthquake Disaster Prevention". Ministry of Public Works and Settlement Government of Republic of Turkey, Ankara.

BS 6349: Part 1. (1984). "British Standard Code of practice for Maritime structures". British Standards Institution.

Gaytwaite, John W. (1981). "The Marine Environment and Structural Design". Van Notrand Reinhold Company, New York.

Gaytwaite, John W. (1990). "Design of Marine Facilities for Berthing, Mooring and Repair of Vessels". Van Notrand Reinhold Company, New York.

Hsu, Tang H. (1984). "Applied Offshore Structural Engineering". Gulf Publishing Company, Houston.

Küçük, M. (2000). "Forces at the Piled Water Structures And Computer Aided Design Criteria", I.T.U. Natural Science Institute. Master Thesis, Istanbul.

Ordemir, I. (1984). "Foundation Engineering". Middle East Technical University Department of Civil Engineering. Ankara.

Quinn, Alonzo Def. (1961). "Design and Construction of Ports and Marine Structures". Mc Graw-Hill Book Company Inc. New York.

Sahin, A. (2001). "Comparison with Mode-Superposition Methods at Symmetric Structures". I.T.U. Natural Science Institute. Master Thesis, Istanbul.

Sumer, B. Mutlu and Fredsoe, J. (1997). Hydrodynamics Around Cylindrical Structures World Scientific Singapore, London.

Tsinker, Gregory, P. (1996). "Handbook of Port and Harbor Engineering". Chapman Hall International, Thomson, Publishing, New York.

U.S. Army Coastal Engineering Research Center (1984). Shore Production Manual, Volume II, Chapter 6, through 8. Government Printing Office, Washington D.C.

Author index